Nanomaterials and their Interactive Behavior with Biomolecules, Cells and Tissues

Authored by

Yogendrakumar H. Lahir

Department of Biophysics,
University of Mumbai,
Mumbai 400098
India

&

Pramod Avti

Department of Biophysics, Research 'B' block,
Postgraduate Institute of Medical Education and Research,
Chandigarh 160012
India

Nanomaterials and their Interactive Behavior with Biomolecules, Cells and Tissues

Authors: Yogendrakumar H. Lahir & Pramod Avti

ISBN (Online): 978-981-14-6178-1

ISBN (Print): 978-981-14-6176-7

ISBN (Paperback): 978-981-14-6177-4

need for a court order if at any point you breach any terms of this License Agreement. In no event will any delay or failure by Bentham Science Publishers in enforcing your compliance with this License Agreement constitute a waiver of any of its rights.

3. You acknowledge that you have read this License Agreement, and agree to be bound by its terms and conditions. To the extent that any other terms and conditions presented on any website of Bentham Science Publishers conflict with, or are inconsistent with, the terms and conditions set out in this License Agreement, you acknowledge that the terms and conditions set out in this License Agreement shall prevail.

Bentham Science Publishers Pte. Ltd.
80 Robinson Road #02-00
Singapore 068898
Singapore
Email: subscriptions@benthamscience.net

BENTHAM SCIENCE

CONTENTS

Foreword

This book comprises ten chapters; each chapter elucidates specific aspects of nanotechnology, nanoscience, and the basic concepts involved during their multifaceted interactions with and within biological systems and biomolecules.

This presentation elaborates on the introductive remarks on nanotechnology and nanoscience. There is a brief discussion on the definition of nanomaterials, scope, and applications in different fields with emphasis on biological sciences and materials sciences. The interactive behavior of nanomaterials relates to their types and nature. The successful applications of nanomaterials enormously depend on their degree of biocompatibility and bioavailability in the biosystem and at the site of the interface. The physicochemical parameters like inter and intramolecular bonding, hydrophobicity, interactive forces, surface charge, and composition of nanomaterials are well illustrated with suitable examples and supported by the references. When nanomaterials encounter a biosystem, the cellular components like glycocalyx, cell membrane, cytoskeleton, act as the first site of the interface. These components influence the interplay and uptake of the nanomaterials. These entrants form conjugates with ligands, proteins, and cause their effects, and interfere with cellular functioning. Nanomaterials undergo internalization involving phagocytosis, endocytosis. These materials show exclusive interactive behavior with proteins and this depends on the structure of protein, zeta-potential, and nature of binding. This behavior intervenes in cellular physiology and the structure. Protein microchip technology is very useful for analyzing different analytes. The internalized nanomaterials interact with the genetic materials (DNA and RNA) in a biosystem and cause changes in their geometry, physiology, stability, and biophysical aspects. Nanomaterials interact with enzymes *in vitro* and *in vivo*. This feature is used in enzyme technology, enzyme immobilization, biomimetics, and industrial enzymology. The defense mechanism of the biosystem is prone to the impacts of nanomaterials causing immunosuppression or immunostimulation. The nanomaterials intermeddle structurally and functionally with the components of the immune system transforming their roles. Lastly, all these interactive aspects congregate into the wide spectrum of the applications of nanomaterials as detection tools, imaging agents, synthesis, medical implants, and various roles in industries.

A good number of books and reviews report on the specific and selective aspects of nanoscience and nanotechnology. However, there is a need for comprehensive essays that give a consolidated overview of the physical, chemical, biological, biophysical, and molecular aspects of the interactions between nanomaterials and biomolecules, cells, and tissues. This book fulfills this need and offers an incriminated description of the incorporation of basic structural, functional, and physicochemical concepts during the interplays between nanomaterials and biomolecules. The lucid explanation in this book eases the mathematical aspects of the concerned concepts involved. This effort aims to infuse the interest in students, researchers, and foster collaboration through the multidisciplinary approach of nanoscience and nanotechnology in the recent frontiers of biological sciences and nanotechnology.

Each chapter starts with the outline of the chapter, introduction, text, and conclusion; these provide a take-home message of the information contained therein. Throughout the book, suitable examples are presented that support the concepts and are an amalgamation of past and recent research. Lastly, this presentation gives glimpses of the multidisciplinary approach and room to maneuver the various concepts from physical, chemical, applied sciences and technologies, for the betterment of their future applications.

I feel the book will provide a handy but complete reference and review of the nanoworld to students and researchers in this field.

Dr. A.V. Chitre
Former Reader & Head, Department of Chemistry
Professor Emeritus, Sophia College, Mumbai
Adjunct/Visiting Faculty in Biophysics
Department of Biophysics, University of Mumbai
Vidyanagari
Santa Cruz (E), Mumbai 400 098
India

PREFACE

Nanoscience and nanotechnology both have been utilized by a man during the early days of scientific developments under various civilizations, like ancient India, ancient Mesopotamia, ancient Egypt, ancient China, Japan, *etc*. During these eras, there have been scientific developments and materials in nanoforms that might be in use as it is evident from the monuments and the products found during the excavation. Possibly a common man might not be aware of the terms like nanoparticles, nanomaterials, nano synthesis, *etc*., except the specialized craftspersons. These specialized craftspersons might be using different terminology concerning the particular concept of dimension, size, and other properties of the materials in these eras. Nanomaterials are useful in almost all fields of present-day life. These materials have at least one dimension within 1to 100 nm ranges.

The physicochemical features of nanomaterials, mode of their synthesis, duration of exposure, and amount of nanomaterials, *etc*., influence their impacts on the biosystem. Most of the nanomaterials, natural and engineered, both get dispersed in all media and move across almost all types of biological barriers. This ability of nanomaterials exhibits a higher degree of derogative or beneficial interactions. These are of investigatory interest concerning biotic and abiotic components of the environment. These features make them potential agents for their varied applications in industrial, domestic, food-technology, medical, cosmetics, pharmaceutical, and other biomedical fields. Most of the administered nanomaterials get readily disbursed in the biosystem and exhibit a higher degree of metabolic interaction. Such interactions depend on the dose, physicochemical properties of nanomaterials, and cause, either conjugation or dissociation in the interactive biomolecules. The interacting nanomaterials induce changes in the biomolecular conformation, the reactive groups, molecular cross-linkage, hydrophobicity and hydrophilicity, and structural damage, and interrupt cellular functions. The harmful impacts involve micro, macromolecules, cell membrane, cell membrane receptors, cell-organelles, and metabolic pathways. The administered nanomaterials come in contact with the contents of body fluids. The adhesion of nanomaterials onto the cell membrane, even if the biomolecular corona is absent on nanomaterial. The adsorption of proteins on the surface of nanomaterials sharply reduces the adhesion in comparison to the conditions when the nanomaterials are without biological corona. The cellular uptake of the nanomaterials involves two steps: i- initial adherence of nanomaterial to the cell membrane, and ii- internalization of nanomaterial by the cell comprising the energy-dependent pathway.

In most cases, the biomolecules such as proteins, lipids, carbohydrates, *etc*., are present in body fluids of the biosystem. These body fluids include blood, hemolymph, lymph, or any other form of fluid present in the biosystem. The interactions between nanomaterials and biomolecules relate to the specificity of binding ability of biomolecules, the composition of nanomaterial, and their surface physical-chemistry. The effect of the nanomaterial-biomolecular complex formed; generally, the compound formed is with proteins or conjugated proteins that influence the responses of the biosystem. The nanomaterial-protein complex built plays a more significant role in their biodistribution in the biosystem because the protein-nanomaterial-complex formed becomes the identity of the nanomaterials involved within the biological system. Interaction between nanomaterial and biomolecules is a dynamic process. Since proteins are relatively in abundance, they dominate these types of interplay, resulting in the formation of complexes depending on the charge on the surface of protein molecules and nanomaterials. The conformation of the protein at the interface influences the cellular uptake of the nanomaterial. Therefore the interactions between nanomaterial and protein are of great significance in biotechnology and molecular biology.

Adsorption of the proteins on the surface of nanomaterial is a complex process. It is primarily related to (i) - dielectric properties and pH of the medium, (ii)-surface morphology, and surface heterogeneity of nanomaterials and (iii) -the quaternary structure of the protein involved. This phenomenon indicates the existence of the different types of interactions between more significant multimeric proteins, nanomaterial, and small oligomeric proteins.

There is dissociation or binding of proteins present in lower concentrations, but have a higher affinity, influence the separation of conjugated proteins, and this aspect slows the kinetics. Thus the nanomaterials coated with protein can undergo enhanced cellular uptake specifically by macrophages. In most cases, opsonins are present in blood and body fluids. It enhances the ability of macrophages to recognize the surface of the particles entered the biosystem (opsonization). Opsonins like albumin, immunoglobulins, fibrinogen, and compounds of complementary system and apolipoprotein are present in body fluids. They play an active role in the clearance or elimination process in the biosystem. Apolipoproteins are the proteins that bind to lipid.

Lipids are oil-soluble substances like fat and cholesterol and form lipoproteins. These apolipoproteins transport the lipid through the lymphatic circulatory system in vertebrates and hemolymph in the case of invertebrates. The apolipoprotein and phospholipids exhibit amphipathic features having both hydrophilic and hydrophobic components, hydrophilic head and a hydrophobic tail. The apolipoprotein and phospholipids are water-soluble surround lipids and lipoproteins. These interact with enzyme cofactors, exhibit ligand-surface receptors, and low-density lipoproteins (LDL). The distribution of nanomaterials is explicit in cases, like oriented targets, like cancer cells, diseased cells, DNA, RNA, gene, *etc*. Surface-bound molecules like proteins can promote cell-specific uptake of nanomaterials. There are chances that nanomaterials can activate the intracellular signaling pathways. Their dispersibility in air, aquatic, and solid media depend on their nature and specificity.

Any material natural or engineered having nano dimensions and intend to interact with the biological system to evaluate, treat, augment, replace tissue and organ, and have specific functionality. The biocompatibility of such nanomaterials is the first and for most priority for their successful application. The biocompatibility of such materials is the ability to perform appropriate host-response in a specific application. The biocompatibility of nanomaterials depends on the molecular adsorption, mechanical, biophysical, and chemical cellular pathways during their cellular internalization. These interactions are either defensive, interfering oriented targets. Biocompatibility may concern long term or short term specifically for the implanted devices and tissue-engineered devices and conceptually, concerns with cytotoxicity, sensitization, irritation, genotoxicity, implantation, hemocompatibility, carcinogenicity, and biodegradability, *etc*. When biomaterials and hosts come in contact during surgical implants, infusion, injection, extracorporeal circuits, or *in vivo* bioreactor, *etc*., initiate the response.

The processes like material degradation, cell adhesion, mechanical forces generated as a result of the administration as the host response progress with many possibilities like inadequate resolution, clinically relevant effects, either tolerable or non-tolerable, inflammation, hyperplasia, thrombosis, calcification, resolution of the reactions, clinically acceptable results also play a significant role the responses due to these interplays. Biomaterial components like metal ions, polymers, additives, contaminants of nanomaterials on cellular internalization involving specific mechanism, phagocytosis, endocytosis, pinocytosis, *etc*., affect the ambient intracellular environment; this can include material degradation, generation of free radicals like ROS/RNS, cellular damage, alterations in the functionality of cell organelles, interference with apoptotic and necrotic pathways, the passage into nucleus affecting gene expression or gene damage. This process involves material mediators for interactions, chemical structure, elasticity, shape, volume, topology, *etc.*

This presentation is an effort to understand the mechanisms and the involvement of probable parameters along with different concepts, theories, laws, and applied principles of chemistry, physics, biological sciences, and computational simulation. There is a dedicated effort to present the matter precisely, even for those who may not have a mathematical background.

CONFLICT OF INTEREST

The authors confirm that there is no conflict of interest.

CONSENT FOR PUBLICATION

Declared none.

Yogendrakumar H. Lahir
Department of Biophysics
University of Mumbai
Mumbai 400098
India

&

Pramod Avti
Department of Biophysics, Research 'B' block
Postgraduate Institute of Medical Education and Research
Chandigarh 160012
India

ACKNOWLEDGMENT

Authors gratefully express their gratitude to their parents who played pivotal and primary roles in their lives. We are thankful to the research workers, writers, reviewers, whose books, review articles, and research papers have enriched and enhanced our understanding of the subject and have been referred to in this presentation. Further, we are solely thankful to these writers, researchers because their published works have inspired us to undertake this adventure. We also appreciate our students who participated in the discussion during lectures, presentations, seminars and conferences resulting in better clarity from the student point of view about the intricacies of the subject, modes of expression of the subject matter.

YKL extends his gratitude to his teachers namely Mr. H J Lahir (Late), Dr. K P Dhage (Late), Dr. A V Chitre from Sophia College, Mumbai, and Dr (Mrs.) P V Kagwade from C.M.F.R.I., Mumbai Center. The encouragement and departmental support provided by Prof, Dr. P M Dongre, and Head of the Biophysics Department are acknowledged. Dr. A V Chitre and Dr. Sivakami S from the Department of Biophysics, the University of Mumbai are also acknowledged for their critical and encouraging comments and the suggestions on various chapters of this presentation. YKL is also grateful to Mrs. Charu Lata Lahir, Jian, Panchali, Ribhu, and Hemank, for their tolerance, analytical suggestions, encouragement and inspiration during the completion of this adventure. YKL also thankfully appreciates the help of little masters Karan Bali and Suraj Iyer.

PA extends heartfelt gratitude to his Professors namely Professor KL Khanduja, Emeritus Professor, and Professor CM Pathak, Department of Biophysics, PGIMER, Chandigarh for their constant encouragement and support. PA is grateful to his parents and young and supportive minds of Mr. Praveen Avti and Ms. Veena for successfully bringing about the book.

DEDICATION

This work is dedicated to
Lahir Family
&
Avti Family.

CHAPTER 1

Nanoscience, Nanotechnology, Nanomaterials and Biological Sciences

Abstract: Nanoscience and nanotechnology help manipulate or maneuver atoms and molecules to enable them to function at the nanoscale. Nanoscaled materials are the products of nanotechnology, and these are synthesized or fabricated based on specific guidelines. Nanomaterials can interact with most of the biomolecules, cell organelles, and cells, and can move across most of the biological barriers. These materials can readily be functionalized and modified as per the required targets. The modified nanomaterials become convenient tools in several fields of biotechnology, enzyme technology, tissue engineering, *etc.* In these fields, modified nanomaterials act as a vehicle for biomolecules, imaging agents, sensors, probes as diagnostic tools, devices, *etc.* The matters in the bulk form and at the nanoscale level show variable physicochemical properties, thereby, showing multifaceted abilities. These features are responsible for their variety of applications in day to day life as well as in specialized fields.

Keywords: Antifungal Agent, Antimicrobial Agent, Nanomaterials, Nanoscience, Nanotechnology, Sensors, Wootz Steel.

OVERVIEW: NANOSCIENCE, NANOTECHNOLOGY, AND NANOMATERIALS

Nanoscience and nanotechnology are multifaceted aspects of science that provide information about the manipulation or maneuvering of atoms and molecules and enable them to function at the nanoscale. Such products readily interact with cell organelles, cells, and most of the biomolecules. Nanoscience guides to design and formulate nanostructures that ensure their feasible applications in various fields such as biomedicine, biomolecules, biochemical, pharmaceuticals, *etc.* In these fields, nanomaterials are applicable as a cargo vehicle (drugs, biomolecules, gene, *etc.*), imaging agents, sensors, diagnostic devices, *etc.* The industrial applications include electronics, energy storage devices, enzyme technology, tissue engineering, *etc.*

Nanoscience is a science of formation and interactions of nanomaterials. Materials that have at least one dimension within the range of 1nm to 100 nm (nm=one-billionth of a meter) are regarded as nanomaterials. The materials at the nanoscale

Yogendrakumar H. Lahir & Pramod Avti

have different electrical, optical, thermal, and mechanical properties in comparison to their bulk forms. These properties relate atoms and molecules assembly, and interaction at the nanoscale. Nanoscience is a multidisciplinary aspect of science involving principles of material science, physics, chemistry, biological sciences, biotechnology, electronics, quantum mechanism, *etc* [1, 2]. Nano is a prefix used in metrics (metric system), and it represents anything that is one-billionth of some matter in size. It expresses a specific unit that measures mass and time. Materials at this dimension have different properties and behaviors, and both are different in comparison to the respective materials with larger sizes.

The field of nanotechnology has enormous impacts on human life. Nanoscale structures help to store information on 20 nm thick magnetic strips, dirt-resistant and scratch-resistant surfaces, materials that are suitable for tissue regeneration, *etc*. Researchers all over the world are making untiring efforts to explore advanced applications of such materials using basic and applied principles of physics, chemistry, biology, materials science, *etc*. As a result, there has been an enormous development in the field of nanodevices, microscopic development systems, structural and engineering systems, storage of information, computational investigations, biomedical devices, *etc*. The prime focus of nanoscience is on the properties of materials at the nanoscale, and the methodology involved in the synthesis, fabrication, and, assembly of these nanostructures. This science also facilitates the characterization, applications, and functionality of the nanomaterial, nanodevices, *etc*. The observations and study of these wonder materials need very specialized instruments and methodologies that should have the ability to either magnify or detect the products of chemical interactions that produce such nanomaterials in nature or otherwise.

Some of the fine aspects of nanoscience and nanotechnology include bioengineered materials and bionanoscience, quantum confined nanoscale materials, novel tools for nanoscale device patterning, imaging, and characterization, molecular nanoscience and electronic materials, *etc* [3]. James Tour and his coworkers made a nanoscale car, consisting of phenylene ethynylene (oligo), alkynyl axles, and four spherical fullerenes (C60) in 1906. This car moves on the gold surface as the temperature increases, and above 300°C, it moves very fast. (This nanoproduct has the chemical formula $C_{430}H_{274}O_{12}$ and molar mass 5632.769). In 1908, the National Nanotechnology Initiative (nano.guv) published a strategy related to nanotechnology. This document is a general guideline that governs the varied aspects involved in this technology.

Generally, nanostructures are the materials in the form of structural elements (particles), clusters, crystallites, or molecules. These products are in high demand as they have significant academic and industrial applications. There have been tremendous efforts to study their properties and changes concerning their infinitely extended solid form to particle size consisting of countable numbers of atoms. The functions of the nanomaterials depend on their size and physicochemical properties. These parameters are of prime concern during their synthesis and investigations. At the nanoscale, the properties and functionalities of matter, such as electrical, optical and magnetic, *etc.*, change. These features are related and exhibit variations about the changes in their infinitely extended solid form to an excellent particle state. At this state of materials, their atoms are countable. This condition also exists even in the confinement of nanoscaled semiconductors or metal clusters or colloids. The nonmetallic elements, like carbon-based nanomaterials such as fullerene, nanotubes, *etc.*, also exhibit similar behavior. These features make them suitable for their pervasive applications not only in nanoscience but also in other biomedical fields [4].

HISTORICAL   ASPECTS   OF   NANOSCIENCE   AND NANOTECHNOLOGY

One of the earlier established applications of nanoscience and nanotechnology has been reported during 600 BC in India. Indian blacksmiths produced wootz steel, mixing specific ingredients like wood from Cassia auriculata and leaves of Calotropis gigantea, and, other ores from the particular Indian mines. These ingredients were used during the forging process in steel industries resulting in the formation of petite cakes. These tiny cakes are called wootz steel, and, the steel formed from these was wootz steel. During this process, and, related ones, like thermal cycling and cyclic forging, catalytic segregation of elements into a different array was induced [5]. Carbon nanotubes and cementite nanowires were noticed in the microstructures of wootz steel. In ancient India, a sophisticated thermomechanical treatment related to forging and annealing had been in practice. This technique has been applied to refine steel with specific qualities. For this purpose, wootz steel cakes were used. This technique was developed and spread globally. The medieval bladesmiths could use a mineral called cohenite to reduce the brittleness of cementite (having carbon contents of 1-2% wt). Mechanical processing makes microstructure of steel to be fine-grained and superplastic at an appropriate high temperature. The addition of tiny amounts of vanadium, chromium, manganese, cobalt, nickel and other, resulted in specific bonding of cementite during thermo-cycling at temperature lowers than the formation of cementite (around 800°C). Actually, during this treatment, the formation of cementite nanowires takes place at the microstructure level [6, 7]. History of nanoscience and nanotechnology is traced at a much earlier stage. Famous glass, Lycurgus Cup; a product of the 4th century, is known for its dichroic behavior because of the presence of colloidal gold and silver particles in the glass. These fine particles make it appear to be opaque green from the outside, and to be translucent red when lit from the inside. The ceramic glazes have been in use during the 9th-10th centuries by the Islamic and European worlds. The material that constituted the ceramic glazed had silver or copper nano-metal particles. Gold chlorides, along with other metal oxide nanoparticles were used in the making of glass used in the glass windows of the European cathedrals during the 6th to 15th centuries. Also, the decoration of the gold nanoparticles intends to purify the air because of their photocatalytic nature. In 1857, Michael Faraday demonstrated that gold nanoparticles in colloidal form display different colors under specific lighting conditions. Erwin Müller invented the Field emission microscope in 1936, and it could take the images near the range of atomic resolution. The information technology got boost because of the invention of semiconductor transistor by Bardeen J, Shockley W, and Brattain W, in 1947, and ensuring the supportive implementation of electronic devices.

Victor La Mar and Robert Dinegar established the theory related to growing monodisperse colloidal materials in 1950. Controlled fabrication of colloids was achieved based on the concept of monodisperse colloidal materials. This concept is applicable in industries like paper, paint, and thin films. They also have dynamic roles in the treatment of dialysis . The idea of molecular engineering is also applicable in fields like dielectrics, ferroelectrics, and piezoelectric, that was pronounced in 1956 at MIT by Arthur von Hippel. The famous, classic, and path-breaking lecture titled 'there's plenty of room at the bottom' based on the technology and engineering at the atomic scale by Richard Feynman delivered in 1959 at the California Institute of technology.

In 1965, Gordon Moor proposed Moor's Law related to the density of transistors on an integrated chip, their size, and cost of these chips; both are of interest. Professor Norio Taniguchi has claimed that there is a technology concerned with the precision machining of materials within dimension, tolerance, and atomic-scale. In 1974, he named this technology like nanotechnology. Gerd Binnig and Heinrich Rohrer invented a scanning the tunneling microscope' in 1981. This device facilitates the visualization of individual atoms by creating direct spatial images. In the same year, Alexei Eskimov of Russia detected nanocrystalline, semiconducting quantum dots in the matrix of glass, and he has also studied their electrical and optical properties. In 1985, nanomaterials-Buckminster fullerene, also called Buckyball, was discovered by Harold Kroto, Sean O'Brien, Robert Curl, and Richard Smalley, at Rice University. These nanomaterials, *i.e.*, C60, are made of only carbon atoms. Gerd Binnig, Calvin Quate, and Christoph Gerber in 1986 have discovered an Atomic force microscope. This exceptional instrument is capable of viewing, measuring, and manipulating materials up to the nanometer size. Even fundamental forces of nanomaterials are measured. Sumio Lijima in

1991 discovered carbon nanotubes. These tubular nanomaterials are composed of only carbon and specifically exhibit extraordinary degrees of strength, electrical, and conductivity properties. During 1999, and, early 2000, nanoscience and nanotechnology were used in consumer products and appeared in the market. These products included lightweight automobile bumper, and are scratch and dent resistant; a golf ball that flies straight in one direction, stiffer rackets used in tennis, *etc*. The degree of better flex and kick of baseball bat is increased by nanotechnology. Socks treated with silver nanomaterials show antibacterial nature. Consumer products like wrinkle and stain-resistant clothing, clear sunscreen, therapeutic and cosmetics, environmental, health, and safety materials, *etc*., also treated with nanomaterials. Nadrian Seeman and co-workers at New York University have created DNA like robotic nanoscale assembly devices during 2009-2010. The 3D DNA structure was built involving synthetic sequences of DNA crystals. This crystal was programmed to self-assembly. Sticky ends of DNA were used to attain appropriate order and orientation to accomplish the self-assembly of 3D DNA structure. The flexibility and density exhibited by 3D nanoscale components helped the assembly of the parts involved. In due course of time, even the DNA assembly line was fabricated [8 - 16].

CURRENT SCENARIO OF NANOSCIENCE, NANOTECHNOLOGY, AND NANOMATERIALS

The present scenario reflects that nanoscience and nanotechnology have attained the status of the foundation of the overall growth of industrial applications and exhibit exponential growth. For example, pharmaceutical concerns involving nanotechnology have an enormous impact on biocompatibility, the biodistribution of biomedical devices like diagnostic biosensors, bioprobes, bioimaging, theranostics, regulated drug delivery system, tissue engineering, *etc* [17 - 20]. In, food and cosmetic industries, nanomaterials are being used to improve and improvise the production, quality, performance, storage, transportation, packaging, shelf-life, bioavailability, *etc*. Zinc oxide quantum dots, silver nanoparticles are employed to check the growth of food-borne bacteria. Nanomaterials play a role as sensors to check the quality, safety, and also to increase shelf-life of food materials [21 - 24]. Nanomaterials used in consumer products related to almost all aspects of day to day life of a modern man. These include cosmetics, clothing, automobiles, electronic goods, medical appliances, antiviral, antifungal and antibacterial agents, tissue engineering, DNA and enzyme technology, and most aspects of biotechnology. Nanomaterials have their significant roles in biodefense, delivery systems, and sensors; wireless-secured Radio-Frequency-links (RF-Linked) between sensors and the equipment, *etc*. The extensive use of nanomaterials in the various products consumed in today's life leads to a significant concern related to the potential risk to health and safety of life of humans, animals, plants, and the environment. This concern directs the investigations to study their derogative impacts on life and the environment. This aspect leads to the development of nanotoxicity, nanomedicine, and other multifaceted applications of nanomaterials. The nanotoxicity involves the study of potential adverse effects of nanomaterials while nanomedicine is concerned with the identification of risk and beneficial impacts of nanomaterials like a reduction of inflammation, tissue engineering, biomedical devices, *etc* [18, 25 - 29].

VISUALIZATION OF BIOLOGICAL SCIENCES AT NANOSCALE

Nanoscience and nanobiotechnology in all probabilities make it possible to visualize molecular interactions and manipulations occurring during their reaction, interplay, and the dynamics of associating proteins, enzymes, and other micro and macromolecules. In a given biosystem, all biomolecules function by their specific structural integrity and individuality involving a molecular orchestral large complex to ensure specific directional macro and micro-molecular assemblies. All these utilize the micromanipulation technique, microfluidic approach, imaging techniques, *etc*.

A micromanipulation technique encompasses all the fundamental operational and appropriate functional aspects of an optical instrument and various fields of microscopy. The prime interest of this study should be to get acquainted with the tools used and the handling of the samples under the specific considerations concerning the various techniques involved. Micromanipulation works on high magnification and motion provided by varied types of microscopy. The optical magnification gives sharp images of the tool point and the sample under study. For example, one can insert a textile fiber into the interior of the stinging hair of the nettle. This flowering plant belongs to genus-*Utica*; family Urticaceae, having stinging hair that has an opening of about less than 30μm (0.003 mm). These actions are accomplished using a stereoscopic binocular microscope along with 3D viewing.

The microfluidic approach involves the basic principles of engineering, physics, chemistry, biochemistry, biotechnology, and nanotechnology. This also provides multiplexing, automated, and a high degree of clarity throughout the screening process. This technique is suitable for developing the appropriate specific ink-jet print head and DNA-chip technology. The DNA chip is a piece of glass or plastic or silicon substrate, on which the DNA probe (as pieces) affixed in a microscopic array, protein array, and miniature array. In this technique, a multitude of different agents like monoclonal antibodies are attached to the surface of the chip. These are neither reconfigurable nor scalable. The performance of these chips improves by adding hardware appropriately.

The lab-on-a-chip technique facilitates many laboratory functions. Chip is an integrated circuit having a size varying from mm to a few centimeters and features automatically with high throughput screening. In such techniques, the quantity of sample required is very less (less than picoliters). This device is a micro-total-analysis-system, and the method is within the perimeter of the microelectromechanical system (MEMS) [30].

CAN NANOTECHNOLOGY BE CONSIDERED AS COMPLEMENTATION OF MICRO- TECHNOLOGY?

Microtechnology encompasses the techniques involved in investigations related to the matter at the micro-level, *i.e.*, one-millionth of a meter, (10^{-6} or $1\mu m$). Microtechnology illustrates a simple example of increasing the efficacy of the microelectronic circuit. If large numbers of microscopic transistors arranged on a single chip, the resultant product helps to improve the performance, functionality, and reliability of the microelectronic circuit. This micro-device is cost-effective and provides extra space for use. This technique plays a significant role in the fields of information technology, science, and mechanical devices.

Further, the efficiency and performance of mechanical devices increase when the miniaturized and batch fabricated modes are applied. Integrated circuit technology is a suitable option for this purpose. Electronics acts as the brain for most of the advanced systems and the related microelectronic and micromechanical products. Micromechanical devices used as sensors and actuators, and, applied in the automobile as airbags, ink-jet printers, blood pressure monitors, and other display systems. As a result of advancements in this direction, these micromechanical devices are expected to be pervasive like electronics [31, 32].

Discoveries of a scanning tunneling microscope, atomic force microscope, and magnetic force microscope are significant milestones in the field of nanoscience and nanotechnology. These play vital roles in the study of materials involving atomic resolution and manipulation. Atomic force microscope and magnetic force microscope provide information by the feel of the atoms and molecules present on the surface of matter under study. These devices offer a resolution of an area less than a nanometer [16]. Synthesis of nanomaterial is carried out either by the atom-by-atom assembly, *i.e.*, bottom-up approach or from a bulk form of matter to atomic level disassembly, *i.e.*, top-down approach. The top-down approach is in use in micro and nanoelectronics. In this mode, mostly a silicon substrate is disassembled, and some of its parts are removed using physical, chemical, electron or optical lithography. Microlithography and nanolithography relate to the scale and method of the pattern on materials used in lithography. If the range is less than 10 μm, it is microlithography; if the features are smaller than 100 nm, it is nanolithography. The basics of photolithography used in the production of the microchips and the fabrication of microelectromechanical systems devices. The photomask or an article as the master is prepared by photolithography and used in further processing [33, 34]. All these examples suggest that somewhere nanotechnology is the product of the extrapolation of microtechnology.

Nanomaterials are omnipresent. These particles are the products of human and natural activities, such as combustion, volcanic eruptions, dust storms, tornado, domestic dust, and anthropogenic activities, *etc*. Natural nanomaterials like coral, paper, cotton, vertebrate bone, components of the crust of the earth, volcanic ash, scales present on the wings of a butterfly, derivatives of skin, shells of foraminifera, a fine spray of water, *etc*., also come under nano category. The products of industrial or mechanical processes, like, vehicular exhaust, domestic dust, carbon soot, *etc*., also contribute to the nanomaterials. Nanofibers, nanotubes, nanorods, nanoplates, nanoribbon, nanocomposites, *etc*., are either engineered or naturally produced. All these products are physicochemical entities and possess specific physical and chemical properties, like any other entity having mass and energy. These materials follow the concerned principles of physics and chemistry while interacting and behaving in a biosystem or the components of the environment. During these interactions, basic concepts concerning surface energy, plasmonic behavior, scattering of light and electromagnetic waves, *etc*., are followed. These nanomaterials also exhibit Raman scattering, Rayleigh scattering, Mie scattering, Compton scattering, X-ray scattering, *etc*. The nanomaterials also manifest the phenomena of absorption of radiant energy, optical properties depending on their physical parameters. The interactive aspects of these nanomaterials are investigated based on these physical phenomena. Various types of nanomaterials, such as, metal and metal oxide nanoparticles, carbon nanomaterials, dendrimers, quantum dots, *etc*., display different interactions that relate to their size, shape, the surface to volume, ratio, surface energy, *etc*. These interactive behavioral aspects help to understand the interactions of the various nanomaterials with other materials and the components of the biosystem [35 - 39].

Biocompatibility, biodistribution, and bioavailability of the nanomaterials play significant roles to ensure their purposeful and successful applications in the fields of biological sciences, biomedical, biotechnological sciences, material science, surface technology, *etc*. The physicochemical features of nanomaterials and biomolecules, cells, biosystems, and the components of the environment, act as the functional parameters. The nanomaterials are conveniently modified or functionalized different techniques or conjugating molecules as per the set functions. Hydrophilicity and hydrophobicity of nanomaterials, both are concerned with type, the topology of the surface, or the presence of adsorbed molecules on the surface of nanomaterials. These features influence the biocompatibility, biodistribution, and bioavailability of nanomaterials. The cell membrane involves a dynamic mechanism that regulates the cellular amenities and the administered nanomaterials. This mechanism relates to the hydrophobic and the hydrophilic nature of nanomaterials, and the behavior of a cell membrane [40]. The nanomaterials undergo manipulation of their functionalization; during such processes, a specific molecule is attached at an appropriate site on a surface of a particular nanomaterial. The process of functionalization of nanomaterials adds to their ability to carry drug or biomolecule, cancer treatment, diagnostics, theranostics, tissue engineering, molecular biology, and understanding the structural and functional relationship between functionalized nanomaterials and set biological target [41]. There are some strategies to make nanomaterials biocompatible, bioavailable; these include synthesis of specific nanomaterials, increase the degree of stability of nanomaterials, prevent their tendency to agglomerate or aggregate, change of the phase of nanomaterials, using or exchanging specific ligand, silanization, using multifunctional hybrid coating technique [42 - 45]. Even using aerogel composite and opsonization, enhance the biocompatibility and biodistribution of nanomaterials [46, 47].

The nature and behavior of the interplay of nanomaterials, both are unpredictable. These interactions are concerned with size, shape, chemical functionality, surface charge, composition, biomolecular signaling, kinetics, transportation on nanomaterials in cell culture, and experimental animal models [48]. Most of the nanomaterials move across the biological barriers and bind with biomolecules and also with factors or components that inhibit the enzyme activity and components of the immune system of a biosystem. The presence of metallic group and toxic compounds, induce their respective effects during such interactions [49]. Some of the fundamentals of physics that regulate the interaction of nanomaterials include 'quantum mechanism, tunneling effect, quantum biology. The quantum mechanics is concerned with the motion and energy of the atom and its electrons. Nanomaterials have a low dimension, and their mass becomes extremely less, as a result, the effect of gravitational force becomes negligible. In these conditions, electromagnetic force regulates the behavior of atoms, molecules. Since, nanomaterials behave as elementary particles because of their dimensions, follow the wave-particle-duality concept. Nanomaterials under the quantum mechanics display the quantum confinement. Quantum biology is the function of quantum mechanism and theoretical chemistry and is most apt in biological sciences. Most of the biological processes concern the interconversions of energy and chemical transformation; mass being very negligible may not be considered but exists in such cases. The quantum tunneling is affecting the matter at nano or low dimensions; this suitably regulates the movement of atoms, molecules across an energy barrier. The fundamentals of chemistry that involve intermolecular bonding, dispersion, and distribution, adhere to the laws of equilibrium-equation, net charge, play significant functional roles in the interactions of nanomaterials. Concepts that govern the processes of agglomeration, adsorption, dissolution, stable covalent bonding, surface chemistry, nucleophilic and electrophilic affinities, electron distribution, energy transfer, *etc.*, also have their specific roles during nanomaterial interaction. DLVO theory and forces like van der Waals forces along with Keesom, Debye, and London forces (related to dipole concept) act as regulatory forces to accomplish the interactions of nanomaterials [50 - 58].

Administered nanomaterials encounter body fluids and cells in a biosystem. They may undergo protein corona formation, and such products are up-taken without undergoing any activation of cell receptors. These wonder particles have the potential to stimulate and influence cells, cell organelles, and associated components. Stimulus-response base interactions exhibit a wide range of applications in various fields. There is a great challenge for nanotechnology and nanoscience to deal with the unpredictability of the derogative impacts and to evaluate the harmful interactions of nanomaterials using the current methodology of nanotechnology, specifically, interactions between nanomaterials and cells. The glycocalyx is the first cell-associated part that nanomaterials confront after internalization. Pericellular matrix, popularly, called glycocalyx, is a specialized layer located between the extracellular matrix and the cell membrane. This sandwiched layer is structurally and functionally associated with both. Glycocalyx dynamically maintains the equilibrium between the body fluid flowing along and its soluble components. It consists of proteoglycans, glycosaminoglycan, and glycoproteins [19, 59 - 63]. Enzymes or the rheological impacts, *etc.*, can erode glycocalyx. Angiopoietin, with the help of heparanase enzyme, sheds or reduces the thickness of the glycocalyx. This layer covers the adhesion and signaling molecules of the cell membrane, thereby, inhibits attachment, uptake, and translocation of the nanomaterials, but, once the glycocalyx gets degraded or eroded, these activities take place actively [64]. Various types of nanomaterials undergo cellular uptake and can cause deformation of membrane or formation of membrane-bound vesicles or carriers. The most common mode of internalization includes phagocytosis, macro, and micro-pinocytosis, caveolae-mediated and clathrin-mediated endocytosis. These processes involve filopodia, lamellipodia, circular ruffles, bleb formation, *etc.* Flotillin proteins (other name Reggie proteins) require during cellular uptake of a particle smaller than 100nm [65]. Biomolecules, like, cell-fusogenic proteins, cell-penetrating or fusogenic peptide motifs, proton sponge hypothesis, enhanced permeability and retention effect, mass transport, degree of degradability, *etc.*, play their significant roles during internalization of nanomaterials. There are some specific pathways like the classical pathway, C-reactive pathway, lectin pathway, alternative pathway are also involved in cellular uptake. The cytoskeleton plays an active role in various aspects of cellular functionality, like cellular elasticity and plasticity, mobility, adhesion, invasion, proliferation, differentiation, phagocytosis, endocytosis, and exocytosis and responding to all types of stimuli. Carbon-based nanomaterials, metal, and metal oxide nanoparticles, quantum dots, *etc.*, induce biophysical impacts during their interactions specifically at the interface formed at the site of interaction [66].

Once nanomaterials get internalized, they encounter protein because protein biomolecules are present in a relatively higher amount, and, many forms. These participate in every physiological and biochemical process. The interactions between nanomaterials and proteins include sensation, molecular assembly of specific proteins, inter and intracellular communication and, other interactions related to cell and cell organelles. Such interactions and the physicochemical features of nanomaterials and proteins have significant roles in accomplishing, either beneficial or derogative interplay [67]. In a biological system, native protein represents functionally stable natural conformation. Parameters like pH, temperature, elimination of water, exposure to hydrophobic surfaces, metal ions, and elevated shear force can influence the stability of proteins. During the denaturing of proteins, the structure is disturbed, and chemical degradation, oxidation, deamidation, and, hydrolysis based on the peptide-bond, reshuffling, or breaking of the disulfide bond, *etc.*, takes place [68]. The processes like opsonization, the ability to form protein corona, and zeta potential influence the interaction between protein and nanomaterials. Other parameters like size, shape, surface charge on nanomaterials, hydrophilic and hydrophobic nature, the composition of nanomaterials, conformation of the interacting protein, duration of exposure, static and dynamic state of body fluid and temperature, nature of bionanointerface, *etc*, influence the process of formation of protein corona and also protein nanomaterials interaction. There are plenty of applications of protein corona in biomedical, biomolecular, biophysical, and biochemical fields [69]. Interactions between nanomaterials and proteins may cause aggregation and folding of proteins. Such dysfunctions lead to some clinical conditions. The nanoparticles cause or act as artificial chaperons and result in fibrillogenesis or detect intermediate folding. The nanoparticle-protein-corona induces conformational changes in protein and also affects cellular interaction [70]. Biodistribution of nanomaterials can go off the target cells or the tissues. Such unwanted biodistribution of nanomaterials can be the cause of toxicity, the decline in therapeutic efficacy immune-related, or other unwarranted physiological activities [71]. Protein chip is a micro-device fabricated based on DNA microassay. It is helpful to investigate many proteins simultaneously, protein interactions, and their functionalities [72].

The genetic material is a stable and intact bioentity that maintains its structural, functional, phylogenetic integrity at least under normal physiological conditions. This nature of genetic materials is of practical significance because it helps to avoid erroneous genetic configuration in an organism and its offspring. Temperature, pH, ionic strength, density, hydrophobic and hydrophilic nature, the impact of exposure of radiations on the optical properties of genetic material (DNA), *etc*, are the structural and functional limiting factors [73 - 75]. Entropy, elasticity, and stack forces concerning nanomaterials can cause structural and functional fluctuations in genetic materials [76 - 78]. The fundamental intermolecular forces participate during the interaction between nanomaterials and genetic materials. These interactions bring changes in the conformation of DNA; as a result, the dynamic light scattering, zeta potential, of treated and untreated DNA and RNA changes, and these changes reflect on the intensity of the interaction. The non-specific interactions disrupt the existing H-bonds in a short double-stranded DNA and take place in the case of interplay between salts of small metal nanoparticles because this binds non-specifically when oligonucleotides of DNA are melting. It also prevents the hydration of a complementary DNA sequence in a standard buffered solution. This non-specific interaction is size-dependent; as size increases the interaction becomes weaker [79]. Generally, the complex of biomolecules and ligands, act as a suitable template with nanoparticles, and function as a catalyst for specific interaction [80]. The interactions between nanomaterials and DNA and RNA, along with their different conformations, are good options as an agent for non-ionized imaging, therapeutics, diagnostics, delivery system, and RNA technology. Mostly nanomaterials interact at Pi-Pi stacking (π-π stacking) within the strand and H-bonds of the nucleic acids. Highly charged nanoparticles interact with single-stranded nucleic acids and convert them into the compact form [81].

Biological activities are ambiguous and come under a broad umbrella of proteins. Pharmaceutical and pharmacy-based industries practice biocatalytic and enzymatic interactions [82]. Shape, size, charge, hydrophobicity or hydrophilicity and others, of the substrate, nanomaterials, are basic features and incriminate during the formation of enzyme-substrate-complex. Enzymes are substrate specific in nature [83]. The primary functional factors, like activation energy, free energy change, the structure of an enzyme and substrate, are essential for the successful accomplishment of an enzymatic interaction. Enzyme interaction Immobilization of enzyme is a versatile aspect of enzyme technology and adsorption, entrapment, cross-linking, and covalent bonding is necessary processes in this technology. Physical binding of either enzyme or cell, or with the surface of the inert and inorganic supportive matrix are involved during adsorption. Silica gel, calcium phosphate gel, glass, alumina, act as an inorganic support matrix [84, 85]. It is of common observation that the nano-based matrix is advantageous in comparison to the traditional matrix. The nanostructured matrix can elevate the efficacy of biocatalysts, specific areas, mass transfer resistance, and loading of the active enzyme. A matrix consisting of nanomaterials like silica, chitosan, gold, diamond, graphene, and zirconium at the nanoscale, are potentially suitable materials for immobilization of enzyme [86].

There are functional factors, like, the chemical composition of the reactants, electrostatic interactions due to surface charge on the nanomaterials, hydrophobic nature of nanomaterials and lipophilic groups present on nanomaterials affect the interactions, and competitive or non-competitive restricted enzyme activity affect the interactions between nanomaterials and the respective enzymes [87]. Whenever there is a change or alteration in these functional parameters, the enzymatic activity under consideration gets affected or becomes inactive functionally. This situation may not favor its industrial applications and enzyme technology. Enzyme technology inducting nanomaterials plays a significant role in clinical, biotechnology, and therapeutic fields [88].

The immune system functions intriguingly. Any entrant, including nanomaterial, to organisms, has to encounter the immune system. In the human body and vertebrates, the components of blood, such as monocytes, platelets, leukocytes, and dendritic cells in tissues, macrophages in lungs, engulf (phagocytosis) these internalized molecules, living or non-living, both. Plasma proteins, opsonins, and immune-related components also interact with the foreign bodies that enter the biosystem. This behavior of the immune system interferes with the interactions of nanomaterials used as a delivery system, prosthesis, sensors, diagnostic and other medical devices, thereby, affecting their biodistribution, clearance and deviate them from the set target tissue [89]. Among human beings, innate immunity is non-specific, and it is the first one to recognize foreign bodies. The pattern recognition receptors (PRPs) dedicate to this identification and the pro-inflammatory response [90]. Nanomaterials activate the complement system- a component of the innate immune system, significantly and have specificity with respect either to inhibit or to enhance the immune response. The successful fabricated or engineered nanomaterials, as drug-carrying agents and, must be nonimmunotoxic but immuno-compatible. These agents should not get destroyed or get eliminated by the components of the immune system [91]. The same functional factors are also applicable to biomedical implants and biosensors.

Immunomodulation plays an essential role in the success of immunotherapy. The process of immunomodulation is vital because it provides either stimulatory or inhibitory signals to T-cells. Dendritic cells, antigens, and antigen-presenting cells play key roles during immunomodulation. The development of innovative nano-bio-material is the focus during the interactions between the modified nanomaterials and the adaptive and innate aspects of the immune system.

Bionanomaterials display immuno-simulative, neutral, or immunosuppressive interactions depending on their type and of fabrication [92, 93]. Immunostimulation is unintended or inappropriate antigen-specific or non-specific activation of the components of the immune system. Nanomaterials are investigated to evaluate their immunostimulatory potentials concerned with the stimulation of innate or adaptive immune responses. These investigations include cytokine secretion, induction of antibody response involving immunogenicity, and complement immune system. The antigenicity of nanoparticles is still in its infancy. Nanomaterials like nanowires, nanotubes, nanoparticles, cantilevers, micro-nano-arrays, are employed in the diagnostics and biomedical sensors. Biodistribution of nanoparticles is based on affinities, kinetics, and stoichiometry of protein that influences the association and dissociation of proteins with nanoparticles. Fabricated nanomaterials like polymeric nanoparticles, nanoliposomes, nanoemulsions, and solid lipid nanomaterials are suitable for biomedical applications [94 - 96]. These regulate the biological responses to specific nanomaterials. Proteins like albumin, IgG, IgM, fibrinogen, and apolipoproteins interact with iron oxide nanoparticles and form protein corona. Protein binding with the nanoparticles involves pre-incubation with bulk plasma/serum or pre-incubation of an individual protein or attaching specific protein [97]. The selective cellular uptake of nanomaterials is applicable to deliver anticancer drugs to tumors involving the Trojan horse mechanism [98].

CONCLUSION

It is easy to envisage that nanomaterials have enormous potential to solve most of the problems related to energy, medical sciences, biomolecular investigations, biotechnology, genetic engineering, tissue engineering, environmental aspects, *etc*. There are possibilities of the derogative impacts on biotic and abiotic components of the environment, industrial developing fields. Their use must be judicially and cautiously made; other aspects like finances, investments, human profit-making tendencies need careful consideration along with consent from the local, state and, federal government regulatory authorities. There is a need to establish the most suitable and distinctly clear risk control policy. This practice must include activities related to the production, storage, distribution, procurement, management, health of personnel involved, regulatory directions, and specific codes on the small scale and large scale production. Any preferred technology must be applied cautiously, and continuous research and investigations should run parallel to the applications to minimize their negative impacts; thereby, help in the sustenance of biota, and the environment.

REFERENCES

[1]   Hubler A, Osuagwvo O. Digital quantum batteries, energy and information storage in nano vacuum

tube array. Complexity 2010; 15(6): 48-55. https:10.1002/cplx.20306

[2]   Saini R, Saini S, Sharma S. Nanotechnology: the future medicine. J Cutan Aesthet Surg 2010; 3(1): 32-3.
[http://dx.doi.org/10.4103/0974-2077.63301] [PMID: 20606992]

[3]   Barbara P. Nanoscience and nanotechnology center makes rapid progress, The University of Texas at Austin – Chemical Compositions 2001; 1-2.

[4]   Moriarty P. Nanostructured materials. Rep Prog Phys 2001; 64(3): 297.
[http://dx.doi.org/10.1088/0034-4885/64/3/201]

[5]   Schwarz C, Stahl U. Eisen 1901; 209-2011.

[6]   Sanderson K. Sharpest cut from nanotubes sword. NATNEWS 2005.
[http://dx.doi.org/10.1038/news061113-11]

[7]   Reibold M, Paufler P, Levin A A, Kochmann W, Patzke N, Meyer D C. Carbon nanotubes in an ancient Damascus sabre, Nature 2006; 444(286)
[http://dx.doi.org/10.1038/444286a]

[8]   Binnig G, Rohrer H, Gerber C, Weibel E. Surface studies by scanning tunneling microscopy. Phys Rev Lett 1982; 49: 57.
[http://dx.doi.org/10.1103/PhysRevLett.49.57]

[9]   Binnig G, Quate CF, Gerber C. Atomic force microscope. Phys Rev Lett 1986; 56(9): 930-3.
[http://dx.doi.org/10.1103/PhysRevLett.56.930] [PMID: 10033323]

[10] Ijima S. Helical microtubules of graphitic carbon. Nature 1991; 354: 56.
[http://dx.doi.org/10.1038/354056a0]

[11] NNI. National Nanotechnology Initiative 2000.www.nano.gov

[12] Read AM. A brief history of Nano, Mater today 2004; 7(7-8): 55.
[http://dx.doi.org/10.1016/S1369-7021(04)00350-5]

[13] The Royal Society. The Royal Academy of Engineering 2004.www.royalsoc.ac.uk

[14] Derycke V, Martel R, Appenzeller J, Aviouris P. Carbon nanotubes inter and intramolecular logic gates. Nano Lett 2007; 1: 453.
[http://dx.doi.org/10.1021/nl015606f]

[15] Bardosova M, Wagner T. Nanomaterials and nano-architectures, NATO Science for Peace and security Series-C, Environmental security, Tyndall, Univerzita Pardubice, Springer Science Business Media Dordrecht, 2015; ISBN 978-94-017-9920-1 (HB).

[16] Stefan L, Martina L, Peter S, Matej J, Eva M. A brief history of nanoscience and foresight in nanotechnology, In Nanomaterials and nano-architectures, NATO Science for peace and security Series-C, Environmental security, Tyndall, Univerzita Pardubice, Springer Science Business Media Dordrecht, 2015; ISBN 978-94-017-9920-1(HB).

[17] Nie S, Xing Y, Kim GJ, Simons JW. Nanotechnology applications in cancer. Annu Rev Biomed Eng 2007; 9: 257-88.
[http://dx.doi.org/10.1146/annurev.bioeng.9.060906.152025] [PMID: 17439359]

[18] Hulla JE, Sahu SC, Hayes AW. Nanotechnology: history, and future, Human and Environmental Toxicology 2015; 34(12): 1318-21.

[19] Lahir YK. A dynamic component of tissue- Extracellular matrix: a structural and adaptive Approach. Biochem Cell Arch 2015; 15(2): 331-47.

[20] Lahir YK. Graphene and graphene-based nanomaterials are suitable for drug delivery, Chapter-7 in Applications of Targeted Nano-Drugs and Delivery Systems. 50-Hampshire Street, Cambridge, MA 02139: Elsevier 2018; ; pp. 157-90.

[21]   Moraru CI, Panchapakesan A, Huang Q, *et al.* Nanotechnology: a new frontier in food science. Food Technol 2003; 57(12): 24-9.

[22]   Weiss J, Takhistov P, McClements DJ. Functional materials in food nanotechnology. J Food Sci 2006; 71(9): R107-16.
[http://dx.doi.org/10.1111/j.1750-3841.2006.00195.x]

[23]   Jin T, Sun D, Su JY, Zhang H, Sue HJ. Antimicrobial efficacy of zinc oxide quantum dots against Listeria monocytogenes, Salmonella Enteritidis, and Escherichia coli O157:H7. J Food Sci 2009; 74(1): M46-52.
[http://dx.doi.org/10.1111/j.1750-3841.2008.01013.x] [PMID: 19200107]

[24]   Lahir YK. Role and adverse effects of nanomaterials in food technology. J Toxicol Health 2015; 2
[http://dx.doi.org/10.7243/2056-3779-2-2] [PMID: 2]

[25]   Oberdörster G, Maynard A, Donaldson K, *et al.* Principles for characterizing the potential human health effects from exposure to nanomaterials: elements of a screening strategy. Part Fibre Toxicol 2005; 2(8): 8.
[http://dx.doi.org/10.1186/1743-8977-2-8] [PMID: 16209704]

[26]   Maynard A. Nanotoxicologist self assemble, Nanotoxicology 2008. http://2020science/org/2008/09/08/nanotoxicologists-self-assemble/

[27]   Egusquiaguirre SP, Igartua M, Hernández RM, Pedraz JL. Nanoparticle delivery systems for cancer therapy: advances in clinical and preclinical research. Clin Transl Oncol 2012; 14(2): 83-93.
[http://dx.doi.org/10.1007/s12094-012-0766-6] [PMID: 22301396]

[28]   Gu Z, Aimetti AA, Wang Q, *et al.* Injectable nano-network for glucose-mediated insulin delivery. ACS Nano 2013; 7(5): 4194-201.
[http://dx.doi.org/10.1021/nn400630x] [PMID: 23638642]

[29]   Lahir YK. Impacts of fullerene on a biological system. Clin Immunol Endocrinol Metabol Drugs 2017; 4(1): 48-57.
[http://dx.doi.org/10.2174/2212707004666171111351624]

[30]   Bückmann T, Thiel M, Kadic M, Schittny R, Wegener M. An elasto-mechanical unfeelability cloak made of pentamode metamaterials. Nat Commun 2014; 5: 4130.
[http://dx.doi.org/10.1038/ncomms5130] [PMID: 24942191]

[31]   Schaller B. The origin, nature, and implication of Moor's Law 2013.http://research.microsoft.com/en-um/people/gray/Moor_La/html

[32]   Christos T. European Commission; ISBN 978-92-79-28892-0. Nanotechnology: the invisible giant tackling Europe's future challenges, DG Res Inov Ind Tech EUR . 2013.
[PMID: 13325EN2013]

[33]   Grilli S, Vespini V, Ferraro P. Surface-charge lithography for direct polydimethylsiloxane (pdms) micropatterning. Langmuir 2008; 24(23): 13262-5.
[http://dx.doi.org/10.1021/la803046j] [PMID: 18986187]

[34]   Paturozo M, Grilli S, Mailis S, *et al.* Flexible coherent diffraction lithography by tunable phase arrays in Lithium niobate crystals. Opt Commun 2008; 281(8): 1950-3.
[http://dx.doi.org/10.1016/j.optcom.2007.12.056]

[35]   Raster A, Yazdanshenas ME, Rashidi A, Bidoki SM. Theoretical review of optical properties of nanoparticles. J Eng Fibers Fabrics 2013; 8(2): 85-96.https://www.jeffjournal.com
[http://dx.doi.org/10.1177/155892501300800211]

[36]   Chen H, Wang B, Gao D, *et al.* Broad-spectrum antibacterial activity of carbon nanotubes to human gut bacteria. Small 2013; 9(16): 2735-46.
[http://dx.doi.org/10.1002/smll.201202792] [PMID: 23463684]

[37]   Chen Q, Xue Y, Sun J. Kupffer cell-mediated hepatic injury induced by silica nanoparticles *in vitro*

and *in vivo*. Int J Nanomedicine 2013; 8: 1129-40.
[http://dx.doi.org/10.2147/IJN.S42242] [PMID: 23515466]

[38]   Zeng S, Baillargeat D, Ho HP, Yong KT. Nanomaterials enhanced surface plasmon resonance for biological and chemical sensing applications. Chem Soc Rev 2014; 43(10): 3426-52.
[http://dx.doi.org/10.1039/c3cs60479a] [PMID: 24549396]

[39]   Chemistry, LibretextsTM, Surface Energy 2016. https://chem.libretexts.org/Bookshelves/Inorganic_Chemistry/Book%3A-inorganic-Chemistry

[40]   Yu Y, Cui C, Liu X, Petrik ID, Wang J, Lu Y. A designed metalloenzyme achieving the catalytic rate of the native enzyme. J Am Chem Soc 2015; 137(36): 11570-3.
[http://dx.doi.org/10.1021/jacs.5b07119] [PMID: 26318313]

[41]   Subbiah R, Veerapandian M, Yun KS. Nanoparticles: functionalization and multifunctional applications in biomedical sciences. Curr Med Chem 2010; 17(36): 4559-77.
[http://dx.doi.org/10.2174/092986710794183024] [PMID: 21062250]

[42]   Keller AA, Wang H, Zhou D, *et al.* Stability and aggregation of metal oxide nanoparticles in natural aqueous matrices. Environ Sci Technol 2010; 44(6): 1962-7.
[http://dx.doi.org/10.1021/es902987d] [PMID: 20151631]

[43]   Fratila RM, Mitchell SG, del Pino P, Grazu V, de la Fuente JM. Strategies for the biofunctionalization of gold and iron oxide nanoparticles. Langmuir 2014; 30(50): 15057-71.
[http://dx.doi.org/10.1021/la5015658] [PMID: 24911468]

[44]   Schoth A, Keith AD, Landfester K, Rafael ME. Silanization as a versatile functionalization method for the synthesis of polymer/magnetic hybrid nanoparticles with controlled structure. Royal Society Chemistry Advances 2016; 6: 53903-11.
[http://dx.doi.org/10.1039/C6RA08896A]

[45]   Ansar SM, Chakraborty S, Kitchens CL. pH-responsive mercaptoundecanoic acid functionalized gold nanoparticles and applications in catalysis. Nanomaterials (Basel) 2018; 8(5): 339.
[http://dx.doi.org/10.3390/nano8050339] [PMID: 29772775]

[46]   Zhang Y, Hoppe AD, Swanson JA. Coordination of Fc receptor signaling regulates cellular commitment to phagocytosis. Proc Natl Acad Sci USA 2010; 107(45): 19332-7.
[http://dx.doi.org/10.1073/pnas.1008248107] [PMID: 20974965]

[47]   Rulla M, Louisa JHW. Louisa JHW, Synthesis of ZnO-CuO nanocomposites aerogel by sol-gel route Journal of Nanomaterials 2014; 2014: 9.
[http://dx.doi.org/dx.doi.org/10.1155/2014/491817] [PMID: 491817]

[48]   Albanese A, Tang PS, Chan WC. The effect of nanoparticle size, shape, and surface chemistry on biological systems. Annu Rev Biomed Eng 2012; 14(1): 1-16.
[http://dx.doi.org/10.1146/annurev-bioeng-071811-150124] [PMID: 22524388]

[49]   Rahi A, Sattarahmdy N, Heli H. Toxicity of nanomaterials –physicochemical effects. Austin J Nanomed Nanotechn 2014; 2(6): 1034.www.austinpublishing group.com

[50]   Linus AJ, Edgar BW. ISBN: 0486648710; Introduction to the quantum mechanism with an application to chemistry. Courier Corporation 1985. 978048664712.

[51]   Jiang J, Oberdorfer G, Biswas P. Characterization of size, surface charge, and agglomeration state of nanoparticles dispersion for toxicological studies. J Nanopart Res 2009; 11: 77-89.
[http://dx.doi.org/10.1007/s11051-008-9446-4]

[52]   Panitchayangkppn G, Mayes D, Fransted KA, *et al.* Long-lived quantum coherence in photosynthesis complexes at physiological temperature, of the National Academy of Sciences (USA) 2010; 107(29): 12766-70.
[http://dx.doi.org/10.1073/pans.1005484107]

[53]   Sun J, Wang F, Sui Y, *et al.* Effects of particle size on solubility, dissolution rate and oral

bioavailability: evaluation using coenzyme Q10 Int J Nanomedicine 2012; 7: 5733-44.
[http://dx.doi.org/10.2147/IHN.S34365] [PMID: 23166438]

[54] Nelson DL, Cox MM, Michael M. Lehninger Principles of Biochemistry. 6th ed., New York: W H Freeman & Co 2013.

[55] Trixler F. Quantum tunneling to the origin and evolution of life. Curr Org Chem 2013; 17(16): 1758-70.
[http://dx.doi.org/10.2174/13852728113179990083]

[56] Rudyak V. Ya, Krasnolutskii SL, (2014) Dependence of viscosity of nanofluids on Nanoparticles size and nanomaterials. Phys Lett A 2014; 378(26-27): 1845-9.
[http://dx.doi.org/10.1016/j.physleta.2014.04.060]

[57] Baalousha M, Lead JR. Frontiers of Nanoscience; Characterization of nanomaterials in complex environment and Biological media. Amsterdam, Boston, London, New York, Sydney: Elsevier 2015; 8: pp. 78-9. www.elsevier.com

[58] Brookes JC. Quantum dots effects in biology: golden rule in enzymes, olfaction, Photosynthesis and magnetic detection, Proceedings of the Royal Society, A Mathematical Physical and Engineering Science 2017; 473(2201)
[http://dx.doi.org/10.1098/rwpa.2016.0822]

[59] Alphonsus CS, Rodseth RN. The endothelial glycocalyx: a review of the vascular barrier. Anaesthesia 2014; 69: 777-84.

[60] Cooper GM. The Molecular Approach,. 2nd Ed., Sunderland (MA): Sinauer Associates 2015.

[61] Lahir YH. Understanding the basic role of glycocalyx during cancer. Journal of Radiation and Cancer Research 2016; 7: 79-84.
[http://dx.doi.org/10.4103/0973-0168.197974]

[62] Lahir YK. Impacts of metal and metal oxide nanoparticles on reproductive tissues and spermatogenesis. J Exp Zool India 2018; 18(2): 594-608.

[63] Xu GK, Qian J, Hu J. The glycocalyx promotes cooperative binding and clustering of adhesion receptors. Soft Matter 2016; 12(20): 4572-83.

[64] Hull MS, Vikesland PJ, Schultz IR. Uptake and retention of metallic nanoparticles in the Mediterranean mussel (Mytilus galloprovincialis). Aquat Toxicol 2013; 140-141: 89-97.
[http://dx.doi.org/10.1016/j.aquatox.2013.05.005]

[65] Nixon SJ, Carter A, Wegner J, et al. Caveolin-1 is required for lateral line neuromast and notochord development. J Cell Sci 2007; 120(Pt 13): 2151-61.
[http://dx.doi.org/10.1242/jcs.003830]

[66] Wu YL, Putcha N, Ng KW, et al. Biophysical responses upon the interaction of nanomaterials with cellular interfaces. Acc Chem Res 2013; 46(3): 782-91.
[http://dx.doi.org/10.1021/ar300046u]

[67] Kane RS, Stroock AD. Nanobiotechnology: protein-nanomaterial interactions. Biotechnol Prog 2007; 23(2): 316-9.
[http://dx.doi.org/10.1021/bpo060388n]

[68] Dunbar J, Yennawar HP, Banerjee S, Luo J, Farber GK. The effect of denaturants on protein structure. Protein Sci 1997; 6(8): 1727-33.
[http://dx.doi.org/10.1002/pro.5560060813]

[69] Walkey CD, Olse JBB, Song F, et al. Protein corona fingerprinting predicts the cellular interaction of gold and silver nanoparticles, ACS Nano 2014; 8(30): 2439-55.
[http://dx.doi.org/10.1021/nn.4060/q]

[70] Romana P, Shamsi TN, Fatima S. Nanoparticles and protein interaction: Role in protein aggregation and clinical implications/ International Journal of Biological Macromolecules 2017; 94(Part A): 386-

95.
[http://dx.doi.org/10.1016/j.ijbiomac.2016.10.024]

[71]   Ilinskaya AN, Dobrovlskaia MA. Interaction between nanoparticles and plasma proteins: Effects on nanoparticle distribution and toxicity.polymer nanoparticles for nanomedicine, Springer Chem. DOI 2016.https://doi.org
[http://dx.doi.org/10.1007/978-3-319-41421-8_15]

[72]   Melton L. Protein arrays: proteomics in multiplex. Nature 2004; 429(6987): 101-7.
[http://dx.doi.org/10.1038/429101a] [PMID: 15129287]

[73]   Harris NC, Kiang C-H. Defects can induce melting time of DNA, Nanoassemblies. J Phys Chem B 2006; 110(33): 16393-6.
[http://dx.doi.org/10.1021/jp062287d] [PMID: 16913768]

[74]   Smialek MA, Jones NC, Hoffman SV, Mason NJ. Measuring the density of DNA film using U V –VIS interferometry. Physical Rev J 2013; E87IS56060701

[75]   Cox MM, Nelson DL. Lehninger Principles of Biochemistry. 5th ed., New York: W H Freeman and Co 2013.

[76]   Marko JF, Siggla ED. Bending and twisting elasticity of DNA. Macromolecules 1994; 27(4): 981-8.
[http://dx.doi.org/10.1021/ma00082a015]

[77]   Udgaonkar JB. Entropy in Biology, Resonance, 2001. https://ncbs.res.in/sitefiles/jayant-reson.pdf

[78]   Johnson   S.   What   causes   the   double   helix   to   twist   in   a   DNA   molecule? 2017.https://sciencing.com/causes-double-helix-twist-dna-picture-2848.html

[79]   Yang J, Lee JY, Too HP, Chow GM. Inhibition of DNA hybridization by small metal nanoparticles. Biophys Chem 2006; 120(2): 87-95.
[http://dx.doi.org/10.1016/j.bpc.2005.10.011] [PMID: 16303234]

[80]   Heddle JC. Gold nanoparticle biological molecule interactions and catalyst. Catalysts 2013; 3: 683-708.
[http://dx.doi.org/10.3390/catal3030683]

[81]   Nash JA, Tucker TL, Theriault W, Yingling YG. Binding of single-stranded nucleic acid to cationic ligand 2016. https://avs.scitation.org/doi/am.pdf/10.1116/1.4966653

[82]   Rahman SA, Cuesta SM, Furnham N, Holliday GL, Thornton JM. EC-BLAST: a tool to automatically search and compare enzyme reactions. Nat Methods 2014; 11(2): 171-4.
[http://dx.doi.org/10.1038/nmeth.2803]

[83]   Jeeger K E, Eggert T. Enantioselective biocatalysts optimized by direct evolution, Current Opinion in Biotechnology 2004; 15(4): 305-13.
[http://dx.doi.org/10,1016/j.copbio.2004.06.007]

[84]   Brady D, Jordaan J. Advances in enzyme immobilisation. Biotechnol Lett 2009; 31(11): 1639-50.
[http://dx.doi.org/10.1007/s10529-009-0076-4] [PMID: 19590826]

[85]   Razi A, Maryam S. Enzyme immobilization: An overview of nanomaterials as immobilization matrix. Biochem Anal Biochem 2015; 4: 178.
[http://dx.doi.org/10.4172.2161-1009.1000178]

[86]   Cipolatti EP, Josse M, Silva MJA, *et al.* Current status and trends in enzymatic nano immobilization, Journal of Molecular Catalysis B. Enzymatics 2014; 99: 56-67.
[http://dx.doi.org/10.1016/j.molcatb.2013.10.019]

[87]   Vineet K. Environmental Toxicity of Nanomaterials 2018.www.crcpress.com

[88]   Kenz CORDIS (Community Research and Information Service), Project ID-322158/Funder under FP7-PEOPLE; Record no 214317.

[89]   Dobrowolski MA, Aggarwal P, Hal JB, McNeil SE. Preclinical studies to understand nanoparticle

interaction with the immune system and its potential effects on nanoparticles, biodistribution. Mol Pharmacol 2008; 5(4): 487-95.
[http://dx.doi.org/10.1021/mp800032f]

[90] Luo YH, Chang LW, Lin P. Metal-based nanoparticles, and the immune system: activation, inflammation and applications. Biomedical Research International 2015.

[91] Boraschi D, Costantino L, Italiani P. Interaction of nanoparticles with immunocompetent cells: nanosafety considerations. Nanomedicine (Lond) 2012; 7(1): 121-31.
[http://dx.doi.org/10.2217/nnm.11.169]

[92] Fontana F, Shahbazi MA, Liu D, *et al.* Multistaged nano vaccines based on porous silicon acetal Dextran cancer cell membrane for cancer immunotherapy, Advanced Materials 2017; 29(7)1603239
[http://dx.doi.org/10.1002/adma.20 1603239]

[93] Fontana F, Figueiredo P. Bauleth- Ramos T, Correia A, Santos HA, Immunostimulation and immunosuppression: Nanotechnology on the brink. Small Methods 2018; 2(5): 1700347.
[http://dx.doi.org/10.1002/smtd.201700347]

[94] Pan X, Chen L, Liu S, Yang X, Gao JX, Lee RJ. Antitumor activity of G3139 lipid nanoparticles (LNPs). Mol Pharm 2009; 6(1): 211-20.
[http://dx.doi.org/10.1021/mp800146j] [PMID: 19072654]

[95] Eshete M, Bailey K, Nguyen TDT, Santosh A, Choi S-O. Interaction of immune system protein with PEGylated and un-PEGylated polymeric nanoparticles. Advan Nano 2017; 6(3): 103-13.
[http://dx.doi.org/10.4236/anp.2017.63009] [PMID: 63009]

[96] Zamani P, Momtazi-Borojeni AA, Nik ME, Oskuee RK, Sahebkar A. Nanoliposomes as the adjuvant delivery systems in cancer immunotherapy. J Cell Physiol 2018; 233(7): 5189-99.
[http://dx.doi.org/10.1002/jcp.26361] [PMID: 29215747]

[97] Mukherjee SP, Bondarenko O, Kohonen P, *et al.* Macrophage sensing of single-walled carbon nanotubes via Toll-like receptors. Sci Rep 2018; 8(1): 1115.
[http://dx.doi.org/10.1038/s41598-018-19521-9] [PMID: 29348435]

[98] Benget F. Hide and seek: Nanomaterial interaction with the immune system. Front Immunol 2019.
[http://dx.doi.org/10.3389/fimmu.2019.00133]

CHAPTER 2

Nanomaterials and their Behavioral Aspects

Abstract: Nanomaterials exhibit some extraordinary features. These features are the bases for their applications in different fields such as biomedical, pharmaceuticals, communication, warfare, clothing, sports industries, automobiles, *etc*. Reports reflect on their interactions with abiotic and biotic components of the environment. It is very imperative to understand their interactions with biomolecules or related materials. These investigations elaborate on their benefits and, damaging effects; these ascertain their appropriate applications. The concerned reactants may be natural, organic, or inorganic. Nanomaterials interact with components of an environment in a medium like air or water, on the bases of their specific structure and functional groups. During such interactions, the physiological and ecological parameters of the environment also play a significant role. The physicochemical properties of nanomaterials and surface functionalities are due to the specific modifications of the nanomaterials. The hydrophobicity or hydrophilicity of nanomaterials influences their interactions between them and the biological and ecological systems. This chapter deals with the behavior of nanomaterials, parameters, and conditions related to their interaction in a biosystem.

Keywords: Absorption, Applications, Drug delivery Systems, Nanomaterials, Physicochemical Properties, Plasmonic Nanoparticles, Scattering, Surface Energy, Tissues Engineering.

INTRODUCTION AND OVERVIEW

Nanomaterials are present in the environment and influence its abiotic and biotic components. Although the world nano is relatively recent, the term, minimal particulate matter, as earlier conceived, has left its impact on animals, fungi, microbes, and plants. These particles are the products of human and natural activities or processes such as combustion, volcanic eruptions, forest fire, dust storms, tornado, domestic dust, and anthropogenic activities, *etc*. Many misconceptions or miss apprehensions, like bad air, bad/evil spirits, phobia, *etc*., are associated with the nanomaterials. Research about nanomaterials reveals their link with respiratory, cancerous, cardiovascular diseases and mortality, *etc* [1]. During this period, these particles have not been precisely defined, characterized, or classified. Their impacts are also not technically analyzed or understood. Such particles differentiate as coarse, ultra-fine, and very fine-textured particles, (bhasm). These are used in different medical branches like Ayurveda, Homeo-

Yogendrakumar H. Lahir & Pramod Avti

pathy, or Unani Medicine for the treatments. The inception of the nanoscale is helpful in the characterization of nanomaterials, and the related investigations enhance the understanding of mechanisms concerning their interactions with biotic and abiotic components of the environment [1].

Studies related to nanomaterials involve size, shape, and, physical features, and, characterization under the guidance of the principle of materials science and metrology (the science of measuring). Such investigations reflect on the term nanoscale (material with at least one dimension in the range 10^-9 m or 1-100 nm). Studies related to the structural and functional aspects of nanomaterials elaborate on the unique physicochemical, optical, electronic, electrical, and thermal conduction and mechanical properties, *etc*. These understandings lead to the successful micro and macro fabrications and strengthen the concept of nanotechnology, its commercialization, and commoditization [2 - 4].

DEFINING NANOMATERIALS

Usually, a state or a phase of matter is defined to ease its detailed description and for identification or nomenclature. This process is solemnized based on the specific features or properties or the impacts of that matter or state. Defining nanomaterials is relatively an area of active scientific and policy-related debate [5]. Maynard (2011) expressed his view against characterizing the manufactured nanomaterials. Related definitions are helpful in identification, description (at least to some extent) of the mater under consideration to assign them a specific nature or impact as benign or hazardous. Understandably, matter at nanoscale exhibits different specific properties as compared to their corresponding bulk forms. Defining the nanomaterials helps to know about their safety, impacts, and concern precautions against their derogative effects. This topic is subjected to legitimate public concern, adaptation, and political acceptance or response and overall industrial applications. Hence, an appropriate definition is a need, even though nanomaterials show heterogeneous nature. Their category defines nanomaterials, impact on human health, and environmental risk. There are chances that technical definitions based on parameters like size, maybe deficient or insufficient but may be helpful to evaluate the risks involved in their commercial production, investment, and marketing aspects [6 - 8].

The following are the definitions approved by some of the global regulatory authorities.

1-"Nanomaterials are insoluble or biopersistent. These materials are either intentionally manufactured or fabricated with one or more external dimensions or an internal structure with the scale from 1 to 100 nm". This definition is regulatory and is proposed by the European Commission on cosmetics Directives. Nanomaterials include under the cosmetic category [9].

2-"The products under FDA regulation are useful in engineered nanomaterials. The prime parameters include dimension and that should be within 1nm to 100 nm range or the physical and chemical properties or biological effects that change the dimension of the nanomaterial". This definition is a piece of advice, but this agency suggests no formal description. Cosmetics, pharmaceuticals, food, and food packaging materials come under these guidelines [10].

3-"There are mostly only outlines that are advisory definitions related to the nanomaterials that are involved in products. Such materials should be solid at 25°C and atmospheric pressure with particle size within 1nm to 100 nm at least one dimension. The materials should show unique and novel properties because of the size, the engineered particles or the aggregates and agglomerates should be within the range but not greater than 10% by weight and dimension less than 100 nm". These are advisory definitions and applicable to all products excluding cosmetics, pharmaceuticals, food, and food packaging materials [11 - 13].

4-"Nanomaterials means a natural or manufactured active or non-active substances containing such particles, in an unbound or as an aggregate or as an agglomerate and where for 50% or more of the particles in the number, size distribution, one or more external dimensions having the size range 1nm- 100nm. The particles of fullerene, graphene flakes, and single-walled carbon nanotubes and that have one or more external dimensions below 1nm are grouped as nanomaterials" [14].

5-"Substances produced in the nanoparticular state define as substances containing unbound particles or aggregates or agglomerate of those particles where 50% or more of the particles in number, size distribution have one or more external dimensions within the size range of 1nm -100 nm. The definition excludes natural, non-chemically modified substances and those for which the fraction in the 1nm-100nm range is a by-product of human activity". These are the regulatory definitions based on a complex set of exemptions [15].

USES OF NANOMATERIALS

Nanomaterials are of multi-utilities and are in use in varieties of industries. The silver, silicon dioxide, potassium, calcium, iron, zinc, phosphorous, boron, zinc oxide, and molybdenum nanomaterials are applicable in the field of agriculture. Tungsten, disulfide silicon dioxide, boron, clay, titanium dioxide, diamond, copper, cobalt oxide, zinc oxide, boron nitride, zirconium dioxide, γ-aluminum oxide, palladium, platinum, cerium-IV oxide, carnauba, aluminum oxide, silver, calcium carbonate, and calcium sulfonate are useful in the automotive/automobile industries. Silver, titanium dioxide, gold, carbon, zinc oxide, silicon dioxide, clay, sodium silicate, kojic acid, hydroxy acid, *etc.*, are used in the cosmetics either directly or indirectly. Titanium dioxide, silicon dioxide, silver, clay, aluminum oxide, calcium carbonate, calcium silicate hydrate, cerium-IV, aluminum phosphate, *etc.* are used in the various aspects of the construction industry. In electronic goods and electronic industry nanoparticles such as silver, palladium, aluminum, silicon dioxide nanoparticles are involved. In the food industry, nanoparticles like silver, gold, titanium dioxide, zinc oxide, silicon dioxide, copper, zinc, platinum, manganese, palladium, and carbon, *etc.* find their applications in various aspects of food technology [16]. Silver, titanium dioxide, carbon-nanomaterials, manganese oxide, clay, gold, and selenium nanoparticles are useful in environmental sciences. Titanium nanoparticles, palladium, tungsten disulfide, silicon dioxide, clay, graphite, zirconium-IV oxide-yttria (stabilized), carbon, gd-doped-cerium-IV oxide, nickel-cobalt oxide, nickel-II oxide, rhodium, cerium-IV oxide doped with barium strontium titanate and silver nanoparticles, have applications in various aspects of the renewable energy industry. Silver nanoparticles and other nanomaterials like zinc oxide, silicon dioxide, diamond, and titanium dioxide find utility in home appliances. Silver and gold and nanomaterials in the form of hydroxyapatite, clay, titanium dioxide, silicon dioxide, zirconium dioxide, carbon, diamond, aluminum oxide, and yttrium, nanoparticles are beneficial applications in the field of medicine. Tungsten, boron, silver, disulfide zinc oxide, silicon dioxide, diamond, clay, boron nitride, titanium dioxide, γ-aluminum oxide, carbon, molybdenum disulfide, and γ-aluminum oxide, nanomaterials play active roles in the petroleum industry. Even the textile and sports industries make use of nanomaterials to enhance the effectiveness of sports goods. Nanomaterials, like silver, titanium dioxide, gold, clay, carbon, copper sulfate, polyethylene terephthalate, and silicon dioxide are suitable for various modifications to the desired functionalities [17 - 19].

Inorganic nanomaterials like quantum dots, nanowires, nanorods, *etc.*, are familiar examples for their specific optical and electrical features depend on their size and shape, and these features help to fabricate nanomaterials to suit an appropriate set target. As a result, such fabricated nanomaterials become excellent options in the field of optoelectronic. These associate with organic materials based on optoelectronic devices like organic solar cells, because, these nanomaterials are efficient agents and assist the photo-induced processes like electron transfer and energy transfer. This specific feature is related to the particular ratio of the inorganic and organic contents used. Nanocrystals having metallic, semiconductors, or oxide origin, exhibit new mechanical, electrical, magnetic, optical, and chemical properties. These products are applicable in various applications, like quantum dots, chemical catalysts, and catalysts based on nanomaterials, *etc.* These are very useful in biomedical applications, tissue engineering, drug delivery systems, *etc* [18, 20 - 24]. The details of different types of nanomaterials are shown in Figs. (**1-5**).

ENGINEERED NANOMATERIALS	BYPRODUCTS OR INCIDENTAL NANOMATERIALS	NATURAL NANOMATERIALS*
Intentionally Made with Specific Purposes And Properties: Commercially Important; Needed as Functionally Advanced Nanomaterials: Colloidal, Particulate Matter; Carbon Black; TiO_2 etc; (Riedker and Katalgarianakis, 2012; CDC, 2013; Mackevica et al., 2016; Iavicoli et al., 2014)	Produced as By- Products Because of the Mechanical or Industrial Processes, Vehicular Exhaust, Combustions, Domestic Dust, Welding Process, Mining, Intentional Anthropogenic Activities, *etc* Carbon Soot-Fullerenes, Very Fine Dust or Spray *etc*; (Sura and Daniel, 2009; Damia and Merinella, 2012; Richard, 2014; NCRP, 2017)	Naturally Occurring Nanomaterials Like As Components of Biological System: As Components of Crust of Earth (Natural Inorganic): Crystals, Clay, Fumed Silica, Pigments, Photonic Crystals (Opal) As Products of Natural Combustion Like Forest Fire, Volcanic Eruption: Radio-Active Decay As Products of Weathering Process: As Product at Acid Mine Drainage Site *etc*: (Phys.Org, 2013; NCRP 2017)

Fig. (1). Types of nanomaterials based on their occurrence.

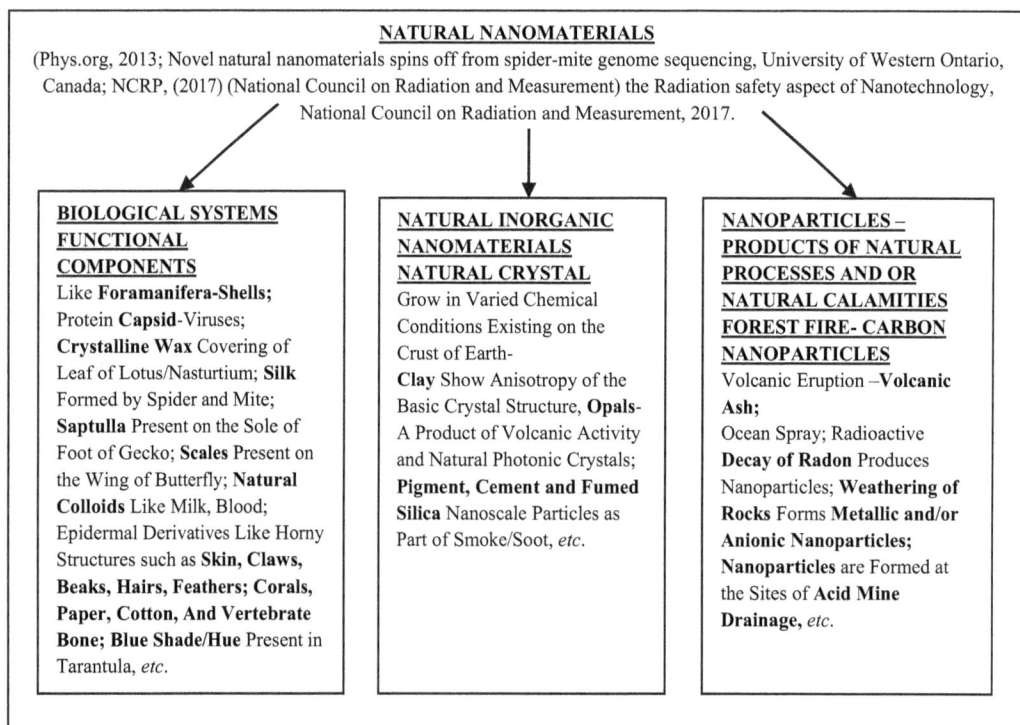

NATURAL NANOMATERIALS

(Phys.org, 2013; Novel natural nanomaterials spins off from spider-mite genome sequencing, University of Western Ontario, Canada; NCRP, (2017) (National Council on Radiation and Measurement) the Radiation safety aspect of Nanotechnology, National Council on Radiation and Measurement, 2017.

BIOLOGICAL SYSTEMS FUNCTIONAL COMPONENTS	NATURAL INORGANIC NANOMATERIALS NATURAL CRYSTAL	NANOPARTICLES – PRODUCTS OF NATURAL PROCESSES AND OR NATURAL CALAMITIES FOREST FIRE- CARBON NANOPARTICLES
Like **Foramanifera-Shells;** Protein **Capsid**-Viruses; **Crystalline Wax** Covering of Leaf of Lotus/Nasturtium; **Silk** Formed by Spider and Mite; **Saptulla** Present on the Sole of Foot of Gecko; **Scales** Present on the Wing of Butterfly; **Natural Colloids** Like Milk, Blood; Epidermal Derivatives Like Horny Structures such as **Skin, Claws, Beaks, Hairs, Feathers; Corals, Paper, Cotton, And Vertebrate Bone; Blue Shade/Hue** Present in Tarantula, *etc*.	Grow in Varied Chemical Conditions Existing on the Crust of Earth- **Clay** Show Anisotropy of the Basic Crystal Structure, **Opals**- A Product of Volcanic Activity and Natural Photonic Crystals; **Pigment, Cement and Fumed Silica** Nanoscale Particles as Part of Smoke/Soot, *etc*.	Volcanic Eruption –**Volcanic Ash;** Ocean Spray; Radioactive **Decay of Radon** Produces Nanoparticles; **Weathering of Rocks** Forms **Metallic and/or Anionic Nanoparticles; Nanoparticles** are Formed at the Sites of **Acid Mine Drainage**, *etc*.

Fig. (2). Natural nanomaterials.

NANOPARTICLES: Any Nanoscaled Object Having All Three Dimensions within Nanoscale but the Longest and Shortest Axes are More or Less Same.
NANOFIBERS: Any Nanoscaled **Solid** Object Having Two Dimensions within Nanoscale.
NANOTUBES: Any Nanoscaled **Hollow** Object Having Two Dimensions within Nanoscale.
NANORODS: Any Nanoscaled Solid Rod like Object Having Dimensions within Nanoscale.
NANOPLATES: Any Nanoscaled Plate like Object with One Dimension within Nanoscale.
NANORIBBON: Any Nanoscaled Flat Object with Two Dimensions within Nanoscale.
NANOFIBERS, NANOTUBES, NANORIBBON, AND NANOPLATES Have One of Their Dimensions is Significantly Larger.
(Nano Forum, (2006) Eighth Nano forum report, Nanotechnology, Nano forum, 2006; ISO/TS 80004-4: 2011, (2011). Nanotechnology –Vocabulary- part-2; Nano-Objects, International Organization of Standardization, 2011; ISO/TS 80004-4: 2011, (2011). Nanotechnology –Vocabulary- part-4; Nano-Objects, International Organization of Standardization, 2011; Klaessig F, Marrapese, M, and Shoji A, (2011). Nanotechnology Standards Nanostructures Science, and Technology, Springer, New York; doi: 10.1007/978-1-4419-7853-0-2.

Fig. (3). Natural and engineered nanomaterials based on their dimension and shape.

NANOCOMPOSITES: Any Nanoscaled Solid Object with One Dimension at Nanoscale and have One or More Physical or Chemical or Collecting Region.
NANOFOAM: Any Nanoscaled Object with either Solid or Liquid Matrix Filled with Gaseous Phase.
NANOPOROUS MATERIALS: Any Material having Nanoscaled Pores or Cavities.
NANO CRYSTAL: Any Nanomaterial that is having Fraction of Crystal Grains.
Note: Quite often **Nanoporous** and **Nanofoam** Materials are not Regarded as Nanomaterials because of their Overall Size is not Regarded at Nanoscale.
Nano Forum, (2006) Eighth Nano forum report, Nanotechnology, Nano forum, 2006; ISO/TS 80004-4: 2011, (2011) Nanotechnology –Vocabulary- part-2; Nano-Objects, International Organization of Standardization, 2011; ISO/TS 80004-4: 2011, (2011) Nanotechnology –Vocabulary- part-4; Nano-Objects, International Organization of Standardization, 2011; Klaessig F, Marrapese, M, and Shoji A, (2011) Nanotechnology Standards Nanostructures Science, and Technology, Springer, New York; doi: 10.1007/978-1-4419-7853-0-2.

Fig. (4). Types of nanomaterials based on their phases of matter.

Matters with nano dimensions have been present in nature much earlier than man could conceive the concept of nano. The width of the DNA molecule is around 3 nm in diameter. Nanostructures present in the wings of a butterfly (morph) and peacock modifies the interaction between the light waves, and, the result is a brilliant blue and green hue. Soap bubbles exhibit iridescence causing varied coloration because the wall of a soap bubble is within the range of nano dimensions. Lotus plants possess some water-resistant nanostructures, *i.e.*, hydrophobic, and water molecules do not adhere to the parts of this plant. An invisible spray of water at the waterfalls or oceanic waves has the nano dimensions. Ultra-fine products of the combustion of fuel (soot) are examples of natural nanomaterials.

TYPES OF NANOMATERIALS BASED ON THE DIMENSIONS Nano Forum, (2006) Eighth Nano forum report, Nanotechnology, Nano forum, 2006; ISO/TS 80004-4: 2011, (2011) Nanotechnology –Vocabulary- part-2; Nano-Objects, International Organization of Standardization, 2011; ISO/TS 80004-4: 2011, (2011) Nanotechnology –Vocabulary- part-4; Nano-Objects, International Organization of Standardization, 2011; Klaessig F, Marrapese, M, and Shoji A, (2011) Nanotechnology Standards Nanostructures Science, and Technology, Springer, New York; doi: 10.1007/978-1-4419-7853-0-2

UNIDIMENSIONAL NANOMATERIALS: These Nanostructures have one Dimension Along with their Cross Section (As Small as Single Atom). Nanotubes and Nanowirs are the Common Examples. These have Mechanical Stability and Used as Template that Maintains their Atomic Configuration.

DIMENSIONAL NANOMATERIALS: These Nanostructures are Two Dimensional Crystalline Matters but have Thickness of One Atomic Layer. Graphene is the Common Example.

MULTIDIMENSIONAL OR BULK NANOMATERIALS: Under this Group Nanocomposites, Nano-Crystalline, Nanostructured Films, Nanostyrctured Surfaces, Box Shaped Graphene are considered. Some of these Show Hollow Nanochannels (25nm & 1nm Thick Wall).

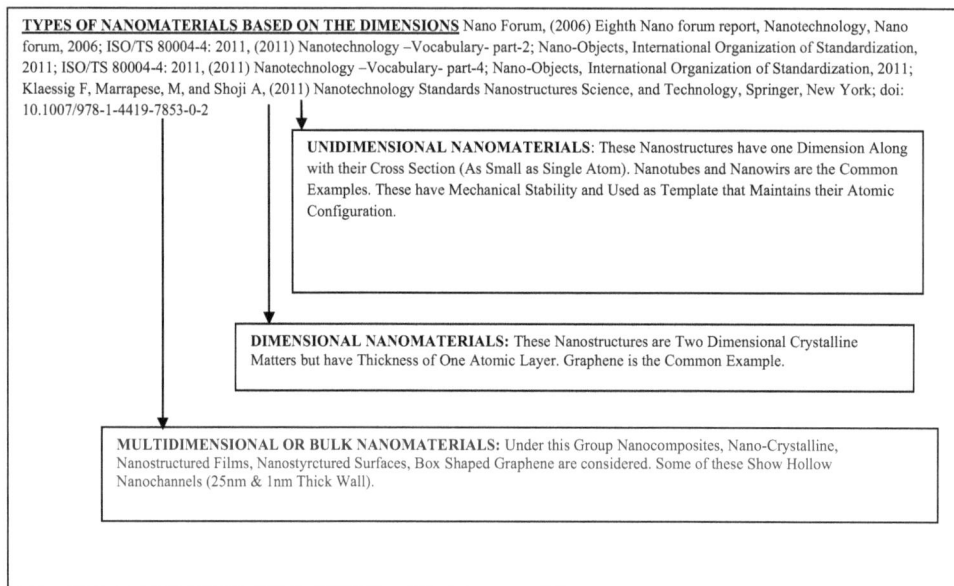

Fig. (5). Types of nanomaterials based on the dimensions.

METAL AND METAL OXIDE NANOPARTICLES AND THEIR PHYSICOCHEMICAL FEATURES THAT IMPACT THEIR INTERACTION OR BEHAVIOR

Metal and metal oxide nanoparticles exhibit particular features, like abundant surface energy and large surface to volume ratio in comparison to their respective bulk forms. The transition between molecular and metallic states represents a specific electronic structure and a transition state. It influences the local density of states, plasmon excitation, short-range order, increased number of kinks, and a large number of low coordinated sites. Such less coordinated forms include corners and edges that have a large number of dangling bonds, consequently, specific and chemical properties. These states can store the excess of electrons [25].

SURFACE ENERGY

The physical properties of nanoparticles, like melting point, vapor pressure, and their reactivity, depending on their high surface to volume ratio. This feature provides a relatively good number of atoms that are available on their surface. Their complex catalyst behavior is also related to the particles available on the surface. In most cases, atoms and molecules at the surface are not well-coordinated, *i.e.*, these are under-coordinated because, when bonds break the energy is spent (released). The atoms at the surface of metallic nanoparticles have a higher level of energy in comparison to their respective bulk forms. The surface energy is interrelated to the availability of atom/s at the surface. This energy is thermodynamically unfavorable; the dangling bonds at the surface of metallic nanoparticles are because of this condition. This condition exists despite the type of bonding such as covalent bonding in metals, ionic bonding in salts, and non-covalent bonding in liquids, like water. One such example is of the beading up of a water droplet on a waxy surface. These small droplets change into smaller spheres, because, the water molecules of droplets act against the force of gravity that tries to flatten them. During this beading process, most of the dangling bonds present on the surface get minimized. In such cases, mostly, hydrogen bonds are involved while in the case of metal or semiconductor particles, the strong covalent bonds break that is present on the surface. There is a possibility that the surface energy of some crystals might be less, because, some of the bonds may be somewhat stronger, *i.e.*, about ¼ of the bond energy. One may consider this relationship as a rule in a broader sense during translating energy per unit area. The surface energy of metal and inorganic salts is around, *i.e.*, 1-2J/m^2 [26].

The bulk form of copper exhibits properties like malleability and ductility, but, copper nanoparticles smaller than 50 nm are one of the superhard materials. Ferroelectric materials having a size of less than 10 nm are not suitable to be used as memory storage because these nanoparticles can switch on their polarization direction involving thermal energy at room temperature. The suspension is related to the interaction of the particles with the solvent; this feature helps in the preparation of a suspension of nanoparticles. This interaction influences parameters like density, resulting in either sinking or floating of the nanoparticles in the solution. The diffusion of nanoparticles is affected significantly by the high surface to volume ratio, specifically, at increased temperature. In the case of larger particles, the sintering gets affected for a shorter duration of time. Finally, density, the flow of nanoparticles, and their tendency to agglomeration can adversely affect their diffusion and distribution.

PLASMONIC NANOPARTICLES

Surface plasmon resonance represents the state of resonant oscillation of the electron involved in the conduction at the interface present between positive and negative permittivity. The incident waves of light induce the state of resonant oscillation at the interface. This phenomenon acts as a functional basis for measuring of adsorption of materials on to planar metal like gold or silver or metal nanoparticles. The developed sensors function on the basic principle of enhanced surface plasmon resonance. These sensors are very efficient tools to sense or detect molecules present within the concentration range of pg/mol. This functional aspect applies to sensors based on metallic and magnetic nanoparticles, carbon-based nanoparticles, latex, and liposome related nanoparticles, *etc*. These sensors are advantageous to detect those analytes that are difficult for detection within the concentrations mentioned [18]. The coupling of a plasmonic nanoparticle is a technique that ensures the fabrication of plasmon nanoparticles involving 3-dimensional cavities of microtubules. The resultant products are useful in manipulating photon-plasmon coupling. This process and the method, both are applied in the field of optical tuning potentials, and, help the understandings of the complexities and mechanisms involved in light-matter interactions [27]. Plasmonic nanoparticles are specific material particles having high electron density. These nanoparticles either couple or associate with plasmonic nanoparticles have far larger wavelengths of electromagnetic waves. This interaction is related to the nature of the interface-the the site of interaction between dielectric metal, medium, and the particles. Pure metals exhibit a specific maximum limit of size and wavelength, and this influences their coupling. Plasmonic nanoparticles exhibit specified scattering, absorbance, and coupling properties. These properties concern with the geometries and their relative positions. Gold nanoparticles are fabricated depending on their ability of self-assembly. The efficacy of their coupling depends on the changes in their average gap width with separation (g) as 1.1nm. The interaction with neighboring particles regulates the plasma coupling. There is a decline in the strength of combining, and, it reflects on either non-local coupling mechanism or mechanism related to quantum mechanical coupling [28].

Dipole and specific oscillation free electrons are the products when electromagnetic waves interact with materials. These oscillations are plasmon. There is a tendency of electrons to migrate to restore the original state. The incident light waves also oscillate and bring about a constant shift towards the dipole. This shifting force makes the affected electrons to oscillate at the same frequency as that of light. This condition of coupling of frequency is applicable only when the frequency of the light wave is either equal to or less than the plasma frequency. If the coupling frequency is higher than the plasma frequency, the condition is called resonant frequency. The process of scattering and absorbance plays a significant role in the specific intensity of a given frequency. Generally, most of the chemical methods of fabrication of such nanoparticles relate to their particular size and the geometries. Their cluster forming nature is responsible for their cluster states, and it represents plasmonic molecules. Special bonding or antibonding between the nanoparticles and the similarity with the molecular orbitals influence the distribution of electrons in these clusters. The polarized light affects or regulates the distribution of electrons the light-waves couple with these electrons. As the geometry of the distribution of electrons of nanoparticles changes, their optical activity and other properties of the system help their alterations. When polarized light gets synchronized the symmetry of the electron within the particles is attained, the dipole moment of the cluster readily manipulated. Such groups can manipulate light waves at the nanoscale level [29].

Metal nanoparticles exhibit an ability in which its free surface electrons get excited when exposed to incident light. The exciting free electrons cause the induction of secondary electromagnetic flow on the affected surface, and this state is plasmon. The surface plasmon resonance is one of the primary functional parameters of the optical behavior of small particles and nanoparticles [30 - 32]. In the visible spectrum, metal nanoparticles exhibit active plasmon resonance, extinction bands. This feature is the cause of deep color or reminiscent of molecular dyes. Plasmon resonance extinction is an analytical technique and helps to investigate molecular interactions. The concerned instruments primarily measure the binding kinetics and affinity of molecular interaction [33]. If the composition of the metal particle stable at a specific constant, the plasmon resonance extinction shifts up to a few hundreds of nm to a particular shape of participating particles or their number density. The particular plasmon resonance of metal nanoparticles indicates the interplay following the classical free electron theory. The spectra of semiconductor particles and metal show their interaction under the influence of the principles of quantum mechanics. The simple electrostatic limits concept regulates the polarizability tendency of metal nanoparticles. These suggest that the nanoscale metal particles may undergo significant changes, even, when their chemical composition is not changed [33]. Spherical gold nanoparticles (40-80 nm) exhibit surface plasmon resonance, and, these nanoparticles are better options to sense signals because these particles show the maximum coupling effect along with surface plasmon resonance. There is a change of phase during the reflection of surface plasmon resonance signal and show a relationship with plasmonic coupling efficacy between gold nanoparticles and sensing thin film [34].

When the bulk material is under the firing electron process, the electrons get excited, and electrons get scattered, the energy becomes transferred into the bulk plasma. The component of the scattering vector directly corresponds to the surface resulting in the formation of a surface plasmon polariton. Electrons and photons can excite surface plasmon polariton (SPP). (SPPs are the electromagnetic waves that have a frequency of the infrared or visible spectrum. These can travel along a metal-dielectric or interface between air and metal). Surface plasmon polariton also gets excited by photon, as both have the same frequency and momentum. There is a difference between the dynamism of free space photon and surface plasmon polariton. It is less in case of free space photon than the surface plasmon polariton. This factor plays a significant role in their dispersion, and both have different dispersion relationship. The free space photon from the air is unable to couple with surface plasmon polariton directly because of this, and the dispersion relationship is different. This difference also prevents the emission of energy from a smooth metal surface or surface plasmon polariton because free space photon is in the dielectric state [34]. Nanomaterials are suitable multifunctional supermolecules. Metal and semiconductor nanoparticles function like potential units for optical and electronic devices and conventional lithographic pattern technology. The metal nanoparticles system is relatively advantageous in comparison to other nanomaterials systems. Metal nanoparticles show energetic plasmon resonance extinction band in-visible spectrum, and this is the reason behind deep or color reminiscent of molecular dyes. Plasmon resonance extinction is one of the analytical techniques for studying the interaction at the molecular level. The related instruments measure the binding kinetics and affinity of molecular interactions. The spectra of semiconductor particles and metal nanoparticles reflect on their interactions. These are well illustrated under the quantum mechanism while the specific plasmonic resonance of metal particles follows classical free electron theory. The simple electrostatic limits explain the polarizability of these nanomaterials. Polarizability is the tendency to form dipoles immediately. This ability plays an essential role in determining the dynamic response of a bond system in response to the external field. This response is the reflection of the internal structure of the molecule or nanoparticles. Polarizability is related to factors like electron density, atomic radii, and orientation of molecules and nanoparticles. Molecular orientation relates to the numbers of electrons. If the number of electrons is enormous than the nuclear charge, this condition shows that there is less control over the distribution of charge on the resultant degree of polarizability and as a result, this ability gets enhanced [33]. The electrical properties at the nanoscale and bulk scale are relatively similar. Surface charge and electron transport at nanoparticles are related to size, dielectric features of the ambient medium. Generally, optical properties are related to the surface chemistry of nanoparticles, but in the case of modified surface chemistry of nanoparticles, they may exhibit affinity to a specific type of analytes in a solution state [32].

PHENOMENON OF LIGHT SCATTERING BY NANOMATERIALS

This aspect is an essential parameter that influences the optical behavior of particles and nanoparticles. The scattering of light is a physical process under which particles change their typical path of movement due to some forms of radiations, such as light, sound, or moving particles. These radiations also cause scattering because of the nonuniformities present or created in the medium in which the affected particles are present. Generally, these deviations in the path of particles follow the universal laws of reflection. Scattering referrers diffuse reflection and specular reflection. Diffused reflection of light and other waves or particles involves the scattering of the incident ray/wave/ray of light from the surface at many angles. An ideal diffused reflection follows the Lambertian reflection, *i.e.*, there is a uniform luminance, when observed from all directions. A specular reflection is a type of regular reflection. It involves reflection/scattering of the wave from a mirror-like surface and angle of incident and reflection is equal, *i.e.*, it follows the laws of reflection [35]. Scattering describes the particle to particle collision involving atoms, electrons, molecules, photons, and other ultra-fine particulate matter. Common examples of a distribution of light phenomena include the dispersal of cosmic rays in the atmosphere of the earth; the collision of particles within particle accelerator, electron collision takes place in fluorescent lamps, in nuclear reactors neutrons collision occurs in a nuclear reactor, *etc.* The particles, bubbles, droplets, show changes in the density of the fluid; the crystallites in polycrystalline solids, the defects present in monocrystalline solids, the roughness of the surface, cells in organisms, and textile fibers in clothes, all of these influence the process of the scattering of light and the concerned theory. The scattering phenomenon plays an essential role in the medical ultrasound technique, semiconductors used to inspect the water quality, monitoring the operation of polymerization, computer-generated images, and free-space communication, *etc.* Sound energy propagates in the form of vibrations having some specific audible wave of pressure, through the medium. Concerning human physiology and psychology, these sounds act as the receptors of acoustic pressure waves having a distinct pitch. These pitches have frequencies within the range of 20 Hz to 20 kHz. Sound waves with a frequency above 20 kHz are the ultrasound while below 20 Hz is infrasound. Hearing abilities among animals vary around frequencies within regular and ultrasound frequencies [36].

Light scattering can be single and multiples. When light incidents on a single particle, in a transparent and homogeneous ambient medium, this mode of dispersion is a single scattering. In this process, neighboring particles do not influence the distribution of light. When these individual scattering particles form a cluster, then the total scattering efficiency is the product of the total of the scattering due to each particle. Single scattering involves only one localized center. It is a random phenomenon, because, it is challenging to locate the single-center causing scattering, and, it is not related to the path of radiation. It also depends on the exact incoming trajectory. This type of light scattering occurs during the firing of the electron at an atomic nucleus. The precise location of an atom about the path of the electron is unknown. It is difficult to measure the size of the concerned atom during the single scattering of light. As a result, the exact route or trajectory of an electron after the collision is unpredictable. This process depicts the probability of distribution [36]. Generally, the particles tend to remain in a cluster or as aggregates. The groups of such aggregates are selected as a representative of single-particle scattering and are used to measure the scattering efficiency. When particles are unevenly distributed or dispersed, this causes multiple scattering. The resultant combined effect of more number of scattering, and it expresses the normal function of the system [37]. The phenomenon of multiple scattering and theory of diffusion both can be combined to enhance their range of application. It is achieved involving when an unambiguous medium and two assumptions. The first assumption is directional broadening, and the second assumption is temporal broadening. Both assumptions worth consideration while studying multiple scattering. In the scattering and diffusion process, the least absorption may take place. It represents the directional broadening. In temporal broadening transport, free path and fractional changes in density are very less than unity.

Further, multiple scattering describes the dual-mode of expression of a particle; first is in the form of quantum, and the second is in the waveform [38]. Specular reflection includes illumination due to incident light and secondary scattering by other particles in the medium [32]. When particles are dispersed relatively more densely in a medium, the particles exhibit multiple scattering. A simplified model of light behavior is studied as chromatic Values in densely dispersed particles [39 - 45].

The optical behavior of small particles follows Mie's theory and other basic concepts related to the electromagnetic response of light. The basics of this theory play a crucial role during single scattering. Specifically, it helps to predict the optical properties of spherical particles in a non-absorbent medium. According to this theory, the efficacy of the light scattering depends on parameters like size, Mie scattering coefficient, and refractive index of the particle, and their refractive index is measured. In the case of a cluster, there is a need to carry some adjustments. The concept of Mie's theory is suitably applicable when the non-absorbent medium is involved while studying optical behavior. The Poynting vector (concerned with electromagnetic waves and the intensities of magnetic, electronic, and transporting energy from one point to another) is applicable during experimenting on scattering is done in an absorbent matrix [32].

The nature of light is also an electromagnetic wave. Maxwell equation regulates the behavior of electromagnetic waves in a broader sense. Based on Maxwell's hypothesis, it is convenient to deal with the basics of electromagnetic waves and also describe the electric field, magnetic field, the intensity of the electric field, and magnetic field. These are regulated and defined under the Gauss's law, Gauss's law for magnetism, Faraday's and Ampere's law respectively. For practical purposes, the Maxwell equation is applicable in the time-harmonic form. Maxwell's equation helps to describe the ordinary and harmonic format. The parameters like the electric field, magnetic induction, electric displacement, magnetic field, current density, change of denseness, time, angular velocity, complex permittivity, phenomenological coefficient, complex index, free index, divergent factor and curl factor, are understood easily using Maxwell's equations [32, 39 - 45].

Elastic and Thermal Scattering

Elastic and thermal scattering, both influence the optical behavior of nanoparticles. When non-polarized light propagates through the atmosphere, its beam vibrates in different polarization planes. When the polarization plane of incident light changes, it results in the process of scattering of light. When the frequencies of the incident and scattered light are identical, then this mode of light scattering is regarded as the Elastic Scattering. This behavior occurs when the absolute temperature is zero. During the process of the Thermal scattering, scattered light gets emitted in all frequencies. The thermal scattering, electrons, neutrons get scattered because of thermal motion. The same is also applicable to the x-rays while these pass through the crystalline lattice. The wavelength of the incident beam influences the phenomenon of light scattering [46].

PHENOMENON OF SCATTERING OF ELECTROMAGNETIC WAVES BY NANOMATERIALS

Electromagnetic radiations represent specific wave/quantum/photon based on the magnetic field, moving through space and carry with them specific radiant energy. These types of radiation involve radio waves, microwaves, infrared, visible, ultraviolet, X-rays and gamma rays, *etc.* and these propagate through a vacuum. The electromagnetic radiations exhibit their velocity in space that is equal to the speed of light, the strength of their electric field (E) is (c) times the power of the magnetic field, the position of the magnetic and electric field are at the right angle to each other and also to the direction of their propagation [47, 48]. The electromagnetic waves are synchronized oscillation of magnetic and electric fields. Electrically heated charged point (particle) induces the influence on acceleration and emits electromagnetic radiation. These emitted waves interact with other charged particles and exert force on the interacting particles. Parameters like energy, momentum, and angular momentum are carried by electromagnetic waves away from their particle (point) source. As per the quantum mechanics, the functional units consist of the photon, an uncharged particle (with zero rest mass). These act as a force of the electromagnetic waves and bring about their interactions with other materials. Quantum electrodynamics describes the interactions of these waves at the atomic level. The quantum impacts of these waves illustrate the transition of electrons into different energy levels (Electromagnetic spectrum, 2019). The energy of one photon can be quantified, and photons with high frequency have high strength in comparison to those that have a lower rate. This relationship obeys Plank's equation, *i.e.*, E=hv; [E is the energy of a photon, v is the velocity of moving photon, and h is the Plank's constant] [49].

Whenever there is an interaction between electromagnetic waves/radiations and particles, scattering takes place, and its pattern depends on the wavelength of radiation and the size of the interacting particles. During this interaction, the local electron distribution on the particles and inhomogeneity entities get disturbed. The disturbed regional electron distribution causes separation of charge within the particle, resulting in an induced oscillation of local dipole moment. Further, during the scattering process, this feature causes periodic acceleration, and that functions as a source of electromagnetic radiation. Since the frequencies of oscillation of the scattered waves are similar to the frequencies of incident waves, the phenomenon is elastic scattering. When these related frequencies are dissimilar, the condition is inelastic scattering. The interaction between electromagnetic radiation and the particles and the electromagnetic scattering exhibits the elastic and inelastic types of scatterings.

Rayleigh scattering and Mie scattering come under the elastic type of scattering while Brillouin scattering, Raman scattering, and Compton scattering, come under the inelastic kind of scattering. There is an inelastic type of X-ray. The thermal energy absorbed to an extent during this interaction, and the combined impact of scattering and absorption causes the extinction of the incident beam.

The underlying basis of the phenomenon of scattering of electromagnetic radiations is well illustrated by involving Rayleigh scattering and Mie scattering. Rayleigh scattering describes the scattering of the small dielectric; nonabsorbing spherical particles while Mie scattering is related to the particles irrespective of their sizes. This concept makes use of geometric limits, precisely, when larger particles are under consideration. Further, the theory of Mie scattering provides a general explanation for scattering due to the spherical particles, and this includes Rayleigh scattering also. This concept is quite tedious to implement. Thus, the theory of Rayleigh scattering remains a better option.

Raleigh Scattering

During Rayleigh scattering, the state of mater, understudy does not change, *i.e.*, there is no change in the quantum state of the interacting material, which includes energy, momentum, or angular momentum. Hence, this process is parametric. The content involved can be either atomic or molecular in nature or both. This scattering is the result of the ability and degree of the polarizability of the particle. The resultant oscillatory electric field of the incident beam acts on the charge of the particle, and they also oscillate at the same frequency. This state of the affected particle behaves like a small radiating dipole during scattering. In this mode of scattering, the size of the particles and the wavelength of the incident radiation (illuminating radiation) are of great significance. When the ratio between the size of particles and the wavelength of incident radiation is higher the intensity of the Rayleigh scattered radiation increases, *i.e.*, an increase of the Rayleigh scattered radiation is directly proportional to the rise in the ratio between particle size and the wavelength of the incident radiation. If there is an increase in the size of the particle around 10% of the wavelength of incident radiation, the principle of Rayleigh scattering does not hold good. At this point, the law of Mie scattering becomes the most suitable option [50, 51].

MIE Scattering

In this type of light scattering, spherical structure as an inhomogeneous state causes scattering of an electromagnetic wave, and its expression is in the form of an infinite series of spherical multipole partial waves. It is Mie Solution, and, it represents an endless series consisting of spherical multipole partial waves. The same applies to the scattering of light by stratified spheres, infinite cylinders, and particles with different geometries [36, 52]. It provides separate equations for radial and angular dependence of the solutions. It describes the scattering process involving particles having a similar size to the incident beam and is suitable for studying scattering due to non-molecular or aerogel particles. These materials are likely to contain spherical particles that come under the perimeter of Mie scattering. The plane of the incident wave and the field of scattering expand into radiating spherical vector wave function while the internal domain into regular spherical vector wave functions and helps in computing the expansion coefficients of scattering [36, 52, 53].

Raman Scattering

The Raman effect it also represents an inelastic type of scattering. When a molecule is excited, a photon moves to a higher level of energy during Raman scattering. Most of the scattered photons exhibit the same frequencies and wavelength (energy) as those incident photons (elastic scattering). A tiny fraction of the scattered photons do not possess the same frequency, and, wavelength as those of scattered photons, *i.e.*, this fraction of photons show inelastic scattering. These photons have lower energy than those of incident photons. This phenomenon predicts the conservation of energy. As a result, the materials under observation either gain or lose power. The photons of incident light belong to visible range, and the profit or loss of energy of materials during this scattering is in vibrational form. There exists a possibility to gain or to lose power in the form of rotational and electronic energy, and it depends on a sample. If the sample is gas, then the rotational kind of energy form is involved, and if the X-ray source is engaged, it entails the electronic type of energy. In this pattern of scattering the absorbed photons reemitted have low power. Thus, there is a difference in the energies between the photons that incident and the photons that are scattered. This difference is equivalent to the energy needed to excite a given molecule to attain a higher vibrational energy level. Practically, high-intensity laser radiation having either visible or near-infrared regions, are made to pass through a test sample. The incident laser beam causes oscillating polarization in the molecule of the sample and excites them to a new virtual energy level. This molecular oscillating polarization gets associated with vibrational or electronic excitation. If it does not get associated, then there is no change in the vibrational state of the affected molecule. Raman scattering phenomenon results in two probable outcomes: (i) the interacting mater absorbs energy, and the photons emitted have lower energy in comparison to the incident photons. This state is called Stokes Raman Scattering, (ii) the interacting mater absorbs energy, and the photons emitted have higher energy in comparison to the incident photons. This state is Anti-Stokes Raman Scattering. The difference of energy between absorbed energy and energy of the emitted photons correlates the difference between two oscillation states (resonant state) of the matter under consideration. This parameter is independent of the absolute energy of the photon [54 - 57].

Compton Scattering

This mode of distribution of light comes under inelastic scattering involves free charged particles. In this case, the frequency of scattered light does not match the frequency of incident light (radiation). When electrons scatter photons, there is a decline in their energy, *i.e.*, their wavelength increases, and it is the Compton Effect. The recoiling electrons get some portion of the energy. During earlier experimentation, the energy of X-ray photon higher in comparison to the binding energy gets involved in the atomic electron; this condition enables the concerned electron to be free. This behavior reflects on situations in which the wavelength of light shifts, and it is the Compton shift. Quite often, charged particles shift part of the energy, and these are concerned with a photon. This shift of energy is called Inverse Compton scattering [58 - 60].

X-Ray Scattering

During X-ray scattering, electrons of the atomic shells participate, and these participating electrons undergo oscillating mode and attain dipole state; as a result, a spherical wave is given out. Thus, the electron of the material plays a prime role in the process of X-ray scattering. The scattered waves inform about nanoscale and atomic disorder, and this interference causes a change in the order;

these changes help to investigate the sample under study. If the sample structure is distinctly/periodically, a discrete, it causes the Bragg Scattering. If the sample under investigation is disordered, then diffused scatting is the result. Like any other electromagnetic wave, an incident X-ray interacts with an atom, an electron cloud (Electron cloud is a collection of charges) forms that move ahead. If these charges re-emit or re-radiate a wave that has the same frequency as that of the incident wave, then, the condition is said to be elastic scattering. Some different effects influence this pattern of scattering. A negligible rescattering among the scattered waves takes place, and this process is secondary scattering. This process follows the principle of the concept of Bragg's Law. The nuclei of atoms also exhibit similar neutron waves, even when unpaired electrons undergo coherent spin, the same processes take place. Whatever the case may be, there can be scattering due to the interaction between cogent spin and unpaired electrons. The fields of scattered or re-emitted waves interfere with each other. There are derogative or beneficial impacts of such interplays. If these waves overlap due to their adding up, that resultant waves give a sharp peak; then this interaction is destructive. If the waves get subtracted, then they provide a weaker peak in comparison to that when these waves are added up. As a result, these waves form a different pattern on a detector or film, and it is named as Diffraction Pattern. This principle applies to the diffraction analysis. Solids (crystals) and liquids give different diffraction patterns. X-ray scattering depends on the specific frequency and angle of incidence. At different incident frequencies and incident angles, give different diffraction patterns [61].

Small-angle X-ray scattering is the most commonly used technique to investigate the dimensions and structure of nanomaterials in samples having solids, powders, and gels phases. This technique is suitable to investigate amorphous, crystalline, semi-crystalline nanomaterials. The small-angle X-ray scattering technique is also helpful to study the colloidal dispersions, surfactants, polymers, biomacromolecules, nanocomposites, membranes, and porous nanomaterials. Comparative analysis of the 2D scattering pattern examined involving small angles X-ray scattering technique is a beneficial technique while investigating the structure of nanomaterials. Small-angle X-ray scattering is conveniently applied to study a dilute biological solution containing protein. This versatile technique is useful to consider the size, shape, folding and unfolding, aggregational behavior, degree of stability, and molecular weight of proteins. This technique is beneficial to study, these aspects under very close to natural conditions of proteins and also the impacts of parameters like concentration, pH, ionic strength and temperature, *etc*. This technique also facilitates the study of a low-resolution molecular shape envelops reconstruction, thereby, provides a piece of complementary information to that obtained by NMR or XRD techniques. Hence, this technique is known as the Biological small-angle X-ray scattering technique. The ultra-small is another technique and is suitable to characterize those samples that contain tiny sized nanostructures with better and higher resolutions. The wide-angle X-rays scattering technique is ideal for the identification, quantification, and measuring sizes of the crystallites present in test samples. It gives 2D wide-angle X-ray scattering pattern. These help to deduce the orientation of crystalline lattice present in anisotropic structures [61].

Brillouin Scattering

This mode of scattering describes the interaction between light waves and particles within a given medium. The refractive index is a prime parameter that plays a significant role in this type of scattering. This feature depends on the material properties of the medium. The refractive index of the transparent material shows deformational changes due to physical stress like compression or distension or sheer skewing, *etc*. During this deformation, the interaction between light radiation and the transmitted light, the frequency, and the energy, *i.e.*, momentum, remains unchanged. Brillouin scattering involves the interaction of electromagnetic waves and one of the three dimensions of crystalline lattice waves (mass oscillation-acoustic oscillation modes-phonons; charge displacements modes in dielectrics-polarons; magnetic spin oscillation modes in magnetic materials-magnons). This scattering exhibit inelastic type of distribution may involve either the stokes process or the anti-stokes process. This shift in energy is the Brillouin shift in the frequencies of photons. This scattering is useful for measuring the voltages, frequencies, wavelength of different atomic chain oscillation types [62, 63].

PHENOMENON OF ABSORPTION OF RADIANT ENERGY BY NANOMATERIALS

The electrons of the interacting materials pick up the energy of photons present in the incident electromagnetic radiation. This energy transforms into the internal energy of the interacting material that acts as an absorber. The transformed energy converts into thermal energy. When an incident beam of electromagnetic radiations travels through a medium, its intensity declines, because the medium absorbs its energy of photon through which it passes. This absorption of energy causes a reduction in the strength of emerging radiation; this process is attenuation [64]. When an atom and a photon of radiation interact with other atoms, it gets excited and moves to a higher energy level within nanoseconds, and later, it falls to either the original energy level or some lower energy level. It is quite common during absorption and scattering. There is a conversion of radiating energy into kinetic energy; this process occurs either during the collision of the particles or oscillation or vibrational energy. Mostly, particles absorb incident light during light scattering, and the absorbed energy changes into another form of energy, mainly as heat energy. [Extinction describes the light absorption and scattering of light by particles. Extinction efficiency of a particle represents the changes for every compartment that is polarized, present in the incident light] [32, 46]. Nanoparticles act like a bridge between bulk material and atomic/molecular structures. The physical properties of bulk materials remain unchanged even if the size changes but not in the case of nanoparticles. This aspect of nanomaterials makes them of scientific and industrial significance. The physical properties heavily depend on the respective size of nanoparticles. These have a high ratio between their surface and volume, the ability to confine electrons and cause quantum impact, *etc*. These features nanomaterials play a significant role in their varied applications. Nanofluids (fluids that contain nanoparticles) exhibit specific behavior concerning the diverse heating scenario. The size of the particles influences the effect of thermal energy even on the melting point [65, 66]. Nanoparticles show a higher degree of absorption of solar radiation in comparison to a thin film of continuous material sheet. The materials enhance the degree of absorption from solar radiation and the modified products, are suitable for various solar thermal applications. When plasmonic nanostars are associated with quantum dot, solar cells, the degree of broadband quantum efficiency enhances. Fabricated solar cells with multi-spiked nanostars elevate the performance of the solar cells. Such modified nanomaterials cause broadband scattering and absorption cross-section. One can make maximum benefit from the total solar spectrum if plasmon nanostars are used because these nanostars need less energy in comparison to quantum dots [67].

IMPACT OF MEDIA ON OPTICAL BEHAVIOR OF NANOPARTICLES

There are two prime conditions for the dispersal of particles, namely, temporal and spatial dispersion. During temporal distribution, the mode of dispersion of particles is related to the time parameter. It exhibits a time-harmonic form of a wave. In the case of spatial dispersion, the additional parameter space also affects this process along with the time parameter. Both parameters influence the polarization of the transmitting electromagnetic field.

Further, during spatial dispersion, the ambient medium around the particles also affects the spatial behavior of the system as an absorbent matter. It is necessary to include the space and the distance among the particle while calculating the flow of electromagnetic field. This mode of flow of electromagnetic field is called the Poynting vector. The scattering phase represents the angular distribution of the Poynting vector in the case of more considerable distances [32, 36, 46].

SCATTERING OF RADIATIONS BY IRREGULARLY SHAPED NANOPARTICLES

Some of the theories related to dispersion of particles may be efficient in obtaining the best and relevant results in a given system, specifically, in the case of irregularly shaped nanoparticles. Effective Medium Theories explains the scattering phenomena due to irregularly shaped particles, which, in turn, is related to the refractive index of an active spherical particle. These considerations improvise excellent and reliable results. The Transition (T) matrix is another technique to study the optical properties of irregularly shaped particles. In this technique, the transition (T) matrix transposes the results, and the calculated results need comparison with the standard spherical particles [32, 46]. Detailed information related to the optical and other related information is available [39 - 45]. (NOTE: the detailed explanation appears to be beyond the scope of this presentation; the interested readers are suggested to refer the concerned literature as mentioned).

APPLICATIONS OF SCATTERING, EXCITATION AND ABSORPTION PHENOMENA

Parameters, like size and distribution of particles concerning volume, number, and intensity of scattering, play their respective roles in the processes of scattering, excitation, and absorption. Zeta potential helps to depict a degree of electrostatic stabilization, and the possibility to reduce or to avoid instability due to the attraction among the particles. This comparative evaluation will facilitate an understanding of the impact of scattering phenomena [68]. One practical aspect of Mie theory explains the distribution of numbers and mass of particles are smaller than 100 nm in size. This distribution relates to the optical features and phase of dispersing medium, and the concerned optical properties include real and imaginary aspects of the refractive index and absorption of light. In such cases, the volume and number means are likely to be smaller in comparison to the mean intensity. There is a possibility that if the size is larger than 100 nm, the fluctuation in the mean size becomes proportional to the different sizes distributed in the dispersion [68]. Another application of Mie theory concerns the transformation of intensity into the volume, and it can be calculated as number distribution. The parameters needed are correct mass, refractive index, and absorption. The distribution of such particles depends on the size of the distributed particles [68].

Stokes shift illustrates the application of absorption of energy during an emission (Fig. **6**). When a molecule or an atom considered as a system, absorbs a photon, it gains energy and enters the excited state. When it emits a photon, it relaxes, *i.e.*, loses energy by emission of a photon and loses energy in the form of heat. Photon can absorb more energy in comparison to the emitted photon, reflecting, that emitted photon has less energy than the absorbed photon. This difference between the energy of the absorbed photon and the energy of the emitted photon represents the Stokes shift. Stokes shift phenomenon depicts the difference between the frequency or wavelength between the position of band maxima of absorption and the emission spectra of the same electronic transition. An Irish Physicist, George C. Stokes manned this phenomenon as the Stokes shift is the result of (i) vibrational relaxation and (ii) dissipation and reorganization of solvent. When energy transfer takes place between photon and a molecule/atom (in a system) the affected molecule or atom enters the vibrational state and emits energy towards the lower frequency. The molecule dissipates energy in the form of heat during the energy transfer between absorbed photon and molecule. There occurs an arrangement in the molecule solution or a solvent.

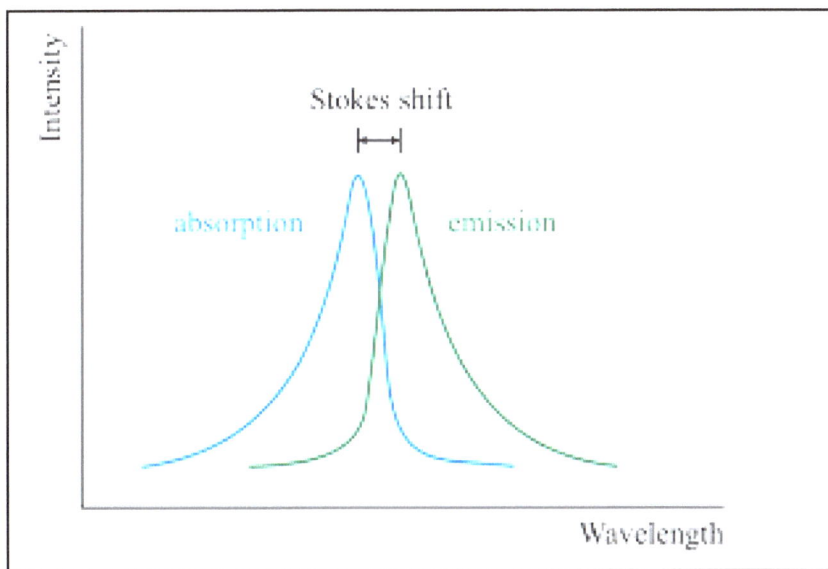

Fig. (6). Stokes Shift.

For example, if a fluorophore Rhodamine and water system form components of an experimental set. A fluorophore is a dipole, and water molecules surround it. When fluorophore enters in the excited state, its dipole moment changes, the water molecules of the medium do not adapt quickly to this change. There is a realignment of the dipole moments during vibrational relaxation, *i.e.*, the dipole of fluorophore and water molecules also realign [69 - 71].

The Antistokes shift is a process in which emitted photon has more energy than the absorbed photon. This additional energy of an emitted photon derives extra power from the dissipation of thermal phonons in the crystal lattice. Phonon represents a quantum of energy or a quasiparticle associated with compressional wave-like sound or a vibration of a crystal lattice. In physics, a phonon is a collective excitation in a periodic, elastic arrangement of atoms or molecules in a condensed matter such as solids and some liquids). Gadolinium oxysulfide nanomaterials doped yttrium acts as an anti-stokes pigment; it is a standard industrial anti-stokes pigment. (Yttrium is a silvery-metallic transition metal element, symbol (Y) and atomic weight 39). When yttrium oxysulfide gets exposure to near-infrared wavelength, it emits energy in the visible range of the spectrum. The emitted photon exhibits more energy than the absorbed photon, and it represents an anti-stokes shift. Photon up-conversion is a process in which the sequential absorption of two or more photons leads to the emission of light at a shorter wavelength in comparison to the wavelength of the excitation radiation. This process acts as an example of conversion on infrared light into visible light [72].

Stokes lines are the representation of particular wavelengths that are present in line spectra. There are dark and bright-line present in the uniform and continuous spectrum. The emission or absorption of light within a narrow frequency range in comparison to the neighboring frequencies forms dark and bright lines. These lines with narrow range help in identifying atoms and molecules [73]. Further, these are associated with fluorescence and Raman scattering. The energy of scattered radiation is less than the incident radiation in the case of Stokes lines.

The energy of the scattered radiation is more than the incident radiations for anti-stokes lines, and Stokes lines are of longer wavelength than the exciting radiation. These are responsible for the fluorescence phenomenon and the Raman Effect. The increase or decrease of energy during excitation depends on the vibrational energy and the spacing of the ground electronic state in a molecule, and acts as the functional principle for the wavenumber of stokes and anti-stokes lines and also acts as the direct measure of the vibrational energies of the molecules. The probability of anti-stokes transition is lower in comparison to that of stokes transition because at room temperature most of the molecules are in the lowest vibrational energy level (Fig. **7**) [74 - 76].

Stokes and Anti – Stokes Scattering

Virtual energy states

Vibrational energy states

4
3
2
1
0

Infrared absorption

Rayleigh scattering

Stokes Raman scattering

Anti - Stokes Raman scattering

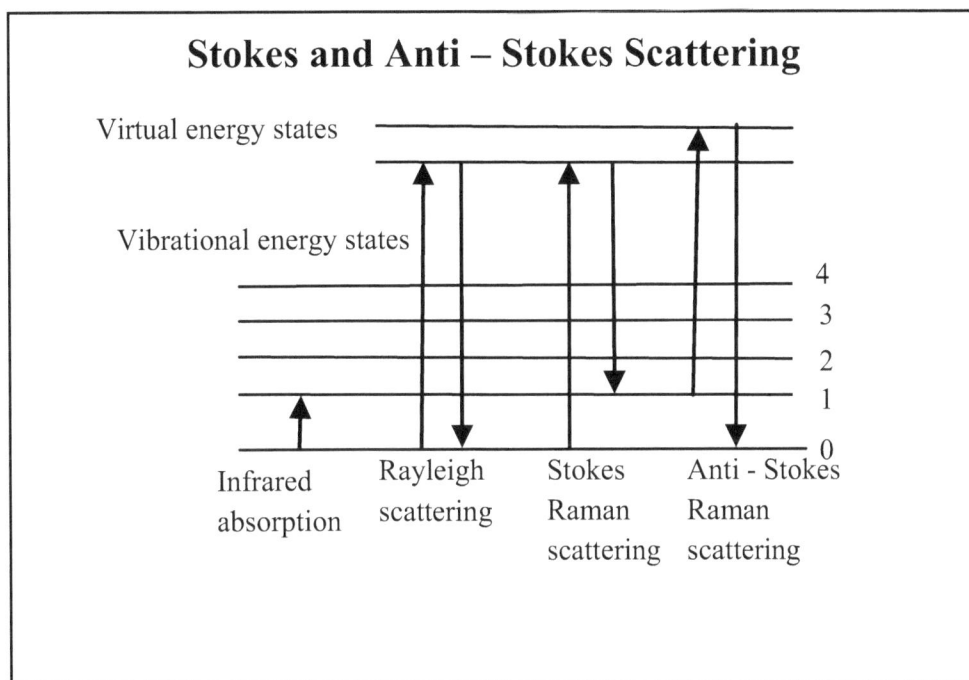

Fig. (7). Raman Scattering.

The Raman Microscopy

This device helps the investigations concerned with molecular identifications involving vibrational, rotational, and other low-frequency modes. Further, it includes molecular optics laser examiners (MOLE) referred to as Raman based microprobes. The scattering phenomenon plays a vital role as the theoretical basis in this technique. When light radiation falls on a molecule, there is an interaction with the electron cloud and bonds of the molecule under consideration. When the incident radiation gets scattered, a photon excites the molecules under investigation from its ground level to a virtual energy level or state or undergo vibrational or rotational state. On relaxation, it returns to a different rotational or vibrational state. There is a difference in the energy, *i.e.*, wavelength or frequency between the original state and the new virtual state. This shift in the frequency of emitted photons and the frequency of the exciting photon is called the Raman Effect (Fig. **8**).

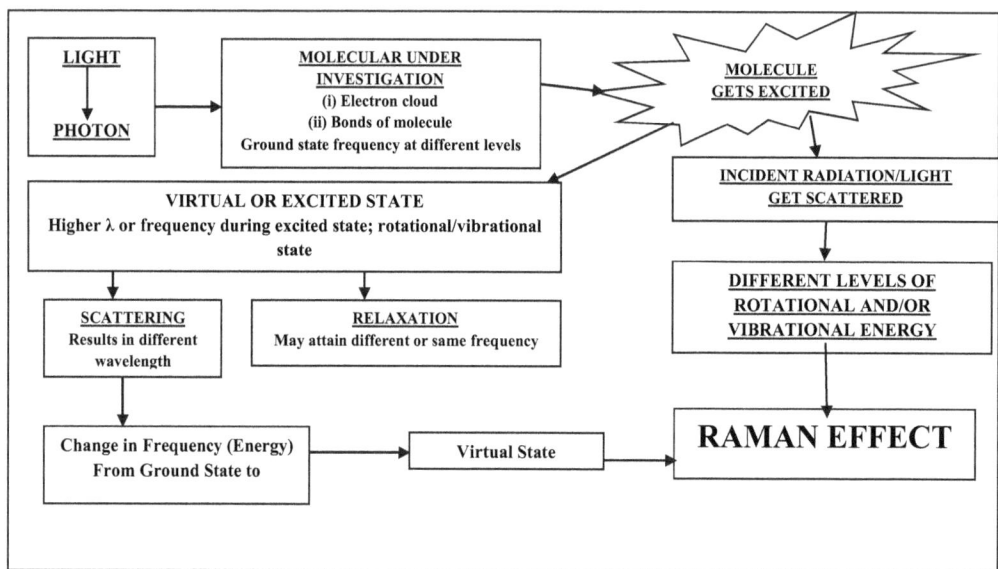

Fig. (8). Schematic representation of the theoretical principle of Raman Effect.

Raman Effect is based on and related to the scattering of the light radiation phenomenon. It should not be muddled with absorption as seen in fluorescence, where a molecule absorbs energy and gets excited to a discrete energy level. When energy transfer takes place between a photon and an atom or ion. It results in an inelastic scattering, but the total energy of the system remains balanced during this process. If the energy of the final vibrational state of a molecule is more than the initial state, the emitted photons shifts to a lower frequency, but the energy of the system remains balanced. A molecule exhibits the Raman Effect and as this change is occurring Polymerization potential or the amount of deformation of electron cloud shows a change concerning vibrational coordination. The intensity of Raman scattering is directly proportional to the amount of polarizability.

The pattern of shift, *i.e.*, Stokes shift or anti-stokes shift, is regulated by the rotational and vibrational state of the molecules under study. This pattern of shift is dependent on the polarizability and differs from infrared spectroscopy when the interaction between a molecule and light involves the dipole moment. Raman shift also includes the typical wave numbers that have units of inverse length. To convert spectral wavelength and wavenumber of a change in the Raman spectrum following formula is applied. Let $\Delta\omega$ represents Raman shift expressed in wavenumber; $\lambda 0$ represent the excitation wavelength; λ represent Raman spectrum wavelength; $\Delta\omega$ (Raman shift number) $= (1/\lambda 0 - 1/\lambda)$ [77 - 81].

Applications of Coherent Antistokes Raman Scattering (CARS)

The Raman scattering technique is useful in procuring chemically selective images because of the coherent anti-Stokes Raman Scattering (CARS) technique. This technique helps to generate a contrast about the vibrational resonance produced while studying a sample. Earlier, this technique was called a three-wave mixing experiment. Begley and his coworkers have changed this and named it as Coherent Anti-Stokes emission in 1974. In this technique, a non-linear optical method involving two laser frequencies is used to excite the sample simultaneously. The frequencies selected in this manner exhibit their differences coincide with the vibrational transition of a sample under study. A strong signal is a result that has higher energy than the excitation beam. This technique allows a selective mapping of the distribution of chemical compounds in a given sample. It is sensitive enough to study live cells and also offers a high 3D resolution. This technique is also suitable for investigations concerning chemistry, physics, and allied branches.

One faces two significant problems during fluorescence microscopy. The first problem is related to the labeling of the sample with a fluorophore. This labeling process is problematic and may change the properties of the test sample. The second problem is related to the photo-bleaching nature of fluorophore. These two parameters interfere actively during the investigation of the sample treated with a fluorophore. Coherent Anti-Stokes Raman Scattering technique handles these problems amicably. The basic principle involved is the use of multiple photons that addresses the molecular vibrations and the production of a signal. This signal emits waves that are coherent with one another; these results in a stronger magnitude than the spontaneous emission produced under Raman Scattering.

This technique is sensitive to vibrations and is non-linear; *i.e.*, the output is not proportional to the input. The related principle can be well illustrated physically involving the classical oscillator model and quantum mechanical model. There is an incorporation of the energy levels of the molecules. Harmonic oscillation is a system when displaced from its equilibrium position experiences a restoring force (F) that is proportional to the displacement (x) (F α x). If Force (F) is the only force acting, then the system is called a simple harmonic oscillator, and it undergoes simple harmonic motion. When a frictional force (damping force) (α), works on to the velocity, then, the system of harmonic oscillator functions as a Damped oscillator/damped oscillation. The harmonic oscillator is the Coherent Anti-Stokes Raman Scattering (CARS), it does not involve a single optical wave but the different frequencies, *i.e.*, $\omega p - \omega s$ between the pump and Stokes beam. This driving mechanism is similar to hearing the low combination tone when striking two different high tone piano keys. Our ear is sensitive to the different frequencies of the high tones.

Similarly, the Raman oscillator is susceptible to the different frequencies of two optic waves. When the frequencies $(\omega p - \omega s)$ approach ωv the oscillator moves very efficiently. At the molecular level, the electron cloud surrounding the chemical bond oscillates vigorously with frequency $(\omega p - \omega s)$. These electronic motions of the sample molecule exhibit periodic modulations as the refractive index of the material. The third laser beam called a probe; beams help to study these periodic modulations. The probe beam propagates the periodically altered medium, and it needs the same intonation. Part of the probe beam formerly at (ωpr) will now get modified into $\omega pr + \omega p - \omega s$, these are observed as anti-Stokes emission. Under certain beam geometrics, the anti-Stokes emission may diffract away from the probe beam and is detected in a separate direction [82].

The explanation related to the energy involved in this technique does not account for quantum mechanical energy. The quantum mechanical strength of the molecule is at ground level initially, *i.e.*, at the lowest energy level. The pump beam excites molecules to a virtual state. A virtual state is not in an eigenstate of an atom or a molecule, and it cannot remain in this state, but at the same time, it does allow transition between otherwise uncoupled real states. The Eigenstate of an atom, molecules, or a crystal, in a dynamic system represents a state with one variable that defines a state like energy or angular momentum that has a determinate fixed value. This state is a physical aspect showing any discrete value from the set of values of total power for a sub-atomic particle confined by a force or a dynamical state that has known state vector or wave function and represents an eigenvector or eigenfunction. This condition belongs to an operator corresponding to a specified physical quantity or energy state). If the Stokes beam is simultaneously present along with the pump beam, then the virtual state acts as an instantaneous gateway to address a vibrational eigenstate of the molecules. Pump beam and stokes beam both together play a functional role during coupling action between the ground state and vibrationally excited state of the molecule at the same time. The molecule is now in the two conditions; at the same time, it resides in a coherent superposition of state. This coherence between these states can be proposed by the probe beam that promotes the system to a virtual state. Again, the molecule cannot stay in the virtual state and will fall back instantaneously to the ground level/state under the emission of a photon at anti-stokes frequency. Thus, the molecule is no longer in a superposition. In the quantum mechanical model, deposited energy in the molecule during the process of the coherent anti-stokes Raman scattering instead of the molecules acting as a medium for converting the frequencies of three incoming waves into the CARS signal. This process is parametric processes. In this process, light interacts with matter to leave the quantum state of material but remains unchanged. Direct consequences are no-transfer of energy, momentum, angular momentum between the optical field, and the physical system. These are, however, related to Coherent Raman Process that takes place simultaneously. It does not deposit energy into the molecule [83 - 86].

APPLICATION OF INFRARED MICROSCOPIC AND SPECTROSCOPIC TECHNIQUES IN UNDERSTANDING THE BEHAVIOR OF NANOMATERIALS

These techniques are instrumental in resolving chemical components in a wide range of materials such as plants, animal tissues, polymers, laminates, geological samples, artwork like paintings, sculptures, *etc*. The electromagnetic spectrum includes the energy range below the UV/VIS range and is called the infrared range. Its range extends from the red end of the visible region to the microwave region, *i.e.*, 700 nm to 25 μm, longer wavelengths of IR radiation limit the spatial resolution. Under this technique, two parameters need specific attention. The first parameter is the signal to noise ratio (S/N ratio); it should be within an acceptable range. It declines as the apertures are reduced to confine the IR beam to a smaller area. The second parameter is the diffraction. The diffraction occurs. When an incident light beam bends around the corners of an obstacle or an aperture and enters into the region of the geometrical shadow of an obstacle, the process of diffraction takes place. This process of diffraction exhibits interference of waves when it encounters an obstacle or silt in its path that is comparable to the size. Similar effects take place when light waves travel through a medium having different refractive indexes. If we confine the aperture for IR to an area of 20-30μm in diameter, these two problems resolve. The higher brightness of the synchrotron source makes the smaller region suitable for the probe with an appropriate S/N ratio.

Thus, the setting of aperture should be appropriate; *i.e.*, it should be smaller than the wavelength of the incident light. The diffraction is controlled involving synchrotron IR source that provides a suitable resolution. In this technique, the infrared wave incident on a molecule and this causes vibrations of higher energy levels. The bonds of the molecules under study are flexible like springs. As a result, the bond vibration is in the form of either stretching or bending, leading to deformation. Infrared (IR) radiation is emitted or absorbed by a molecule. When molecules change its rotational and vibrational movements, either absorption or emission of IR radiation takes place. The energy of IR radiation excites vibrational modes in molecules by changing the dipole moment and converts the frequency suitable for investigation. The energy states of the molecule are related to its symmetry.

IR microscopy examines the absorption and transmission of the photon within the infrared energy range. During infrared microscopy, one records the frequencies, and spectroscopy represents the functional groups that are present in the molecule under study. These are called fingerprints. It is difficult to assign the bands when the number of functional groups increases in a complex molecule. Groups of certain bands regularly appear near the same wavelength, which belongs to specific functional groups. This trend facilitates the diagnosis of the structure of the molecule under study. A slight or significant change in the molecular environment might affect the useful group/s because of the wavenumber associated with the particular functional group that gets affected. By this feature, one can easily differentiate between the C-H vibrations in methylene ($-CH_2-$) and methyl group (CH_3-) [75].

When a molecule, along with its functional groups, is exposed to radiations, like infrared, UV, or NIR (Raman spectroscopy), these molecules get excited due to absorption bands. The spectrometer records and plots against the wavelength. A concluded spectra help to know the chemical nature of a sample under study. The combination of IR and Raman spectroscopy with light microscopy is very useful to analyze small samples or a specific region of the sample because these studies provide chemical and spatial information. The IR spectroscopy works on the principles of absorption; the sample under investigation should be suitable for the microscopic analysis. This analytic technique involves microtomy or cryomicrotomy/cryomicroscopy. For this purpose, the tissue should be attenuated (reduced in thickness) for total reflection (ATR), and this step makes the handling of the sample comfortable. Attenuated entire reflection element can be a crystal having a high reflective index like diamond or germanium. These elements exhibit an absorption of the evanescent wave. Evanescent wave arises at the interface of two media when light moves from highly dense medium to less dense medium under total internal reflection. The location of the sample should be within this evanescent wave (0.5 to 0.2 µm) for investigation [75, 87].

Features That Influence the Behavior of Quantum Dots

Quantum dots are the functional entities of semiconductor technology. These are the tight confines of electrons in semiconductor materials. This feature reflects on their natural existence as a single, discrete bound entity with a specific electron state. Semiconductor technology helps the fabrication of the pool of electrons in a semiconductor material within the range of few hundreds of Angstroms (A°). These fabricated products are Single-electron transistors, quantum dots, zero-dimensional electron gases, or Coulomb islands, *etc*. Metallic electrodes control the valid nuclear charge in these highly potential entities of artificial atoms. These small entities possess a specific number of electrons and the discrete spectrum of frequencies (energy levels). Their electronic configurations exhibit many variations when subjected to two leads; even a single electron can do so. The charge and the energy can be quantized insignificantly small metallic or the semiconductor entity [88, 89]. The process of emission of light from quantum dots is photoluminescence because, during this process, the photon gets excited and moves to a higher energy level. The excited photons at this energy level, relax, fall back to the lower energy level, or reradiate energy or exhibit the recombination. Quantum dots at nano-size exhibit higher energy gaps between the valance and conduction bands in comparison to the corresponding bulk matter. The quantum confinement impact, the quantized (discrete) electronic state and the effects of quantum confinement, are based on their nano size. When there is a change in the atomic structure of a given particle, the energy band gets influenced because it causes a shift in the electronic wave, having a range per the size of the particle. The wavelength boundaries constrain the electrons, and this explains the dependence of their properties on size.

Further, the excitation restricted in all three dimensions. The confinement of energy and the excitation of quantum dots are related to the size and frequency of the light they emit. The small size of quantum dots is responsible for their discrete or quantized state. There are the larger bandgaps due to the effect of quantum confinement, and these distinct energy levels are observable. Further, these quantized energy levels regulate the electronic configuration in the single-molecule, and this has one gap while bulk materials of semiconductors have continuous energy levels within bands [88, 89].

Quantum dots efficiently conduct electric current, and the range of degree of conductivity of metals like copper, silver, and gold, *etc*., restrict this feature, and also the degree of conductivity of non-conductors like glass. These materials are readily excited when light incidents on them or electric current passes through them. The frequencies of emitted radiations are dependent on their size, shape, and constituent components [90]. The size and shape of quantum dots readily influence the dispersion curve, specifically, under the effect of lower frequencies. Even the degree of dispersion and absorption is affected because of mostly non-linearity results with changes in size, shape, and wetting layer [91]. The properties of quantum dots like structure, electronic and optical, depend on their size and orientation or assembly pattern such as superlattice, colloidal crystals, and super crystals, *etc*. There is a dipole-dipole interaction between the interacting particles; there is a transfer of energy involving ambient nanocrystals. Electronic-tunneling also occurs in the nearby nanocrystals. The overall impact results in dark and photon-conductivity [92].

Features that Influence the Behavior of Carbon Nanomaterials-Carbon Nanotubes, Fullerene, and Graphene

Carbon nanotubes are tubular structures and made of only carbon. These tubular structures have diameter 1-3 nm but are very long. These nanomaterials are around 100 times stronger and 1/6 timeless in weight in comparison to steel. These are efficient electrical and thermal conductors, but these features are dependent on their configuration. These can also act as semiconductors and extraordinary strength and stiffness. The ratio between length and diameter is around 132,000,000:1 ratio [93]. Their unique properties make them preferred options in the fields of nanotechnology, materials science, optics, electronics, biomedical science, *etc*., These have multiple uses in the products of material science, sports goods, like a baseball bat, golf clubs, automobile parts, *etc*., because of their extraordinary abilities concerning thermal, electrical conductivity and mechanical capacities [94, 95].

Carbon nanomaterials involve van der Waal Forces and more specifically π-π stacking. These π-π stacking and interactions refer to attractive noncovalent interactions amongst aromatic rings. These interactions commonly occur during the stacking of nucleic bases of DNA and RNA molecules, folding of proteins, synthesis based on the directives from the template, materials science, and molecular recognition, *etc*. Among proteins, π-stacking associated with the isolated amino acid dimmers. The orientation in protein represents parallelly displaced π-stacking. Aromatic amino acids like phenylalanine, tyrosine, histamine, and tryptophan, exhibit the stabilized interactions and these present at a greater distance than the average radii of van der Waals forces. Carbon nanotubes exhibit a chemical bonding specific orbital hybridization because the chemical bonding of carbon nanotubes involves completely sp^2 hybrid carbon atom. These sp^2 hybrid carbon atoms are mostly similar to those present in graphene but much stronger than those found in alkanes and diamonds [96]. It is imperative to measure the strength of carbon nanotubes for their successful applications in materials science and other uses. Scanning electron microscope incorporates nanostressing because it helps to measure the tensile strength of carbon nanotubes, specifically, multiwalled carbon nanotubes. Stresses around 270 to 950 Giga Pascal can break the multiwalled carbon nanotubes. Varied fragments like nanoribbon, changed wave pattern, partial radial collapse, *etc*., are formed at the site of broken carbon nanotubes. The advent of carbon nanotubes with their high degree of stiffness and strength has given new dimensions to tensile strength and elastic modulus [97]. Carbon nanotubes exhibit low density (1.3 to 0.4g/cm³) and specific strength to the tune of 48,000 kN.m.kg⁻¹ in comparison to the strength (154 kN.m.kg⁻¹) of high carbon steel. It is tedious to measure the extraordinary mechanical properties of carbon nanotubes because of either ineffective or inappropriate or inadequate preparation of the test sample and deduced information.

The quantum-mechanical estimation, the chemical treatment, and the high-resolution imaging techniques help investigate the numbers of fractures in a shell, the chirality of the outer shell, and the cross-linking present within the outer shell, *etc*. Computational simulation reflects on the defect on load sharing between the shells or exterior, while the various irradiations influence this type of simulation [98]. Carbon nanotubes as an individual shell exhibit high strength and weak shear interaction amongst adjacent shells. In multiwalled carbon nanotubes, adequate force reduces significantly, and under high electron, irradiations rectify them. The high electron irradiation enhances the cross-linking among inner shells and tubes, resulting in elevating the strength of these nanomaterials up to 60 GPa. Carbon nanotubes are not competent to withstand compression due to their hollow nature and high ratio of length to diameter, and buckles under torsional, compression, and bending stress. The degree of thermal conductivity elevates in carbon nanotubes, and it occurs along the nanotube [99, 100].

Varied fragments like nanoribbon, changed wave pattern, partial radial collapse, *etc*., are formed at the site of broken carbon nanotubes. The advent of carbon nanotubes with their high degree of stiffness and strength has given new dimensions to tensile strength and elastic modulus [97]. Carbon nanotubes exhibit low density (1.3 to 0.4g/cm³) and specific strength to the tune of 48,000 kN.m.kg⁻¹ in comparison to the strength (154 kN.m.kg⁻¹) of high carbon steel. It is tedious to measure the extraordinary mechanical properties of carbon nanotubes because of either ineffective or inappropriate or inadequate preparation of the test sample and deduced information.

The degree of thermal conductivity elevates in carbon nanotubes, and it occurs along the nanotube. This feature is considered to be of Ballistic type of conduction as this conduction involves the transport of charge carriers. These carriers offer the least or negligible electrical resistivity and include scattering. When there is no scattering, this conduction takes place, because the electrons follow Newton's second law of motion, at non-relativistic speed in the absence of the scattering. Materials exhibit resistivity are related to the distribution of moving electrons due to the impurities called defects, present in the materials under study reflecting the influence on the mean free path. Ballistic transport is of frequent occurrence in the nanosized metal nanowires (Kondo and Ohnishi, 2001). At room temperature, thermal conductivity along the axis of the single-walled carbon nanotubes is around 3500 $W.m^{-1}.k^{-1}$ in comparison to that of copper-element with excellent thermal conduction (385 $W.m^{-1} k^{-1}$) [101]. Thermal conduction in the case of single-walled carbon nanotubes (SWCNT) at room temperature across the axis, *i.e.*, in the radial direction is about 1.52 $W.m^{-1} k^{-1}$. This value is quite close to the thermal conductivity of soil [102 - 106]. When nanotubes are in mesoscopic assemblies like film or fibers' the thermal conduction is around 1500 $W.m^{-1} k^{-1}$. In vacuum and air, the thermal conduction is around 2800°C and 750°C, respectively. Defects present in materials of crystals, strongly influence the thermal properties of carbon nanotubes. These are the results of the scattering of phonons that elevates the rate of relaxation of phonon, reduction in the mean free path consequently declined thermal conductivity. Stone-Wales defects are common in crystallography. These induce changes in the π-bonded carbon atoms. This change causes rotation by 90° to the midpoint of these bonds [102 - 106].

Concerning the electrical properties of carbon nanotubes, these nanomaterials behave as moderate semiconductors materials. Under different conditions, they act as either metallic or semiconducting materials. The carbon nanomaterials do not exhibit semimetallic behavior, because in such nanomaterials, the degenerate point shifts away from k-point; (k-point is the wave vector in the first Brillouin zone and represents Brillouin zone sums). The degenerate point is a point where the π-bonding band joins the antibonding π-band; at this point, the energy comes to zero. The shifting of the degenerate point takes place because of the hybridization of σ and π antibonding bands [107, 108].

The curvature effects influence the electrical properties of carbon nanotubes having a small diameter. This factor also affects their metallic and semiconductor nature and may not follow the regular behavior pattern. Some of the single-walled carbon nanotubes may act as metallic materials when they have features of semiconductors. The zigzag chiral single-walled carbon nanotubes have a minimal diameter and may behave as metallic materials. Carbon nanotubes as components of a given composite material are good options as interconnecting and conducting promoter components. This feature is applicable in the case of highly conducting electrical wires. Parameters, like undesired current saturation under-voltage, resistive nanotubes to nanotubes junctions and the impurities of the materials, *etc.*, are significant functionally because these parameters tend to lower the degree of the electrical conductance of the resultant mesoscopic nanotubes [109]

Carbon nanotubes are regarded as unidimensional conductors because electron moves only along the axis of the nanotubes. In single-walled carbon nanotubes maximum electrical conductance is expressed as 2G0; (G0 =2e2/h). It is the conductance in a single Ballistic quantum channel (Ballistic conduction also referred to ballistic transport). In this mode of transport, charge carriers have negligible electrical resistivity. In most cases the charge carriers are electrons. During the charge transportation, the electrons get scattered due to the presence of defects in the medium through these charge carriers are transported [109].

Generally, there is a rational approach towards the metallic and semiconductor nature of nanotubes concerning their topographic structural aspect, the sophistication of tight bindings, and other physical parameters like curvature, inter-tubular interaction, and the defects from topological aspects [110]. Intrinsic superconductivity related to carbon nanotubes is debatable [111 - 114].

The optical properties of carbon nanotubes are suitably applicable in phenomena like absorption, photoluminescence-fluorescence, and Raman spectroscopy. These applications help to investigate the characteristics of carbon nanotubes. These types of studies facilitate the fabrication and modifications of carbon nanotubes and carbon nanoparticles to enable them to meet the set targeted applications. Carbon nanotubes are among the most suitable options as nanoscale light emitters, infrared wavelength interaction, and optical communication. These features depend on their metallic and semiconductor nature and their range of diameters. The non-linear optical properties and electro-optic features of carbon nanotubes are applicable in laser, high field emitters, and exhibit saturable absorber and modulators for electro-optics [115, 116].

Single-walled carbon nanotubes are one of the promising options in the field of optoelectronics. These nanomaterials are very much suitable for investigating the processes of one-dimensional excitation because of their one-dimensional nature. Their one-dimensional character provides secure quantum confinement and facilitates the formation of photo-created electron-hole pairs, which is exciton. (Each exciton represents hydrogenic Coulomb-bound electron-hole pair [117 - 119]. Earlier, an ideal one-dimensional system can bind energy of excitons as infinite, but this binding energy was observed to be around hundreds of meV [120 - 122]. Single carbon nanotube, an ideal one-dimensional system is good enough to investigate energy binding. They exhibit some specific inter-band gaps because of their structural aspects like diameter, chiral angle, and the electronic format, (*i.e.*, metallic and semiconducting nature). These features of single-walled carbon nanotubes make them a powerful and useful tool to study such features and their distribution in a study sample.

Sometimes, the momentums of the photon are comparatively much less than the energy of the Bloch electrons; under such conditions, approximate conservation of total linear momentum does not allow the Horizontal transition. As a result, one observes only vertical changes. In most of the solids, the total linear momentum is conserved, and this ensures only vertical transition among solids. In the case of single-walled carbon nanotubes, only equal linear energies move along their axis [123]. There exists a relationship between valance and conduction sub-band and the angular momentum of the axis of the tube [123]. The angular momentum of the carbon nanotubes plays a significant role during absorption and emission. If polarized light is parallel to the axis of the tube, it does not exhibit angular momentum. This type of transition occurs when equal angular momentum exists, but, if the position of polarized light is at a right angle to the axis of the tube, then, it influences the angular momentum and relates to the circumference of the tube. The perpendicular position of polarization permits the transition involving a change in angular momentum. There is a strong impact of depolarization when the field is perpendicular, and there are chances that absorption can be suppressed causing difficulties in observing optical transition [115 - 117, 119, 123, 124].

Fullerene is a very reactive and useful carbon nanomaterial because it has specific physicochemical properties. Fullerene is a very capable scavenger for reactive oxygen species (ROS). Fullernol [C60 (OH18)] is a hydrophilic form, and it has an excellent affinity for electrons. This feature makes it a potential scavenger for free radicals. The degree of scavenging becomes high using sodium salt of Hexa (sulfobutyl) fullerene. In this form, there are more addends per C60 cage in comparison to the fullernol [125, 126]. There are many double bonds and lowest uncoupled molecular orbitals present in the molecules of fullerene that facilitates the acceptance of electrons [127]. These provide sites for the attachment of hydroxyl radicals. As a proton approaches the vicinity of the pentagon of fullerene (1°A), the proton shifts inside the fullerene structure and positions at a space of 1.1°A, this behavior reflects on the possibility of absorbing a maximum of 6 protons [128, 129]. A single molecule of fullerene accommodates around 34 methyl radicals and can interact with many superoxide radicals [127].

Fullerene also exhibits a specific photocatalytic activity. Under UV radiation, this activity coordinates with its ability to form radicals like singlet oxygen ($1O_2$), superoxide radical (O_2^-) at the surface when subjected to U V radiation. The fullernol a derivative of fullerene destabilizes and interact with noncovalent and divalent counterions (Na^+, Ca^{++} and Mg^{++}), and these change their behavior. Such behavioral changes include non-responsive nature to transmembrane pressure and breaking of aggregated fullernol. This process helps the passage of the derivatized molecules or particles [130]. Fullerene exhibits a delocalized conjugated structure and a tendency to accept an electron. During electron acceptance, fullerene induces initiation of the rapid photo, induce charge separation also charge recombination at a slower rate [126, 131, 132]. Fullerene increases the quantum efficiency and transfer of the charge. In the case of fullerene, the conduction band gets lowered in comparison to the potential reduction band. Possibly this is because the photon-generated electrons move from the conduction band of fullerene towards that of TiO_2 molecule as these two interact under visible light. The photocatalytic activity of fullerene is proportional to the concentration of fullerene and CdSe [126, 133]. Graphene is another allotropic form of carbon, and it is a subunit of graphite.

The term graphene originates from graphite, and polycyclic hydrocarbon, *i.e.*, the term -graph is from graphite and suffix -ene is from polycyclic aromatic hydrocarbon, both after amalgamate and the term graphene (graph+ene = graphene). The prime differences between metals, conducting, and the non-conducting matter is the difference in the energy band of electron between them. These energy bands may be empty or may have energy gaps in metals and conducting materials but not in non-conducting materials [134 - 136] Interactions between graphene and other molecules depend on the specific structural aspect and the duration available. Structurally graphene is a single sheet consisting of carbon atoms in hexagonal lattice rendering it a honeycomb appearance. The structural stability of graphene is because of the compact arrangement of carbon atoms and the hybridization state of sp^2 orbitals. This hybridization is because of the combination of orbital s, px, and py resulting in the formation of σ-bonds and final pz orbital electron contributing to π-bond. During transitional conditions, π-bonds hybridize π-bonding and anti π bonding orbital. Structurally this state is a 2D allotropic form of carbon with sp^2 orbital that binds with each other with a molecular bond length of 0.142 nm. This state exhibits absorption of light at π α 2.3% of white light. This state is capable of participating in spin transport [136, 137]. The prominent features of the C-C bond in graphene include a bond length as 1.42 A°; the atomic radius of resonance is 0.71 A°. The bond length is twice the atomic radius. The area occupied by resonance bond atomic radius is 4.3X4 A°. There are four resonance bonds and four single bonds that are responsible for the six free bonds about six-ring carbon atoms configuration. This structural configuration is related to the six other rings of graphene. The center of the hexagon of graphene is of spherical with 0.71 A° radius which is similar in dimension to the radius of resonance of a bond of about 0.71 A° [136, 137].

There are odd electrons available as a pool representing a broad structural pattern for graphene. These electrons play a significant role during the interaction of graphene because these are involved in covalent bonds formation and specific π-electrons. The rate of interaction of graphene is directly related to the rate of involvement of the free electrons of radicals. The electrons related to π bond do not correlate when placed on the same locus while electrons of the radical correlate significantly during the said interaction. In these situations, the electrons have their spin and are away from each other [138, 139]. The chemical functionality of graphene is more effective in comparison to the physical feature [136, 138] Graphene and graphene-based nanomaterials are very reactive, and their interactions are related to the high specific surface area and π-π stacking. The hydrophilic form of graphene elevates the degree of its physiological interactions, the degree of biocompatibility, and the dispersibility many folds in comparison to its hydrophobic form. The chemistry and other properties, like mechanical, tribological (related with design, friction, wear, and lubrication of interacting surfaces in relative motion) and corrosion impacts of this nanomaterial readily undergo modifications as per the need [136, 139 - 141].

Features that Influence the Behavior of Dendrimers

The dendrimers are nanomaterials resembling tree-like branching and exhibit 3D spherical morphology. These show repeated dichotomous systematic branches ranging from few to excessive branching. These nanomaterials are synthesized as per the need or set targeted or predetermined functions. The dendrimers are spherical polymeric molecules and formulated as carriers for drugs or any cargo that gets trapped in its structural spaces. Dendrimers have a large surface area about their respective volumes. The space present between its branching can be conveniently is used for loading the molecules/cargo to be carried in the biological system. The surface to volume ratio may extend up to $1000 m^2/g$ [142]. Dendrimers exhibit specifically polymeric, monodisperse perfect symmetrical structure. Dendrimers are considered either low molecular weight or high molecular weight species. The smaller molecular species are dendrons, and dendrimers with high molecular weight are called polymer brush because of the hyperbranched polymers.

In most cases, the properties of dendrimers are dependent on the functional group present on these polymeric molecules. Dendrimers have improved physicochemical properties. These properties depend on the mode and controlled synthesis. Dendrimers can be functionalized at their core components as well as at their peripheral zone because of their unique architecture. The incorporation of a specific group within the interior of the dendrimer relates to parameters like the design of monomer, mode of synthesis, properties of the type of backbone of a dendrimer, *etc*. During such activities, an effort is made to adhere to the covalent modifications [143]. The Olefin metathesis facilitates the internalization of the small guest molecules successfully within the inner zone in dendrimers. Metathesis: -a chemical reaction involves different kinds of molecules exchange parts forming other types of molecules and ruthenium-based Grubbs Catalyst). In this case, alkylated pyrene, and imidazolidinone gets cross-metathesis [144].

Dendrimers are the most suitable drug carriers because their internal units can be tuned or manipulated to load drugs or cargo. The external surface can be attached to a ligand that is suitable for the specific target without affecting its particular stability. This manipulation of dendrimer enables them to either avoid or deceive the mononuclear phagocyte system (a component of the immune system) of the recipient.

Some terminology and nomenclature are used based on their structural and chemical nature. These polymers are the products that involve incoherent polymerization of monomers, and either a monomer or a functional group is incorporated. These dendrimers have facile synthesis where a well defined finite structure is the result, and because of this, the degree of mimicking and their utility enhances. The role of dendrimers as a molecular tool and biotechnical potential in the biomedical field is well evident [145]. The process of polymerization results in their linear fold and branched configuration. The branching can be undefined or well defined. The linear polymers show intermediate physiochemical properties between the hyperbranched polymers and linear polymers. The type of polymerization depends on the nature of structural conformation [146].

Dendrigraft (forms of dendrimers) have well-defined branching with specific molecular structure and maintain their monodisperse nature. The dendrigraft polymers belong to a class in which highly branched macromolecule are incorporated but do not have a core component. Dendrigrafts are similar to dendrimers that are formulated with well defined molecular structure and remain in monodisperse. These differentiate into linear polymers chain and attached copolymer chains. This arrangement may be repeated, resulting in a hyperbranched motif with a specific number of associated polymers. It may not appear to be a palm tree-like appearance [145].

Dendron/Dendric is the wedge form of dendrimer that shows dichotomous branching but does not have a core component. Dendron-a polymeric way is useful in dendrimer synthesis and involves a parallel mode of synthesis. These dendritic wedges are available commercially and assembled covalently and non-covalently. These forms of dendrimers are the Fréchet-type dendron. These types of dendrons act as a skeleton for the poly-benzyl ether that is a hyperbranched form. The attached groups are excellent sites for functionalization and also serve as a polar group to induce hydrophilicity even though these are hydrophobic [145]. Hyperbranched dendrimer form of dendrimers exhibits the molecular structure with well-defined branching. These forms are the result of the final association of dendron /dendric wedge along with the core component; thus, most of such types evolve into dendrimer [145].

Generations of dendrimers are the branching/hyper-branching starting from core up to the periphery give a similar structure between the focal/branching points. The numbers of branching point/focal points are the number present between the core and the outermost (periphery) layer. If there are five layers of focal points, the dendrimer is said to be having five generations, and it represents G-5 dendrimer. If a dendrimer consists of polypropylene imine and poly amidoamine, these are having five generations, and then it is abbreviated as G-5-PPI [145]. The core is the central component to which the first generation is attached. The core component is zero generation (GO). There is no focal point in the core because hydrogen substituent is considered to be focal points, *e.g.*, in case of polypropylene mine dendrimer (G-5-PPI) 1, 4-di amido butane is a core component, and G-5 poly amidoamine (G-5PAMAM) ammonia is the core component [145]. Concerning dendrimers half-generation term is used concerning dendrimers, and it is represented by specific functional groups during the synthesis of dendrimers, specifically about amino-terminated dendrimers and applies to the biological system. Shell is the zone which differentiates into the outer shell and inner shell. This region extends between the core component and the space of generations. The outer shell is a space/zone between the last and peripheral focal point/branching point and the surface. The inner shell extends between the core zone and the outer shell [145]. The distinct most focal points represent pincers just before the surface of the dendrimer. Generally, the number of pincers is half the number of groups attached to the surface; two surface groups attach to one focal point [145].

Properties of dendrimers are dependent on their structural diversity; hence, they may show variations [147]. Dendrimers are specific macromolecules and have a low molecular weight that exhibits wide varieties of properties like solubility, toxicity, and fixation, removal of active group or materials, and distribution within the system. Among dendrimers, polymerization is a controllable process because of this fact; their size and molecular mass are controlled or regulated during their synthesis. These exhibit specific molecular architecture. This feature enables dendrimers to exhibit highly improved physical and chemical properties in comparison to the traditional linear polymers [148]. Dendrimers have a large surface area about their respective volume between its branching. This space accommodates molecules as cargo for the biological system. Lower generation Dendrimers that have fewer generations may not be tightly packed in the spherical structures but have an extensive surface area concerning the volume. This feature facilitates the loading of drugs or any guest molecule. The surface to volume ratio may extend up to $1000m^2/g$ [142].

The dendrimers exhibit hydrophilicity and behave stable, compact spherical structures in solution. This feature affects their rheological properties and shows a lower degree of viscosity. The numbers of chain-ends present on dendrimers are responsible for their degree of solubility, miscibility, and higher reactivity. The intrinsic viscosity increases with the increase of the molecular mass, but up to the 4th generation after that, it declines polymerization in the dendritic pattern) while in linear polymerized macromolecules/classic polymers the intrinsic viscosity increases with the increase of molecular mass. The nature of the surface groups present on the dendrimers plays a vital role in determining the degree of solubility of dendrimers. The hydrophilicity of surface groups makes dendrimers soluble in the polar solvent while hydrophobic surface groups make them soluble in non-polar solvents. Such nature of the surface groups also affects the chemical reactivity of dendrimers [148].

Dendrimers exhibit globular shape at least after the 4th and 5th generations. Despite their more or less compact nature, there are enough spaces within the dendrons. These spaces are suitable to lodge macromolecules like drugs, dyes, fluorophores, *etc*., thereby, enhance their utility as a drug carrier, detector using fluoresce technique, or other chemicals as per the need. There is a possibility of trapping a macromolecule like the Rose Bengal/ p-nitrobenzoic acid within the dendrimer Box made of poly (propylene) imines and consists of 64 branches up at periphery. The interaction between the amino acid L-phenylalanine at the surface of the dendrimer, a shell is formed at the edge, after lodging the macromolecules. The outer shell of the said dendrimer is subjected to hydrolysis to release the gust molecule (cargo) [149, 150]. The dendrimer box opens photochemically, and to achieve this, the 4th generation polypropylene imine dendrimer having 32end groups modifies with azobenzene groups. The azobenzene groups undergo complete, irreversible photo-isomerism having E-isomer and Z-isomer. These forms interchange at 313 nm and 254 nm, respectively. E-isoform changes into Z-isoform at 313nm and Z-isoform change into E- isoform at 254nm. The photochemical types of dendrimers have the potential to encapsulate the macro-molecules like eosin-Y. Z-isomers with 4th generation have a better capability to hold guest molecules than E-isomers [151 - 153]. Formulated dendrimer with absorbing dyes attached at the periphery and mode of the energy of light transfer to the chromophores placed at the core. The absorption spectra of complete macromolecule are specifically broad because the chromophores occupy a wide range of wavelengths [153]. When the energy transfers to the core chromophore, it converts into narrow emission of core chromophore. This ability of light-harvesting increases with an increase in the numbers of generation and with the rise of generation numbers, the number of peripheral chromophores also increases [151 - 153].

The cationic dendrimers have amine and imine at their terminals and act as hemolytic and cytotoxic agents. These dendrimers have cationic groups at low pH (acidic pH). Their degree of toxicity is related to the number of generations and the number of surface groups present. PAMAM dendrimers having 2nd, 3rd, and 4th generations and react with the membrane proteins of red blood cells (erythrocytes) and change their conformations. These result in their toxic impact [148, 154]. Anionic dendrimers have carboxylate groups at their surface. When these dendrimers along with phosphate-buffered saline and PAMAM they cause aggregation among red blood cells, but nucleated cells like fibroblasts from the Chinese hamsters remained more or less unaffected [148, 155].

Nanomaterials exhibit specific physicochemical properties like the advantage of the large area, the ability to carry molecules, avoid their degradation, and increase the duration in blood circulation, relevance interaction, and enzyme-like impact within the biosystem. As a result, nanomaterials find wide applications in biological sciences, medicine, and biotechnology [156].

CONCLUSION

There is a need for an appropriate nomenclature and definition of the nanomaterials. The classification and proper description help normalize the understanding related to the precautions and particular applications. The different nanomaterials cause their respective impacts on the biosystems and their components. These depend on their physicochemical properties. The fabrications, modifications, and judicial and selective applications of these nanomaterials will lead to the sustenance of biota and abiotic components of the environment. This practice will also ensure their appropriate uses and reduce their derogative impacts.

REFERENCES

[1]   Buzea C, Pacheco I, Robbie K. Nanomaterials and nanoparticles: sources and toxicity. Biointerphases 2007; 2(4) MR17.
[http://dx.doi.org/ 10.1116/1.2815690]

[2]   Hubler AW, Osuagwa O. Digital quantum batteries, energy and information storage in vacuum tube arrays. Complexity 2010; 15(5): 49-55.
[http://dx.doi.org/10.1002/cplx,20306]

[3]   McGovern C. Commoditization of nanomaterials. Nanotechn Persp 2010; 6(3): 155-78.
[http://dx.doi.org/10.4024/N15GO10A.ntp.06.03]

[4]   Eldridge T. Achieving industry integration with nanomaterials through financial markets 2014.www.nanotech-now.com/columnd/?articles=825

[5]   Maynard AD. Don't define nanomaterials. Nature 2011; 475(7354): 31.
[http://dx.doi.org/10.1038/475031a] [PMID: 21734686]

[6]   International Council of Chemical Association (ICCA). Core elements of a regulatory definition of manufactured nanomaterials 2010.http://www.icca-chem.org/ICCAdocs/oct-2010_ICCA-core-elements-of-a-Regulatory- definition-of-manufactured-nanomaterials.pdf

[7]   Bleeker EAJ, de Jong WH, Geertsma RE, *et al.* Considerations on the EU definition of a nanomaterial:

science to support policy making. Regul Toxicol Pharmacol 2013; 65(1): 119-25.
[http://dx.doi.org/10.1016/j.yrtph.2012.11.007] [PMID: 23200793]

[8]   Boverhof DR, Braman CM, Butala JH, *et al.* Comparative assessment of nanomaterials definition and safety evaluation considerations. Reg Toxi Pharmacol 2015; 73(1): 137-50.
[http://dx.doi.org/10.1016/j.yrtph.2015.06.001]

[9]   European Commission Regulation (EC) No 1223/2009 of the European Parliament and of the Council of 30 November 2009 on cosmetic products. Official Journal European Union (EN). 52(1): 59-209. L342.

[10]   US FDA (FDA). Considering Whether an FDA-regulated Product Involves the Application of Nanotechnology: Guidance for Industry: Draft Guidance, US Food and Drug Administration 2011.http://www.fda.gov/RegulatoryInformation/Guidances/ucm257698.htm

[11]   US EPA. US EPA, Nanotechnology White Paper, US Environmental Protection Agency, 2007.http://www.epa.gov/osa/pdfs/nanotech/epa-nanotechnology-whitepaper-0207.pdf EPA 100/B-07/001.

[12]   US EPA. U S Environmental Protection Agency Proposed significant new use rules on certain chemical substances. 76(249): 81441-7.

[13]   US EPA. U.S. Environmental Protection Agency, Chemical substances when manufactured or processed as nanoscale materials; TSCA Reporting and recordkeeping requirements, 2014. Fed Regist 2014; 80(65): 18330-18342X.

[14]   EC European Commission Regulation (EU) No 528/2012 of the European Parliament and of the Council of 22 May 2012 concerning the making available on the market and use of biocidal Products. Official Journal European Union (EN) 2012; 1-123.: L167.

[15]   Belgium Royal Decree Concerning the Placing on the Market of Substances, Produced in Nanoparticular State Federal Public Service Health, Food Chain Safety, and Environment, Kingdom of Belgium.

[16]   Lahir YK. Role and adverse effects of nanomaterials in food technology. J Toxicol Health 2015; 2.
http://www.hoajonline.com/journals/pdf/2056-3779-2-2.pdf
[http://dx.doi.org/Doi:10.7243/2056-3779-2-2]

[17]   Klaessig F, Marrapese M, Shoji A. Nanotechnology Standards Nanostructures Science, and Technology. New York: Springer 2011.

[18]   Zeng S, Baillargeat D, Ho HP, Yong KT. Nanomaterials enhanced surface plasmon resonance for biological and chemical sensing applications. Chem Soc Rev 2014; 43(10): 3426-52.
[http://dx.doi.org/10.1039/c3cs60479a] [PMID: 24549396]

[19]   Zhang Y, Leu Y-R, Aitken RJ, Riediker M. Inventory of engineered nanoparticles containing consumer products available in Singapore retail market and likely hood of release into the environment. Int J Environ Res Public Health 2015; 12(8): 8717-43.
[http://dx.doi.org/10.3390/ijerph120808717] [PMID: 26213957]

[20]   Salata O. Applications of nanoparticles in biology and medicine. J Nanobiotechnology 2004; 2(1): 3.
[http://dx.doi.org/10.1186/1477-3155-2-3] [PMID: 15119954]

[21]   Kessler R. Engineered nanoparticles in consumer products understanding a new ingredient, Environmental Health Perspective. 2011; 119: pp. (3)A120-5.
[http://dx.doi.org/doi:10.12 89/ehp.119-a120]

[22]   Garbayo E, Estella-Hermoso de Mendoza A, Blanco-Prieto MJ. Diagnostic and therapeutic uses of nanomaterials in the brain. Curr Med Chem 2014; 21(36): 4100-31.
[http://dx.doi.org/10.2174/0929867321666140815124246] [PMID: 25139519]

[23]   Kerativitayanan P, Carrow JK, Gaharwar AK. Nanomaterials of engineering stem cells Responses. Adv Healthc Mater 2015; 4(11): 1600-27.

[http://dx.doi.org/10.1002/adhm.201500272] [PMID: 26010739]

[24]   Gajanan K, Tijare SN. Applications of nanomaterials, Materials to-day: proceedings. 2018; 5: pp. (1)1098-6.
[http://dx.doi.org/doi.org/10.1016/j.matpr.2017.11.187]

[25]   Mirela D. Metallic properties nanoparticles 2009. www.ung.si/sstanic/teachingseminar/2009/2009 1214-Dramar-MetNP.pdf

[26]   Chemistry, Libretexts™, Surface Energy 2016. https://chem.libretexts.org/Bookshelves/Inorganic_ Chemistry/Book%3A-inorganic-Chemistry

[27]   Yin Y, Wang J, Lu X, *et al. In situ* generation of plasmonic nanoparticles for manipulating photon-plasmon-coupling in microtubules coupling. ACS Nano 2018; 12(4): 3726-32.
[http://dx.doi.org/10.1021/acsnano.8b00957] [PMID: 29630816]

[28]   Chen T, Pourmand M, Feizpour A, Cushman B, Reinhard BM. Tailoring plasmon coupling in self-assembled one dimensioned gold nanoparticles chain through simultaneous control of size and gap separation. J Phys Chem Lett 2013; 4(13): 2147-52.
[http://dx.doi.org/10.1021/jz401066g] [PMID: 24027605]

[29]   Chuntonov L, Haran G. Trimeric plasmonic molecules: the role of symmetry. Nano Lett 2011; 11(6): 2440-5.
[http://dx.doi.org/10.1021/nl2008532] [PMID: 21553898]

[30]   Englebienne P, Hoonacker A, Verhas M. Surface plasmon resonance: principles, methods and application in biomedical science. Spectroscopy (Springf) 2003; 17: 255-73.
[http://dx.doi.org/10.1155/2003/372913]

[31]   Pokrant IS. The crystal structure of TiO_2 nanoparticles. Microscope Microanatomy 2008; 14(2): 354-64.
[http://dx.doi.org/10.1017/S1431927608082159]

[32]   Raster A, Yazdanshenas ME, Rashidi A, Bidoki SM. Theoretical review of optical properties of nanoparticles. J Eng Fibers Fabrics 2013; 8(2): 85-96.
[http://dx.doi.org/10.1177/155892501300800211]

[33]   Feldheim DL, Foss CA. Metal nanoparticles: Synthesis, Characterization and Applications. New York: Marcel Dekker Inc. 2002.

[34]   Zeng S, Yu X, Law W-C, *et al.* Size dependence of gold nanoparticles enhanced surface plasmonic resonance based on differential phase measurement Sensors and Actuators B: Chemical 2013; 176: 1128-1133 .

[35]   Scott MJ. Photoelectric sensors and controls, selection and application. CRS Press 1988; p. 29.

[36]   Bohren CF, Hoffman DR. Absorption and scattering of highly small particles. New York, USA,: Wiley-Interscience 2010.

[37]   Gonis A, Butler WH. Multiscattering in solids. Springer 1999.

[38]   Harrison D. Complementarity and Copenhagen interpretation of quantum mechanics, UPSCALE Department of Physics, University of Toronto, 2002. retrieved 2008-06-21.

[39]   Mie G. Beitrage zur medien, speziell kolloidalar mettallosungen. Ann Phys 1908; 25: 377-445.
[http://dx.doi.org/10.1002/andp.19083300302]

[40]   Mie G. The optics of translated version – P Newman Translation Chicago. National Translatory Center 1978.

[41]   Cheng DK. Introduction to electromagnetic. USA: Addison –Wesley Publishing Co 1983.

[42]   Popovic Z, Popovic BD. Introductory electromagnetic. New Jersey, USA: Prentice-Hall Inc 2000.

[43]   Harrington RF. Time-Harmonic electromagnetic fields. IEEE, Press 2001.

[http://dx.doi.org/10.1109/9780470546710]

[44]   Kokhanovsky AA. Light Scattering Reviews 5: single light scattering and Radiative transfer. New York: Springer-Berlin Heidelberg 2010.
[http://dx.doi.org/10.1007/978-3-642-10336-0]

[45]   Ingrosso C, Panniello A, Comparelli R, Curri ML, Striccoli M. Colloidal inorganic nanocrystals based nanocomposites: functional materials for micro and nanofabrication. Materials (Basel) 2010; 3: 1316-52.
[http://dx.doi.org/10.3390/ma3021316]

[46]   Mishchenko MI, Travis LD, Lacis A. Scattering, absorption, and extinction of light by small particles. New York: NASA Goddard Institute for space studies 2004.

[47]   Browne M. Physics for engineering and science. 2nd Ed. New York: McGraw Hill/Schaum 2013; p. 427.ISBN: 978-0-07-161399-6.

[48]   Purcell M. Electricity and magnetism. 3rd ed., New York: Cambridge University Press 2013.ISBN: 978-1-107-01402-2 .
[http://dx.doi.org/10.1017/CBO9781139012973]

[49]   Tipler PA. Physics for scientists and engineers, Vol-1, Mechanics, oscillations and waves Thermodynamics. MacMillan 1999.

[50]   Barnett CF. Some applications wavelength turbidometry in the infrared. J Phys Chem 1942; 46(1): 66-75.
[http://dx.doi.org/10.1021/j150415a009]

[51]   Young AT. Rayleigh scattering. Appl Opt 1981; 20(4): 533-5.
[http://dx.doi.org/10.1364/AO.20.000533] [PMID: 20309152]

[52]   Sttratton JA. Electromagnetic Theory. New York: McGraw Hill 1941.

[53]   Hahn DW. Light scattering theory, (PDF). USA: University of Florida 2009.

[54]   Raman CV. A New Radiation. Indian J Phys 1928; 2: 387-98.

[55]   Keith JL, Meiser JH. Physical Chemistry. Benjamin/Cummings 1982.

[56]   Long DA. The Raman Effect, John-Wiley and Sons Ltd, 2002.

[57]   Singh R, Raman C V. C. V. Raman and the discovery of the Raman effect. Phys Perspect 2002; 4(4): 399-420.
[http://dx.doi.org/10.1007/s000160200002]

[58]   Christillin P. Nuclear Compton Scattering. J Phys Nucl Phy 1986; 12(8): 837.
[http://dx.doi.org/10.1088/0305-4516/12/9/008]

[59]   Griffith SD. Introduction to elementary Physics. Wiley 1987.
[http://dx.doi.org/10.1002/9783527618460]

[60]   Moore CT. Observations of the transition from Thomson to Compton scattering in optical multiphoton interactions with electrons 1995.www.lle.rochester.edu/media/publications/thesis/moore.pdf

[61]   Roessle M. Basic X-ray scattering Fachhochsule Lübeck, Lübeck University of Applied Science. 2009.https://www.embl-hamberg.de/biosaxa/courses/embo2012/slide/x-ray-scattering-Basics-roessle.pdf

[62]   Feinberg Evgenii l. The forefather (about Leonid Issakovich M). Phys Uspekhi 2002; 4(1): 81.
[http://dx.doi.org/10.1070/PU2002v045n01ABEH00126]

[63]   Meassures RM. Structure monitoring with fiber optic technology. San Diago, California, USA: Academic Press 2001.

[64]   West W. Absorption of electromagnetic radiation (Eastman Kodak Company, Rochester, New York) Access Science. McGraw Hill 2013.

[http://dx.doi.org/doi:10.1036/1097-8542.001600]

[65]   Buffat PH, Borel JP. Size effect on the melting temperature of gold particles Physical Review 1975 13(6): 2287.
[http://dx.doi.org/10.1103/PhysRevA.13.2287]

[66]   Hewakuruppu YL, Dombrovsky LA, Chen C, *et al.* Plasmonic "pump-probe' method to study semi-transparent nanofluids Applied Optics 2013; 52(24): 6041-50.2013;
[http://dx.doi.org/https://10.1364/AO.006042]

[67]   Wu J, Yu P, Sasha AS, *et al.* Broadband efficiency enhanced in quantum dots solar cells coupled with multi spiked plasmonic nanostars. Nano Energy 2015; 13: 827-35.
[http://dx.doi.org/10.1016/j.nanoen.2015.02.012]

[68]   Nobbmann U, Morfesis A. Light scattering and nanoparticles Materials Today. 2009; 12: pp. (5)52-4.
[http://dx.doi.org/10.1016/S1369-7021(09)70164-6]

[69]   Gispert JR. Coordination Chemistry. Wiley-VCH 2008.

[70]   Brennan MC, Herr JE, Nguyen-Beck TS, *et al.* Origin of size-dependent Stokes shift in $CsPbBr^3$ perovskite nanocrystals. J Am Chem Soc 2017; 139(35): 12201-8.
[http://dx.doi.org/10.1021/jacs.7b05683] [PMID: 28772067]

[71]   Sahi S, Magill S, Ma L, *et al.* Wavelength-shifting properties of luminescence nanoparticles for high energy particle detection and specific physics process observation. Sci Rep 2018; 8: 10515.
[http://dx.doi.org/10.1038/s41598-018-2874-y] [PMID: 10515]

[72]   Hassel M. Schafer, Unconversion of nanoparticles. Angew Chem Int Ed 2011; 50: 5808-29.
[http://dx.doi.org/10.1002/anie.201005159]

[73]   Peach G. Theory of the pressure broadening and shift of spectral lines. Adv Phys 1981; 30(3): 367-74.
[http://dx.doi.org/10.1080/00018738100101467]

[74]   Atkins P, de Paula J. Journal of Physical Chemistry. 8th ed., W H Freeman 2006.

[75]   Atkins P, de Paula J. Elements of physical chemistry. 5th ed., Oxford: Oxford University Press 2009.

[76]   Kitai A. Luminescent Materials and applications. John Wiley and Sons 2008.

[77]   Chou KC. Identification of low-frequency modes in protein molecules. Biochem J 1983; 215(3): 465-9.
[http://dx.doi.org/10.1042/bj2150465] [PMID: 6362659]

[78]   Chou KC. Low-frequency vibrations of DNA molecules. Biochem J 1984; 221(1): 27-31.
[http://dx.doi.org/10.1042/bj2210027] [PMID: 6466317]

[79]   Urabe H, Tominagago Y, Kubota K. Experimental evidence of collective vibrations in DNA double helix Raman Spectroscopy. J Chem Phys 1983; 78: 5957-39.
[http://dx.doi.org/10.1063/1.444600]

[80]   Urabe H, Sugawara Y, Ataka M, Rupprecht A. Low-frequency Raman spectra of lysozyme crystals and oriented DNA films: dynamics of crystal water. Biophys J 1998; 74(3): 1533-40.
[http://dx.doi.org/10.1016/S0006-3495(98)77865-8] [PMID: 9512049]

[81]   Carey PR, Chen Y, Gong B, Kalp M. Kinetic crystallography by Raman microscopy. Biochim Biophys Acta 2011; 1814(6): 742-9.
[http://dx.doi.org/10.1016/j.bbapap.2010.08.006] [PMID: 20797452]

[82]   Dhar L, Roger SA, Nelson KA. Time-resolved vibrational spectroscopy in the impulsive limits. Chem Rev 1994; 94(1): 157-93.
[http://dx.doi.org/10.1021/cr00025a006]

[83]   Begley RF, Harvey AB, Byer L. Coherent Anti-Stokes Raman Spectroscopy 1974; 25(7): 387-97.

[84]   Ellis DI, Goodacre R. Metabolifinger printing in a diagnosis, biomedical application of IR and Raman

Spectrometry. Analyst (Lond) 2006; 131(8): 875-85.
[http://dx.doi.org/10.1039/b602376m] [PMID: 17028718]

[85]   Hellerer T, Axäng C, Brackmann C, Hillertz P, Pilon M, Enejder A. Monitoring of lipid storage in Caenorhabditis elegans using coherent anti-Stokes Raman scattering (CARS) microscopy. Proc Natl Acad Sci USA 2007; 104(37): 14658-63.
[http://dx.doi.org/10.1073/pnas.0703594104] [PMID: 17804796]

[86]   Luis M, Pablo M, Jaun F, *et al.* Ultra-broadband and generation of multiple concurrent non-linear coherent interaction in random quadratic media. Appl Phys Lett 2013; 103: 1011101-101105.

[87]   Hardwood LM, Moody CJ. Experimental organic chemistry: principle and practice. Illustrated Ed., Wiley-Blackwell 1989.

[88]   Kastner MA. Artificial atoms. Phys Today 1993; 46(1): 24.
[http://dx.doi.org/10.1063/1.881393]

[89]   Ashoori RC. Electrons in artificial atoms. Nature 1996; 379: 413-9.
[http://dx.doi.org/10.1038/379413a0]

[90]   Sabaeian M, Ali KN. Size-dependent inter sub-band- optical properties of dome-shaped InAs/GaAs quantum dots with wetting layer. App Phys 2012; 51(18): 4176-85.
[http://dx.doi.org/10.1364/AO.51.004176]

[91]   Ali KN, Sabaean M, Sahoori M, Fallahi V. Kerr nonlinearity due to inter sub-band transition three-level InAS/GaAs quantum dots: impact of a wetting layer on dispersion curve. J Opt 2014; 16055004
[http://dx.doi.org/10.1088/2040-8978/16/5/055004]

[92]   Murray CB, Kagan CR, Bawendi MG. Synthesis and characterization of monodisperse Nanocrystals and close-packed nanocrystals assemblies. Annu Rev Mater Res 2000; 30(1): 545-610.
[http://dx.doi.org/10.1146/annurev.metsci.30.1.545]

[93]   Nano A. Carbon nanotubes, definition, properties, industry application and possible environmental concern 2005.www.azonano.com/article.apex&article

[94]   Wang X, Li Q, Xie J, *et al.* Fabrication of ultralong and electrically uniform single-walled carbon nanotubes on clean substrates. Nano Lett 2009; 9(9): 3137-41.
[http://dx.doi.org/10.1021/nl901260b] [PMID: 19650638]

[95]   Gullapalli S, Wong MS. Nanotechnology: A guide to nano-objects. Chem Eng Prog 2011; 107(5): 28-32.

[96]   McGaugheyey GB, Gagne M, Rappe AK. Pi-stacking interactions alive and well in proteins. J Biolog Chem 1998; 273(25): 15458-63.
[http://dx.doi.org/10.1074/jbc.273.25.15458]

[97]   Yu M-F, Lourie O, Dyer MJ, Moloni K, Kelly TF, Ruoff RS. Strength and breaking mechanism of multiwalled carbon nanotubes under tensile load. Science 2000; 287(5453): 637-40.
[http://dx.doi.org/10.1126/science.287.5453.637]

[98]   Peng B, Lacoscio M, Zapol P, *et al.* Measurement of near ultimate strength for MWCNT and irradiation-induced cross-linking improvement. Nature-Nanotechnology 2008; 3: 626-31.
[http://dx.doi.org/10.1038/nnano.2008.211]

[99]   Jensen K, Mickelson W, Kis A, Zettl A. Buckling and kinking force measurements on individual MWCNT. Physics Review-B 2007; 76: 19-.
[http://dx.doi.org/https://10.1103/PhysRevB.76.195436]

[100]   Filleter T, Bernal R, Li S, Espinosa HD. Ultra strength and stiffness is cross-linked hierarchical carbon nanotubes bundles. Adv Mat 2011; 23(25): 2855-60.
[http://dx.doi.org/org/10.1002/adma.2011.100547]

[101]   Pop E, Mann D, Wang Q, Goodson K, Dai H. Thermal conductance of an individual single-wall carbon nanotube above room temperature. Nano Lett 2006; 6(1): 96-100.

[http://dx.doi.org/10.1021/nl052145f] [PMID: 16402794]

[102] Sinha S, Barjami S, Germano I, Schwab A, Muench G. Off-axis thermal properties of Single-Walleded carbon nanotubes film. J Nanopart Res 2005; 7(6): 651-7.
[http://dx.doi.org/10.1007/s11051-005-8382-9]

[103] Thostenson ET, Li C, Chao T-W. Nanocomposites in context. Compos Sci Technol 2005; 65(3-4): 491-516.
[http://dx.doi.org/10.1016/j.compscitech.2004.11.003]

[104] Mingo N, Stewart DA, Broid DA, Srivastava D. Phonon transmission through defects in carbon nanotubes from first principles. Phys Rev B 2008; 77: 033418.
[http://dx.doi.org/10.1103/PhysRevB.77.033418]

[105] Brayfindley E, Irace EE, Castro C, Karney WL. Stone-Wales rearrangements in polycyclic aromatic hydrocarbons: a computational study. J Org Chem 2015; 80(8): 3825-31.
[http://dx.doi.org/10.1021/acs.joc.5b00066] [PMID: 25843555]

[106] Kozio K, Janas D, Brown E, Hao L. Thermal properties of continuously spun carbon Nanotubes Fibers Physica E: Low dimensional system and Nanostructures 2017; 88(): 104-8.
[http://dx.doi.org/https://doi,org/10.1016/j.physe.2016.12.011]

[107] Lu X, Chen Z. Curved π-conjugation, aromaticity and related chemistry of small fullerene and single-walled carbon nanotubes, Chemical Reviews 2005; 105(10): 3643-96.
[http://dx.doi.org/10.1021/cr030093d]

[108] Laird EA, Kenneth FK, Steele GA, *et al*. Quantum transport in carbon nanotubes. Rev Mod Phys 2015; 87(3): 703.
[http://dx.doi.org/10.1103/RevModPhys.87.703]

[109] Vasylenko A, Tamie W, Poulov C, *et al*. Encapsulated nanowires boosting electronic transport in carbon nanotubes. Physical Reviews-B 2017; 95(12) arX10:1611.04867; .
[http://dx.doi.org/10.1103/PhysRevB.95.121408] [PMID: 121408]

[110] Charlier JC, Blasé X, Roche S. Electronic and transport properties of nanotubes. Rev Mod Phys 2007; 79(2): 677-732.
[http://dx.doi.org/10.1103/RevModPhys.79.677]

[111] Tang ZK, Zhang L, Wang N, *et al*. Superconductivity in 4 angstrom single-walled carbon nanotubes. Science 2001; 292(5526): 2462-5.
[http://dx.doi.org/10.1126/science.1060470] [PMID: 11431560]

[112] Takesue I, Haruyama J, Kobayashi N, *et al*. Superconductivity in entirely end-bonded multiwalled carbon nanotubes. Phys Rev Lett 2006; 96(5): 057001.
[http://dx.doi.org/10.1103/PhysRevLett.96.057001] [PMID: 16486971]

[113] Bockrath M. Carbon nanotubes: the weakest links. Nat Phys 2006; 2(3): 155-6.
[http://dx.doi.org/10.1038/nphys252]

[114] Lortz R, Zhang Q, Shi W, *et al*. Superconductivity characteristics of 4-A carbon nanotubes-zeolite composite. Proc National Acad Sci. 73-3.

[115] O'Connell MJ. Carbon nanotubes: properties and applications, CRS, . Boca Raton 2006.

[116] Nanot S, Thompson NA, Kim HG, *et al*. Single-walled carbon nanotubes, in Springer Handbook of Nanomaterials. Ed Vajtai., Berlin, Heidelberg: Springer-Verlag 2013.
[http://dx.doi.org/10.1007/978-3-642-20595-8_4]

[117] Ajiki H, Ando T. Aharonov-Bohm effects in carbon nanotubes. Physica B 1994; 201: 349-52.
[http://dx.doi.org/10.1016/0921-4526(94)91112-6]

[118] Oga WT, Takagahare T. Interbandnd absorption spectra Somerfield factors of the one-dimensional the electron-hole system. Phys Rev B Condens Matter Mater Phys 1991; 43: 14325-8.
[http://dx.doi.org/10.1103/PhysRevB.43.14325]

[119] Urya S, Ando T. Cross Polarized excitation absorption in carbon nanotubes with Aharonov-Bohm flux. Physic Review – B 761154202007;

[120] Knox RS. Theory of excitation of solid-state, Physics. New York: Academic Press 1963; 5.

[121] Elliot RJ, Loudon R. Theory of absorption edge in semiconductor in a high magnetic field. J Phys Chem Solids 1960; 15: 196-207.
[http://dx.doi.org/10.1016/0022-3697(60)90243-2]

[122] Maultzsck J, Pomraenka R, Reich S, *et al.* Excitation binding energies in carbon nanotubes from two-photon photoluminescence. Physics Reviews-B 2005; 72: 241402.
[http://dx.doi.org/10.1103/PhysRevB.72.241402]

[123] Ando T. Theory of electronic states and transport in carbon nanotubes. J Phys Soc Jpn 2005; 74: 777-817.
[http://dx.doi.org/10.1143/JPSJ.74.777]

[124] O'Connell MJ, Bachilo SM, Huffman CB, *et al.* Band gap fluorescence from individual single-walled carbon nanotubes. Science 2002; 297(5581): 593-6.
[http://dx.doi.org/10.1126/science.1072631] [PMID: 12142535]

[125] Jeng US, Lin TL, Chang TS, *et al.* Comparison of the aggregation behavior of water-soluble hexa(sulfobutyl)fullerene and polyhydroxylated fullerene for free radicals scavenging activity. In Koutsouko PG, Eds Trends in colloid and interface Science, X V,. Berlin, Heidelberg, : Springer 2001; 118.

[126] Lahir YK. Impacts of fullerene on a biological system, Clinical Immunology, Endocrinology and Metabolic Drugs 2017; 4(1): 48-57.
[http://dx.doi.org/doi.org/10.2174/13816128113199990614]

[127] Bakry R, Vallant RM, Najam-ul-Haq M, *et al.* Medicinal applications of fullerenes. Int J Nanomedicine 2007; 2(4): 639-49.
[PMID: 18203430]

[128] Chistyakov VA, Smirnovava YO, Prazdnova EV, Soldatov AV. Possible mechanism of fullerene C60 antioxidant action, Bio Med Res Int 2013; vol-2013. Article TD 2013; 821498.
[http://dx.doi.org/10.1155/2013/82/821498]

[129] Bozdaganyan ME, Orekhov PS, Shaytan AK, Shaitan KV. Comparative computational study of the interaction of C60 fullerene and tris-malonyl-C60-fullerene isomers with lipid layer: relation to their antioxidant effect, PLoS one 2014; 9: 7.
[http://dx.doi.org/10.1371/journal.pone.0102487] [PMID: e102487]

[130] Jassby D, Chae SR, Hendren Z, Wiesner M. Membrane filtration of fullerene nanoparticle suspensions: effects of derivatization, pressure, electrolyte species and concentration. J Colloid Interface Sci 2010; 346(2): 296-302.
[http://dx.doi.org/10.1016/j.jcis.2010.03.029] [PMID: 20381057]

[131] Davis JJ, Hill H, Kurz A, Leight AD, Safronov A. Aqueous electro-chemistry of C60 and ethanol of fullerene. Journal of Electrochemistry 1997; 429: 7.
[http://dx.doi.org/10.1016/S0022-0728(97)00068-5]

[132] Meng ZD, Zhu L, Choi J-G, Park CY, Oh WC. Preparation, characterization and photocatalytic behavior of WO3-fullerene/TiO_2 catalysts under visible light. Nanoscale Res Lett 2011; 6(1): 459.
[http://dx.doi.org/10.1186/1556-276X-6-459] [PMID: 21774800]

[133] Meng ZD, Zhu L, Ye S, *et al.* Fullerene modification CdSe/TiO_2 and modification of photocatalytic activity under visible light. Nanoscale Res Lett 2013; 8(1): 189.
[http://dx.doi.org/10.1186/1556-276X-8-189] [PMID: 23618055]

[134] Boehm HP, Setton R, Stump PE. Nomenclature and terminology of graphite intercalation Compounds. Carbon 1986; 24(2): 241-5.

[http://dx.doi.org/10.1016/000-623(86)901126-0]

[135] Bobroff J, Bouquet F, Georbig M-O, Funch J N. Animation of graphene atomic aspects, The University of Paris, Sud et CNRS, UPMC, CNRS. 2015.

[136] Lahir YK. Graphene and graphene-based nanomaterials are suitable for drug delivery. Chapter: 7 in Applications of Targeted Nano-Drugs and Delivery Systems 50 Hampshire Street, Cambridge, MA 02139;: Elsevier 2018; pp. 157-90. ISBN: 978-0-12-814029-1.

[137] Heyrovska R. Atomic structure of graphene, benzene, and methane with bond length as sums of the single, double and resonance bond radii of carbon 2008.https://arvix.org/pdf/0804.4086.

[138] Sheka EF. The uniqueness of physical and chemical natures of graphene: their coherence and conflict, Moscow (Russia) People's friendship university of Russia 2013. 117198https://arxiv,org/pdf/1312.1189 2013.

[139] Tashiro K. Molecular theory of mechanical properties of crystalline polymers, Progress in Polymer Science 1993; 18(3): 377-435.1993;
[http://dx.doi.org/https://doi,org/10.1038.nature11458]

[140] Yang L, Zhang L, Webster TJ. Carbon nanostructures for orthopedic medical applications. Nanomedicine (Lond) 2011; 6(7): 1231-44.
[http://dx.doi.org/10.2217/nnm.11.107] [PMID: 21929458]

[141] Li M, Lin Q, Jia Z, *et al.* Graphene oxide/hydroxyapatite composite coating fabricated by the electrophoretic technique, nanotechnology for biological applications. Carbon 2014; 67: 185-97.
[http://dx.doi.org/10.1016/j.carbon.2013.09.080]

[142] Alper J. Rising chemical "stars" could play many roles. Science 1991; 251(5001): 1562-4.
[http://dx.doi.org/10.1126/science.2011736] [PMID: 2011736]

[143] Hecht S. Functionalizing the interior of dendrimers, synthetic challenges, and applications, Journal of Polymer Science Part A; Polymer Chemistry 2003; 41(8): 1047-58.
[http://dx.doi.org/doi:10.1002/pola] [PMID: 10643]

[144] Catherine O, Fréchet JMJ. Incorporation of functional guest molecules into an Internally functionalizable dendrimers through olefin metathesis. Macromolecules 2005; 38(15): 6276-84.
[http://dx.doi.org/10.1021/ma050818a]

[145] Boas U, Christensen JB, Heegaard PMH. 2006.http://www.springer.com/978-0-85404-852-6

[146] Emric T, Fréchet JMJ. Curr Opin Colloid Interface Sci 1999; 4: 15.
[http://dx.doi.org/10.1016/S1359-0294(99)00008-4]

[147] Abbasi E, Aval SF, Akbarzadeh A, *et al.* Dendrimers: synthesis, applications, and properties. Nanoscale Res Lett 2014; 9(1): 247.
[http://dx.doi.org/10.1186/1556-276X-9-247] [PMID: 24994950]

[148] Klajnert B, Bryszewska M. Dendrimers: properties and applications. Acta Biochim Pol 2001; 48(1): 199-208.
[http://dx.doi.org/10.18388/abp.2001_5127] [PMID: 11440170]

[149] Jansen JF, de Brabander-van den Berg EM, Meijer EW. Encapsulation of guest molecules into a dendritic box. Science 1994; 266(5188): 1226-9.
[http://dx.doi.org/10.1126/science.266.5188.1226] [PMID: 17810265]

[150] Jansen JFGA, Meijer EW. The dendritic box: Shape-selective liberation of encapsulated guests. J Am Chem Soc 1995; 117: 4417-8.
[http://dx.doi.org/10.1021/ja00120a032]

[151] Archut A, Azzellini GC, Balzani V, Cola LD, Vögtle F. Towards photoswitchable Dendritic hosts. Interaction between azo-benzene-functionalized dendrimers and eosin. J Am Chem Soc 1998; 120: 12187.
[http://dx.doi.org/10.1021/ja9822409]

[152] Gilat SL, Adronov A, Fréchet JMJ. Light harvesting and energy transfer in novel convergently constructed dendrimers. Angew Chem Int Ed Engl 1999; 38(10): 1422-7.
[http://dx.doi.org/10.1002/(SICI)1521-3773(19990517)38:10<1422::AID-ANIE1422>3.0.CO;2-V]
[PMID: 29711576]

[153] Adronov A, Gilat SL, Fréchet JMJ, *et al.* Light harvesting and energy transfer in laser-dye-labeled poly(aryl ether) dendrimers. J Am Chem Soc 2000; 122: 1175-85.
[http://dx.doi.org/10.1021/ja993272e]

[154] Roberts JC, Bhalgat MK, Zera RT. Preliminary biological evaluation of polyamidoamine (PAMAM) Starburst dendrimers. J Biomed Mater Res 1996; 30(1): 53-65.
[http://dx.doi.org/10.1002/(SICI)1097-4636(199601)30:1<53::AID-JBM8>3.0.CO;2-Q]   [PMID: 8788106]

[155] Malik N, Wiwattanapatapee R, Klopsch R, *et al.* Dendrimers: relationship between structure and biocompatibility *in vitro*, and preliminary studies on the biodistribution of 125I-labelled polyamidoamine dendrimers *in vivo*. J Control Release 2000; 65(1-2): 133-48.
[http://dx.doi.org/10.1016/S0168-3659(99)00246-1] [PMID: 10699277]

[156] Tsekhmistrenko SI, Bityutskyy VS, Tsekhmistrenko OS, *et al.* Enzyme-like activity of Nanomaterials, Regulatory Mechanism in Biosystems. 2018; 9(3): 469-76.
[http://dx.doi.org/10.15421/021870]

CHAPTER 3

Biocompatibility and Bioavailability of Nanomaterials Outline

Abstract: Biocompatibility, biodistribution, and bioavailability are essential aspects of those nanomaterials that are used in the field of biological, biomedical, and biotechnological sciences. These are applicable like agents for the drug delivery system, biomolecules, biomedical applicants, biosensors, theranostics, *etc*. These aspects are intricately interdependent and play prime roles in successful applications of nanomaterials. The physicochemical features of cell, biosystem, and nanomaterials play a significant part in these processes. The nanomaterials can be modified or functionalized by various techniques or by conjugating with a variety of molecules that have specific functional groups or phase transfer of the nanomaterials. The highest degree of biocompatibility of the nanomaterials is attained by minimizing the cytotoxic, genotoxic, and other derogative impacts of nanomaterials with respect to the physiology of a biosystem. Bionanomaterials should be hemocompatible, histocompatible, and cytocompatible for their successful performance. Nanomaterials are functionalized or modified suitably to achieve the selected performances. This aspect needs to alter the physicochemical properties, the surface topography of the nanomaterials that permit the smooth functioning of fabricated nanomaterials. In this chapter, the biocompatibility of nanomaterials, strategies involved, probable pathways along with some examples have been reviewed. This will provide an overview of these significant aspects related to the interaction between nanomaterials and the biosystem.

Keywords: Bioavailability, Biocompatibility, Biodistribution, Biodispersibility, Functionalization of nanomaterials Hydrophilicity, Hydrophobicity, Wettability.

BIOCOMPATIBILITY, BIODISTRIBUTION, AND BIOAVAILABILITY OF NANOMATERIALS IN BIOSYSTEM: AN OVERVIEW

Biocompatibility is the ability of any material to accomplish specific functions safely within a biosystem [1]. Biocompatible materials are considered to exhibit non-interfering behavior towards the biological system. The non-interfering response is related to non-thrombogenicity, non-allergenicity, non-carcinogenicity, and non-toxicity [2]. Biocompatibility is either long term or short term, explicitly concerning tissue-engineered and biomedical devices. The biomedical devices are used to evaluate, treat, augment, and replace tissue/organ and to either rectify or support the functional aspects of the organ or tissue. The

Yogendrakumar H. Lahir & Pramod Avti

surfaces of the bionanomaterials act as an interface when these interact with biomolecules, like proteins, and cell, and cell organelles. Under these conditions, biocompatibility relates to the contextual concepts like cytotoxicity, sensitization, irritation, and genotoxicity, the success of implants, hemocompatibility, carcinogenicity, tissue engineering, biodegradability, and biodistribution. Williams suggested that there is nothing like biocompatibility, but it is the response of the biosystem towards the structure and functional components of administered nanobiomaterials [3].

Properties like hydrophilicity and hydrophobicity depend on the nature of the surface of the nanomaterials and the medium. The surface topography or presences of adsorbed, crystalline, or amorphous molecules also play significant roles in biocompatibility, biodistribution, and bioavailability of nanomaterials in a biosystem. Hydrophilic molecules containing groups like COO^-, OH^-, NH^{-2} or SO_3H^-, influence the hydrophilic nature of nanomaterial, while groups like CH^{-3}, $CH_2\text{-}CH^{-2}$, chains, and rings of hydrocarbons, elevate the degree of hydrophobicity of nanomaterials [4].

Biological membranes exhibit attractive and repulsive responses towards water molecules. The compositional material of the layer and the corresponding surface chemistry plays a significant role during the interaction with water. One aspect of this interaction is wettability. Wettability of a material is related to a particular surface property that yields a unique value for that material. The value of the surface tension of a substance is an essential factor when wettability is under consideration. Wettability of a given substance in the specific liquid helps to measure the contact angle between a solid surface and a droplet on that surface. The surface tension is a product of internal forces that exist between two different materials. When two materials, say liquid droplet and solid surfaces, come in contact with each other and form an interface or a boundary. The force of surface tension is due to the tendency of all materials to reduce their surface area in response to the unbalanced state among the intermolecular forces. These forces are active at the point of contact between the two materials. The significant fundamental comparison of surface tension helps to understand the principle of wettability. Generally, liquids having lower values of surface tension readily spread on materials (liquids/solids). Liquids having higher amounts of surface tension do not show this tendency. This feature of a material affects its hydrophilic and hydrophobic behavior. When the multi-component solution is under consideration, its expression relates to the surface tension that depends on solubility complications of the said solution [5, 6].

The nature of a membrane and the wettability of any materials play a significant role in the degree of biocompatibility, biodistribution, and bioavailability of nanomaterials in a biosystem. Particles that foul in the aqueous medium tend to be hydrophobic in behavior. Colloids, starch, metal colloids, complex-ion aggregates, groups of molecules, liquid-solid; and liquid-liquid suspensions or environment exhibit hydrophobic behavior. Proteins have positive and negative charges but also have a hydrophobic region and behave like hydrophobic materials. Other hydrophobic materials include clays, silicates, alumina, ferric hydroxide, oil particles, paraffin, surfactants, and greases, *etc* [5, 6].

Particles that foul in the aqueous medium tend to be hydrophobic in behavior. Colloids, starch, metal colloids, complex-ion aggregates, groups of molecules, liquid-solid; and liquid-liquid suspensions or environment exhibit hydrophobic behavior. Proteins have positive and negative charges but also have a hydrophobic region and behave like hydrophobic materials. Other hydrophobic materials include clays, silicates, alumina, ferric hydroxide, oil particles, paraffin, surfactants, and greases, *etc* [5, 6]. Generally, hydrophobic particles tend to cluster or group. This behavior lowers the interfacial free energy, *i.e.*, surface tension. There is a formation of bigger spherical particles because the spherical shape has the minimum surface area, and its exposure is limited to the hydrophilic environment. Utilizing modified surface chemistry that functions like the hydrophilic component is a useful technique to regulate the process of fouling of membrane [7, 8]. The probable pathways of intracellular biocompatibility of nanomaterials are shown in Fig. (**1**).

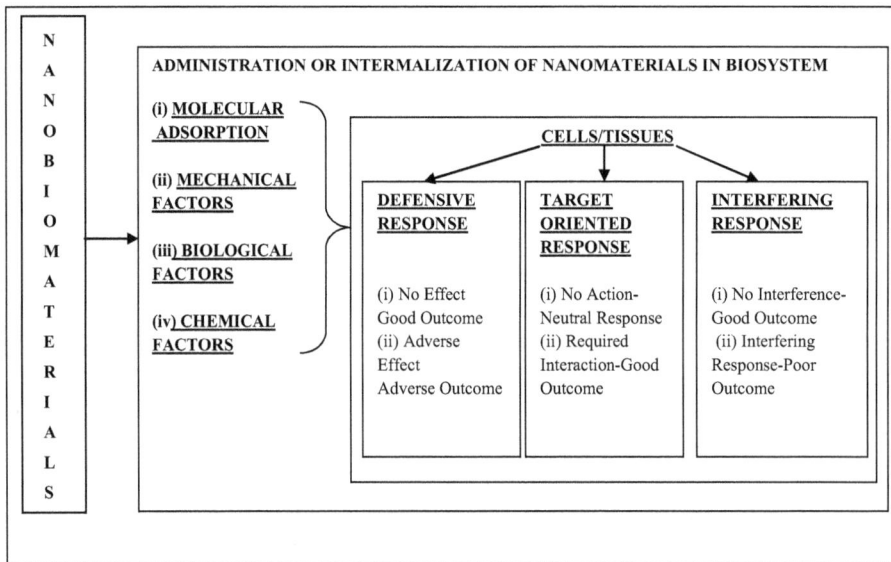

Fig. (1). The probable pathways of intracellular biocompatibility of nanomaterials.

STRATEGIES THAT AFFECT BIOCOMPATIBILITY OF NANOMATERIALS

The term biocompatibility refers to the capability of a nanomaterial to execute a suitable host response in a given biological system. Pristine and modified nanomaterials exhibit different biocompatibility. Magnetic nanoparticles exhibit an excellent range of biocompatibility in biosystems because these have biomedical applications as an active catalytic, imaging agents, and physiologically active components. There exists a powerful mechanism that maintains the cellular amenities for various metabolites that sustain cellular homeostatic state and administered nanomaterials are no exceptions. Magnetite and hematite nanoparticles induce cellular stress in pheochromocytoma in the rat. The Higher degree of apoptosis is the result because of ferric oxide (Fe_2O_3) while ferrous oxide (Fe_3O_4) is physiologically cooperative. Nanoparticles of Ferrous oxide (Fe_3O_4) induce the activity of catalase enzyme action. This enzyme regulates processes like cellular oxidative stress, apoptosis, and cellular autophagy. This interaction is quite competent and capable to check or nullifies the derogative effect of Fe_2O_3. The natural catalase enzyme facilitates the effect of Fe_2O_3 during the progress of the said interaction. The catalase enzyme technique is useful to investigate the pathways of biocompatibility and the fabricated theranostics nanomaterials [9]. Terminal oxidase catalase enzyme reduces four electrons to form oxygen and produces water molecules; the energy released meets the energy needed in the formation of adenosine triphosphate molecule. Most of the designed catalyst, the mimic terminal oxidase enzyme does not exhibit similar efficacy. Recently, the developed oxidase enzyme in myoglobin shows a reduction rate ($52s^{-1}$) involving oxygen. This product shows the efficacy close to the activity of the cytochrome oxidase enzyme. During this process, electrostatic interaction between a functional model of oxidase from the sperm whale and its corresponding native redox counterpart takes place. There is a higher degree of electron transfer rate (400 folds) and helps to understand a tunable process like electron transfer, and functional aspect of enzyme activity [10].

Cell-based strategies for treating ailments in the medical field depend on the cellular monitoring responses and their intensities of rates, cellular delivery, and behavior. The devices involved in the cell-based treatment are dependent on real-time, quantitative evaluation of cell behavior like migration, distribution, viability, differentiation, and cellular fate. These aspects exhibit the role of medication, the state of treatment, deciding dose, cell-choice-optimization, and delivery route. X-ray computed tomography provides cellular tracking, and, it also shows relatively better capabilities for the spatial and temporal resolution to pinpoint a target site. This strategy helps to understand the pathway of nanomaterials related to biocompatibility [11].

FUNCTIONALIZATION OF NANOMATERIALS

The functionalization of nanomaterials is essential, and it influences and regulates the degree of biocompatibility of nanomaterials. This technique involves the addition of the desired functional group/s and capabilities to nanomaterials under consideration. This technique helps the manipulation of nanomaterials to attain a specific degree of biocompatibility and involves prudent and calculative changes in the surface chemistry of the concerned materials [12]. Recent studies suggest that functionalization is about the recognition of specific moieties that ensures their functional efficacy, activity, increased uptake, and reduced side effects in a biosystem. There is a need to develop a suitable methodology that facilitates the conjugation of the group, biochemical moiety that results in appropriate biocompatible and readily dispersing product. Such functionalized nanomaterials become more suitable for drug and molecular delivery, cancer treatment, diagnostics, theranostics, tissue engineering, molecular biology, and understanding the structural and functional relationship between functionalized nanomaterials and set targets [13]. The manipulation involves in functionalization is accomplished by attaching or deleting a molecule to the surface of specific nanomaterial. The modes of functionalization require chemical bonding like covalent or ionic bonding or adsorption, *etc.* The various combinations involving nanomaterials and molecules exhibit the high potentials of applications. The essential aspects of the related success and optimal performance depending on the choice based functionalization process in accordance to a biosystem involved. The physicochemical modifications are a necessary aspect of the specific applications of the final product. This is further strengthened by the chemical stability, an appropriate adjustment that enhances its degree of dispersion and biocompatibility. Such products find their applications in the fields of biotechnology, biomedical sciences, bio-robotics, biosensors, as adjuvant agents in vaccines, biomarkers for diagnostics, and treatment [14].

The processes of surface functionalization of nanomaterials exhibit restricted behavior in different solvents thereby reflect on the limited applications of surface functionalization. Modifications on the surface of nanomaterials facilitate the enhancement of their suitability for varied applications. The changes depend on the regulatory parameters like the material of nanoparticles; surfactant molecules used to shape and the growth of the product. These changes are accomplished involving the stronger binding of the ligand to the crystal facets. Composite particles with domains of different materials like heterodimers of cadmium and iron/platinum (Cd and Fe/Pt), cobalt, and cadmium-selenium-Co/CdSe) possess fluorescent and magnetic properties which are useful to fabricate nanomaterials.

STABILIZATION AGAINST AGGREGATION OR AGGLOMERATION OF NANOPARTICLES

The nanoparticles tend to aggregate or agglomerate. This tendency of nanoparticles hinders their structural and functionality because particle size changes and also the groups present on the surface. The ligand molecules that bind to the surface of nanoparticles during their synthesis regulate the growth and prevent their agglomeration. The repulsive forces amongst particles exist because of electrostatic repulsion, steric exclusion, or hydration layer formed on their surface, *etc*. Steric exclusion takes place when a molecule of solute in water possesses a larger hydrodynamic radius than the water molecules and causes deficiency of the solute molecules, specifically, in the zone of a second solute molecule that has some hydrophilic surface. The choice of appropriate ligand for use depends on the particle system that includes the concentration of the material and the type of solvent [15].

During initial steps, the ligand molecule binds to the surface of the nanoparticle and involves some attractive interactions like chemosorption, electrostatic interaction, or hydrophobic or hydrophilic effect. In most cases, the head of the ligand molecule participates in such communications. The solubility in polar or aqueous medium of ligand and the solvent is related to their nature and charge, and the polar or charged ligand molecules get solubilized in polar and aqueous solvents. Nanoparticles with polar ligand molecules like hydrocarbon chains are soluble in a polar organic solvent like hexose, toluene, or chloroform. Specific amphiphilic ligand molecules like polyethylene glycol possess amphiphilic properties and nanoparticles conjugated with such ligands, or other ligand molecules dissolve in solvents with intermediate polarity [15].

State of aggregation of nanoparticles, their rate of sedimentation, and electrophoretic mobility of nanomaterials in the aqueous medium depend on ionic strength, presence of natural organic matter, but there is no effect of pH. The physical organic matter gets adsorbed on the nanoparticles in nature and the environment. These nanoparticles behave as reductive agents for aggregation and stabilization under varied conditions. Their specific electrophoretic mobility is related to the change from reaction due to the diffusion of restricted/limited aggregation [15].

USE AND EXCHANGE OF LIGAND AND NANOMATERIALS

The use and exchange of ligand attachment to the nanomaterials help to improve the degree of solubility of nanomaterials. The molecule of a ligand can be changed that has been attached previously to a given nanomaterial. The new atom or molecule of the ligand should have a stronger binding capacity in comparison to the already attached ligand molecule. This provision improves not only the solubility of the final product but also renders new properties, functionality to the resultant products. Thus, this provision facilitates the designing and fabrication of the desired nanomaterials to achieve the set target.

Surface ligands that cover the metal core play many roles like passivating the crystal facets; determining the shape and growth influencing the size and the stability of colloids. The shape of the particles depends on the selected ligand and its length. Some of the factors related to the stability of nanoparticles in dispersion state are size, shelf life, chemical nature, ω-functionalities, and use of ligand. These parameters play a significant role in dispersion and rendering biocompatibility to the product. Ligands influence the electrical properties, the morphology of the particles, and the nature of the interface of the nanoparticles participating in the interactions [16]. The nanomaterials having surface ligand and specific end-functionalized groups, maintain their specific geometrical aspects, electro-optical properties, the stability of colloids, and range of solubility. The ligands can bring about cross-linkage and self-assembly during the interaction between nanoparticles and solid surfaces [16 - 20]. Ligands facilitate the stabilization of the dispersion of nanoparticles by setting inter-particle space. The ligands deposit on nanomaterials and result in the insulating barrier and do not allow making metallic contact with the deposited layers, thereby reducing the electrical conductivity [21].

Ligands help in the synthesis of silver nanoparticles in the form of seed by reducing silver salts. The reactants like acetyl trimethyl ammonium bromide, ascorbic acid are involved, and silver nitrate acts as a reducing agent. Single-crystalline gold nanoplates are formed in an aqueous medium using tetrachloroaurate, PVP (polyvinyl pyrrolidone), and sodium citrate as a reducing agent. A low concentration of sodium citrate creates many nuclei that are very small while remaining gold ions get attached to the lateral sides. This method results in the formation of a narrow shaped triangle or hexagonal nanoplates. Nanoplates are the result if the concentration of PVP is less and the numbers of nuclei formation are also less. These nanoplates are larger, and if the PVP is absent, then the irregular or rounded nanoplates are the results. The capping ligand regulates the stability of colloids, affects the shelf-life, and the functionality of the nanomaterials. Ligand exchange is essential to ensure the formulation of nanomaterials regarding the suitable solvents involved [22, 23]. The exchange of ligand provides a better option to replace one ligand for better functionality. For example, ethanedithiol or ammonium thiocyanate replaces an insulating ligand oleylamine on gold nanoparticles, the stability of the nanomaterials increases [16, 20, 21].

ROLE OF PHASE TRANSFER PROCESS DURING THE BEHAVIOR OF NANOMATERIALS

Phase transfer-a process in which a reactant is shifted from one phase to another phase during the reaction. Mostly ionic reactants get dissolved in the aqueous phase (polar), but these do not dissolve in the organic phase (non-polar). If a phase transfer catalyst, *i.e.*, detergent or surfactant is added to the reactions then these insoluble ionic reactants behave as a soluble reactant. The phase transfer reflects on the efficacy of this process. These processes enhance the rate of reaction, facilitate the velocity of the reaction, elevate the yield, and significantly reduce the production of by-products. There is a reduction in the use of dangerous and expensive solvents. These help to dissolve the reactants in either of the phases. This process diminishes the use of costly raw materials and also the formation of waste during the reaction. This process is one of the most preferred options in the field of green synthesis technology. The phase transfer process not only restricted to hydrophobic and hydrophilic interactions but also applicable to the reactions involving a change of phases of matter like solid, liquid, or gas. When solvents having functional groups like carboxyl acid, ammonium ions, sulphonic acid, *etc.*, are used the solubility of monolayers of chemisorbed ions in polar solvent gets elevated. This aspect plays a significant role in increasing the solubility and dispersibility of metallic ions like gold in the aqueous phase [24].

Nanomaterials need either aqueous or organic systems for their synthesis. Nanomaterials in these systems exhibit specific features that are suitable for a particular functionality. The change of phase from polar to non-polar and vice-versa occurs during the synthesis of nanomaterials and depending on the applications of nanomaterials synthesized, the structure of core-shell of bimetallic nanoparticles, catalytic abilities of nanoparticles and replacement interactions under the organic system. Semiconductors and metallic nanoparticles exhibit the functional reversibility of polar and non-polar phases. Such features are of many applications like depositing noble metals on semiconductor nanomaterials [25]. This type of strategy enhances solubility; the dispersibility of the final product in a suitable medium may be biochemical or biological in nature [25].

Colloidal nanoparticles that are synthesized using organic solvents are a suitable option as biologically compatible materials. These products are transferred to the aqueous solution, thereby making them compatible with the physiological media of a biosystem. This practice improves the scope of applications in biomedical, theranostics, and biomolecular fields. Hydrophobic nanoparticles or nanomaterials are converted into the compatible form using some specific solvent or a material system like polymeric blends. This step elevates the degree of improvisation of hydrophobic nanomaterials in the biosystems. Basically, during phase transfer process strategies like exchange and modification of ligands are carried out. These manipulations facilitate the addition of extra layers of materials that bring about the stabilization of nanomaterials and rendering them relatively more suitable options in their varied applications [26].

The nanomaterials which have protection with aqueous phase protein, transform into fluorescent nanoclusters. This conversion takes place in a reverse micellar medium containing silver nanoparticles and bovine serum albumin protected silver nanoclusters. Basic pH is used to induce negatively charged protein capping and formation of protein protected nanoclusters. This interaction accomplishes the transfer of nano matter from the aqueous phase to the cationic surfactant phase. This surfactant is a mixture of hexane: hexanol: water in the ratio (16:2:16). This mixture acts as a reverse micelle phase [26]. Further, this transfer did not change the original fluorescent characteristics of the basic nanoclusters. The mixture of these nanomaterials coexists in the aqueous phase and gets uniformly distributed, retaining the high intensity of fluorescence [26]. Mercaptoundecanoic acid makes the synthesized gold nanoparticles more prone to functionalization. These exhibit their capability for aggregation, redispersion, and phase transfer from the aqueous phase to the organic phase depending on the specific pH. When these nanoparticles are in the aggregated stage, the pH of the solution is around 4, and during their phase transfer process it is about 9. Induction of ion pair formation between the amine group present in octadecyl amine, carboxyl group of mercaptoundecanoic acid and hydrophobic alkyl chain present in octadecyl amine facilitate the phase transfer from aqueous to organic phase of nanomaterial [27].

SILANIZATION

The basic concept of silanization involves the change of ligand and helps to form the first layer of silanes* on to the surface of nanoparticles. The first layer of polymeric cross-linked inorganic silica shell deposits on to the particles can be further derivatized depending on the set requirements of applications [28, 29]. [*Silanes are colorless gaseous compounds of silica and hydrogen with a sharp repulsive smell like CH_3COOH. These have strong reducing properties and are spontaneously inflammable ($Si-H_4$).]

This strategy modifies the core materials and considered as an inorganic polymer of nanomaterials/nanoparticles. This method of functionalization helps to encapsulate the inorganic nanoparticles through mini-emulsion-polymerization. It is crucial to functionalize the surface of inorganic materials because this helps to convert their hydrophilic nature into hydrophobic nature. This feature increases their degree of compatibility with the polymers [30].

Some of the inorganic compounds are in use for various purposes, for example, silica as filler material, for enhancement of the mechanical strength, Titania (titanium), and magnetite as contrast agents in case of magnetic resonance imaging. These inorganic compounds absorb UV absorption. In most cases, when inorganic materials are in use, these neither attain the expected degree of colloidal stability nor biocompatibility [28 - 31]; but if polymeric materials and dispersion polymerization or seeded emulsion polymerization, encapsulate these inorganic nanoparticles, these nanoparticles become useful. These techniques face problems like structural limitations such as polymer particles get surrounded by inorganic particles or polymer shells surrounds inorganic particles [32 - 35]. The synthesis of polymer-inorganic nanocomposites involving the mini-emulsion polymerization method depends on the factors like the effect of initiator, the concentration of surfactant, monomer system, and the surface features of inorganic nanoparticles under consideration. Such factors play a regulatory role in the morphology of the resultant nanocomposites [32, 36].

NANOMATERIALS AS AGENTS TO DEVELOP MULTIFUNCTIONAL HYBRID COATING FOR SCRATCH AND CORROSION RESISTANT SURFACES:

This technique is of multiple applications in the field of engineering, biomedical, biological sciences, biotechnology, *etc*. This technique involves the activation of sites on the surface of nanomaterial to optimize for hydrophobicity, hydrophilicity, and conductivity of the concerning nanomaterials. Surface modification is of significance in the composites and is accomplished by (i) liquid-phase oxidative treatment and (ii) plasma-assisted coating.

Wet oxidation is the other name for liquid-phase oxidation. It is a type of hydrothermal process involving oxidation of dissolved or suspended materials in an aqueous medium. Oxygen acts as an oxidizer agent. H_2O_2, HNO_3, and ammonium bicarbonate also act as oxidative agents. If air is used instead of oxygen, then this process is called wet air oxidation. The oxidative interaction takes place within the range of temperature, *i.e.*, above the average boiling temperature of the water and below the critical point of water (above 100°C to 374 °C). The complete system is under pressure, and this avoids excessive evaporation of water. Thus, energy consumption due to the latent heat of vaporization is controllable. This aspect helps the retention of water during the process of oxidation and prevention of oxidation of compounds under the conditions, but the temperature and pressure at dry conditions do not favor the set interaction [37, 38]. Plasma-assisted coatings are a modified version of a chemical vapor deposition technique under vacuum. This technique helps to form solid material products that show high quality, better, and improved performance. Wafer after its exposure to either one or more volatile precursors acts as a suitable substrate.

These interact or get decomposed on the substrate provided, giving the expected deposit. Volatile by-products are eliminated if formed by gas flow from the reaction chamber [39]. Based on this technique, organic plasma polymer-coated copper nanoparticles, polymerization with ethylene glycol, hydroxyl methanol, and ethanol are good options. The product formed shows increased reactivity and improved the degree of dispersion between solid and liquid phases. The contact angle between the surface (copper nanoparticle) and ethylene glycol and ethane exhibits hydrophilic and hydrophobic features, respectively [40]. The degree of anti-corrosion properties of nanomaterial elevates by the use of nanoparticles like SiO_2, TiO_2, ZnO, Al_2O_3, Fe_2O_3, nano aluminum, nano titanium, *etc*. Hybrid coating as nano-coating is formed on a metal substrate using a specific electrolyte. Such materials can block the U V radiations [41].

ROLE OF AEROGELS IN NANOMATERIALS AND NANOTECHNOLOGY

Aerogel is a porous, very light product having very low density and low thermal conductivity. It is a product of the gel in which a gas replaces the liquid component. There are different names of aerogel like frozen smoke, solid smoke, solid air, solid cloud, and blue smoke, *etc*., depending on the translucent and light scattering nature of the gel [42].

Aerogels are three-dimensional materials that possess ultra-high surface area because their porous skeleton or framework wraps its interior. These products have high applications as it provides a potentially available surface for the attachment of molecules and nanomaterials. Aerogels offer enormous scope in the process of surface functionalization. For example, silica aerogel provides an extensive surface area inside the aerogel; aerogel having 0.1 g/cm^3 density has an area equivalent to 750 m^2 area. This area is around 10% of the soccer field. Thus, the aerogels provide a very high surface to volume ratio. Different materials can be used as a skeleton (backbone) of aerogel and provide enormous chemical flexibility. Frequently aerogel is made through ambient temperature using bench-top processes. This technique refers to a method used in a research laboratory to carry out authorized treatment of waste. It deals with the handling and disposal of chemicals as proposed guidelines by the National Research Council. According to a general basic concept that describes aerogel is the number of molecule or nanomaterials that accommodate as per the need. The final product is competent, efficient, and biocompatible and has reliable functionality [43 - 45].

These aerogel types of nanomaterials exhibit three-dimensional forms and are among the most preferred options for varied applications. Hence, there is a need to look for an effective strategy that facilitates hydrogel with set applications. For example, the gold hydrogel is a network using a three-dimensional arrangement induced by gold nanoparticles and dopamine. The aerogel, a final product, is having an extensive surface area and an enormous degree of porosity after supercritical drying. The material consists of nanowires having a diameter with 5-6 nm. This material offers a better method of oxidation of some of the small molecules that have been suitably functionalized. This material and mode are of great utility in electrocatalysis. The Silica aerogel and its building blocks, hydroxy-gels with poly (hexamethylene-di-isocyanate) orient as the cross-linking arrangement is of common use. Similar methods make this silica gel a light weighted material with extra strength that does not collapse even if it comes in contact with the fluid [46].

Chitosan hydrogel shows excellent sorption ability. It is produced involving the interaction between propylene glycol, suitable acid, or amino acids as an agent to establish cross-linkage under microwave radiations. Chitosan is a polymeric molecule, derived from chitin as a result of deacetylation. This polymeric product possesses specific features and acts as an antibacterial and antioxidant agent. It exhibits homeostatic action and better biodegradability, lower toxicity, and higher tolerance [47 - 51]. Processes, like methylation, carboxylation, alkylation, and sulfonation, and precise degree of deacetylation functionalize chitosan. If the specific value of functionalization exceeds the limit, its properties become different. Polyethylene glycol, polyvinyl alcohol, polylactic acid, polyamide-6, and polyvinyl pyrrolidone blend chitosan. This blended form elevates its degree of bioactivity and affinity for biological tissue [52 - 55]. Zinc, an oxide-copper oxide aerogel, exhibits very active catalytic and sensing abilities and is the product of treatment involving epoxide addition sol-gel technique. This form of aerogel is porous, having a particle size ranging between nano to micrometer consisting of zinc-copper composite aerogel in a molar ratio of 50:50-90:10. Annealing of the aerogel at 400°C converts it into a material whose phase is changed, and it has a highly crystalline nature. ZnO-CuO aerogel possesses a nanosized network of about 25-550 nm range. ZnO-CuO aerogels are most suitable as a material with better catalytic and sensing abilities. Hence, this material is the preferred option as agents for catalytic and sensing devices. These features reflect on its biocompatibility and suitable choice for the biosystem [56].

OPSONIZATION

Opsonins are the components of the blood vascular system that enhance the uptake of coated materials. Mostly this uptake is brought about by macrophages of the reticuloendothelial system and natural killer cells. A particular molecular structure is a result when opsonins attache to the surface of nanomaterials. The immune system of a biosystem easily recognizes these resultant specific molecular structures and influences the route of internalization of nanoparticles and their biodistribution also. Some of the common opsonins present in the blood are albumin, immunoglobulins, fibrinogen, components of the complement system, and apolipoproteins (proteins that bind lipids-oil-soluble substances such as fat and cholesterol to lipoprotein [57, 58]. Lipoprotein consists of a water-insoluble component, and the presence of amphipathic molecules present in lipoprotein makes them water-insoluble and acts as a detergent. One of the most common functions of detergent is related to solubility involving membrane. This process brings the transformation of the membrane into micelle or other suitable forms [59]. Apolipoprotein and phospholipids surround the lipid and lipoprotein, make it soluble in water. These also act as enzyme cofactor; accept ligand cell receptor and low-density-lipoprotein. Another functional aspect of opsonins is related to the active clearance of the nanoparticles from the biosystem [58, 59].

SOME EXAMPLES OF NANOMATERIALS EXHIBITING BIOCOMPATIBILITY, BIODISTRIBUTION, AND BIOAVAILABILITY IN BIOSYSTEM

Biocompatibility, biodistribution, and bioavailability of materials and nanomaterials are intricately interconnected. Their interconnection is the bases of the application of nanomaterials in the field of biomolecular, biomedical, theranostics, drug, biomolecule, and genetic materials delivery, *etc.* Factors affecting these processes also influence components of blood or body fluids in the tissue, the pressure of interstitial fluid in the tissue, contents of the stroma of the tissue, and display functional significances. Physicochemical features of nanomaterials like size, shape, surface charge, chemistry, and type of modification of the surface of nanomaterials affect their biocompatibility. Ability to bind with ligand or degree of PEGylation, *etc.*, and composition of cargo, location of tissue or organ in the body of the biosystem, pretreatment status of the recipient also affect biocompatibility [60]. Nanomaterials exhibit diversity in the structure; physicochemical features that enable them to interact with subcellular structures and biomolecules, and exhibit biocompatibility. Hence, these features facilitate the movements of nanomaterials across biological barriers, to carry a specific payload of cargo and release it at the target site. There are chances of appropriate or probable restrictions in their clearance from the biosystem. If the surfaces of nanomaterials are modified or functionalized or coated with materials that elevate their biodistribution and clearance from the biosystem, they can overcome such restriction. The materials used for this purpose can be neutral polyethylene glycol and other similar molecules. Such a stealth layer on nanomaterials may reduce the impact of opsonization or enable its surface charge to get modified by providing a suitable charge system. For this purpose, even a Zwitterionic system is a good option. All these modifications enhance their biodistribution, bioavailability and to a greater extent their clearance from biosystems also. Further, these improvisations render these nanomaterials more suitable for the applications in the fields of theranostics, imaging, and sensing modalities [61].

Rapid clearance of administrated and circulating nanoparticles is a prime issue during their biodistribution. Factors like blood residence time, specific accumulation of nanoparticles in the organs, interactions between nanoparticles and biological barriers, some tunable features such as composition, size, core properties, modification of surface including chemical, charge, and ligand conjugation, *etc.*, play a significant role in such processes. Factors like the half-life of circulating nanoparticles in blood and body fluids, low degree of non-specific uptake, declined degree of opsonization and elevate the level of accumulation in a specific tissue, *etc.*, these influence biodistribution of nanomaterials. Biological barriers act as impediments during biodistribution in a given biosystem [62, 63]. Commonly, biological-barriers in a biosystem include cellular and humoral components of the immune system and mucosal barriers. Such biological-barriers are overcome through diffusion into and out of the tissues, extravasation, and escape from hepatic clearance. The composition of the nanomaterials can be modified to enable them to have a longer circulating time. PEGylated nanomaterials reduce the degree of adsorption of blood proteins, opsonization, and non-specific tissue uptake in a biosystem [62, 63].

Chemically modified micelles also influence biodistribution. N-(2hydroxyethyl) methacrylamide) oligoacetate (mPEG-b-p (HEMAm-Lac (n) and core-cross linked biodegradable thermo-sensitive micelles help in the synthesis of nanomaterials. The synthesized materials are useful as test nanomaterials to investigate biodistribution under the influence of temperature in an aqueous medium. These micelles exhibit critical micelles temperature above which these either distort or disintegrate. Non-cross linked micelles undergo rapid disintegration while core-cross linked micelles maintaining their integrity even when the temperature of the aqueous medium becomes either above or below the critical micelles temperature. Core-cross linked micelles exhibit a relatively higher degree of stability. When these two forms of cross-linked micelles undergo hydrolysis of the lactate side chain, the core-cross linked micelles behave as a stable component in comparison to non-cross linked micelles. Core-cross linked micelles exhibit much- improved circulation and biodistribution in contrast to non-cross-linked micelles. Even non-cross linked micelles may not avoid opsonization, engulfment by macrophages while core-cross-linked micelles show favorable biodistribution, better chemical integrity, and access to the set target [64, 65].

Nanomaterials exhibit variations in their shape and dimensions. Quite a good number of them have the same dimensions similar to those of biological molecules. Nanomaterials interact with biomolecules, cell organelles, and cells. Smaller nanoparticles can reach most of the parts of the biosystem and can quickly move across most of the biological barriers [63, 64]. Nanomaterials in the pristine form or as conjugates, suspension or colloidal forms are likely to get internalized in the biosystem. Their interactions within biosystems may be beneficial, neutral, adverse, and toxic. These interactions are likely to render physicochemical, physiological, biochemical impacts that may or may not be the cause of structural or functional dysfunction in the biosystem.

Adverse or derogative or toxic impacts of nanomaterials depend on their physicochemical features like size, shape, surface charge, chemistry, composition, stability, duration of exposure, and the nature of the cells, tissue, or biosystem. Biocompatibility acts as the bases of biomedical applications, drug and biomolecule, and gene delivery systems, medical imaging, and other diagnostic devices. There is a need for a very rigorous evaluation of the factors leading to the security of the targeted tissues in the biosystem. During this evaluation, these parameters are put to stringent analytical standards to judge the suitability of these nanomaterials for various biomedical and other applications. Hemocompatibility and cytocompatibility are significant aspects of biocompatibility. Hemocompatibility relates to the safe journey of fabricated or modified nanomaterials through the blood vascular system or circulatory system and reaches the designated target. Such nanomaterials do not cause any damage to the components of the blood vascular system and can move across most of the biological barriers without being affected adversely and without causing harm to the blood components. Hemocompatibility of nanomaterials indicates that these do not interference with hemolysis, standard functionality of leukocytes, and do no cause biochemical fluctuations in the biochemical aspects of serum/plasma [66]. Cytocompatibility of nanomaterials indicates the structural and functional integrity of cellular components, cells, extracellular matrix, and tissue of a biosystem [67].

Mostly physical sequestration of nanoparticles or materials takes place in the liver and spleen and by passive entrapment of these materials at the fenestrated region of capillaries in the liver and spleen. These organs are the significant sites of biotransformation and immune clearance. The enzymatic transformation under Phase-I and Phase-II modes play a vital role in detoxifying or elevating the toxicity of given materials [68, 69]. Some biological effects, like the signaling pathway, *etc.*, may get stimulated when nanoparticles interact with cell, cellular, and extracellular environment. Such interactions depend on the dynamic physicochemical characteristics that associate with nanomaterials. It is essential to understand the mechanism involved during such interaction and help the rational designing of biocompatible nanomaterials that exhibit set biodistribution. Most of the biomedical applications focus on the delivery system, diagnostic devices, and implants to ensure a successful treatment of the targeted tissue [70]. Nanomaterials undergo designing, fabricating, or formulation to suit as agents for drug carriers or as the nano biomedical device. Such nanoproducts should ensure biocompatibility, biodistribution, and biodegradation in the biosystem and should avoid or deceive the mononuclear phagocyte system (MPS). Phagocytes, Kupffer's cell or dendritic cells residing in liver and spleen, *etc.*, should not clear the fabricated nanomaterials. These are the prime sites of the elimination of any foreign matter from the biosystem. The surface modifications carried out on the nanomaterials should render it stable or resistant to the interaction of proteins, lipids, and impacts of pH, electrolytes, *etc.*, present in the body fluids of a biosystem.

Further, the final product should exhibit a suitable degree of permeability to reach the target cell or tissue, and enhanced permeability and retention (EPR) processes also influence permeability. Thus, the nanoproducts should have an optimized degree of EPR and should be able to carry an appropriate payload of the drug or cargo and release it as per the required doses [71].

Almost all features of nanomaterials either directly or indirectly affect their uptake and biodistribution within a biosystem. Protein, lipid, and recognition by opsonins, immunoglobulins, reticuloendothelial system, biotransformation in liver, and spleen, hinder the biodistribution of nanomaterials in a biosystem. Modifications of interfaces of interacting nanomaterials can inverse biodistribution in a biosystem [72 - 75]. Modifications of the nanomaterials reduce the impacts of protein and other biomolecules in the biosystem [76]. This mode inhibits the agglomeration of nanoparticles and pulmonary uptake. Nanomaterials like CNTs when functionalized with $-SO_3H$, -COOH and other negative charged groups conveniently escape hindrances caused due to the circulatory system and remain in circulation for a much longer duration. Even quantum dots also exhibit similar behavior [72, 77 - 79]. The shape is another parameter that plays an essential role in the biodistribution of some of the nanomaterials [75]. If the form of the nanomaterials is specially modified, then their degree of biodistribution can be enhanced, like rod-like and spherical nanostructures remain in circulation for comparatively longer duration in comparison to other shapes of nanomaterials. Nanomaterials with a higher length/diameter ratio show a tendency to stay in circulation for a relatively long period [75]. SiO_2 nanoparticles having a higher ratio could escape accumulation in the liver and spleen to a higher degree [33]. The cells of the reticuloendothelial system readily trap larger sized nanoparticles and do not permit them to accumulate in large quantities in the skin, fat (adipose tissue) and muscles. Fullerene (C60) and CNTs when coated with the tween-80, showed better distribution in comparison to the uncoated ones [72, 75]. Renal clearance is another aspect of successful biodistribution and reduced toxicity. Quantum dots having a lesser diameter than 5 nm get cleared readily than those having a larger diameter [80]. Carbon nanomaterials like carbon nanotubes, carbon nanodiamonds, and graphene, *etc.*, are good options for theranostics, biomedical and imaging, drug delivery, and photothermal therapy. The photothermal aspect is significant for targeting specific cells or tissues, *i.e.*, cancer cells or tissues. Such applications indicate the level of biocompatibility, biodistribution, and bioavailability of carbon nanomaterials in a biosystem. Carbon nanomaterials have tremendous ability to get functionalized to become suitable for the desired applications, may it be imaging, carrier of drugs or biomolecules, *etc.*, even regulated the controlled release of their cargo amicably. Carbon nanomaterials like carbon nano-onions, carbon nano-horns are relatively more biocompatible and get conveniently biodistributed and become readily bioavailable in a biosystem. Some of them may be functionalized or modified involving cycloparaphenylenes, or their ability of size-dependent fluorescence emission is useful for specific targets in the biomedical field and theranostics [81].

Surface active agents (SAA) exhibit a specific ability to alter the path of nanomaterials during internalization or otherwise. Linear alkylbenzene sulfonate is a better choice for silica magnetite modified cationic surfactant aggregates. The limits of deviation are within the range (0.8 -1.9 µg/L) [82]. The Anionic surfactant template-mesoporous silica (AMS) and the co-structure directing agent (CSDA) are used to establish an interaction between the surfactant head group and silica species. The pairing effect between CSDA and the surfactant results in the formation of a regular array of the organic group. It relates to the stoichiometry and geometric arrangement of the surfactant, which produces functionalized materials with a uniform distribution [83]. Unique phosphopeptides as the surface-active agent (SAA) play a significant role in the synthesis of iron nanoparticles. This mode of interaction exhibits a dramatic effect of the rational designing of the size and shape of iron-iron oxide core-shell nanoparticles prepared in one pot synthesis using sodium borohydride reduction of iron salt. These products exhibit a better range of biocompatibility in comparison to the untreated iron salts [84]. The protein molecules are functionalized nanomaterials and these nano molecules are very effective biochemical for detecting different analytes in the test samples and within biosystem [85].

CONCLUSION

Biocompatibility reflects on the capability of bionanomaterials to accomplish their desired functions in biosystems concerning with medical therapy, biomolecular and biochemical investigations or industrial applications without interfering with the physiology and biochemistry of a biosystem. These interactions do not elicit any undesirable impact on the host. These nanomaterials cause beneficial cellular or tissue response reflecting on the specific and optimum performances. The nanomaterials are modified and functionalized so that these do not interrupt the normal biochemical and physiological functioning. Hemocompatible evaluates the hemolytic studies; the impact on blood coagulation indicates the degree of hemocompatibility of nanomaterials. The target drug delivery system involves superparamagnetic iron oxide, dendrimers, mesoporous silica particles, noble metal nanoparticles, carbon-based nanomaterials. These do not cause derogative impacts on cell or tissue of a biosystem. The basic principles related to administration, metabolism, degradation, and clearance, are maintained during the biocompatibility phenomenon. There appears to be vast scope for such investigation to ensure appropriate and judicial use of the nanomaterials.

REFERENCES

[1]   Williams DF. The William Dictionary of Biomaterials. Liver Pool UK: University Press 1999.

[2]   Keel D. Polymers for biomedical applications: improvements of the interface Compatibility, In the book, Advances in polymer science. Heidelberg: Springer-Verlag 2000; 149: pp. 1-57.

[3]   Williams DF. There is no such thing as a biocompatible material. Biomaterials 2014; 35(38): 10009-14.https://dx.doi.org/10.11-16/j.biomaterials.2014.08.035
[http://dx.doi.org/10.1016/j.biomaterials.2014.08.035] [PMID: 25263686]

[4]   Voet D, Voet J. Biochemistry. 4th ed., Hoboken: John Wiley, and Sons 2011.

[5]   Zeaman LJ, Zydney AL. Microfiltration and ultrafiltration: principles and applications. New York: Marcel Dekker Inc 1996.

[6]   Perry RH, Green DH. Perry's chemical engineering handbook. 7th ed., McGraw Hill 1997.

[7]   Chandler D. Interfaces and the driving force of hydrophobic assembly. Nature 2005; 437(7059): 640-7.
[http://dx.doi.org/10.1038/nature04162] [PMID: 16193038]

[8]   Kukizaki M, Wada T. Effects of membrane wettability on the size and size distribution of microbubbles formed from shirasu-porous-glass membrane. Colloids Surf A Physicochem Eng Asp 2008; 317(1-3): 146-54.
[http://dx.doi.org/10.1016/j.colsurfa.2007.10.005]

[9]   Wang L, Wang Z, Li X, *et al.* Deciphering active biocompatibility of iron oxide nanoparticles from their intrinsic antagonism. Nano Res 2018; 11(5): 2745-55.
[http://dx.doi.org/10.1007/s12274-017-1905-8]

[10]   Yu Y, Cui C, Liu X, Petrik ID, Wang J, Lu Y. A designed mettalloenzyme achieving the catalytic rate of the native enzyme. J Am Chem Soc 2015; 137(36): 11570-3.
[http://dx.doi.org/10.1021/jacs.5b07119] [PMID: 26318313]

[11]   Kim J, Chhous P, Hsu J, *et al.* Use of nanoparticles contrast agents for cell tracking with computed tomography. Biocon Chem 2017; 28(6): 1581-97.
[http://dx.doi.org/10.1021/acs.bioconjchem.7b00664]

[12]   Sperling RA, Parak WJ. Surface modification, functionalization and bioconjugation of colloid

inorganic nanoparticles, Philosophical Transactions A. Phil Trans R Soc A. 368: 1333-83.
[http://dx.doi.org/10.1098/rsta.2009.0273]

[13]   Subbiah R, Veerapandian M, Yun KS. Nanoparticles: functionalization and multifunctional applications in biomedical sciences. Curr Med Chem 2010; 17(36): 4559-77.
[http://dx.doi.org/10.2174/092986710794183024] [PMID: 21062250]

[14]   Helio R, Randow PVC, Vile DN, Andrade LM, *et al.* Surface modification and application of nanomaterials in biotechnology. Academic J Poly Sci 2018; 1(2): 555556.
[PMID: 555556]

[15]   Keller AA, Wang H, Zhou D, *et al.* Stability and aggregation of metal oxide nanoparticles in natural aqueous matrices. Environ Sci Technol 2010; 44(6): 1962-7.
[http://dx.doi.org/10.1021/es902987d] [PMID: 20151631]

[16]   Kanelidis I, Kraus T. The role of ligands in coinage-metal nanoparticles for electronics. Beilstein J Nanotechnol 2017; 8: 2625-39.
[http://dx.doi.org/10.3762/bjnano.8.263] [PMID: 29259877]

[17]   Xia Y, Xiong Y, Lim B, Skrabalak SE. Shape-controlled synthesis of metal nanocrystals: simple chemistry meets complex physics? Angew Chem Int Ed Engl 2009; 48(1): 60-103.
[http://dx.doi.org/10.1002/anie.200802248] [PMID: 19053095]

[18]   Sardar R, Funston AM, Mulvaney P, Murray RW. Gold nanoparticles: past, present, and future. Langmuir 2009; 25(24): 13840-51.
[http://dx.doi.org/10.1021/la9019475] [PMID: 19572538]

[19]   Barriere C, Piettre K, Latour V, *et al.* Ligand effects on the air stability of copper nanoparticles obtained from organometallic synthesis. J Mater Chem 2012; 22(5): 2279-85.
[http://dx.doi.org/10.1039/C2JM14963J]

[20]   Fafarman AT, Hong SH, Caglayan H, *et al.* Chemically tailored dielectric-to-metal transition for the design of metamaterials from nanoimprinted colloidal nanocrystals. Nano Lett 2013; 13(2): 350-7.
[http://dx.doi.org/10.1021/nl303161d] [PMID: 23215159]

[21]   Fafarman AT, Hong SH, Oh SJ, *et al.* Air-stable, nanostructured electronic and plasmonic materials from solution-processable, silver nanocrystal building blocks. ACS Nano 2014; 8(3): 2746-54.
[http://dx.doi.org/10.1021/nn406461p] [PMID: 24484271]

[22]   Ah CS, Yun JY, Park HJ, *et al.* Size-controlled synthesis of machinable single-crystalline gold nanoparticles. Chem Mater 2005; 17(22): 5558-61.
[http://dx.doi.org/10.1021/cm051225h]

[23]   Reiser B, González-García L, Kanelidis I, Maurer JHM, Kraus T. Gold nanorods with conjugated polymer ligands: sintering-free conductive inks for printed electronics. Chem Sci (Camb) 2016; 7(7): 4190-6.
[http://dx.doi.org/10.1039/C6SC00142D] [PMID: 30155064]

[24]   Sastry M. Phase transfer protocols in nanoparticles synthesis. Curr Sci 2003; 85(12): 1735-45.

[25]   Yang J, Lee JY, Ying JY. Phase transfer and its applications in nanotechnology. Chem Soc Rev 2011; 40(3): 1672-96.
[http://dx.doi.org/10.1039/B916790K] [PMID: 21120233]

[26]   Sahu DK, Pal T, Sahu T. A new phase transfer strategy to convert protein capped nanomaterials in uniform fluorescent nanoclusters in the reverse micellar phase. ChemPhysChem 2018; 19(17): 2153-8.
[http://dx.doi.org/https://doi.org.10.1002/cphc.201800191]

[27]   Ansar SM, Chakraborty S, Kitchens CL. pH-responsive mercaptoundecanoic acid functionalized gold nanoparticles and applications in catalysis. Nanomaterials (Basel) 2018; 8(5): 339.
[http://dx.doi.org/10.3390/nano8050339] [PMID: 29772775]

[28]   Jordan A, Wust P, Scholz R, *et al.* Cellular uptake of magnetic fluid particles and their effects on

human adenocarcinoma cells exposed to AC magnetic fields *in vitro*. Int J Hyperthermia 1996; 12(6): 705-22.
[http://dx.doi.org/10.3109/02656739609027678] [PMID: 8950152]

[29]   Wang H, Xu P, Meng S, *et al.* Polymethylmethacrylate/silica/titania nanocomposites with greatly improved thermal and ultraviolet-shielding properties. Polym Degrad Stabil 2006; 91(7): 1455-61.
[http://dx.doi.org/10.1016/j.polymdegradstab.2005.10.008]

[30]   Schoth A, Keith AD, Landfester K, Rafael ME. Silanization as a versatile functionalization method for the synthesis of polymer/magnetic hybrid nanoparticles with controlled structure. Royal Society Chemistry Advances 2016; 6: 53903-11.
[http://dx.doi.org/10.1039/C6RA08896A]

[31]   Mori Y, Kawaguchi H. Impact of initiators in preparing magnetic polymer particles by miniemulsion polymerization. Colloids Surf B Biointerfaces 2007; 56(1-2): 246-54.
[http://dx.doi.org/10.1016/j.colsurfb.2006.11.023] [PMID: 17196799]

[32]   Zou H, Wu S, Shen J. Polymer/silica nanocomposites: preparation, characterization, properties, and applications. Chem Rev 2008; 108(9): 3893-957.
[http://dx.doi.org/10.1021/cr068035q] [PMID: 18720998]

[33]   Huang X, Li LL, Liu T, *et al.* The shape effect of mesoporous silica nanoparticles on biodistribution, clearance, and biocompatibility *in vivo*. ACS Nano 2011; 5(7): 5390-9.
[http://dx.doi.org/10.1021/nn200365a] [PMID: 21634407]

[34]   Mahadevan AR, Ashjari M, Makoo AB. Preparation of poly/styrene-methyl methacrylate/ SiO_2 composite nanoparticles *via* emulsion polymerization: An investigation into the compatibilization. Eur Polym J 2007; 43(2): 336-44.
[http://dx.doi.org/10.1016/j.eurpolymj.2006.10.004]

[35]   Colver PJ, Colard CA, Bon SAF. Multilayered nanocomposite polymer colloids using emulsion polymerization stabilized by solid particles. J Am Chem Soc 2008; 130(50): 16850-1.
[http://dx.doi.org/10.1021/ja807242k] [PMID: 19053450]

[36]   Asua JM. Mapping the morphology of polymer-inorganic nanocomposites synthesized by mini-emulsion polymerization Macromolecular Chemistry, and Physics 2014; 215(5): 458-64.
[http://dx.doi.org/10.1002/macp.201300696]

[37]   Mishra V, Mahajan V, Joshi J. Wet air oxidation. Ind Eng Chem Res 1995; 34: 2-48.
[http://dx.doi.org/10.1021/ie00040a001]

[38]   Bhargava SK, Tardia J, Prasad J, Akolekar DB, Grocott SC. Wet oxidation and catalytic wet oxidation. Ind Eng Chem Res 2006; 45: 1221-58.
[http://dx.doi.org/10.1021/ie051059n]

[39]   Woehri N, Ochedowaski O, Gottlieb S, Shibasaki K, Schulz S. Plasma enhanced chemical vapor deposition of graphene on a copper substrate. AIP Adv 2014; 4(4): 1-9.
[http://dx.doi.org/10.1063/1.4873157] [PMID: 047128]

[40]   Tavares J, Swanson EJ, Coulombe S. Plasma synthesis of coated metal nanoparticles Tailored for dispersion. Plasma Process Polym 2008; 5(8): 759-69.
[http://dx.doi.org/10.1002/ppap.200800074]

[41]   Cao G, Wang Y. Nanostructures and Nanomaterials -- Synthesis, Properties, and applications. World Scientific Publishing 2011.
[http://dx.doi.org/10.1142/7885]

[42]   Pekala RW. Organic aerogels from the polycondensation of resorcinol with formaldehyde. J Mater Sci 1998; 24(9): 3221-7.
[http://dx.doi.org/10.1007/BF01139044]

[43]   Jones S. Jet propulsion laboratory-Pasadena, California 2002. https://stardust.jpl.nasa.gov/newa/news93.html

[44]   Compendium of chemical terminology The Gold Book 2nd Ed., 2006. ISBN: 0-96785500-9-8.
        [http://dx.doi.org/https://doi,org/10.1351/goldbook]

[45]   Aegerter MA, Leventis N, Koebel MM. Aerogel Hand Book. Springer Publishing 2011.ISBN978--
        -4419-7477-8 .

[46]   Wen D, Liu W, Haubold D, *et al.* Gold aerogels: three dimension assembly of nanoparticles and their
        use as electrolytic interfaces. ACS Nano 2016; 10(2): 2559-67.
        [http://dx.doi.org/10.1021/acsnano.5b07505] [PMID: 26751502]

[47]   Yang J, Chen J, Pan D, Wan Y, Wang Z. pH-sensitive interpenetrating network hydrogels based on
        chitosan derivatives and alginate for oral drug delivery. Carbohydr Polym 2013; 92(1): 719-25.
        [http://dx.doi.org/10.1016/j.carbpol.2012.09.036] [PMID: 23218359]

[48]   Zheng H, Tang C, Yin C. Oral delivery of shRNA based on amino acid modified chitosan for
        improved antitumor efficacy. Biomaterials 2015; 70: 126-37.
        [http://dx.doi.org/10.1016/j.biomaterials.2015.08.024] [PMID: 26310108]

[49]   Kaya M, Baran T, Merten A-O, *et al.* Extraction and characterization of chitin and chitosan with
        antimicrobial and antioxidant activities from cosmopolitan orthoptera species, (Insecta). Biotechnol
        Bioprocess Eng; BBE 2015; 20(1): 168-79.
        [http://dx.doi.org/10.1007/s12257-014-0391-z]

[50]   Moreno-Vásquez MJ, Valenzuela-Buitimea EL, Plascencia-Jatomea M, *et al.* Functionalization of
        chitosan by a free radical reaction: Characterization, antioxidant and antibacterial potential. Carbohydr
        Polym 2017; 155: 117-27.
        [http://dx.doi.org/10.1016/j.carbpol.2016.08.056] [PMID: 27702495]

[51]   Radwan-Proglowaska J, Piakowaski M, Janus L, Bogdal D, Matysek D. Biodegradable, pH responsive
        chitosan aerogels for biomedical applications. Royal Society of Chemistry Advances 2017; 7(52):
        32960-5.
        [http://dx.doi.org/10.1039/C6RA27474A]

[52]   Young SN, Won HP, Ihm D, Hudson SM. Effects of the degree of deacetylation on thermal
        decomposition chitin and chitosan nanofibers. Carbohydr Polym 2010; 80(1): 291-5.
        [http://dx.doi.org/10.1016/j.carbpol.2009.11.030]

[53]   Abdul KHPS, Chaturbhuj KS, Adnan AS, *et al.* A review of chitosan-cellulose blends and nano-
        cellulose reinforced chitosan biocomposites: properties and their application, Carbohydrate Polymers
        2016; 150(): 216-26.
        [http://dx.doi.org/10.1016/j.carbpol.2014.10.071]

[54]   Usman A, Zia KM, Zuber M, Tabasum S, Rehman S, Zia F. Chitin and chitosan based polyurethanes:
        A review of recent advances and prospective biomedical applications. Int J Biol Macromol 2016; 86:
        630-45.
        [http://dx.doi.org/10.1016/j.ijbiomac.2016.02.004] [PMID: 26851360]

[55]   Ahmadi Majd S, Rabbani Khorasgani M, Moshtaghian SJ, Talebi A, Khezri M. Application of
        Chitosan/PVA Nano fiber as a potential wound dressing for streptozotocin-induced diabetic rats. Int J
        Biol Macromol 2016; 92: 1162-8.
        [http://dx.doi.org/10.1016/j.ijbiomac.2016.06.035] [PMID: 27492559]

[56]   Rull M, Louisa JHW. Synthesis of ZnO-CuO nanocomposites aerogel by sol-gel route. J Nanom 2014;
        2014: 9.
        [http://dx.doi.org/10.1155/2014/491817]

[57]   Roos A, Xu W, Castellano G, *et al.* Mini-review: A pivotal role for innate immunity in the clearance
        of apoptotic cells. Eur J Immunol 2004; 34(4): 921-9.
        [http://dx.doi.org/10.1002/eji.200424904] [PMID: 15048702]

[58]   Zhang Y, Hoppe AD, Swanson JA. Coordination of Fc receptor signaling regulates cellular
        commitment to phagocytosis. Proc Natl Acad Sci USA 2010; 107(45): 19332-7.

[http://dx.doi.org/10.1073/pnas.1008248107] [PMID: 20974965]

[59] Lichtenberg D. Dopatowaski, Kozolov MM, Phase boundaries in a mixture of membrane-forming amphiphiles and micelle-forming amphiphiles. Biochim Biophys Acta 2000; 1508(1-2): 1-19.https://doi.org [BBA].
[http://dx.doi.org/10.1016/S0304-4157(00)00004-6] [PMID: 11090815]

[60] Ernsting MJ, Murakami M, Ray A, Li S-D. Factors controlling the pharmacokinetics. Biodistribution and intratumoral penetration of nanoparticle. J Control Release 2013; 172(3): 782-94.
10.16/j.jconrel

[61] Arvizo RR, Miranda OR, Moyano DF, *et al.* Modulating pharmacokinetics, tumor uptake and biodistribution by engineered nanoparticles. PLoS One 2011; 6(9): e24374.
[http://dx.doi.org/10.1371/journal.pone.0024374] [PMID: 21931696]

[62] Alexis F, Pridgen E, Molnar LK, Farokhzad OC. Factors affecting the clearance and biodistribution of polymeric nanoparticles. Mol Pharm 2008; 5(4): 505-15.
[http://dx.doi.org/10.1021/mp800051m] [PMID: 18672949]

[63] Lahir YK. Impacts of metal and metal oxide nanoparticles on reproductive tissues and Spermatogenesis. J Exp Zool India 2018; 18(2): 594-608.

[64] Huang X, Brittain WJ. Synthesis and characterization of PAMA nanocomposites by suspension and emulsion polymerization. Macromolecules 2007; 34(10): 3255-60.
[http://dx.doi.org/10.1021/ma001670s]

[65] Rijcken CJ, Snel CJ, Schffelers RM, van Nostrum CF, Hennink W E. Core-Cross-linked thermosensitive polymeric micelles: synthesis, characterization and *in vivo* studies. Biomaterials 2007; 28(36): 5581-93.
[http://dx.doi.org/10.1016/j.biomaterials.2014.03.070]

[66] Shi J. Biocompatibility of synthetic nanomaterials, Sweden, 2012.www.reproprints.se

[67] Li X, Wang L. Biocompatibility and toxicity of nanoparticles and nanotubes. J Nanom 2012; 2012: 19.
[http://dx.doi.org/10.1155/2012/548389] [PMID: 548389]

[68] Sevior DK, Pelkonen O, Ahokas JT. Hepatocytes: the powerhouse of biotransformation. Int J Biochem Cell Biol 2012; 44(2): 257-61.
[http://dx.doi.org/10.1016/j.biocel.2011.11.011] [PMID: 22123318]

[69] Lahir YK. Principles and applications of Toxicology. Bangalore ,India: See Kay Publications 2013.ISBN: 978-81-924169-5-3 .

[70] Naahidi S, Jafari M, Edalat F, Raymond K, Khademhosseini A, Chen P. Biocompatibility of engineered nanoparticles for drug delivery. J Control Release 2013; 166(2): 182-94.
[http://dx.doi.org/10.1016/j.jconrel.2012.12.013] [PMID: 23262199]

[71] Liu W, Chaix A, Gary-Bobo M, *et al.* Stealth biocompatibility of Si-based nanoparticles for biomedical applications. Nanomaterials (Basel) 2017; 7(10): 288.
[http://dx.doi.org/10.3390/nano7100288]

[72] Deng K, Jia G, Wang H, *et al.* Translocation and the fate of multiwalled carbon nanotubes *in vivo*. Carbon 2007; 45: 1419-24.
[http://dx.doi.org/10.1016/j.carbon.2007.03.035]

[73] Liu J-H, Yang S-T, Wang H, Chang Y, Cao A, Liu Y. Effect of size and dose on the biodistribution of graphene oxide in mice. Nanomedicine (Lond) 2012; 7(12): 1801-12.
[http://dx.doi.org/10.2217/nnm.12.60] [PMID: 22830500]

[74] Chang X-L, Yang S-T, Xing G. Molecular toxicity of nanomaterials. J Biomed Nanotechnol 2014; 10(10): 2828-51.
[http://dx.doi.org/10.1166/jbn.2014.1936] [PMID: 25992420]

[75] Chang X-L, Ruon L, Yang S-T, *et al.* Quantification of carbon nanomaterials *in vivo*: direct stable

isotope labeling on the skeleton of fullerene C60. Environ Sci Nano 2014; 1: 64-70.
[http://dx.doi.org/10.1039/C3EN00046J]

[76]   Yang ST, Fernando KA, Liu JH, *et al.* Covalently PEGylated carbon nanotubes with stealth character *in vivo*. Small 2008; 4(7): 940-4.
[http://dx.doi.org/10.1002/smll.200700714] [PMID: 18574799]

[77]   Deng X, Yang S, Nie H, Wang H, Liu Y. A generally adoptable radiotracing method for tracking carbon nanotubes in animals. Nanotechnology 2008; 19(7): 075101.
[http://dx.doi.org/10.1088/0957-4484/19/7/075101] [PMID: 21817626]

[78]   Yang ST, Guo W, Lin Y, *et al.* Biodistribution of pristine single-walled carbon nanotubes *in vivo*. J Phys Chem C 2007; 111(48): 17761-4.
[http://dx.doi.org/10.1021/jp070712c]

[79]   Schipper ML, Iyer G, Koh AL, *et al.* Particle size, surface coating, and PEGylation influence the biodistribution of quantum dots in living mice. Small 2009; 5(1): 126-34.
[http://dx.doi.org/10.1002/smll.200800003] [PMID: 19051182]

[80]   Choi HS, Liu W, Misra P, *et al.* Renal clearance of quantum dots. Nat Biotechnol 2007; 25(10): 1165-70.
[http://dx.doi.org/10.1038/nbt1340] [PMID: 17891134]

[81]   Bartelmess J, Quinn SJ, Giordani S. Carbon nanomaterials: multi-functional agents for biomedical fluorescence and Raman imaging. Chem Soc Rev 2015; 44(14): 4672-98.
[http://dx.doi.org/10.1039/C4CS00306C] [PMID: 25406743]

[82]   Pena-Pereira F, Duarte RM, Trindade T, Duarte AC. Determination of anionic surface active agents using silica coated magnetite nanoparticles modified with cationic surfactant aggregates. J Chromatogr A 2013; 1299: 25-32.
[http://dx.doi.org/10.1016/j.chroma.2013.05.040] [PMID: 23773585]

[83]   Han L, Che S. Anionic surfactant templated mesoporous silicas (AMSs). Chem Soc Rev 2013; 42(9): 3740-52.
[http://dx.doi.org/10.1039/C2CS35297D] [PMID: 23081758]

[84]   Peltier R, Siah WR, Grant VMW, *et al.* Novel phosphopeptides as a surface-active agent in iron nanoparticles synthesis. Aust J Chem 2012; 65: 680-5.
[http://dx.doi.org/10.1071/CH12168]

CHAPTER 4

Physicochemical Aspects that Influence the Interactive Behavior of Nanomaterials

Abstract: Nanomaterials have occupied ubiquitous status in present-day life. Nanotechnology has become the backbone for technical aspects of energy-storing, communication industries, domestic, health, and safety, *etc.* Interactions and behavior of nanomaterials are the primary concern among the related research fraternity. The main focus is on the mechanisms involved in the interactions and the responses of nanomaterials concerning abiotic and biotic components of the environment during the pertinent research. The interactions and behavior of nanomaterials follow the basic principles of physics, chemistry, material science, biological sciences, *etc.* Nanomaterials abridge the atomic and molecular state of the matter and the respective bulk forms. In such interactions, the quantum mechanics and tunneling effect, parameters like, inter and intramolecular binding forces, hydrophobicity, and hydrophilicity, net charges, *etc.*, have functional significance. Nanomaterials exhibit the ability to get precisely designed as per the assigned functions. As a result, such nanomaterials act as preferred options in different fields like vehicles for cargo and diagnostic tools, *etc.* In this chapter, the functional roles of the physicochemical parameters and related forces are reviewed regarding the behavior of nanomaterials in the biosystem.

Keywords: Coulomb Forces, Electrostatic Forces, Hydrophilicity, Hydrophobicity, Intermolecular Bonding, Magnetic Properties of Nanomaterials: Quantum Mechanism, Optical Properties of Nanomaterials, Tunneling Effect.

INTRODUCTION AND OVERVIEW

Currently, nanomaterials and nanotechnology act as the backbone in current industrial, biomedical, and academic research scenarios because nanomaterials are in use for most of the day to day life products. The behavior of nanomaterials in the biosystem is ambiguous and needs investigation concerning mechanisms involved in nanomaterials, and biomolecules, and cellular components. Nanomaterials have the advantage of being designed as per the required assigned functions to enhance avidity for interactions with a specific target such as biomolecules and cellular components. These materials interact in a derogative or helpful manner within a biological system based on their unique reactivity. The probable derogative impacts are on biomolecules, cells and tissues, and potential

Yogendrakumar H. Lahir & Pramod Avti

interferences during these interactions. Natural and engineered nanomaterials pose challenges regarding the uncertainty about their health-hazardous potential and approvals from various regulatory authorities at different levels, *etc.*

Nanotechnology ensures appropriate designing, fabrication, and synthesis of nanomaterials with maximum precision so that engineered nanomaterials perform the designated functions. The features of engineered nanomaterials, such as surface properties, charge modifications, size, and shape, enable them to interact with the target moieties. The half-life of nanomaterials is affected while they are in circulation within a biosystem. This parameter implies that such nanoparticles and adducts formed during their interaction require prolonged duration for clearance from the biosystem.

The mechanism and severity of the interactions and behavior of nanomaterials are unpredictable, thereby making it essential to understand the mechanism involved. The current detailed investigations are concerning size, shape, chemical functionality, surface charge, composition, and biomolecular signaling, kinetics, transportation, and toxicity in cell culture and experimental animal models [1]. Properties like chemical composition or surface modification, hydrophobicity, hydrophilicity, and presence of lipophilic groups, *etc.*, of nanomaterials, play a significant role in their interactions and behavior. The ability of nanomaterials to move across most of the biological barriers and to bind with biomolecules specifically, factors or components that inhibit enzyme activity and immune system are the main points that need attention. The presence of metallic group and toxic compounds induces their respective effects during such interactions [2]. Features, like dimensions, shape, the tendency to agglomerate, crystallinity, surface coating, *etc.*, of the nanomaterials, have relatively more cytological capabilities. In a given biological system, nanomaterials having a spherical, tubular, rod, and needle-like shapes, *etc.*, exhibit kinetics of deposition and adsorption [3].

NATURAL AND ENGINEERED NANOMATERIALS

Some of the common examples of natural nanomaterials are present in nature. Nanostructures present in the wings of a butterfly (morph) and peacock modify the interaction between the light waves, and the result is a brilliant blue and green hue. Soap bubbles exhibit iridescence causing varied coloration because the wall of a soap bubble is within the range of nano dimensions. Lotus plants possess some water-resistant nanostructures, *i.e.*, hydrophobic, and water molecules do not adhere to the parts of this plant. An invisible spray of water at the waterfalls or oceanic waves has the nano dimensions. Fine products of the combustion of fuel (soot) are examples of natural nanomaterials. Natural nanomaterials are produced in nature by natural processes like combustion, volcanic eruption, mining activities, forest fires, anthropogenic physical processes, *etc*. The ambiguous behavior of nanomaterials raises doubts about the validity of their use. Engineered nanomaterials are appropriately designed, fabricated as per the requirements; this reduces the real enigma. Aspects related to derogative or beneficial effects, appropriate utility, half-life, and the clearance form biosystem, *etc*., and the approvals from a competent authority, are primarily considered during their applications and research. These parameters should get administrative support while designing and fabricating nanomaterials for the set targeted tissue, moiety, or functions. Nanomaterials exhibit unique and intermediate dimensions between atoms and molecules, and the corresponding bulk materials. This feature has a significant role during the behavior of nanomaterials and their applications in various fields like biological, biomedical, theranostics, and other industrial applications [4].

SOME FUNDAMENTALS RELATED TO PHYSICS THAT AFFECT THE BEHAVIOR OF NANOMATERIALS

The behavior of nanomaterials follows some of the fundamentals of physics. Thus, it is appropriate to enumerate some of the prime fundamentals.

Quantum Mechanics

This mechanism comes quite handy in explaining or understanding the behavior of nanomaterials because the principles involved are relatively strictly applicable to atoms and molecules. Surface to volume ratio gets enhanced at the nanoscale. The quantum mechanism is related to the motion and energy of atom along with the electrons. Since nanomaterials are low dimensional materials, their mass becomes extremely less, and as a result, the gravitational force comes to a negligible level. Under this condition, electromagnetic force becomes the regulatory parameter in controlling the behavior of atoms, molecules, and nanomaterials. Nanomaterials are elementary particles with nano size or elementary particles, and negligible mass exhibits the wave-particle duality concept. Under these conditions, such particles exhibit a wave-like nature, and it may be relatively in a distinct manner. Electrons show wave behavior, and their wave function shows their probable position. Nanomaterials exhibit quantum mechanism and manifest quantum confinement, *i.e.*, nanoparticles of specific metal have electrons restricted within the particular space. This condition of electrons does not exist in the corresponding bulk metal form. Furthermore, electrons exist at a discrete energy level; this is evident in the case of quantum dots where the impact of quantization of energy is displayed [5, 6].

Tunneling Effect

This effect describes the passage of a subatomic particle through a potential barrier. When a sub-atomic particle passes through a potential barrier, it is unable to cross over under normal or classical mechanical conditions. This phenomenon is the basis for the physical process, such as nuclear fission. It is also the functional base for tunneling diode, quantum computing, scanning electron microscopy, *etc.* Tunneling current develops when electrons try to move across a barrier that they are unable to cross over under normal conditions and to do so need extra energy [7]. The impact of the quantum tunneling process has its influence on the materials having nano and smaller dimensions and regulates the elementary particles, atoms, and molecules to move across an energy barrier. These particles or materials do not have the required energy to do so. These conditions influence the physical impacts of materials under consideration, and its practical applications and functions are related to the photon tunneling mode. The quantum tunneling phenomenon helps to trace the origin and evolution of life [7 - 10].

The tunneling effect of quantum mechanics is the functional basis of Scanning Tunneling Microscopy. An energy barrier, *i.e.*, energy potential, is compared to the wavelength of the particle under the action, and it is essential for the existence of a tunneling effect. An electron quantum can exhibit tunneling effect when the electron has lower/less energy that exists in the other region of the energy barrier, which is forbidden. The sample under observation is above absolute zero, and the molecules always exhibit movements because of their kinetic energy. At the nanoscale, this movement is very less in comparison to the bulk size of the matter and shows either no or least impact. The nanoparticles, during their actions and their interaction, exhibit random kinetic motions, which can be compared with Brownian movements of particles [7 - 10].

Quantum biology relates to the applications of quantum mechanisms and theoretical chemistry in biological sciences. Many biological phenomena involve the interconversions of energy and chemical transformations. Furthermore, during such physiological processes, biomolecules undergo chemical interactions, adsorption of light/energy, the formation of excited electron state and transfer of the excitation energy, transfer of an electron, and proton (as H^+) at least during photosynthesis, olfaction, cellular respiration, conduction of nerve impulse, *etc.* Most of the biological processes do follow the fundamentals of chemistry and physics [11]. Quantum tunneling involves electron tunneling and proton tunneling. Biochemical redox interactions, like photosynthesis, cellular respiration, and enzymatic or catalytic interactions, involve the electron tunneling phenomenon. The spontaneous DNA mutational interactions follow the basics of the proton tunneling process [10]. When protons defy difficulties during quantum tunneling, it causes spontaneous DNA mutations referred to as a proton tunneling [12]. Quantum tunneling also plays a significant role during aging and the progress of cancer [13].

The functioning of the Scanning Tunneling Microscope depends on the concept of quantum tunneling. (When the samples under study, top, and surface of the sample, understudy should be nearby, a voltage difference develops). Passage of electrons gets established that permits movement of electrons through the tunnel *via* a vacuumed zone formed between them. As a result, the functional tunneling current is developed (between the tip position and the sample). Other factors affecting this process are the voltage applied and the local density of the states. The information is collected while the current is monitored at the position of the tip across the surface scanned and projected as an image. This concept and technique help to a greater excellent resolution of the sample to the tune of 0.1 nm lateral resolution and 0.01nm depth resolution. Overall, tunneling effect/concept reflects on the stability and behavior of the nanomaterials. The process of quantum tunneling has a significant functional role during senescence and the advancement of cancer [13, 14].

Chemical aspects regulate the life processes (prebiotic chemistry) occurring in a biological system of the environment and involve geothermal (heat) energy and functionalized nanomachines. This phenomenon plays an effective role in molecular and biological evolution and also in the prebiotic chemical regime [10].

SOME FUNDAMENTALS OF CHEMISTRY THAT INFLUENCE THE BEHAVIOR OF NANOMATERIALS

The behavior of nanomaterials follows some of the fundamentals of chemistry. Hence, it is appropriate to review some of the related prime chemical fundamentals.

Inter and Intramolecular Bonding

During the interactions of nanomaterials, all types of chemical bondings participate. Intramolecular bonding brings about the changes in the structural aspects of a biochemical and chemical molecule like ionic bonds, covalent bonds, metallic bonds, *etc*. Intermolecular bonding is related to physical interactions but does not bring change in the chemical structure of the interacting molecules. Ion-ion and ion-dipole interactions, van der Waals forces and bonds, hydrogen bonds, hydrophobic bonds, repulsive forces, and steric repulsions, come under intramolecular bonding and forces. Macromolecules are the products of many molecules held as one unit, and these units assume a three-dimensional structural aspect. This association involves intramolecular bonding. Hydrogen and van der Waals bonds incriminate during intramolecular bonding. These result in the formation of the relatively bigger molecule. Hydrogen and van der Waals bonds are relatively weaker bonds, but when present in large numbers, these have overall higher energy impact [15 - 17].

Large Surface Area

Nanomaterials have a large surface area and weak forces; the three-dimensional structure of a macromolecule incorporates intermolecular bonds that have a precise and specific biological function. The irreversible change affects the quaternary structure of the molecule resulting in the loss of its proper and accurate biological function, and this phenomenon occurs during the denaturation proteins, loss of enzymatic activity, *etc* [18].

Hydrophobicity

Hydrophobicity has a significant functional role during interactions of nanomaterials. The hydrophobic effect or interaction of nanomaterials with biological materials follows the principle of entropy. It is a property of non-pola--polar molecules like oil that tries to form aggregates in water. This tendency reflects on the exclusion of non-polar molecules in an aqueous medium. This behavior gets initiated, when highly dynamic hydrogen bonds are present between the molecules of water in the liquid state. Polar groups, like OH^- present in methanol, do not exhibit hydrophobic impact [19]. (For this topic, refer earlier chapters).

With the rise in ubiquitous applications of nanomaterials in most spheres of life and industries, there has been an enormous increase in the production of these materials and reflects on the elevated utility of nanomaterials and increased possibilities of interactions at bio-nano-interface. This bio-nano-interface includes nanomaterials-biomolecules, nanomaterials-cell organelles, nanomaterials-cell membranes, nanomaterials-extracellular matrix, nanomaterials-biological fluids, *etc*. These interactions involve bio-physicochemical and colloidal forces. In all possibilities, the biomolecules present in a biosystem may exhibit processes such as phase transformation, the release of free energy, reconstruction, dissolution at nano-bio-interface, *etc*. These are likely to induce a predictive and developmental correlation between the structure and activity of nanomaterials. This specific correlation depends on the size, shape, surface chemistry, nature and coating, and functionalization of nanomaterials, biochemical along with the physiological aspects of bio-nano-interface. These aspects influence the behavior of nanomaterials, their safe use, and impacts in a biosystem. Their uptake, distribution, transportation to the specific site within cells and tissues also reflect on the behavior of nanomaterials involving their interaction within a biosystem. During such transit, parameters like half-life, interactions with biomolecules, cell organelles, signaling system, and metabolic pathways, and their elimination from the biosystem all relate to their behavior.

FACTORS AFFECTING INTERACTIONS OF NANOMATERIALS

Dispersion of nanomaterials in fluid media and parameters like size, shape, surface properties, charge on the surface, state of aggregation, crystallinity, potential to generate reactive species, *etc.* influence the interaction of nanomaterials. These parameters, if not suitably modified and regulated, have the potential to cause derogative impacts.

Ability of Nanomaterials To Get Distributed and Dispersed in Media

Natural nanomaterials have a specific range of size, shape, chemical composition, *etc*. These parameters are related to their distribution and movements within an aquatic and aerial medium. In the present contest, the specifically aquatic medium is essential because water is an integral component of a biosystem. The size and shape of nanomaterials are related to the mode of their formation or fabrication or synthesis. There are more chances of the creation of nanomaterials having different sizes within a given range. This group of nanomaterials is likely to exhibit different distribution behavior in a medium. Larger sized nanoparticles tend to settle down while smaller ones remain suspended in the medium. Thus, rates of settling down are related to the size of nanomaterials and the nature of the medium. The settling force depends on gravity and buoyancy [20, 21].

If F is settling force, Vx is the volume of nanoparticles, ρx is the density of nanoparticles, and ρm is the density of the medium,

F= Vx ρx g (gravity); F= − (ρmVx g) (buoyancy); F= Vx g (ρx − ρm) settling force.

In due course of time, the concentration of suspended nanoparticles changes because larger nanoparticles settle down and do not remain suspended in a medium. The rate of settling and degree of suspension of nanoparticles in a given medium affects the transport of nanomaterials. The settling velocity is the terminal velocity (Vx). It is affected by the setting force and viscosity (η) of fluid (resistance of nanoparticles offered by fluid medium):

If r = radius of nanoparticles, g = gravity; Vx = 2/9 X r2g/η X (ρx − ρf)

Particles sized within nano-dimensions (less than 100 nm) behave more or less like particles performing Brownian Movements. These movements also resemble the movements (velocity) of particles caused due to thermal effects in a given fluid medium [20, 21]. Suspension of nanoparticles in a given volume of fluid may not behave like a true solution because the distribution of nanoparticles does not exhibit an equilibrium state. The suspension system follows the kinetics related to the settling mechanism and shows apparent solubility, *i.e.*; some part of it is called a non-solution state. Following equation represent the apparent solubility of nanoparticles:

If K as $= [X] f / [X] s$

During initial stages, nanomaterials in the medium behave within the thermal kinetic range, and then these conditions function towards the determination of equilibrium concentration of nanoparticles in a given solution.

Under the equilibrium condition, If k = Boltzmann's constant, T = absolute temperature, h = linear measure of separation of nanoparticles, $[X]s$ = concentration of non-suspension. At saturation state, the non-suspension is likely to cause no real impact on the degree or amount of suspension [20, 21].

The Boltzmann equation is likely to provide the probable concentration of total suspended nanoparticles at equilibrium aqueous state $[X]aq$

Concentration of total suspended nanoparticles $[X] aq = e\ Vx\ g\ (\rho x - \rho m)/2Kt$ .0.012

Application of Equilibrium Equation

Equilibrium equations relate with the size, concentration of nanoparticles along with the shape of the temporal stability of the heterogeneous solution. If the size of the nanoparticles fixed the resultant suspension should be stable. When there are no external forces on the solution and nanoparticles are agglomerated, then the resultant solution is metastable. This equilibrium also helps the prediction of concentration in aqueous medium of nanoparticles depending on their physical properties. Materials at nano dimensions exhibit apparent solubility (Kas) and the suspension is metastable. It follows the distribution function (f) by the volume of nanoparticles in a suspension state. The dynamic time-course-chemical reaction is a useful parameter to predict the stability of nanoparticles in a suspension state, and predict the transportation of nanomaterials and their exposure to the receptor in the biosystem [20, 21].

Net Charge on Nanomaterials

This aspect is critical as it is related to the change in ion distribution, and the dipole distribution of the components of the solvent, and the immediate next nanoparticles in the vicinity. In a medium, counter ions make a layer adjoining the charged surface of nanoparticles. The charged surface of nanoparticles develops a layer of counter ions. All these, together form a layer of the charged surface along with a layer of counter ions charge. The resultant layer formed is called the stern layer. Nanoparticles are in a constant state of Brownian's movements in a given medium along with this stern layer on them. Under normal conditions, the ions of the stern layer should balance the surface charge on the nanoparticles. If there is a difference between these two, it is stern potential. This difference is a differential movement in charge that makes the stern potential to interact with other components of the system, and it results in sheer electromagnetic force. This force is called the Zeta potential (ζ-potential). For a more straightforward understanding, one may say that ζ-potential is a net charge of nanoparticles present in the suspension.

FORCES AFFECTING THE BEHAVIOR OF NANOMATERIALS IN BIOSYSTEM

Some physical forces influence the interactions between nanoparticles, biomolecules, and components of the biosystem. This interactive behavior also depends on the size, shape, and other physical features of nanomaterials. During these interactions, modifications in the functional aspect of size or other characteristics of nanomaterials may occur in the medium. The concerned medium is either related to a biological entity or the environment. Some physical features of nanomaterials are likely to change by the extent of agglomeration, adsorption, and dissolution, *etc.*, when present as a suspension state in the medium. Following are the concepts that regulate these changes:

1. The stable covalent bonds stabilize the internal structure of nanomaterials. This process ensures the insolubility of the restructured nanomaterials in the biological medium, at least during the initial phase of the interaction.
2. In most cases, nanomaterials exhibit their chemical effects based on their surface chemistry. The surface chemistry is a resultant functional aspect of the composition, structure, and functionalization of the nanomaterials.
3. Nanomaterials prefer to exhibit weaker nucleophilic and electrophilic affinities, as such, strong affinities, are likely to affect their stability. Nanomaterials respond to the weaker ions and van der Waals interactions.
4. Factors like energy transfer, electron distribution, and architecture of the surface of nanomaterials help to predict their behavior in a biological system. This behavior is possible, specifically, when partial charge sharing, excitation, quenching, occur in a heterogeneous number of nanomaterials and a medium.

Nanomaterials are useful, reactive, and their interactions are dependent on their surface chemistry to the greatest extent. Generally, those interactions that are related to the surface chemistry of nanomaterials, their mechanism follows the DLVO theory. This theory, proposed in 1941, is a combined concept of Derijaguin B, Landau L, Verny E, and Overbeek, (DLVO). These researchers suggested that this theory explains the process of aggregation of the aqueous dispersion. This concept also describes the role of forces involved during charged surfaces interacting through a liquid medium. It also explains the combined effects of van der Waals attractions and electrical repulsions caused due to the double layer of counter ion' The DLVO concept provides summated forces acting in a given aqueous/liquid medium.

If FT = total force field; FR = electrostatic repulsive forces; Fvdw = van der Waals; FS=forces of solvency.

Then this summation is as FT = FR + Fvdw + FS.

Nanomaterials possess low inherent inertia and high Brownian's velocity during charged surfaces interactions. These forces become a significant driving force in the case of nanomaterials in a given liquid medium; otherwise, these forces are typically weak.

Some basic concepts are related to nanomaterials, colloid, solution, and particles. The solution is a mixture of molecules-molecules that are around 1 nm in diameter; the suspension is a mixture of particles-particles that are larger than 1000nm; colloid/colloid dispersion is a mixture of colloid-particles having a diameter ranging between 1nm to 1000 nm. Colloid suspensions have two components-particles and the dispersion medium. Common examples of colloids in our day to day life include mayonnaise, milk, butter, gelatin, jelly, muddy water, plaster, colored glass, paper, and smoke. The dimensions of nanoparticles are within 1nm to100 nm and in most cases remain in a state of suspension until and unless agglomeration takes place among these nanoparticles.

In colloidal suspension, the dispersed particles remain suspended in the dispersion medium and show Brownian's movements (zigzag movements or random movement of particles caused by the collision in the medium). Colloid exhibit Tyndall Effect, (*i.e.*, an intense light becomes visible due to colloidal particles and appears as a column of light). Colloidal particles do not settle down to the bottom in a medium. There is some similarity between colloids and nanoparticles [22]. His proposal deals with the size of nanomaterials within the 1-100 nm range, and colloids are smaller than 10 nm in size. Nanomaterials exhibit homogenous molecular composition are around 15% size dispersion, and, these cause less polydispersity for nanocluster around nanoparticles. The colloids are poorly defined compositions and are less than 15% sized dispersion. Nanomaterials are reproducible synthetically. Their synthesis depends on size, shape, and composition, but the colloids are not. These exhibit uncontrolled morphological and compositional aspects. The physicochemical properties and catalytic activities can be reproduced while colloids have features that are not reproducible. Nanomaterials can be solubilized in polar/non-polar solvent depending on the use of a suitable stabilizing agent while colloids are soluble in polar solvents. Nanomaterials have a clean surface and surface of colloid possess surface adsorbed species like -OH, -X, *etc.* Despite these aspects, the DVLO concept is applicable, at least in a broader sense.

Role of Electrostatic Forces [Coulomb Forces (Fs)] During The Behavior of Nanomaterials

The electrostatic force is either attractive or repulsive between two charged interacting atoms, molecules, and moieties. The cause of charge between the interacting atoms, molecules, and moieties is due to the stability in the valency of the atoms, molecules or moieties at the site of interaction or medium, *e.g.*, ionizable salts. In these cases, there are relative electron affinity differences between two interacting anions and cation. The separation of charge is related to low energy conformation. Whenever there is a transfer of energy between neutral and charged moieties, this energy is called ionization energy. The energy of ionization and the status of valency participate in the electrostatic repulsive forces. The electrostatic (Coulomb forces) forces between the internalized nanomaterials undergo fluctuations because these forces depend on the types and the physicochemical features of the nanomaterial.

The repulsive force acts through a minimal distance or range, and, this force increases many folds when the distance between the positive and the negative charges becomes very small. This force reduces when the gap between these two increases. Another aspect of this concept includes repulsive-negative and attractive force-positive and, the net effect between the two. All these features are the functions of interatomic distance (r). When an atom moves closer to another interacting atom, an attractive force increases, *i.e.*, the interatomic gap reduces and potential energy becomes less and indicates the establishment of a stable and stronger bond. When these atoms are very close to each other, their electron clouds overlap, this causes repulsion between the two atoms, and the force exerted is called repulsive force [23]. Since the force of repulsion is a strong short-range force, the net force declines because of interatomic space increases. If (r) represents interatomic space or distance and if r = r0, then the net force reduces to zero and so is the potential energy (it may become minimum). This (r0) is considered to be bonding length, and it signifies and represents the equilibrated interatomic distance. At this state, bonding between the two interacting atoms is most stable.

When potential energy is zero, it represents the strength of the bond; it is also the energy is required to break this bond or separate the two interacting atoms. A stage when atoms are closer than r0 and repulsive force increases substantially resulting in the involvement of potential energy of the interacting atoms [23]. Coulomb energy is due to the Coulomb forces. During the agglomerative reaction, these are repulsive. The reasons for this nature are similar and like nanoparticles attain the same type of charge. The density of charge depends on the respective size. When dipole conditions emerge the charge can arise, and the dipole aligns between the external surface and the interior of the nanoparticles under question. The steric state of atom or nanoparticle or molecule can affect the interaction [24 - 26].

If x and x′ are the two points, q is the charge on nanoparticles on particle x and x′, ε0 is the electric constant ($8.8548782X^{-}12$ C2 $N^{-}1m^{-}2$), εs is the dielectric constant and E(C) x and x′ is the potential energy of the system [24, 25]. Then one can calculate potential energy, according to Langel, 2005 [25] as [E(C) x and x = qx.Qx′/4π.z.ε0.εs]

Van der Waals FORCES (vdW)

The interactions between molecular moieties are called van der Waals interaction, and the forces involved are van der Waals forces (vdW) [27]. These forces cause attractive interactions between the molecules, atomic groups, or moieties but not due to the covalent bonds or electrostatic interactions, *i.e.*, charge present on the interacting molecules or neutral molecules. Theoretically, between two identical bodies in a given medium, the vdW force will always be attractive in nature and positive. Between two different entities, vdW may be either attractive or repulsive (attractive force is a positive and repulsive force is a negative [28 - 32]. The vdW force is always positive between two condensed bodies in vacuum and air, but in a solvent, medium vdW reduces highly. This force (vdW) can be positive, negative, or zero, depending on the conditions of interaction [33].

The energy of van der Waals forces in a medium along with nanomaterials expresses the weak attractive and repulsive forces existing therein. The attractive and repulsive forces are present between the surfaces of nanoparticles and the medium, surface of nanoparticles, and the components of the medium. The net forces depend on the difference between the molecular distances and between the interacting molecules. When the atomic gap exhibits repulsive force, the forces show maximum attraction and are exothermic. If there is a change in the free energy of the system, it is due to the product of enthalpy and adsorption. The difference between the forces of repulsion and attraction and between the interacting nanomaterials is approximated using a modified Lennard-Jones relationship. The unmodified version of the Lennard-Jones relationship is applicable suitably in case of interaction between molecules in a medium [25]. If $E(w)x$ x′ is the inter-nanoparticle potential energy between two particles x and x′; z is the distance between particles x and x′; z0 is the effective thermodynamic distance at which potential energy E(w) is the same as enthalpy of adsorption (-H α); n is the number of binding sites on nanoparticles or nanomaterials under question.

Then inter-nanoparticle potential energy E (w) x x′ = 4πnz0. Hα [1/45 (z0/z) 9–1/6 (z0/z) 3] [25].

Size, The toxicological behavior of nanomaterials depends on their surface charge and agglomeration state of nanoparticles (Jiang *et al.*, 2009). The viscosity of nanofluids increases when the size of the nanomaterials is less. This viscosity also relates to the nature of the nanomaterials [34].

The interacting molecular moieties exhibit attraction or repulsion [25]. The vdW forces actively participate when uncharged atoms, molecules, or moieties interact. These forces participate during the cohesion of condensed phases, adsorption of gases on the physical manner, and attraction between two macroscopic objects in the absence of forces of repulsion. The vdW forces actively participate during the aggregation of fine particles and coagulation of matter or tiny structures in a dispersed colloid mixture [33]. The processes like adhesion, surface tension in a film related to nanostructures, nanosystems, nanofilms, self-assembly of nanowires, and long-range interaction in nanoscience involve van der Waals forces [35 - 38]. There is an active role of van der Waals forces during physical adhesion [36]. The ability of materials to get wet also depends on the involvement of van der Waals forces. The van der Waals forces and theory of dispersion take part during the surface melting of ice [39]. During the formation of free-standing films in a soap bubble, the van der Waal forces play a significant role [40]. During flocculation and deflocculation in the colloidal system, van der Waals forces play a regulatory role [41]. These forces affect the properties of gases, liquids during the process of wetting and formation of thin-film [42]. The van der Waals forces accomplish the interaction between oxidized polystyrene surface and colloid surface in a system. Hodges measured the van der Waals force using atomic force microscopy (AFM) during the interaction on the surfaces of the polymer [43, 44]. The van der Waals forces are involved during the docking of the protein-protein process [45]. The computational investigation related to binding modes of peptide and non-peptides inhibitors revealed the involvement of van der Waals forces, and molecular dynamics simulation confirms it [14].

Keesom Forces, Debye Forces, and London Dispersion Forces are Under Van Der Waals Forces

Although van der Waals forces appear to be simple forces and accomplish several interactive processes. Forces like Keesom, Debye, and London dispersive forces come under the umbrella of van der Waals forces.

Keesom forces are due to Keesom interactions. William Hendricks Keesom studied these interactions. These interactions are weak van der Waals interactions and do not take place in an aqueous solution having electrolytes. Keesom forces are temperature-dependent and originate due to the attraction between moieties having permanent dipoles. The molecules are in the state of constant rotation, and these molecules do not get fixed into a specific place or position. This rotational and locking state affects the dipoles of these molecules. The energy of Keesom interaction is proportional to the inverse 6th power of the distance.

The induction force or polarization capacity represents Debye forces. Peter J W Debye studied this aspect of the interaction. This force is the result of interactions between permanently and constantly rotating dipoles and also due to the ability of atom or molecule to polarize. Polarization takes place on the induction of dipole forces, and also when a molecule having permanent dipoles repels the electrons of other molecules. A molecule with a permanent dipole induces a dipole in similar ambient molecules and these results in mutual attraction between them. Even atoms exhibit Debye forces and Debye interactions. Induced dipoles can freely shift or rotate around a non-polar molecule. These forces are not dependent on temperature. This induction is an attractive interaction between a molecule with permanent dipole and others with the induced dipole. This interaction represents a Debye force. The forces between induction and induced interaction are weaker in comparison with Keesom forces.

London Dispersion Force (LDF) is also Dipole-induced dipole interactions. These forces are related to the dispersibility of molecules and exhibit fluctuating dipole-induced dipole interactions. Polarization is due to polar molecule or repulsion of negatively charged electron and electron clouds among non-polar molecules. This aspect reflects that London interactions are concerned with random fluctuations in the electron density in the electron cloud. Those atoms which have many electrons exhibit London dispersion force in comparison to those atoms having lesser electrons. London dispersion forces are most suitable because almost all materials can be polarized. The Principles of London dispersion forces govern the atom-atom interaction. These forces are universal in application. London dispersion forces are weaker intermolecular forces and are due to quantum induced multiple polarizations in molecules. Even molecules having multipolar moments exhibit these forces. Correlated movements in the electrons between the interacting non-polar molecules also show London Dispersion Force. These correlated movements are due to the repulsion between the fast movements of the electrons and results in the redistribution of electron density between the interacting molecules. This behavior of an electron causes instantaneous dipole formation. Thus, LDF is the functional basis of all chemical groups and also in condensed matter. LDF is relatively weaker in comparison with ionic and hydrogen bonds and forces. The behavior of LDF enhances the degree of polarization of molecules, at least in those molecules that have significant and dispersed electrons cloud. When the surface contact of materials increases, these forces become stronger.

Dispersibility and Solubility of Nanomaterials

The force of solubility and degree of dispersibility play significant roles during the dispersion and distribution of nanomaterials in biological and other fluids. The differences between the levels of free energy (Gibbs free energy) of pure solvent and the solute (in current case nanomaterials) are the main factors that affect the forces of solubility [25]. Gibbs free energy (G) is the available free energy of the solubility. According to Gibbs, the highest amount of mechanical work obtained from a given quantity of a substance in a given initial state without increasing its total volume or allowing heat to pass or from external bodies except such as at the close of the process remain in their initial condition [46]. This concept is aptly applicable in the case of nanomaterials and their relevant medium of dispersion either environmental or biological. When nanomaterials disperse in aquatic medium (hydrosol), the binding site of water molecule attaches to the individual nanomaterial, and this links to other water molecules: $X+H_2O \cdot H_2O \rightarrow X+H_2O \cdot H_2O$. Thus, there exists an energy difference (ΔG) between $X + H_2O \cdot H_2O$ and $X + H_2O \cdot H_2O$. This difference is the free energy of the resultant hydrosol. The thermodynamic benefit of the hydrosol or the product of any nanomaterials dispersion in a medium is useful if the value of ΔG is less than zero (0), *i.e.*, $\Delta G < 0$. This condition results in the spontaneous dispersion of the nanomaterials. The source of the force of energy difference is related to solubility or dispersibility of nanomaterials in a suitable medium (solvent). The dispersibility of the nanomaterials is an analyzable and quantifiable parameter [46, 47]. The dispersibility of nanomaterials plays a significant role during their formulation as per the need concerning the biological fluid medium. This feature also ensures access to the target tissue. The suitably functionalized surface of nanomaterials is a feasible option to carry the pharmaceutical products successfully. Further, this technique also ameliorates the dispersion or solubility of comparatively less hydrophilic or non-polar nanomaterials [47, 48].

Influence of Size of Nanomaterials During their Interactions

The size of nanomaterials is the fundamental parameter, and it accounts for during synthesis, designing, functionalization, and interactions of nanomaterials. The sizes of nanomaterials influence their uptake, distribution, interaction, and impact of a biosystem. Quite often, smaller and larger seized nanoparticles find enormous applications in academic experimentation, research, and other commercial fields. There are forces like interacting and unbinding forces that become active during the uptake of such nanomaterials. These forces facilitate the uptake of gold nanoparticles having a size ranging diameter between 4 to 17 nm. It is observed that the larger size of nanoparticles gives different impacts [49]. The use of gold nanoparticles in imaging and diagnostic components is well evident. Gold nanoparticles are stable, safe, and are not toxic [50], but there are conflicting reports in the recent past related to derogative impacts in the biosystem. Au nanoparticles (average diameter of 18 nm) with their surfaces capped with citrate interact with human leukemic cells for three days. These gold nanoparticles did not show a toxic impact. The precursors of the modified Au nanoparticles may be toxic. Au NPs with 0.8-15 nm size after stabilized with triphenylphosphine interact with connective tissue fibroblast, epithelial cells, macrophages, and melanoma cells. The observations indicate, the sensitivity concerning nanoparticles under consideration, specifically with 1.4 nm sized nanoparticles. IC50 values are within 30 to 56 μM with 1.4 nm size nanoparticles. AuNPs with 15 nm diameter and coated with Tauredon – gold thiomalate, interact with these cells, and their impact is nontoxic. Nanoparticles with 1.4 nm diameter cause necrosis in these cells while Au NPs with 1.20 nm result in apoptosis leading to cell death [51]. AgNPs (10, 100 nm), molybdenum (MoO_3, 30, 150 nm), aluminum (Al, 103 nm), iron oxide (Fe_3O_4, 47 nm) and Titanium dioxide (TiO_2, 40 nm) exhibit toxic cellular responses. These responses are more derogative than those caused by bigger particles like cadmium oxide (CdO, 1μm), manganese oxide (MnO_2, 1-2 μm), and tungsten (W, 27 μm). The toxicity in BRL-3-A rat liver cells is in terms of the function of mitochondria (MTT), leakage/activity of LDH, glutathione (GSH), mitochondrial membrane potential and reactive species of oxygen (ROS) in response to 24 hours exposure. The lower doses of Fe_3O_4, 47 nm, Al, 103nm, MoO_3, 30, 150 nm, TiO_2, 40 nm, (10-50 μg/ml) show mild effects while higher doses (100-250 μg/ml) did cause a higher degree of cellular toxicity. Among all the nanoparticles, Ag nanoparticles exhibit a higher degree of toxicity, but MoO_3 NPs are moderately toxic while Fe_3O_4, Al, MnO_2, and W, are the least toxic. Relatively larger particles of Ag NPs (with 5-50 μg/ml concentration) decrease mitochondrial function to a greater extent. The result of exposure of respective NPs, having the abnormal size to the cells under study, shows cellular shrinkage or irregularity in the shape of the cells. Among all the nanoparticles studied AgNPs are relatively toxic. This toxicity is in the form of depletion and decline in GSH level, mitochondrial membrane potential, and relative enhanced ROS level [52].

Gold and silver nanoparticles having a coating of antibodies undergo internalization through a membrane and this process involves binding and activation of membrane receptors and subsequent proteins. This interaction of nanoparticles is size-dependent. Nanoparticles within a 2-100 nm size range change essential cellular signals meant for basic cellular functions. The maximum intensity of impact is in the case of nanoparticles with 40, and 50 nm size. The nanoparticles within this range are not suitable for the medical/therapeutic applications, and these may lead to derogative conditions [53]. Gold and silver nanoparticles coated with antibiotics are capable of controlling the membrane receptor related to internalization. The size of nanoparticles is a useful parameter and plays an active role in binding, activation of a receptor on the membrane. Nanoparticles within the nanoscale range, 2-100 nm can influence cellular signaling processes related to the essential cellular functions but can mediate the biological effects of the cell [54].

The effects of the size of NPs well suspended in uniform mesoporous silica NPs (range of diameter 38-28- nm) on the HeLa-cell line undergo cellular uptake. Mesoporous silica nanoparticles with 50 nm size exhibit maximum cellular uptake concerning the HeLa-cell line. The green fluorescence technique is helpful to study such types of studies [55]. Platinum nanoparticles (4-8 nm size) capped with polyvinyl alcohol (PVA) enter human cells because of diffusion. These nanoparticles cause more damage to DNA and increased localization in cells of S-phase of the cell cycle, apoptosis, and genotoxic stress. The activities related to cellular tumor protein, p53-a tumor suppressor, becomes irregular as a result of this interaction. When polyvinyl-pyrrolidone (PVP) coated iron-oxide nanoparticles having a diameter ranging between 30-120 nm interact with hepatic lesions *in vivo*, these iron oxide nanoparticles behave as biocompatible and show a high degree of crystallinity. The cellular uptake of these modified nanoparticles is in proportion to their size. Nanoparticles with 30 nm diameters, size, and particle with 100nm hydrodynamic diameter exhibit a higher degree of cellular uptake and show a relatively higher degree of distribution in comparison to those iron oxide nanoparticles coated with PVP [56].

Physicochemical features of nanoparticles influence the pharmacokinetic behavior of modified forms; modification of iron oxide nanoparticles with PVP-10-37 (size 100 nm) elevates their efficiency. Uptake of anatase, rutile (mineral forms of TiO_2), and TiO_2 nanoparticles follow the principle of kinetics and is dependent on the size of hard agglomeration but not on the actual size of TiO_2 nanoparticles [57]. The uptake of nanoparticles induces the degree of intracellular oxidative stress, the proinflammatory response in A549 lung epithelial cells in response concerning the size TiO_2 nanoparticles. These observations relate to the static light scattering (SLS) technique, Raman microspectroscopy (μ-Raman), and TEM. These techniques confirm the location of size-specific Au NPs uptake within the cell and those reaching cell organelles and nucleus [57]. On internalization, gold nanoparticles having 2.4 nm sizes concentrate in the nuclei of CO5-1 Line while 5.2 nm nanoparticles and nanoparticles with 8.2 nm-sized partly reach the cytoplasm. Gold nanoparticles with 16 nm and then bigger size than this do not show internalization. The range used was 2.4 to 8.9 nm [58].

There exist lacunae in the understanding of the mechanism involved in the interaction between nanoparticles and the biological molecules and the components of the biosystem. This mechanism is related to time and length scale concept, in connection to inter and intracellular entry, distribution of nanoparticles (NPs) within a specific type of cell and the cellular uptake of NPs [59]. There is a need to make an effort to investigate this mode of internalization using fluorescence carboxyl-modified polystyrene nanoparticles having sizes within range 40-2μm in the cells of different cell lines like HeLa, A549 epithelial cells, 1321 Astrocytes, HCMEC D3 endothelial cells and murine RAW 264.7 macrophages. Study on a semi-quantitative approach using confocal microscopy, flow cytometry, to check uptake and subcellular distribution of model NPs, shows that uptake of NPs and internalization is size-dependent in the entire cell lines studied. Even the same type of nanoparticles but having different sizes exhibit different rates and amounts of uptake. HeLa cell line shows the minimum rate of intake of nanoparticles while phagocytic cell line (RAW 264.7) the maximum; the rate of intake and internalization is higher in the case of smaller sized NPs in comparison to larger ones. Among fluoresceine-labeled silicon nanoparticles (32 nm and 83 nm) only nanoparticles with 32 nm-sized nanoparticles localize in cell and nucleus while nanoparticles with 83 nm size could not enter the Caco-2 cells (heterogeneous human epithelial colorectal adenocarcinoma cells). These nanoparticles of both sizes are non-cytotoxic.

Further, the rate of migration is higher in the case of silicon NPs with 32 nm sizes in comparison to 83 nm-sized nanoparticles [60]. Gold NPs smaller than 10 nm (referred to as ultra-small NPs) exhibit better penetration and localization in comparison to those Au PNs a having bigger size (10 nm) in the cells of breast cancer, multicellular tumor spheroids, and tumors in mice. Gold NPs with tunable size from 2-15 nm and tiopronin (Thiola) surface coatings and the charge, had better penetration. (Tiopronin (trade name Thiola) is a cancer drug). The smaller NPs can penetrate deeper into the cells of tumor spheroid, while larger nanoparticles could not penetrate [50]. Aqueous CdTe quantum dots having thiol ligand with 2-7 nm as diameter are suitable as image probes for living cells (A-

427 cells- lung cells under adenocarcinoma). The surface characteristics of CdTe QDs with thiol ligand affect adsorption on the cell membrane, and their subsequent internalization exhibits their impact concerning their concentration and distribution dynamics. The cellular uptakes of thiol ligated CdTe QDs escape the trapped environment and get distributed in large quantities in the cytoplasm of the cells used as test materials. This behavior of treated QDs is dependent on the hydrodynamic diameter and the ability to escape the ambient environment [61].

The oral bioavailability of quaternary ammonium salt di-dodecyl-dimethyl-ammonium bromide (DMAB) can improve if modified nanoparticles having a uniform size range 50nm to 300nm involving Deserno's model, are used [62]. The nature of the cell membrane is fluid, and, it has to carry out some critical functions like cell recognition, receive nutrition, and select proteins to meet its requirements. The fluid nature of the membrane makes it difficult to accomplish the life-sustaining activities. It becomes very tedious to investigate such cellular functional aspects. These membranes related problems can be solved not only functionally but also dynamically at the nanoscale with the use of computer programming [63]. This model is suitable to study physicochemical properties and cellular uptakes. The emulsion solvent diffusion method appears to be appropriate to investigate coumarin-6(C-6) loaded poly-(Lactide-co-glycolide) PLGA nanoparticles. There is an involvement of forces like electrostatic forces, hydrophobic forces, bending and stretching forces, during these interactions with the experimental cells model like human epithelial colorectal adenocarcinoma cells (Caco-2) and human colon adenocarcinoma cells (HT-29). When the modified surface of PLGA with DMAB (Di-dodecyl dimethyl-ammonium bromide), the uptake of NPs under investigation gets remarkably improved and exhibit optimum cellular uptake with optimal particles size (100 nm) and indicates the sized dependent cellular uptake [62].

Similarly shaped nanomaterials but having different sizes also shows a varying degree of derogative behavior. Longer TiO_2 fibers are more damaging in comparison to smaller ones (15μm). These longer fibers nanomaterials cause inflammatory activities, the release of inflammatory cytokines that involve cathepsin-β-mediated mechanism. Longer nanomaterials fibers (TiO_2 nano-belts) behave like fibers of asbestos or silica with macrophages [64].

Nanoparticles interact with cells like histocytes and blood macrophages (reticuloendothelial cells). These nanomaterials move across the cell membrane and exhibit varying degrees of bioaccumulation and biodistribution in tissues such as lymph nodes, bone marrow, spleen, adrenal, kidney, liver. Their size is an effective parameter concerning their interaction with immune activation and renal clearance. When this interaction is delayed, it results in increased duration in the circulation. This factor is useful in those nanoparticles that are used for therapeutic purposes. The hydrophilic nanoparticles, within 10-100 nm range in size, can reduce activation of mononuclear phagocytes but these do not undergo renal filtration [65]. Treatment of thioether ligands with well-defined platinum and palladium metal nanoparticles (within the size range 1-5 nm) enhances their size regulation and stabilization. Their interactions with fluorophore are dependent on the size, and so is the energy transfer. These particles exhibit reduced catalytic effectiveness about environmental pollutants due to the storage of multiple electrons [66]. PLGA linked nanoparticles are relatively more biocompatible and biodegradable; when such nanoparticles are functionalized with cationic groups, their cellular uptake increases because electrostatic interactions among them also increase. The electrostatic interaction is involved between the cationic surface and the negative charge of the cellular lipid members [67].

At the subcellular level, the degree of distribution and trafficking of quantum dots among macrophages depends on their size. Smaller sized quantum dots readily reach the target like the histones of DNA. The distribution of gold nanoparticles with a size range of 2.4 to 89.0 nm show variable distribution. Nanoparticles with 2-4 nm size reach the nucleus, 5.2 and 8.2 nm size are partially localized within cytoplasm while 16 nm and above sized do not undergo internalization at all [58].

Influence of Shape of Nanomaterials on their Interactive Behavior

Natural and engineered nanomaterials exhibit various shapes such as planes, tubes, rings, spheres, fibers, or irregular shapes. These variations in forms induce the impacts of the interaction of the nanomaterials concerning cell membranes and extra and intracellular components. The forms of nanomaterials are a physical parameter and influence the interactions and behavior of nanomaterials in a biosystem. Each form has its specific effects with varying degrees of interplay and impact. Spherical nanoparticles with appropriate dimensions exhibit convenient cellular internalization and cause no or least adverse effects in the biosystem as compared to a rod-like or fiber shaped nanoparticles. Processes such as endocytosis, phagocytosis, caveolin formation, enhanced permeability, and retention effects (EPR effect), *etc.*, are the various modes of internalization. The shape of the nanomaterials varies within a wide range of different types of nanomaterials. Carbon nanomaterials exhibit a wide range of shapes like tubular, hollow balls, sheet, agglomerated forms, forming conjugates with biomolecules, *etc.* Nanomaterials that have spherical, cubical, rectangular, rod-like, crystalline shapes show a change in the area exposed during interaction [4]. Different designs of nanoparticles exhibit a different rate of uptake by phagocytes [68].

The composition of nanomaterials is another factor that affects the interaction and behavior of nanomaterials. Quantum dots incorporated with heavy metals like cadmium have the potential to cause health hazards. Quantum dots either coated or enclosed in nanoshells are relatively less toxic. The size and surface coatings on nanomaterials influence the rate and mode of internalization. Polymeric nanoparticles internalize differently and with a different degree in Caco-2 (human colon adenocarcinoma cells), *in vitro* [69]. Polymeric nanoparticles coated with poly-lactic-co-glycolic acid, fluorescent polystyrene nanoparticles of standard size, and polyvinyl alcohol, and it Vit-E, exhibit different behavior. Nanoparticles coated with Vit-E exhibit 1.4 times elevated rate of uptake in comparison to those coated with polyvinyl alcohol. This uptake is 4 to 6 times higher in contrast to non-coated nanoparticles [69]. Mammalian cells exhibit a different rate of uptake of Different shapes of gold nanoparticles that exhibit different rates of uptake in mammalian cells. Spherical nanomaterials exhibit 375% to 500% increased cellular uptake while other forms of nanomaterials show the appreciable rate of uptake and cause mild to substantial toxic and derogative impacts [68].

The different shapes of nanomaterial make contact with membrane at a specific angle; this angle and the normal formed at the point of contact play an essential role during the initial stage of internalization. During later stages, the vector, *i.e.*, angle at which the mean direction of tangents drawn to the target, contour between the initial contact point and the centerline of the target also affects the interactions and cellular internalization of nanomaterials [70]. The angle between the membrane and the initial contact depends on the shape of the nanomaterial. Phagocytosis of nanoparticles depends on their size and shape while the rate of phagocytosis is associated with the angle and velocity of internalization, (the distance moved by the membrane of macrophage to complete the process of phagocytosis-the factor divided by the time needed to complete the process of phagocytosis). The velocity of phagocytosis decreases when the initial contact angle increases [70].

The impacts of nanomaterials also relate to the duration of exposure, concentration, size, and shape of the nanomaterials, and these nanoparticles ameliorate processes like attachment, proliferation, expression of osteogenic gene ALP/DNA, protein/DNA, and mineralization of human adipose-derived stem cells. Further, the shaped based nanomaterials show fluctuations in protein content in the tissue, and proteins that induce bone formation are also affected [71]. The various crystalline structures of TiO_2 and different shapes of TiO_2 exhibit different derogative and toxic behavior, like oxidative DNA damage, lipid peroxidation, and formation of micronuclei, *etc* [71]. Graphene oxide nanosheets and its treated forms with fetal bovine serum both exhibit cytotoxic behavior and show protein corona mediated cytotoxic interaction. Low concentrations of graphene oxide GO (1%) show concentration-dependent toxic action on cells; at a concentration (10%) this behavior gets intensified. The mitigation of toxic and derogative impact is probably due to the interactions between flat graphene sheets on the cell membrane that result in physical damage to the cell membrane. Further, the nanocrystalline forms of graphene oxide and uric acid result in morphological destruction to the structural aspects of cell and tissue along with inducing inflammation [72].

TiO_2 nanoparticles having different crystalline structures exhibit varied derogative interactions and impacts. Rutile TiO_2 nanoparticles result in oxidative damage of DNA, the formation of micronuclei, lipid peroxidation, while anatase TiO_2 crystals induce the formation of ROS [73, 74]. Plate-shaped silver nanoparticles are more toxic to fish cells in comparison to other shapes like spheres and wires because of the presence of surface defects. Defective crystalline structures and nanoplates induce oxidative stress because they produce toxic superoxides. The target cells pretreated with N-acetylcysteine (antioxidant molecule this acts as a protective agent) there is no damage. The surface defects on the nanoplates are responsible for releasing respective ions. The High-Resolution Transmission Electron Microscopy (HRTEM) establishes the process of release of ions. Such defects are stacking faults and point defects, and quite often, these defects are present in good numbers in respective crystalline structures. During formulation or engineering, the specific nanomaterials the shape aspect is considered seriously. Carbon-based nanomaterials like carbon nanowires, multi-walled carbon nanotubes (MWCNTs), and single-walled carbon nanotubes (SWCNTs) different exhibit shapes. Of these shapes, only carbon nanofibers show the least toxic behavior [75]. Among graphene and nanotubes, graphene can cause a higher degree of toxic expression in tissues like mouse keratinocytes, bone marrow formation, and adversely affected the cell to cell communication during embryogenesis [76]. Antibacterial silver nanoparticles are the standard components of consumer products. Silver nanomaterials having shapes like nanowires, nanocubes, spherical, *etc.*, cause a decline in growth (5.3% and 39.6%) in the case of growing roots of *Folium multiform* [77]. The FDA has permitted the use of nano amorphous silica nanoparticles as a food additive, but not the silica in crystalline shape, and crystalline forms are carcinogenic [78]. These observations reflect on the role of the shape of nanomaterials in different products.

Influence of Surface Properties of Nanomaterials During the interactive behavior

Nanomaterials follow the basics of surface science during their interactive behavior. Surface science deals with the physical and chemical features of the materials in totality. The surfaces of nanomaterials function and permit the flow of the materials or energy across the intersurface, initiator, or terminator of a chemical reaction precisely, about catalytic activity, and, can hold or release other matter (specifically therapeutic drugs). This behavior of nanomaterials is related to the nature of their surface and obeys the basics of surface science [79]. The surface and interface of nanomaterials represent the boundary between two phases interacting with nanomaterials. At the nanoscale, the total volume remains the same, but the collective surface area becomes highly increased. The surface science explains the role of chemical groups present on the interface of the material, and specific features, like catalytic reactivity, electrical resistivity, adhesion, gas storage, chemical reactivity of the nanomaterials, *etc.* There is an existence of a significant proportion of atom on the surface of nanomaterial. This feature can alter some of the physical properties like a melting point at the nanoscale. The melting point in bulk materials is comparatively higher than in the nanomaterial of the same matter. The atoms present on the surface are more available for the interaction in comparison with the corresponding bulk material. The energy needed to hold these surface atoms is less in the case of nanomaterials than the energy for holding intermolecular forces in bulk state, because of this feature the melting point in nanomaterial gets lowered [80 - 82].

Influence of the Surface Charge on Nanomaterials During their Interactive Behavior

Charge on the surface of nanomaterials is responsible for the electrical potential on the surface and their stability in their suspension state. This feature of nanomaterials plays a significant and functional role during their interactive behavior in a biological system. Phenomena related to nanomaterials, like the stability of nanomaterials in suspension, the potential energy of nanomaterials due to attractive, repulsive forces and solvent attributions, correlate to the charge present on the surface of nanomaterials. The related attractive and repulsive forces influence significantly the movements of nanomaterials that have a more considerable distance between them. These movements resemble Brownian movements, solvent attributions, and are also concerned with the interaction between the solvent and the nanomaterials [21]. To visualize the stability of nanomaterials in suspension, one needs the information related to the features of the surface of nanomaterials such as coating or adsorbed species on the surface, defective sites on the surface, a layer of oxide, acidic and basic groups, *etc.* All these parameters contribute to the surface charge, charge density, and total net charge on nanomaterials. The value of Zeta potential (ζ) provides information about the surface charge and to measure zeta potential directly is difficult. The Zeta potential is the potential difference between the potential at the shear plane. The shear plane is present between the stern layers and a layer, which is very close to the surface of nanomaterials. It is considered to be set or fixed and the diffused layer, a layer formed by the cloud of positive and negative charge. Since the zeta potential is difficult to measure; one can calculate zeta potential experimentally by following the concept of the Henry equation. The parameters needed for this equation are: movement of a particle in an applied electric field-electrophoretic mobility $\mu\epsilon$; ϵ is dielectric constant; (fa) is the relative functional size of nanomaterial; (k) is the Debye length (a range within $1.0 - 1.5$); (η) is the viscosity: $\mu\epsilon = 2\epsilon\zeta/3\eta \; X \; f\,(ka)$........Henry equation.

The potentiometric titration technique is useful to find zeta potential, and it relates to H^+ and OH^- activity and activity about the non-specific ion adsorption that is having changes in H^+ and HO^- activity. This technique of titration is helpful to investigate the zeta potential of heterogeneous systems that include colloid, emulsions, and nanomaterials. The system contains solids that have an extensive surface area. This technique is quite useful to find zeta potential on the surface of particles under various conditions. In this technique, the isoelectric point is an essential parameter and indicates the pH value at which the zeta potential is around zero. At an isoelectric point, the colloid tends to be unstable, and nanoparticles get coagulated or flocculated. Acids and bases act as a titrating agent in this titration technique. The potential titration technique helps to determine the optimum dose of surfactant to attain stabilization and flocculation in the heterogeneous system. Other technologies involved to find the zeta potential, are micro-electrophoresis, electrophoretic light scattering, and electro-acoustic phenomenon [83].

Five cell lines that are generally considered as a portal of entry or systemic cellular target for engineered nanoparticles, namely macrophage (RAW 264.7), epithelial cells (BEAs-2B), human micro endothelial cells (HMEC), hepatoma (HEPA-1) and pheochromocytoma (cells of neuroendocrine tumor of the medulla of the adrenal gland) (PC-12) and NH_2 labeled polystyrene (PS) nanospheres of 60nm diameter. The observations from this technique indicate that the nanomaterials are highly toxic to RAW 264.7, BEAs -2B but HEMC and HEPA-1, PC-12 are found to be resistant to particle injury. The death pathway in the case of RAW 264.7 follows caspases activation mode, and in BEAs cells cytotoxic response is necrotic. The deposition of the particles in the cytosol of other cells activate uptake of mitochondrial Ca^{++}, and it is possible to suppress cell death by using cyclosporine-A (CsA). Further, HEPA-1, HMEC, and PC-12 cells pick up the cationic particles, but no change in the degree of permeabilization, enhanced Ca^{++} flux, and damages to mitochondria are very distinct. These observations indicate the significance of the cell specificity uptake mechanism and the different pathways involved [83].

Positively charged nanoparticles and microparticles internalize readily at a faster rate while negatively charged nanoparticles are not. There are cases when negatively charged particles like carboxyl coated quantum dots (QDs- 655-COOH) internalize relatively readily in comparison to those that are positively charged polyethylene glycol (PEG) amine coated PEG QDs [84, and 85]. Negatively charged, neutral particles and microparticles remain in blood circulation for more time in comparison to positively charged nanoparticles. Thus, one may conclude that those nanoparticles which remain in blood circulation for longer duration appear to be suitable candidates as delivery agents [86, 87].

Influence of Composition of Nanomaterials on their Behavior and Interaction

The composition of nanomaterials plays a significant functional role during their interaction. There is an optimal structural aspect of every nanomaterial that reflects on its reactivity, their adverse and beneficial impacts, *etc*. Generally, if the optimal structural aspect changes intentionally or due to the interaction with biomolecules; there is a possibility of different behavior of nanomaterials which can be toxic or derogative or neutral. The optimal structural aspect is essential in the case of engineered nanomaterials. During the formulation and fabrication, some specific organic chemicals or moieties involve enhancing surface reactivity of nanomaterials. These are the probable sites of responsiveness and the formation of reactive oxygen species.

Quantum dots (QD) find applications in biological and biomedical fields. The different types of coatings are added to QDs to carry out the specifically assigned function. Optimal functional status concerning the optimal structural aspect of the nanomaterial must be envisaged while it is being fabricated or synthesized. The surface composition of nanomaterials, such as discontinuous crystal planes in the case of nanocrystals and material defects in the case of nanoparticles, *etc*., are the probable sites of the generation of reactive species of oxygen. Quantum dots have the potential to cause health hazards these incorporate heavy metals like cadmium and coated or enclosed in nanoshell. Otherwise, quantum dots in their pristine form are relatively less toxic and partly biotolerant. Very often, nanoparticles are coated with polymeric organic molecules to enhance their reactive effectiveness. The poly-lactic-co-glycolic acid nanoparticle fluorescent polystyrene nanoparticles of standard size coated with polyvinyl alcohol and Vit E, use as polymeric nanoparticles. The nanoparticles coated with Vit-E, exhibit 1.4 times higher uptake in comparison to those overlay with polyvinyl alcohol. This uptake is 4 to 6 times higher in contrast to non-coated nanoparticles [69]. This aspect reflects on the significance of appropriate overlay on quantum dots and their applications in biomedical, biomolecular, and biotechnological fields.

Influence of Optical Properties of Nanomaterials on their Interactive Behavior

In general, optical properties show a close relationship with the electrical and electronic properties of the materials. Optical properties are the reflections of the interactivities between electromagnetic radiations and matter. An incident electromagnetic ray is either reflected, transmitted, refracted, or absorbed. The incident ray, in all probabilities, may get, either scattered or absorbed. The scattering causes relatively less effect, but the absorption of incident rays activates processes such as electronic, vibrational, rotational, and transitional effects. There may be re-emission of absorbed energy. When the absorbed energy may dissipate in heat form, the process is dissipative absorption. The electron transition causes changes in atomic orbitals, molecular orbitals, or change in the band (energy range and bandgap, *i.e.*, energy difference). When electromagnetic ray or wave falls on a matter having atomic spices, it induces an electron cloud oscillation, and in turn, creates dipole oscillations and emits radiation that is having a similar frequency in all directions. The process of scattering and absorption are frequency inter-dependent [88]. Origin of color is the result of different mechanisms like absorption, emission, reflection, transmission, scattering (blue sky), dispersion (prism), interference (oil films on water), wings of a butterfly, *etc*. The readable surface on a compact disc has a spiral track wound tightly so that it diffracts incident light within a full visible spectrum; it is an example of an optical impact. The color-flame test of sodium – a golden or yellow-colored flame affirms the atomic and electronic transition [88]. A plasmon is explicitly formed during the interaction between electromagnetic radiation and metal involving absorption.

Incident visible electromagnetic radiations cause plasmon oscillation, and because of these, the incident radiations absorb while bulk metals behave like opaque materials in the visible region. Metals like gold, platinum, aluminum, and silver when interacting with electromagnetic radiations exhibit frequency regions such as low frequency ($\omega\tau < 1$) – beam penetrate metals for short-depth (δ) skin deep below the surface; high frequency ($\omega\tau \gg 1$) – visible and ultraviolet range (reflection $\omega < \omega p$, plasma frequency) and $\omega < \omega p1$. In this case, metal becomes or behaves like non-absorbent – transparent dielectric matter, (τ is the time between two consecutive collisions, ($\tau \sim 10^{-}14$ s) [88].

When ionic crystals interplay with optical phonon (light) they exhibit absorption and reflection in the infrared region. Semiconductor compounds such as gallium arsenate have partial ionic features in their bond and also show absorption and reflection in the infrared region. When the energy of an incident photon is higher than the bandgap, then it results in the absorption of a photon. An exciton formed when material and it can shift energy without shifting the charge. The excited state is capable of moving through the lattice without shifting the charge. The impurities in the matter bind the exciton. Exciton is a contiguous state of one electron and one electron-hole, these are attracted to each other by electrostatic and Coulomb forces, because of the presence of an electrically neutrally quasiparticle in insulators. When a semiconductor absorbs a photon, it results in an exciton. Bulk metals absorb a visible range of electromagnetic radiation. The thin metal films can transmit partially because, in the thin metal film, there is insufficient material to absorb the radiation. Gold film (10s of nm) thick is partially transparent. There exists a dominance of surface plasmon, quantum confinement effect, in addition to insufficient material among nanomaterials. Optical absorption and optical emission shift to higher energy when the size of the quantum dots declines. The process of decline is relatively more prominent in the case of semiconductors in comparison to metals. In the nanosized scale, metal nanoparticles are likely to develop a bandgap, *i.e.*, semiconductors or insulators cause. When size decreases, the electron gets restricted to the process of showing confinement effect and leads to an increase in bond level, and that gets quantized. The energy level spacing enhances when the dimension decreases [88]. [*Plasmon is a collective longitudinal excitation of quantized conductive electron oscillation in gas. A plasmon is a quantum of plasma oscillation. Plasmons are like quantized microscopic vibration in solid. Plasma is a state of matter similar to gas, but the atomic particles are charged rather not neutral. The plasmons are the collective oscillation of free electrons, and bulk plasmons are a longitudinal oscillation of gas electron].

Influence of Magnetic Properties of Nanomaterials During their Interactive Behavior

The materials are grouped as ferromagnetic, paramagnetic, and diamagnetic depending on their degree of magnetization and magnetic susceptibilities. Ferromagnetic materials show high vulnerability towards the magnetic field, *e.g.*, iron, cobalt, and nickel; diamagnetic materials have weak negative sensitivity towards magnetic field *e.g.*, copper, gold, and silver; the paramagnetic materials show attraction and magnetic properties under the influence of the external magnetic field. However, in the absence of an external magnetic field, these properties are not functional, *e.g.*, magnesium, molybdenum, lithium, and tantalum. Primarily, the paramagnetic properties of the material are dependent on the presence of some unpaired electrons. These electrons realign along with the path set due to the external magnetic field. When the paramagnetic materials are under the influence of the uniform magnetic field, these materials get adjusted in such a manner so that their longest axis lies in the direction of the magnetic field while short-axis lies perpendicular to the longest axis. Under the influence of the non-uniform magnetic field, the electrons move to the stronger zone from the weaker zone. Paramagnetic materials exhibit better magnetic permeability because these substances allow magnetic lines of force to pass across them so that the magnetic induction is within the paramagnetic system. These materials are affected weakly (not forcefully) by the external magnetic field, indicating that their susceptibility is lower or less but positive. These particles or materials follow Curie's law. Curie's law states that the magnetization of the paramagnetic materials is directly proportional to the magnetic field applied, and magnetization is inversely proportional to the temperature of the material under investigation.

Superparamagnetic nanoparticles have around 50 nm in size and single magnetic domain, exhibit superparamagnetism. These materials exhibit magnetic susceptibility like ferromagnetic materials under the influence of the external magnetic field but in its absence. The behavior of superparamagnetic material depends on their ability to share the magnetic properties of ferromagnetic materials and paramagnetic materials. The superparamagnetic materials exhibit a short relaxation time with time range between 10^{-9} to 10^{-10} s. During this time, their degree of magnetization reduces to zero. The magnetic moment of superparamagnetic substances gets oriented or disoriented when the external magnetic field is applied and removed alternately. There is a loss and regain of the magnetic domain during the change of frequency. The loss of energy causes a rise in the temperature of the ambient environment. This feature is considered as magnetic susceptibility and is very high in superparamagnetic nanoparticles. This unique feature of nanoparticles made from superparamagnetic substances plays a significant role in their applications in the biomedical fields like the repair of tissue by local healing, detoxification of biological fluids. These superparamagnetic nanoparticles carry drugs, biomolecules, and genes, and deliver them in a controlled manner magnetically. These superparamagnetic nanoparticles act as contrast agents as these are preferred agents to label the cells under study specifically during magnetic resonance imaging [89]. Metallic magnetic nanoparticles are in use for technical purposes because of their higher magnetic moment, and in the biosystem act as phosphoric agents. These also act as oxidizing agents at least up to some extent. Thus, this feature makes them unsuitable for use in the biomedical field even in a colloidal form otherwise;

oxides of iron such as maghemite and magnetite are beneficial in the biomedical field [89].

The physical and chemical properties of magnetic nanoparticles primarily depend on the method of formulating and their chemical structure. Particles within the range of 1-100 nm in size are likely to exhibit superparamagnetic features. Magnetic nanomaterial/particles have magnetic elements such as Fe, Ni, Co, and their chemical compounds/constituents. The nanomagnetic materials are within the size range of 5-500 nm. Magnetic materials within a 0.5-500 micrometer size range are considered to be magnetic nanobeads or magnetic nanoparticle clusters. Superparamagnetic oxide nanoparticles, like ferrite nanoparticle clusters (around 80 superparamagnetic nanoparticles) beads and coated with a silica shell, exhibit specific features, namely, (i) a higher degree of chemical stability, (ii) a restricted size distribution, (iii) a higher degree of colloidal stability because these do not agglomerate magnetically, (iv) their magnetic moment can be regulated for nanoparticle cluster size, and, (v) ability to retain their superparamagnetic nature. Ferrite particles have a size having less than 128 nm, do not get self-agglomerate, and act as superparamagnetic particles. These particles show magnetic behavior in the presence of an external magnetic field [90]. The surface of ferrite particles can be modified using silica, silicone phosphoric acid, and surfactants. The coatings of suitable biocompatible materials on superparamagnetic nanoparticles, and, small core-shell with the conformation of SPION, and coating of sodium oleate, enhance their functional and analytic efficacy. The magnetic dipole-dipole interaction varies per the experimental conditions. SPION coated with polymeric starch matrix shows the primary agglomerated increase in size up to 36 nm. Methoxy-poly-ethylene-glycol coated SPION exhibit an increase in cluster size up to 120 nm. This feature enhances the degree of application of SPION in the biomedical field for imaging and functional investigations, like magnetic resonance imaging, tissue engineering, drug-delivering, *etc* [91]. The number of magnetic nanoparticle clusters is an important parameter, and one has to manipulate according to the suitability of these magnetic nanoparticles in biomedical applications. In single nanocluster, about 67 magnetic nanoparticles are packed, and these depend on the influence of the magnetic interactions within the nanoparticles present in the core component. The annealing technique is useful to attain this mode of functionality. Further, annealing does not alter the size and phase of nanomaterials/nanoparticles. Thus, it is difficult to control the clustering of ferric nanoparticles; their magnetic moment either increases or decreases and is converted into magnetic nanobeads [66, 92].

Generally, oxide of iron (ferrites) like maghemite and magnetite are inert and do not participate in the molecular interactions, because these do not form a covalent bond with most of the functionalized molecules. When metallic magnetic nanoparticles having metallic core are treated for gentle oxidation and after treating them with surfactants, polymer and precious metals are conveniently exploited in the biomedical field, for example under oxygen environment Co nanoparticle form anti-ferromagnetic layer of CoO [90]. Magnetic nanoparticles coated with this layer of silica shell having many functional groups involving covalent bonds exhibit more reactivity. The covalent bonds establish between the organic-silane and silica shell during oxidative interactions [79]. A coat of thin layer of silica increases the degree of reactivity of the surface of iron oxide magnetic nanoparticles. A technique involving hydrolysis and polycondensation of tetraethyl orthosilicate in the presence of an alkaline catalyst (NH_3 or KOH) and ethanol solution to form a stable suspension of citric acid-coated nanoparticles is suitable for this purpose. This model induces the formation of heterogeneous nucleation of the silica layer with less than 2 nm thickness [79, 90].

Further, the reduced thickness in layer forms a homogeneous monolayer of –Si, –OH, on to the nanoparticles and it enhances the degree of surface reactivity, but the ability of magnetization reduces by 20%. One may graft the (3-aminopropyl) triethoxysilane on to the surface of nanoparticles to increase their interaction with amino acids group/nm_2. Superparamagnetic nanoparticles linked with biomolecules activate the surface of such nanoparticles. These are very useful as detection tools for biomedical and biotechnological fields such as magnetic separation for biomolecules and cells, multiple detections based on biosensors, to increase magnetic resonance imaging contrast, as agents for magnetofection (a technique used to separate nucleic acid), targeting a tissue involving magnetic nanoparticles, *etc*. The surface of nanoparticles has the potential to bind with different cargos like proteins, drugs, and fluorescent molecules, molecules that increase permeability, and biocompatibility. For such types of interactions, strong covalent bonds are required to set up between the molecules and the functionalized surface of nanoparticles and ensure the efficacy of a set targeted function. Chemically synthesized nanocrystals that have cobalt core, a shell of cobalt oxide, and onto this shell gold shell are suitably added to accomplish such interactions [93]. Highly magnetic metallic face-centered cubic cobalt nanoparticles having 20-60 nm diameters are helpful in biomedical and biotechnological fields, *etc*. These fabricated nanomaterials get protection from oxidation because a 1 nm thick layer of cobalt oxide is present on them [93].

The impact of the ambient environment on the nanomaterials of magnetization relaxation is helpful to study their dynamic magnetic behavior concerning aggregation state. The Quasi-static magnetization curves do not exhibit degradation of the nanoparticles during their interaction within the biosystem on a specific time scale. The location, incubation time, type of cell line and coating on nanoparticles do not influence this behavior [94]. During the interactions between nanomaterials and the biomolecules within the biosystem or otherwise their respective ambient and microenvironment, mechanisms involved and their physicochemical properties are very significant. There are possibilities of a backlash of their applications due to the controlled modifications, and functionalization of nanomaterials [95].

CONCLUSION

The physicochemical properties of nanomaterials make them useful options for investigating catalytic and biological activities. They exhibit enhanced electrical conductivity, specifically in association with ceramics, and, magnetic composites, and. elevated electric resistance in metals. Mechanical hardness, the toughness of metals and alloys can be made relatively better as per the proposed target. The elasticity and superplasticity of ceramics can also be improved. These features make nanomaterials suitable for the applications in instrumentation, biomedical and biotechnological fields. Magnetic traits are modified to bring about superparamagnetic behavior and enhanced magnetic coercivity concerning the specific size. Coercivity is the intensity of the magnetic field required to decline the magnetization of a ferromagnetic material to zero after it has reached saturation or its ability of a ferromagnetic material to resist any change in the magnetization. Optical features affect the optical absorption, fluorescence frequencies, and elevate the degree of conductor crystals. These features promote permeability, the ability to move across the biological barriers, and maintain suitable biocompatibility. Further, these nanomaterials can bring sterical changes either in nanomaterials or in biomolecules to ensure their transportation within the cell, tissue, or biological systems. Thus, these minute particles help to understand the various mechanisms involved in natural living phenomena, environmental complexities, and above all, to understand the bio-physicochemical cycles occurring between abiotic and biotic components of the environment.

REFERENCES

[1] Albanese A, Tang PS, Chan WC. The effect of nanoparticle size, shape, and surface chemistry on biological systems. Annu Rev Biomed Eng 2012; 14(1): 1-16.
[http://dx.doi.org/10.1146/annurev-bioeng-071811-150124] [PMID: 22524388]

[2] Rahi A, Sattarahmdy N, Heli H. Toxicity of nanomaterials –physicochemical effects, Austin Journal of Nanomedicine & Nanotechnology, 2014; 2(6): 1034.

[3] Roberts SM, James RC, Williams PL. Principles of toxicology – environmental and industrial applications. 3rd ed., New Jersey: John Wiley and Sons, Inc 2015.

[4] Filipponi L, Daima HK. nanoyou.eu/attachments/188_Module_1_chapter -4-proofread.pdf

[5] Leggett AJ. Macroscopic quantum system and the quantum theory of measurement. Prog Theor Phys Suppl 1980; 69: 80-100.
[http://dx.doi.org/10.1143/PTPS.69.80]

[6] Linus AJ, Edgar BW. Introduction to the quantum mechanism with an application to chemistry.

Courier Corporation 1985.: 978048664712.

[7]   Taylor J. Modern Physics for Scientists and Engineers Prentice Hall. 2004. ISBN: 978-0-13-8057 1502.

[8]   Davies PCW. Quantum tunneling time. Am J Phys 2005; 73(1): 23.
        [http://dx.doi.org/10.1119/1.181.0153]

[9]   Panitchayangkoon G, Hayes D, Fransted KA, *et al.* Long-lived quantum coherence in photosynthetic complexes at physiological temperature. Proc Natl Acad Sci USA 2010; 107(29): 12766-70.
        [http://dx.doi.org/10.1073/pnas.1005484107] [PMID: 20615985]

[10]   Trixler F. Quantum tunneling to the origin and evolution of life. Curr Org Chem 2013; 17(16): 1758-70.
        [http://dx.doi.org/10.2174/13852728113179990083] [PMID: 24039543]

[11]   Brookes JC. Quantum dots effects in biology: golden rule in enzymes, olfaction, and photosynthesis and magnetic detection, Proceedings of the Royal Society, A Mathematical Physical and Engineering Science, 2013; 473(2201).

[12]   Matta CF. Quantum biochemistry: Electronic structure and biological activity. Weinheim: Wiley-VCH 2014.

[13]   Cooper WG. Roles of evolution, quantum mechanics and point mutations in origins of cancer. Cancer Biochem Biophys 1993; 13(3): 147-70.
        [PMID: 8111728]

[14]   Chen J, Zhang D, Zhang Y, Li G. Computational studies of difference in binding modes of peptide and non-peptide inhibitors to MDM2/MDMX based on molecular dynamics simulations. Int J Mol Sci 2012; 13(2): 2176-95.
        [http://dx.doi.org/10.3390/ijms13022176] [PMID: 22408446]

[15]   Margenau H, Kestner N. Theory of intermolecular forces, International Series of Monograph in Natural Philosophy. Pergamon Press 1969.

[16]   Nelson DL, Cox MM, Michael M. Lehninger Principles of Biochemistry. 6th ed., New York: W H Freeman & Co 2013.

[17]   Lindh U. Biological functions of the elements.Solinus, Olle, Essential of Medical geology. Revised Ed., Dordrecht: Springer 2013.
        [http://dx.doi.org/10.1007/978-94-007-4375-5_7]

[18]   Yu DP, Zalgaller VA, Kudryavtsav LD. Area.Hazewinket M, Encyclopedia of Mathematics. Springer Scientist Business Media B V/Kluwer Academic Publishers 2001. (Original work published in 1994)

[19]   Tanford C. The hydrophobic effect. New York : Wiley 1980.

[20]   Atkins P, Paula J. Physical Chemistry. Oxford Press 2006.

[21]   Baalousha M, Lead JR. Frontiers of Nanoscience; Characterization of nanomaterials in complex environment and Biological media. 8; 2015.

[22]   Bradley F. Material Chemistry, Springer Science +Business Media B V. Dordrecht, Heidelberg, London, New York: Springer 2011; p. 476.

[23]   Zhen G, Tan L. Fundamentals and applications of nanomaterials. Boston, London: Artech House 2009.

[24]   Langel W. Computer simulation of the surface.Handbook of theoretical and computational nanotechnology. Forschungszentrum Karlsruhe, Germany 2005; Vol. 1: pp. 1-54.

[25]   Seller K, Mackay C, Bergson LL, *et al.* Nanotechnology and the environment.

[26]   Jiang J, Oberdorfer G, Biswas P. Characterization of size, surface charge, and agglomeration state of nanoparticles dispersion for toxicological studies. J Nanopart Res 2009; 11: 77-89.
        [http://dx.doi.org/10.1007/s11051-008-9446-4]

[27]   London F. The general theory of molecular. Trans Faraday Soc 1937; 33: 8b-26.
         [http://dx.doi.org/10.1039/tf937330008b]

[28]   Hamaker HC. The London-van der Waals attraction between spherical particles. Physica 1937; 4: 1058-72.
         [http://dx.doi.org/10.1016/S0031-8914(37)80203-7]

[29]   Derjaguin BV. A theory of heterocoagulation: interaction, adhesion, of dissimilar particles in the solution of electrolytes. Discuss Faraday Soc 1954; 85-98.
         [http://dx.doi.org/10.1039/df9541800085]

[30]   Visser J. On Hamaker constants: A comparison between Hamaker constant and Lifshitz-van der Waals constant. Adv Coll Inter 1972; 3: 331-63.
         [http://dx.doi.org/10.1016/0001-8686(72)85001-2]

[31]   Fowkes FM. Surface, and interfaces. New York: Syracuse University Press 1967; 1: p. 199.

[32]   van Oss CJ, Omenyi SN, Neumann AW. Negative Hamaker coefficient Li-phase separation of the polymer solution. Colloid Polym Sci 1979; 257: 737-44.
         [http://dx.doi.org/10.1007/BF01474103]

[33]   Leite FL, Bueno CC, Da Róz AL, Ziemath EC, Oliveira ON. Theoretical models for surface forces and adhesion and their measurement using atomic force microscopy. Int J Mol Sci 2012; 13(10): 12773-856.www.mdpi.com/journal/ijms
         [http://dx.doi.org/10.3390/ijms131012773] [PMID: 23202925]

[34]   Rudyak V. Ya, Krasnolutskii SL, Dependence of the viscosity of nanofluids on nanoparticles size and nanomaterials. Phys Lett A 2014; 378(26-27): 1845-9.
         [http://dx.doi.org/10.1016/j.physleta.2014.04.060]

[35]   Bruch LW. Evaluation of van der Waals forces for atomic force microscopy Phys Rev–b, 2005; 72
         [http://dx.doi.org/10.1103/PhysRevB.72.033410]

[36]   Leite FL, Borato CE, da Silva WT, Herrmann PS, Oliveira ON, Mattoso LH. Atomic force spectroscopy on poly(o-ethoxyaniline) nanostructured films: sensing nonspecific interactions. Microsc Microanal 2007; 13(4): 304-12.
         [http://dx.doi.org/10.1017/S1431927607070262] [PMID: 17637080]

[37]   Nicolosi V, Vengust PD, Sanvito S, *et al.* Observations on van der Waals drove the self-assembly of most nanowires in too low symmetry structure using aberration-corrected electron microscopy, Advanced Mater, . 2007; 19: pp. 543-7. V A, (1970).

[38]   French RH, Pasegian VA, Podgornik R, *et al.* Long-range interactions in nanoscale science. Rev Mod Phys 2010; 82: 1887-944.

[39]   Elbaum M, Schick M. Application of the theory of dispersion forces to the surface melting of ice. Phys Rev Lett 1991; 66(13): 1713-6.
         [http://dx.doi.org/10.1103/PhysRevLett.66.1713] [PMID: 10043288]

[40]   Ninham BW, Parsegian VA. van der Waals forces across triple layer films. J Chem Phys 1970; 52: 4578-87.
         [http://dx.doi.org/10.1063/1.1673689]

[41]   Adhesion forces between surfaces in liquids and condensable vapors, Surface Science Research 1992; 14: 109- 159.

[42]   Elbaum M, Lipson SG. How does a thin wetted film dry up? Phys Rev Lett 1994; 72(22): 3562-5.
         [http://dx.doi.org/10.1103/PhysRevLett.72.3562] [PMID: 10056231]

[43]   Hodges CS. Measuring forces with the AFM: polymeric surfaces in liquids. Adv Colloid Interface Sci 2002; 99(1): 13-75.
         [http://dx.doi.org/10.1016/S0001-8686(02)00003-9] [PMID: 12405400]

[44]   Lubrasky GV, Mitchell SA, Davidson MR, Bradley RH. van der Waals interactions in a system involving oxidized polystyrene surface, Colloids Surf A. Physiochem Eng Asp 2006; 279: 188-95.
[http://dx.doi.org/10.1016/j.colsurfa.2006.01.002]

[45]   Wang CX, Chang S, Gong XQ, Yang F, Li CH. Progress in scoring functions of protein- protein docking. Wuli Huaxue Xuebao 2012; 28: 751-8.
[http://dx.doi.org/10.3866/PKU.WHXB201202022]

[46]   Gibbs JW. A method of geometrical representation of the Thermodynamics properties of substances by means of surface. Trans Conn Acad Arts Sci. 1873; pp. 382-404.p-400.

[47]   Sun J, Wang F, Sui Y, *et al.* Effects of particle size on solubility, dissolution rate and oral bioavailability: evaluation using coenzyme Q10. Int J Nanomedicine 2012; 7: 5733-44.
[http://dx.doi.org/10.2147/IHN.S34365] [PMID: 23166438]

[48]   Hu J, Johnston KP, Williams RO III. Nanoparticle engineering processes for enhancing the dissolution rates of poorly water soluble drugs. Drug Dev Ind Pharm 2004; 30(3): 233-45.
[http://dx.doi.org/10.1081/DDC-120030422] [PMID: 15109023]

[49]   Shan Y, Ma S, Nie L, *et al.* Size-dependent endocytosis of single gold nanoparticles. Chem Commun (Camb) 2011; 47(28): 8091-3.
[http://dx.doi.org/10.1039/c1cc11453k] [PMID: 21687845]

[50]   Huang K, Ma H, Liu J, *et al.* Size-dependent localization and penetration of ultra-small Au NPs in cancer cells, multicellular spheroids and tumor cells *in vivo.* ACS Nano 2012; 6: 4483-93.
[http://dx.doi.org/10.1021/nn301282m] [PMID: 22540892]

[51]   Yu P, Sabin N, Leifert A, *et al.* Size-dependent cytotoxicity of AU nanoparticles. Small 2007; 3(11): 1947-9.
[http://dx.doi.org/10.1002/smll.200700378]

[52]   Hussain SM, Hess K L, Gearhart JM, Geiss KT, Schlager JJ. In vitro toxicity of nanoparticles in BRL 3A rat liver cell, Toxicity in vitro, 2005; 19(7): 975-83.
[http://dx.doi.org/10.1016/j.tiv.2005.06.034]

[53]   Jiang W, Kim BYS, Rutka JT, Chan WCW. Nanoparticle-mediated cellular response is size-dependent. Nat Nanotechnol 2008; 3(3): 145-50.
[http://dx.doi.org/10.1038/nnano.2008.30] [PMID: 18654486]

[54]   Wang J, Betty YS. Kim, Rutka JT, Chan WCW, Nanoparticles mediated cellular response in size-dependent. Nat Nanotechnol 2008; 3: 145-50.
[http://dx.doi.org/10.1038/nnano.2008.30] [PMID: 18654486]

[55]   Lu F, Wu SH, Hung Y, Mou CY. Size effect on cell uptake in well-suspended, uniform mesoporous silica nanoparticles. Small 2009; 5(12): 1408-13.
[http://dx.doi.org/10.1002/smll.200900005] [PMID: 19296554]

[56]   Huang J, Bu L, Xie J, *et al.* Effects of NPs size on cellular uptake and liver MRT with polyvinyl pyrrolidone coated iron-oxide NPs. ACS Nano 2010; 4(12): 7151-60.
[http://dx.doi.org/10.1021/nn101643u] [PMID: 21043459]

[57]   Andersson PO, Lejon C, Ekstrand-Hammarström B, *et al.* Polymorph- and size-dependent uptake and toxicity of TiO$_2$ nanoparticles in living lung epithelial cells. Small 2011; 7(4): 514-23.
[http://dx.doi.org/10.1002/smll.201001832] [PMID: 21265017]

[58]   Oh E, Delehanty JB, Sapsford KE, *et al.* Cellular uptake and fate of PEGylated gold nanoparticles is dependent on both cell-penetration peptides and particle size. ACS Nano 2011; 5(8): 6434-48.
[http://dx.doi.org/10.1021/nn201624c] [PMID: 21774456]

[59]   dos Santos T, Varela J, Lynch I, Salvati A, Dawson KA. Quantitative assessment of the comparative nanoparticle-uptake efficiency of a range of cell lines. Small 2011; 7(23): 3341-9.
[http://dx.doi.org/10.1002/smll.201101076] [PMID: 22009913]

[60]   Schubbe S, Schumann C, Cavolius C, *et al.* Size-dependent location and quantitative evaluation of the intracellular migration of silica NPs in Caco-2 cells. Chem Mater 2012; 24(5): 914-23.
        [http://dx.doi.org/10.1021/cm2018532]

[61]   Dong C, Irudayaraj J. Hydrodynamic size-dependent cellular uptake of aqueous QDs probed by fluorescence correlation of spectroscopy. J Phys Chem. 2016; 116: pp. 12-132.
        [http://dx.doi.org/10.1021/jp305563p]

[62]   Xu A, Yao M, Xu G, *et al.* A physical model for the size-dependent cellular uptake of nanoparticles modified with cationic surfactants. Int J Nanomedicine 2012; 7: 3547-54.
        [http://dx.doi.org/10.2147/IN.S32188] [PMID: 22848178]

[63]   Benedict JW, Desreno M. Membrane composition-mediated protein-protein interaction. Biointerphase 2008; 3: FA117.
        [http://dx.doi.org/10.1116/1.2977492]

[64]   Hamilton RF, *et al.* Length-dependent TiO_2 nanomaterials, Toxicity, and bioactivity 2009; 6: 35.
        [http://dx.doi.org/10.1186/1743-8977-6-36]

[65]   Conde J, Doria G, Baptista P. Noble metal NPs application in cancer. J Drug Del 2012; 12p.
        [http://dx.doi.org/http://dx.doi.org/10.1155/2012/751075] [PMID: 751075]

[66]   Liyana AW. Size-dependent of metal NPs with fluorophore and semiconductors Dissertations (PhD) 2014;   235   http://scholarworks.wmich.ed/dissertation/235
        http://scholarworks.wmich.ed/dissertation/235

[67]   Gossmann R, Langer K, Mulac D. New perspective I the formulation and characterization of didodecyl dimethyl ammonium bromide (DMAB) stabilized poly (lactic-co- glycolic acid) (PLGA) NPs. PLoS One 2015; 10(7): e0127532.
        [http://dx.doi.org/10.1371/journal.pone.0127532] [PMID: 26147338]

[68]   Chithrani BD, Ghazani AA, Chan WC. Determining the size and shape dependence of gold nanoparticle uptake into mammalian cells. Nano Lett 2006; 6(4): 662-8.
        [http://dx.doi.org/10.1021/nl052396o] [PMID: 16608261]

[69]   Win KY, Feng SS. Effects of particle size and surface coating on cellular uptake of polymeric nanoparticles for oral delivery of anticancer drugs. Biomaterials 2005; 26(15): 2713-22.
        [http://dx.doi.org/10.1016/j.biomaterials.2004.07.050] [PMID: 15585275]

[70]   Champion JA, Mitragotri S. Role of target geometry in phagocytosis. Proc Natl Acad Sci USA 2006; 103(13): 4930-4.
        [http://dx.doi.org/10.1073/pnas.0600997103] [PMID: 16549762]

[71]   Ispas C, Andreescu D, Patel A, Goia DV, Andreescu S, Wallace KN. Toxicity and developmental defects of different sizes and shape nickel nanoparticles in zebrafish. Environ Sci Technol 2009; 43(16): 6349-56.
        [http://dx.doi.org/10.1021/es9010543] [PMID: 19746736]

[72]   Hu W, Peng C, Lv M, *et al.* Protein corona-mediated mitigation of cytotoxicity of graphene oxide. ACS Nano 2011; 5(5): 3693-700.
        [http://dx.doi.org/10.1021/nn200021j] [PMID: 21500856]

[73]   Gurr JR, Wang AS, Chen CH, Jan KY. Ultrafine titanium dioxide particles in the absence of photoactivation can induce oxidative damage to human bronchial epithelial cells. Toxicology 2005; 213(1-2): 66-73.
        [http://dx.doi.org/10.1016/j.tox.2005.05.007] [PMID: 15970370]

[74]   Petkovic J, Zequra B, Stefanovic M, *et al.* DNA damage and alteration in expression of DNA damage responsive gene introduced by TiO_2 in human hepatoma HepG2-cells. Nanotoxicology 2015; 5(3): 341-53.
        [http://dx.doi.org/10.3109/17435390.2010.507316] [PMID: 21067279]

[75]   George S, Lin S, Thomas CR, *et al.* Surface defects on plate-shaped silver nanoparticles contribute to its hazard potential in a fish gill cell line and Zebra fish embryos. ASC Nano 2012; 6(5): 3745-59..
[http://dx.doi.org/10.1021/nm204671v]

[76]   Li X, Liu W, Sun L, *et al.* Effects of physicochemical properties of nanomaterials on their toxicity. J Biomed Mater Res Part A 2014; 103(7): 2499-507.
[http://dx.doi.org/10.1002/jbm.a.35384]

[77]   Gorka DE, Osterberg JS, Gwin CA, *et al.* Reducing environment toxicity of Ag NPs through shape control. Envir Sci Techn 2015; 49(16): 10093-8.
[http://dx.doi.org/ 10.1021/acs.est.5 b01711]

[78]   Dong KY, Georgia CP. Nanomaterials, development, and applications, CRC Press. Boca Raton, London, New York: Taylor and Francis Group 2014.ISBN- 13:978-1-4398-7641-1.

[79]   Slavko K, Darko M, Stanislov C, Miha D. Producing ultrathin silica coating on iron-oxide nanoparticles to improve their surface reactivity. Journal of Magnetism and Magnetic Materials 2010 ; 322(12): 1847-53.
[http://dx.doi.org/10.1016/j.jmmm,2009.12.038]

[80]   Connor EE, Mwamuka J, Gole A, Murphy CJ, Wyatt MD. Gold nanoparticles are taken up by human cells but do not cause acute cytotoxicity. Small 2005; 1(3): 325-7.
[http://dx.doi.org/10.1002/smll.200400093] [PMID: 17193451]

[81]   Slavko K, Miha D, Darko M. Controlled surface functionalization of silica-coated magnetically nanoparticles with terminal amino and carboxyl groups. J Nanopart Res 2011; 13(7): 2829-41.
[http://dx.doi.org/10.1007/s11051-010-0171-4]

[82]   Frohlich E. Role of surface charge in cellular uptake and cytotoxicity on medical nanoparticles. Int J Nanomedicine 2012; 7: 5577-91.
[http://dx.doi.org/10.2147/IJN.s.36111]

[83]   Xia T, Kovochich M, Liong M, Zink JI, Nel AE. Cationic polystyrene nanosphere toxicity depends on cell-specific endocytic and mitochondrial injury pathways. ACS Nano 2008; 2(1): 85-96.
[http://dx.doi.org/10.1021/nn700256c] [PMID: 19206551]

[84]   Zhang LW, Montiero-Riviere NA. Mechanism of QDs nanoparticles cellular uptake. Toxicology Science 2009; 110: 138-55. http://refhub.elsevier.com/S0048-9697 (16)31150-0/rf0400

[85]   Jiang, *et al.* Specific effects of surface amines on polystyrene nanoparticles in their interaction with mesenchymal cells. Biomacromolecules 2010; 11(3): 748-53. http://refhub.elsevier%2Ccom/ S0048.9697(16)31150-0-/rf%200165

[86]   Kim B, Han G, Toley BJ, Kim CK, Rotello VM, Forbes NS. Tuning payload delivery in tumour cylindroids using gold nanoparticles. Nat Nanotechnol 2010; 5(6): 465-72.http://refhub.elsevier.com/S0048-9697(16)31150-0/rf0175
[http://dx.doi.org/10.1038/nnano.2010.58] [PMID: 20383126]

[87]   Feliu N, Hühn J, Zyuzin MV, *et al.* Quantitative uptake of colloidal particles by cell cultures. Sci Total Environ 2016; 568: 819-28.
[http://dx.doi.org/10.1016/j.scitotenv.2016.05.213] [PMID: 27306826]

[88]   Zhang JH. Optical properties of nanomaterials 2009.home.iitk.ac.in/anandh/MSE694/course MSE694/NPTEL-optical%20properies

[89]   Roman Silvia. mappingnorance.org/2014/03/05/superparamahnetic-nanoparticles-and-the- separation-problem/#comments2014.

[90]   Lu A-H, Salabas EL, Schüth F. Magnetic nanoparticles: synthesis, protection, functionalization, and application. Angew Chem Int Ed Engl 2007; 46(8): 1222-44.
[http://dx.doi.org/10.1002/anie.200602866] [PMID: 17278160]

[91]   Kim DK, Mikhaylova M, Gurrero G, Martin PH, Vioux A. Anchoring of phosphonate and phosphinate coupling molecules on Titania particles. Chem Mater 2003; 15(8): 1617-27.

[http://dx.doi.org/10.1021/cm021349j]

[92]   Marin T, Slavko K, Marko J, Darko H, Darko M. Magnetic properties of novel super- paramagnetic iron oxide, nanocluster and their peculiarity under annealing treatment. Appl Surf Sci 2014; 322: 255-64.
[http://dx.doi.org/10.1016/apsusc.2014,09.181]

[93]   Johnson SH, Johnson CL, May ST, Hirsch S, Cole MW, Spanier JE. Co @CoO@Au core- multi-shell nanocrystals. J Mater Chem 2010; 20(3): 439-43.
[http://dx.doi.org/10.1016/j.apsusc.2014.09.181]

[94]   Fortes Brollo ME, Hernández Flores P, Gutiérrez L, Johansson C, Barber DF, Morales MDP. Magnetic properties of nanoparticles as a function of their spatial distribution on liposomes and cells. Phys Chem Chem Phys 2018; 20(26): 17829-38.
[http://dx.doi.org/10.1039/C8CP03016B] [PMID: 29923574]

[95]   Carlota A-S, Tabata N, Pabio J-V, *et al.* Interactions of nanoparticles and biosystems: Microenvironment of nanomaterials and biomolecules. Nanomaterials (Basel) 2019; 9: 1365.
[http://dx.doi.org/10.3390/nano9101365]

CHAPTER 5

Interaction Between Nanomaterials and Glycocalyx, Cell Membrane, Cytoskeleton, Cell Organelles and Tissues

Abstract: Biosystems are responsive to almost all types of stimuli. These stimuli are in the form of fluctuations in their internal and external environments. Abiota, biota, and nanomaterials are interactive components of the environment. These units exhibit a wide range of reactivity because of their respective physicochemical, biomolecular, biochemical, and biophysical features. Biosystem is a complex unit of biota and these are acellular, cellular, unicellular and multicellular structurally and functionally in nature. Cell being the structural and functional unit of the biosystem, is a well-organized structure exhibiting wide variety, nature, and functions that bring about the sustenance of the biosystem. Nanomaterials are some of the most desired novel materials to be used as agents to carry drugs, as a component of biomedical aids, diagnostic tools, biomedical imaging, *etc*. Inter-actions between nanomaterials and the biosystem are very ubiquitous and at the same time ambiguous. Most of the physicochemical properties of nanomaterials play significant roles and cause impacts on interacting materials. These materials are inorganic, organic, or living. Cellular uptake of nanoparticles is a common phenomenon and has a wide range of applications in the field of nanomedicine from cell tracking, cellular to molecular imaging, disease targeting, drug/gene delivery, diagnosis, and therapy. Nanoparticle-based diagnostic or therapeutic applications are mainly attributed to the various methods of functionalization and localization in the cellular and subcellular compartments. The pre-requisite for the nanoparticle-based therapeutic applications mainly involves the mechanisms of the uptake of the nanoparticles which also determines the fate of these nanoparticles for effective efficacy. Applications of nanomaterials are dependent on the regulated interaction between biota and abiota of the environment. The wide range of functionality of nanoparticles is because of their physicochemical properties, ability to get modified or formulated readily, as per the need, and the greatest flexibility among the adaptability of biota. The potential use of nanomaterials may be the cause of their derogative impacts on abiota and biota. It is one of the prime concerns during development, formulation, and applications in varied fields to provide insight into the interactions between nanomaterials and the biosystem. One must understand the intricacies of their interactions within biosystems.

Yogendrakumar H. Lahir & Pramod Avti

Keywords: Biocompatibility, Cellular-uptake, Cell membrane, Cytoskeleton, Cellular-uptake, Caveolae-mediated endocytosis, Clathrin-mediated endocytosis, Enhanced permeable and retention effect, Glycocalyx, Internationalization of nanoparticles, Phagocytosis, Proton sponge effect.

OVERVIEW – NANOMATERIALS AND BIOSYSTEMS

Interaction between nanomaterials and biosystems is very unique. Most of the physicochemical properties of nanomaterials have their impacts on interacting material may it be inorganic, organic, or living beings. Applications of nanomaterials are dependent on the regulated interaction between biota and abiota of the environment. Biosystems are basically very responsive to the external as well as internal stimulations and the fluctuations. Let us label abiota, biota, and the nanomaterials as the reactive components of the environment. Biological reactive components and the nanomaterials exhibit a wide range of reactivity because of their physicochemical, biomolecular, biochemical, and biophysical features. This overall complex phenomenon needs a thorough understanding to make the best of the situation. In the last few decades, nanoscience and nanotechnology have made enormous progress and made nanomaterials as most suitable options for almost all aspects of industries, food technology, agriculture, pharmaceutics, cosmetics, clothing/garment, military ware fare, chemical technology, *etc.*, the list seems to be endless. This range of functionality of nanoparticles is dedicated to their physicochemical properties, ability to get formulated and modified as per the need, and the highest degree of adaptability of biota. Increasing concern about the potential use of nanomaterials may be the cause of their derogative impacts on abiota and biota. To provide insight into the interactions between nanomaterials and biosystems, one must understand the intricacies of their interactions with biosystems.

Biosystem itself is a complex unit of biota. This unit is acellular, cellular, unicellular and multicellular structurally and functionally in nature. The cell is the structural and functional unit of the biosystem is a well-organized structure exhibiting wide variety, nature, and functions that bring about the sustenance of the biosystem. The cell is bounded by a cell membrane that is enveloped by glycocalyx externally and strengthened by the cytoskeleton internally. The component of the cell membrane is in communication with external as well as internal environment. Within the cell, cell-organelles are organized depending on the type of the cell involving endoplasmic reticulum, cytosol, cytoskeleton, and the ambient physiological fluid in and around each cell organelles and the cell itself. The interaction between nanomaterials and cells, tissue seems to be very crucial in nature.

Nanomaterials are some of the most desired novel materials to be used as agents to carry drugs, as a component of biomedical aids, diagnostic tools, biomedical imaging, *etc*. This reflects on the probable biocompatibility of varied nanomaterials to the biosystem. Thus nanomaterials should be biocompatible and should exhibit fairly good biodistribution. These aspects play major roles during designing and formulation of natural nanomaterials and also the engineered nanomaterials to accomplish the set target, may it be a cancerous cell, diseased tissue, and should not harm the normal or nontargeted tissues. Currently, there are many types of nanoparticles that are synthesized and being used for a variety of biological and biomedical applications. Among them to name a few are the iron oxide nanoparticles, gold and silver nanoparticles, quantum dots, polymeric nanoparticles, lipid-based nanoparticles, carbon-based nanoparticles, *etc*. Most of these nanoparticles are prepared either by a bottom-up approach or a top-down approach. The most important aspect of the synthesized nanoparticles is that they are immiscible in water and have very less water solubility due to which they cannot be used directly for the biological applications due to a variety of changing physiochemical properties such as their aggregation in the solution or at physiological pH which leads to toxicity, or aggregation based macromolecular formation losing their inherent properties for which they are synthesized. This ultimately leads to non-biocompatibility and cause toxicity when used for any biological applications. Therefore, the nascent synthesized nanoparticles cannot be directly used for the biological applications unless they are surface modified to make them more biocompatible. The surface of nanomaterials, polymer, and bulk materials are different. In the dry state, there is minimal surface energy. There is a shift of groups referred to as the group mobility. The non–polar groups move to the phase boundary formed with air while under aqueous conditions, the polar groups move to the phase of the boundary, *i.e.*, at the periphery. The surface modifications approaches include either covalent modification or non-covalent modification. In case of covalent modification either the small molecule carbohydrates, surfactants, proteins, DNA, RNA, lipids or any synthesized molecule is linked covalently to enhance their water solubility at the physiological conditions.

Compounds having low molecular weight move towards the phase boundary or away from it. As a result, there is a change in the properties of the materials under consideration. On administration of synthetic and engineered nanomaterials may face some of the conditions in the biosystem. The first component of the biosystem that comes in contact with the administered nanomaterials is its body fluids. The biochemical components like proteins, lipids, and related molecules present in these body fluids interact with biomaterials either physically or chemically. In blood, the interaction between nanobiomaterials and protein involves adsorption leading to the formation of a layer. This reflects on the hemocompatibility. But this interaction depends on the nature of the protein and the nature of the interacting surface properties of biomaterials. The biomaterials may be used implants (fixed/localized) in a specific organ, tissue, or as biosensor, either for a short or long duration. This biomaterial may face acute, intense, or chronic exposure. Some of the common nano-biomedical aids include orthopedic devices, dental applications, artificial pancreas, cardiovascular-related devices, or any nanomachines, *etc.* The physical, as well as the chemical nature of the biomaterial along with the stability of the interacting materials, locations/sites, and biophysical and biochemical features, play important roles in these interactions. The nanobiomaterials may face degradation or erosion due to the chemical or biochemical harshness like pH, changes in concentrations, enzymes, *etc.* Quite often, inflammation or changes in the cyto-homeostatic condition may occur.

The multifaceted utilities of nano-materials show some restrictions. The unique physicochemical features of nanomaterials are the causes of beneficial as well as derogative impacts on biosystems and the environment. These materials have dimensions similar to most of the cell organelles and the potential to interfere with structural aspects and vital functional aspects of the cell. This feature is acting as a fulcrum for their adverse or toxic impacts on the biosystem. Natural and engineered nanomaterials add to this challenge because of. During the experimentation stage during their formulation and fabrication, this challenge becomes difficult. Related computational studies can potentially minimize this challenge. Despite several advancements related to the safety of these materials, there are relatively limited reports on the clinical toxicity, indicating the probability of adverse biological interactions if not toxic, leading to an unwanted outcome. Studies are reporting on the dangerous effects of nanomaterials. Nanomaterials like metal and a metal oxide, quantum dots, fullerene, and fibrous nanomaterials cause derogative impacts like chromosomal fragmentation, DNA strand breaks, point mutation, oxidative DNA adduct and changes in gene expression. The nanomaterials and other materials can readily cross over almost all biological barriers [1]. All these potentials enhance the chances of toxicity or adverse impacts in the biosystem and its components. The current technology is relatively deficient in evaluating the degree of toxicity and the unconfirmed report related to toxicity. The toxicity of nanomaterials has become tangential studies in most of the fields. Fig. (**1**) represents overall bio-nano interactions and nanoengineering processes and applications.

NANOBIOINTERACTIONS AND NANO ENGINEERING PROCESSES AND APPLICATIONS				
SYNTHESIS & CHRACTERIZATION:	**EXPOSURE TO BIOLOGICAL SYSTEM:**	**MEASURED BIOLOGICAL RESPONSES:**	**DATA BASE AND SIMULATION:**	**ENGINEERING NANO DEVICES:**
Size, Shape, Surface Chemistry, Surface Reactivity, Aggregation/ Agglomeration	Mammals, Mammalian Cells, Blood and Blood Cells, Mice, Rats, Plants	Toxicity, ROS Generation, Biodistribution, Tumor Targeting, Clearance	Create Data Base, Modeling, Simulation	Diagnostics, Therapeutics, Optical Devices, Durg Carriers, Cosmatics

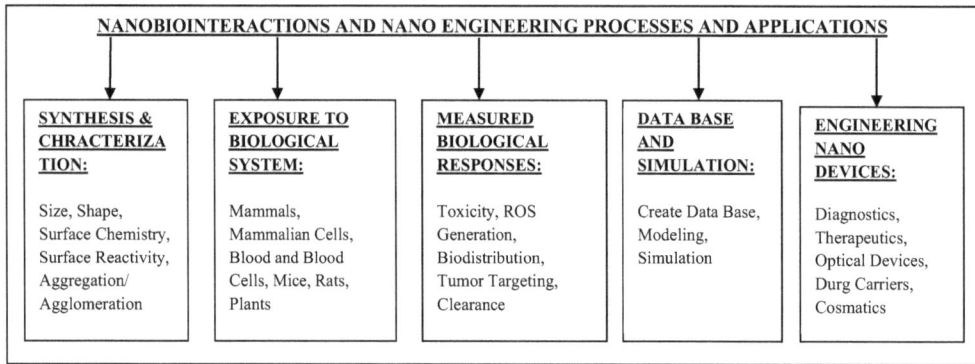

Fig. (1). Nanobiointeractions and nano engineering processes and applications.

The primary challenges concerning nanoscience, nanotechnology, and nanomaterials are to induce imaginative, innovative, precise, and safe technology for biota, and the environment. To meet these challenges, one needs (i) an appropriate, workable, economical instrument to evaluate environmental exposure because of nanomaterials. (ii) Simple, reliable, reproducible assessment of systemic toxicity of the nanomaterials. (iii) Choice of suitable experimental models to study the potential physical, biochemical, bimolecular, and physiological impacts of natural and engineered nanomaterials. The observations obtained should be extrapolated with a human being if the model is non-mammalian, (iv) The survey of related strategies to the identification of risk, its management, and the resultant outcome and relevance to the safety of the environment and the products [2]. Advances in the fields of health care and diagnostics concerning nanotechnology range from nano delivery, artificial organs, and devices, to clinical and therapeutic applications are available. Nanoscience and nanotechnology have the potential to address the global issues concerning diseases, poverty, food, potable water, fertile land for agriculture, and cultivation. These potentials are related to microbiomes, biological sensing, imaging, tissue reconstructions, gene expression, delivery and technology, DNA and RNA technology, delivery with precise intervention in a living system. Similarly, energy production and storage offer enormous activity and promises. There is a great need to develop a scale up to meet the commercialization and industrial applications [3]. Different fabricated and natural nanostructures find various applications in the energy, health, and environmental fields. There is a specific relationship between structure, activity, and the performance of nanomaterials. This relationship is a critical value because it helps to achieve the appropriate, precise design to get suitable results. Thus creating, the formulation of nanostructures is a prime challenge. There is a strong need to either find or develop a core component, *i.e.*, fundamental nanostructure base on the functional set target that meets the requirement. There should be a continuous supply of such building blocks for the fabrication of sophisticated and purposeful nanodevices. At the same time, structural-activity- relationship and controllable targeting synthesis must sustain. Another aspect is to find or visualize new routes/modes of coalescence of such products that must be economical, environmentally feasible have to be searched so that the scalable production is maintained. Evaluation of systemic toxicity and environmental feasibility appear to be the prime challenges [4].

SOME SPECIAL MODES THAT FACILITATE INTERACTIONS BETWEEN NANOPARTICLES AND CELL MEMBRANE

To establish interplay between nanomaterials and the biomolecules, and other cellular components, cellular internalization, and contact with the cellular membrane is a crucial phase. Features of nanomaterials like nature and surface charge of nanomaterials, physical, cellular processes, cell membrane components like receptors, ligands, ion channels have functional significance. These all parts facilitate the process of internalization directly or indirectly.

Cell Fusogenic Proteins (CFPs)

In a biological system, membrane fusion is of frequent occurrence, and it involves specific membrane fusion proteins. Membrane fusion is conventional under membrane biology. In this process, two distinct lipid bilayers merge involving their hydrophobic components resulting in the formation of one interconnected membrane. There is the formation of the aqueous bridge *via* which the contents of the two fusing vesicles mix. Common membrane fusion proteins are Gp 41-Glycoprotein 4, a subunit present in the protein envelop complex of retrovirus (HIV) transmembrane, SNARE- (SNA+RE =SNA – snap, RE-receptor). Primarily these facilitate the fusion of vesicles and also mediated fusion of the membrane of vesicles and target membrane-bound compartments like lysosome. These are common functional aspects in presynaptic vesicles and presynaptic membrane of the neuron [5]. These should not be confused with fusion proteins or chimeric proteins [a fusion protein or chimeric protein constitutes a minimum of two domains. Different genes encode these proteins. These genes are connected so that these are transcribed and translated as one unit forming one peptide]. Membrane fusion proteins are quite common among viruses and the protein that causes fusion of membrane. This protein exists as homotrimers. Structurally it consists of two major domains, the extended helical coiled-coil domain and other a fusogenic peptide. The helically coiled-coil region helps the fusogenic peptide to maintain its geometrical and structural aspects to ensure successful fusion with the host cell membrane.

Further, optimal fusion associates the trimerization of the fusogenic peptide. It works on the hypothesis that dictates the oligomerization of fusogenic peptide enhances the process of membrane fusion, and during this process, the lipid bilayer is subjected to localized destabilization [6]. Some specific modes accomplish the translocation of nanomaterials. These modes help in the natural translocation of biomolecules without damaging the cell membrane. Viruses individually possess cell-fusogenic peptides (CFP). These can penetrate the cell membrane without forming any mechanical structures like pores or whole *etc*. Cell fusogenic peptides are molecular vectors involve during internalizing hydrophilic cargoes and as a potential biological therapeutic tool to carry target protein or drug into the specific cell. Nanoparticles, when conjugated with natural cell-penetrating motifs, can transduce cell membranes effectively into cytosol or nucleus [7].

During the study of the silencing of an oncogene, small interfering RNA (siRNA) is used as a functional mediator for the post-transcriptional gene silencing process –RNA interference-(RNAi). Gene silencing complex molecule must be free from the endosomal membrane to reach the cytosol-the location of RNA interference. This strategy accomplishes the release and mimicking of endosome and fusion with the membrane of the host cell. Such approaches are useful to study the functioning of the fusion domain in the case of the influenza virus. The effect of influenza derived fusogenic peptide (di1NF-7) on gene silencing and the efficacy of small interfering RNA that target epidermal receptors for growth factor and K-ras oncogene [8]. Cytosis is one of the prime pathways for cellular uptake of exosomes for cell signaling and as drug delivery. Interaction between the exosomal membrane and cellular receptors like intracellular adhesive molecules, lymphocyte function-associated antigen, phosphatidylserine, *etc*., binding to T-cell immunoglobulin domain and mucin domain protein-1 and help to bind exosome and host cell membrane. The membrane of the exosome and endsome is a lipid bilayer. This lipid bilayer acts as a biological barrier. Exosomal cargo moves across the barrier to reach cytosol to accomplish their intracellular functions, and these are present copiously in human body fluids like blood, urine, saliva. Thus there is always extreme competition among them for cellular uptake. The disease affected and related cells secret exosomal contents, thereby facilitating the progress of the disease and also inter and intracellular communication [9] (Fig. **2**).

Fusogenic Peptide [HA$_2$]; Binding Protein from Sea Urchin	Green Fluorescent Protein

Fusogenic Peptide and Green Fluorescent Protein form Complex and this Complex Associates with Nanomaterial

Fusogenic Peptide and Green Fluorescent Protein Formed Complex Binds and Fuses with Plasma Membrane Depending on the pH
[It Does Not Influence the Cellular Viability and Proliferation]
And Enters Cell
Finally Releases the Nanomaterial in Cytoplasm

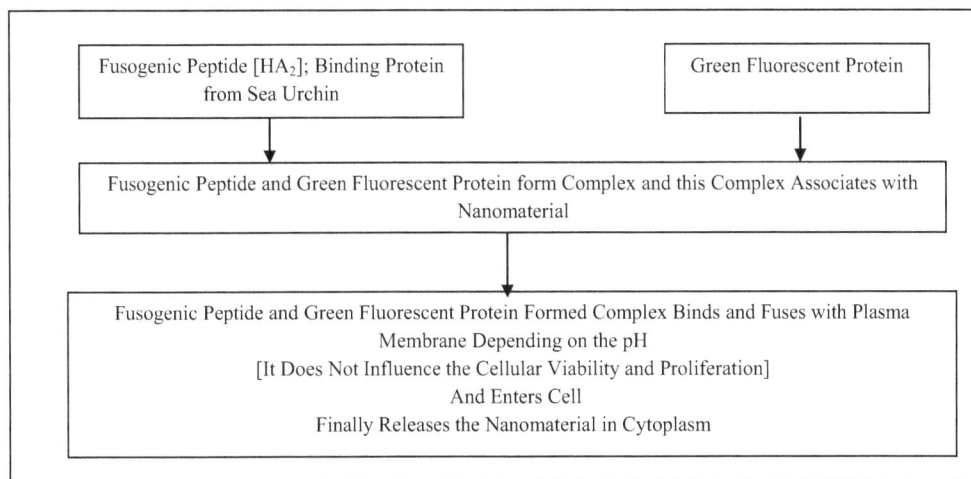

Fig. (2). Flow chart showing the role of fusogenic proteins in cellular uptake of nanomaterials.

Cell Penetrating Peptides (CPP)

Cell-penetrating peptides are in use for intracellular delivery of various cargoes *in vivo* and *in vitro*. Modifications of nanoparticles with cell-penetrating proteins need either a covalent or non-covalent approach in combination with liposomes and micelles that are favorable options for the loading of drugs, DNA, *etc*. There has been wide biodistribution due to the non-selectivity of cell-penetrating peptides. Cell-penetrating peptides carry the cargo of different sizes ranging from nanoscaled to small biomolecules. These cargo molecules get chemically bonded with the peptide *via* covalent bonding or non-covalent bonding/interactions. Endocytosis is a general route for the internalization of cell-penetrating peptides. There are three classes of cell-penetrating peptides, (i) first classes is polycationic-CPP and its peptide contains most of the positively charged amino acids like lysine or arginine, (ii) second class-amphipathic CPP, the peptide of this class includes alternately arranged non-polar and charged hydrophobic amino acids and (iii) the peptides of the third class consists of only hydrophobic amino acids, these are required for cellular uptake [10].

Different cell-penetrating peptides show variety in size, amino acid sequences, and charge they carry but all of them are capable of moving across the plasma membrane and deliver their cargo into the cytoplasm and to the specific organelles. Cell-penetrating peptides translocate through cell membranes either by direct penetration or internalization involving endocytosis or by forming a transitory structure and are an open field for biomedical and biomolecular research because cell-penetrating peptides are beneficial options as carriers for varied biomolecules [11 - 13]. Cell-penetrating peptides can be conveniently synthesized, functionalized, and characterized. These get conjugated with biomolecules readily and carry them to the interior of cells. Cell-Penetrating peptides are passive and non-selective, but when explicitly functionalized and chemically modified, act as one of the most suitable and sufficient drug/ biomolecules/gene vectors/carriers. These nanostructures succeed in reaching the set target [14]. Zinc sulfide nanoparticles doped with manganese (Mn) exhibit a higher degree of cytocompatibility, a lower degree of ROS generation, and a faster rate of clearance from the system. These features are suitable for a successful therapeutic carrier. Coated forms of modified cell-penetrating peptides like PEN-having origin from homeodomain of Drosophila transcription factor; pVEC with cadherin and R9 with delivery application genes are good options to study loading capacity for drugs. Among these modified forms of a cell-penetrating peptide, Mn-doped ZnS nanoparticles are tested to investigate the comparative efficacy for loading and to deliver paclitaxel drug of the three varieties R9. Mn: ZnS nanoparticles exhibit better therapeutic efficacy to carrying capacity, reaching the set target (in this case, can be breast, ovary, and cervical cancer cells). This combination showed co-loading efficiency, probably due to minimal interaction between paclitaxel drug and IR-780 dye (related to detection and diagnosis). The combination R9 was more toxic to cancer cells in comparison to the remaining two cell-penetrating peptides modified forms paclitaxel-loaded nanoparticles [15] (Fig. **3**).

Proton Sponge Hypothesis and Nanomaterials

This hypothesis follows the pH buffering effect. This phenomenon is common among some of the cationic polymers having buffering capability over a wide range of pH. Such polymers possess protonatable secondary and tertiary amine groups whose pH is very near to the pH of the endosome-lysosome.

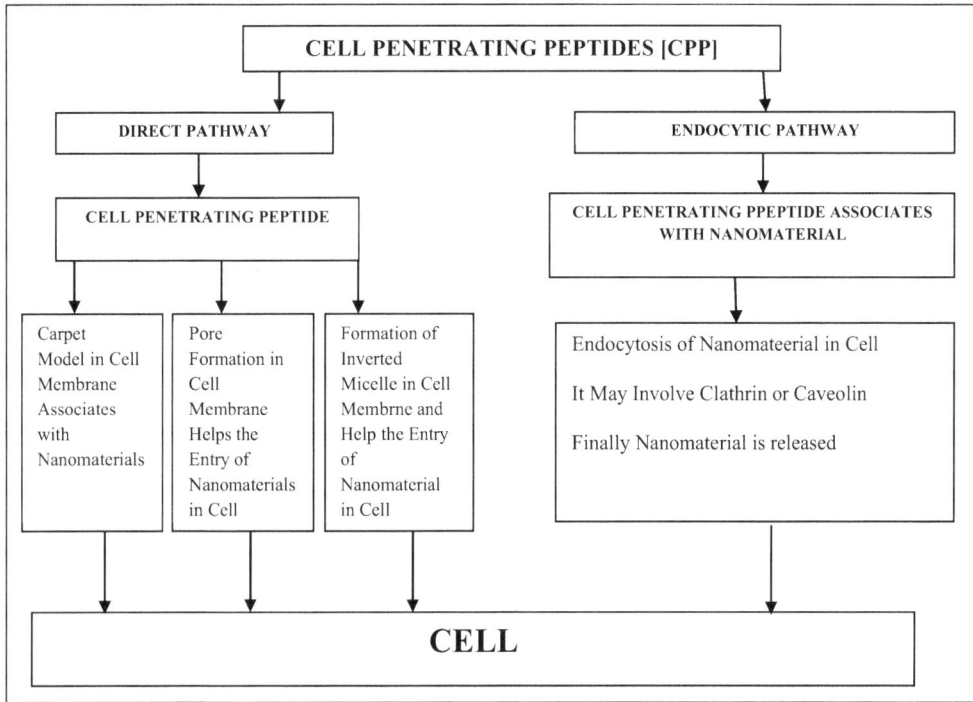

Fig. (3). Flow Chart Representing Role of Cell Penetrating Peptide During Cellular Uptake of Nanomaterials.

Polyplexes enter the cell through endocytosis, and these get enveloped in the endosome. Membrane-bound ATPase proton pumps actively to translocate protons into the endosome. The polymers become protonated and then resist the acidification of endosomes. Hence more protons will be pumped into endosomes continuously to lower the pH. The proton pumping action is followed by passive Cl^- ion entry, thereby increasing ionic concentration and water influx. High osmotic pressure causes the swelling and consequently, ruptures of endosome releasing contents. Nanoparticles can be fabricated involving pH-sensitive and redox-sensitive gatekeepers systems. These nanoparticles readily enter the tumor in a passive mode based or bind with different ligands present on the target cells or tissue. There is an increase in their uptake and controlled availability of the cargo present in the circulation [16]. Dendrimers can transfer biomolecules like DNA, RNA, oligonucleotides, *etc.*, and this ability relates to the buffer and acidification of endosomes. The proton sponge hypothesis helps in rupturing the endosomes to can be made to burst and the release of nucleic acids. Buffering dendrimers induce osmotic pressure, and this technique is conveniently used to break the endosomes to release their cargo. The DNA/vector complex establishes because proton sponge results in instability; this helps to form a compound of DNA and vector, and the release of the contents takes place. Different dendrimers like second-generation polyamidoamine, polyethyleneimine, and poly-1-lysine having a different composition of tertiary amines, are capable of binding plasmid DNA and condensing it. These reduce acidification and increase Cl⁻ accumulation. As a result, swelling induces in the related endosomes. There may be some undermined factors that do not result in the bursting of the endosome. The dendrimer understudy results in bursting the endosomes. One of the factors, pKa, plays an active role in this process. Amines having high pKa when present in dendrimers get readily protonated before it enters endosome; if the endosome gets acidified, no significant change in the rate of protonation takes place [16].

Mesoporous silica nanoparticles exhibit higher loading capacity, ease of getting a modification of the surface, its pore size and nature can be readily tuned as per the need. These features make mesoporous silica nanoparticles suitable agents to ensure controlled drug release. The internal stimuli such as decline pH, reducing ambient environment, or enzymes can be incorporated in mesoporous nanoparticles to ensure controlled drug release. These nanoparticles undergo endocytosis using endosomes. Generally, the environmental tumors have low pH as a result of hypoxia, and this feature makes it easy to find the target site. Psudorotaxan encircled by β-cyclodextrin, tannic acid, polymer, and lipid coating are some of the examples that activate the capping system. A block copolymer having positive charges of artificial amino acids and oleic acid block are suitable agents for capping and as endosomal releasing agents. When these endosomes are protonated, the pore blocking components are either degraded or dissociated, resulting in the release of cargo [17 - 25] (Fig. **4**).

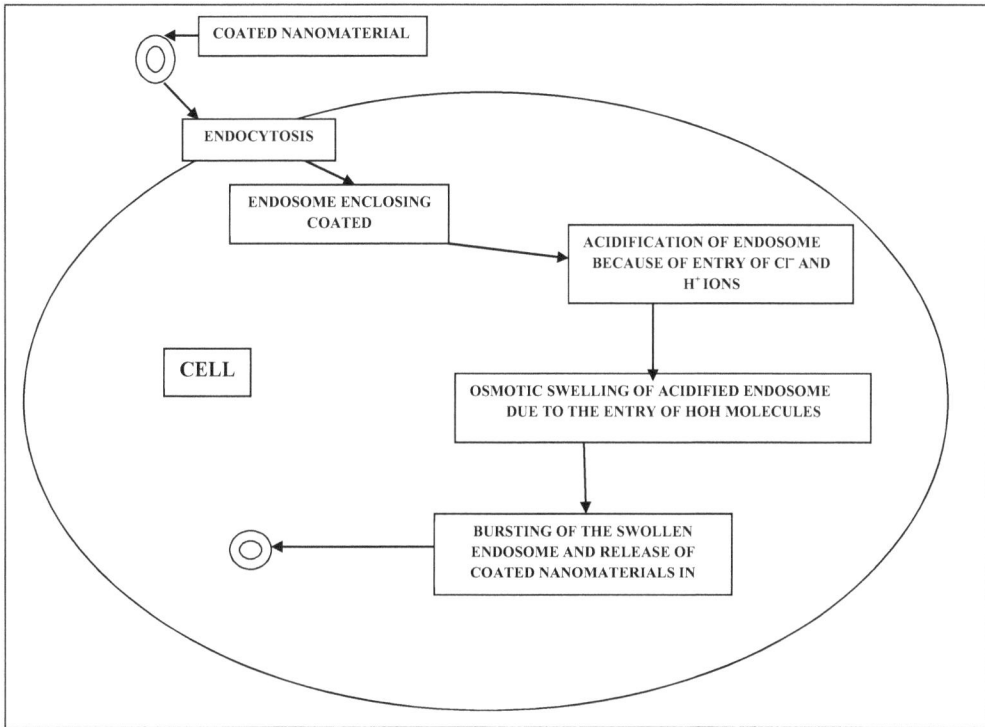

Fig. (4). Flow Chart Depicting Proton Sponge Effect.

BEHAVIOR OF NANOMATERIALS

Under the influence of enhanced permeability and retention (EPR) process, some of the molecules of the particular size typically liposomes, nanomaterials, and macromolecular drugs a trend to get accumulated in the tumor tissue in comparison to the extent of their accumulation in the healthy tissue. This process becomes more effective because of vascular endothelial factors, growth factors in the case of nanoparticles. Since in the affected tissue (tumor tissue), the lymphatic system is not very prominent, there is relatively much poor drainage of the materials from it. EPR is effective in delivering nanomaterials, liposomes to cancer tissues [26], and influences proteins biomolecules like enzymes, growth factors, some hormones, and cytokines. These undergo degradation and become weak because of physicochemical and biological stability, immunogenicity, *etc.*, may exhibit restricted potential benefits. Protein modifies to reach the intracellular target, but differential membrane permeability or release from endosomes restricts their biological activity. Various strategies help to accomplish therapeutic applications [27].

The process of enhanced permeability and retention has been reported to be one of the prominent bases for the delivery of cancer drugs. It is considered to be a simple, safe, and effective method for cancer therapy. Because of this phenomenon ensures the long duration of nanomaterials in circulation and their access to the target. Thus this phenomenon is suitable for the translocation of nano-drug carriers. Further, this process is ineffective for the free drug delivery system during the respective clinical trials. The physiological states of tumor tissue play significant roles during the distribution of the drug. The vulnerable physiological conditions of cancer tissue include (i) non-uniform vasculature, (ii) high tumor interstitial fluid pressure, and (iii) poor flow of blood within the tumor. The observations from the EPR method during the clinical trials of drug distribution cannot be extrapolated in the case of a human being, at least when the experimental model is a non-mammal. The EPR phenomenon has limited applications and applies to case-by-case [28].

Nanosized drugs exhibit some of the specific features like better loading capacity, better maintenance of payload, reduced degradation, a higher degree of regulated release rate and the increased surface area to facilitate an enhanced degree of conjugating target ligand and better biodistribution. Further, the nanosized drug readily gets internalized *via* permeable vasculature of the tumor and does not arrive quickly cleared or drained because of the weak lymphatic drainage system present in tumor tissue. Thus the EPR process becomes prevalent in such cases. The EPR process can be modified to ascertain the interaction, duration, and the amount of drug to enhance much-needed effectiveness. If tumor blood flow modulates the efficacy of the EPR process increases; this will elevate the blood flow and pressure. The modulated form of vasculature and stroma of the tumor tissue increases the permeability of vascular tissue and helps to manipulate the stroma of the tumor. All these steps facilitate to attain appropriate dose, duration, distribution to ensure the killing of cancer cells [29].

Intratumoral drug delivery and distribution are the major hurdles during the treatment of tumors and using ultrafine iron oxide nanoparticles (size below 5 nm with core size 3.5nm) these hurdles minimize. This sized iron oxide nanoparticles readily translocate from the vasculature of a tumor to the stroma of tumors in comparison to the larger sized nanoparticles. The elevated acidic interstitial medium of tumor tissue disrupts the internalization of nano-drugs. By elevating the rate of entry, reducing the rate of drainage, improvised delivery, and distribution of the nano-drug under consideration, elevated acidity reduces. Magnetic resonance imaging and involving fluorescent technique using dye-labeled nanoparticles exhibit better bright T-1 contrast during the early stages of experimentation and after 1 hour and 24 hours contrast dark T-2 reflected on the extended permeability and retention of the nano-drug in the tumoral tissue [30]. The fabrication of nanoparticles involves pH-sensitive and redox-sensitive gatekeepers systems, and these help in convenient internalization in tumor cells and tissue. It includes passive mode related to enhanced permeability and retention effect or can be conjugated with different legends to facilitate their uptake by the target cells/tissue. The PEGylation enhances the circulatory life and their availability in the circulatory system [24].

Enhanced permeability and retention effect or process plays a significant role during the internalization of nano-drug carriers in the cancer tissues. The factors like (i) degree of vascularization in tumor and degree or extent of angiogenesis; this includes a varied degree of porosity and the size of pore present in the blood vessels; (ii) presence of interstitial fluid in tumoral tissue, its pressure influences the internalization of nano-drug and nano-carrier; (iii) the role of extracellular matrix and dense cellular population in tumoral tissue specifically around blood vessels act as a barrier to the uptake of nano-drug. By manipulating these mentioned factors, improve the permeability and retention effect and drug delivery in cancer tissue improves.

PEGylation does not appear to be a good option because it interacts with opsonins and also with nano-carriers and cell surface; (iv) non-uniform vascularization of blood and its flow impede the homogenous distribution of nano-drug within tumoral tissue. These factors result in impairment of the efficacy of cancer drugs. The approach of targeting the specific tumoral cell/tissue is reported to be an ideal way to tackle this issue. One such method is passive targeting. This approach utilizes the differences existing between extravasation and retention of nano-carrier and nano-drug that remain in circulation for a longer time. The other approach is called active targeting. It involves the functionalization and modification of the surface of the nano-carrier to act suitably to identify tumor cells. By selecting specific interaction between ligands like peptides, nucleic acids, sugar molecules, antibodies, density and location of ligands, the ratio between ligands to polymeric materials, the chemistry of end group of ligand/s, *etc.*, influence the active targeting process The interaction between these ligands and the available reactive receptors present on the surface of nano-carrier [31].

ROLE OF MASS TRANSPORT OF NANOMATERIALS IN BIOSYSTEM

Nanomaterials are in use in almost all aspects of the life of biosystems, including food, cosmetics, medical, therapeutics, pharmaceuticals, *etc.* These materials exhibit mass transport in a given biological system because of their physicochemical characteristics. The biosystem provides the most suitable medium for them to get distributed. This tendency is likely to either accelerate or delay the intended action of the nano-based product in the biosystem. These materials negotiate most of the biological barriers [1], and exhibit reactivity with micro and macro biomolecules like antibodies, aptamers (oligonucleotides, small peptide molecules, *etc.*), peptides, bioswitches, and can form conjugates with biomolecules. These features also play a significant role in their dispersion and distribution in a biosystem. These materials can escape the mononuclear phagocyte system (MPS), reticuloendothelial system, epithelial membrane, complex network of blood vessels, and abnormal blood flow. Sometimes the conjugates formed with bionanomolecules may find it difficult to negotiate some biological barriers. Such conjugates include enzyme degradation, molecular efflux pump; ionic efflux pumps [32]. Nanoparticles like liposomes are biocompatible, spherical vesicles, with either single or multi bilayers and a wide range of compositions. These nanomaterials undergo convenient functionalization concerning the specific ligand and protect the cargo in a biosystem [33, 34]. Albumin-bound nanoparticles can utilize albumin related endogenous pathways and carry hydrophobic consignments in the biosystem *via* a blood circulatory system. Albumin tends to bind with hydrophobic molecules involving non-covalent reversible binding; thus, take without causing solvent-based adverse effects [34, 35]. The properties like biocompatible nature, ease of surface modification, light absorption, light scattering nature, fluorescence behavior, surface-enhanced Raman Scattering, *etc.*, render them suitable for multiple applications in diagnostics, therapeutics, imaging, *etc.*, *e.g.*, gold nanoparticles [34, 36]. Polymeric nanoparticles are the products of biocompatible and biodegradable polymers. These are formulated using block copolymers having a different degree of hydrophobicity. Such polymeric nanoparticles get easily assembled as core-shell micelle formulation that also in an aqueous medium. Such polymers are capable of carrying hydrophobic, hydrophilic molecules and proteins, nucleic acids, and medium-sized drugs as these nanoparticles have much-improved biodistribution in the biosystem [34, 37]. Among polymeric nanoparticles, dendrimers are very facile carriers because these can upload small molecules within inter and intra branch spaces *via* chemical linkage, H-bonds, and hydrophobic effects. Their peripheral surface is easily modified, functionalized with those chemical groups that readily conjugate with the target molecules, ligands, and groups present on the recipient cell or tissue of the biosystem [34, 38].

Superparamagnetic iron oxide (SPION) as ferumoxide (120-180nm) and ferucarbotran (60 nm) is a better option as nanoparticles to investigate apoptosis and the expression of a gene. SPION functionalize with those ligands that have magnetic, optical, radionuclides affinity. As a result, this specific SPION acts as a diagnostic and investigatory agent, and it comfortably distributes is biosystem [34, 39]. Quantum dots exhibit a wide absorption band. Further, their emission is as a bright color, having longer life duration, high efficiency, and better stability against photobleaching. These nanomaterials have well defined and a better degree of biodistribution, hence they are very beneficial tools for biomedical, optical and imaging, and as diagnostic agents [34, 40].

SIGNIFICANCE OF DEGRADABILITY OF NANOMATERIALS IN BIOSYSTEM

The degradability of nanomaterials after internalization is an essential parameter that affects their interaction and clearance within the biosystem. There is every chance that non-degradable nanomaterials find it difficult to get cleared from biosystems, especially when these miss the interaction with the mechanism of clearance. In such conditions, these nanomaterials start getting accumulated in cells, tissues, or organs. These accumulated nanomaterials may act as the causes of a derogative impact on the recipient. When biodegradable nanomaterials internalize naturally or intentionally, and if their degraded forms or adducts do not release from biosystems, then these interfere with the normal functioning of the biosystem.

Nanomaterials are developed, produced, applied, and modified concerning their ability to get biodegraded. The biodegradation of nanomaterials relates to the biochemical and chemical interactions at the interfaces and the sites of such interplay. These all reflect on the behavior of nanomaterials in the biosystem. The formation of protein corona is the most common interactions between nanomaterials and protein biomolecules. The physicochemical features of corona formed also deviate during metabolic interconversions in a biosystem either due to the presence of biomolecules and the components present in nanomaterials. The enzymes such as peroxidase, lysosomal hydrolases, and isoforms of CYP450 influence the biodegradation of the internalized nanomaterials [41].

Polymeric nanomaterials act as an option for the formulation of biotechnological and pharmaceutical products because these polymeric compounds are derived or evolved from biodegradable products. These products can act as a target, release the drug in a controlled manner, and act as well as versed diagnostic agents. In this category, the most common examples are dendrimers and liposomes [42]. Nanomaterials like liposomes, carbon nanotubes, dendrimers, polymeric micelles, polymeric conjugates, and polymeric nanomaterials are extensively used in chemotherapy as these reach their target and get biodegraded to release their cargo at their respective sites. There is no approval for such useful nanomaterials from FDA [43].

Oxidative enzymes related to inflammatory cells such as myeloperoxidase, eosinophil-peroxidase, lactoperoxidase, hemoglobin, and xanthoxidase, *etc.*, play a significant role in biodegradation. Peroxynitrite pathway relates to the macrophages and NADPH oxidase and NO synthase-triggered oxidative based interactions also play a substantial role in the biodegradation of nanomaterials. In addition to oxidative enzymatic interactions, some cells participate in the interaction between foreign particles like microglial cells, myeloid-derived suppression cells, and help in the biodegradation of nanomaterials [41].

Azido coumarins and one catechol derivative both are excellent mediators for redox reactions. These elevate the degree of catalytic activity of enzymes like horseradish peroxidase and xanthine oxidase. These enzymes are very active in hepatic tissue, even in the case of mammalian liver. Azido purine horseradish peroxidase enzyme readily degrades the functionalized single-walled carbon nanotubes with azido coumarins and catechol derivative. The degree of degradation is much higher in the case of the horseradish peroxide enzyme in comparison to that of the xanthine oxide enzyme [44]. Magnetic nanoparticles incorporate with stem cells as components of the tissue and undergo degradation at the intracellular level and easy to detect using imaging. This degradation takes place at a single endosome scale or level. When iron contents release, there is no influence in the iron homeostasis of the tissue [45].

INFLUENCE OF BIOPHYSICAL ASPECTS ON INTERACTION BETWEEN NANOMATERIALS AND THE COMPONENTS OF BIOSYSTEM

Membranes in an aqueous environment exhibit attractive and repulsive responses towards the water. The materials composition of the cell membrane and the corresponding surface chemistry determines the reaction with water affect wettability. The wettability of any materials is a characteristic surface property that yields a unique value for each substance. The amount of surface tension helps to find the wettability of material by specific liquid involving the measurement of the angle of contact between a solid surface and a droplet of a liquid on the surface. Surface tension represents the internal force due to the state of unbalance in molecular forces that occurs when two different materials such as liquid droplets on a solid surface are brought in contact with each other forming an interface or boundary. All substances show a tendency to reduce their surface area in response to the state of unbalance among molecular forces. These forces are active at the point of contact within the interface. The comparison between the values of surface tension of different substances indicates the degree of hydrophobicity or hydrophilicity. Nanomaterials also follow the same principle. Liquids with lower values of surface tension generally, spread on the solid-liquid materials than those having higher values for surface tension. The surface tension of the multi-component solution can differ significantly in their behavior due to solubility complications [46, 47].

IMPORTANCE OF CHARACTERISTIC MEMBRANE WETTABILITY

Particles that foul in aqueous media tend to be hydrophobic. Particles like colloids, starch, and metal colloids, complexion aggregates, a grouping of molecules with a diameter ranging 1.0 1.7 mm, liquid-solid and solid-liquid suspended matter, *etc.*, exhibit hydrophobic behavior in an aqueous medium. Although proteins there have positive and negative charges, they also have hydrophobic regions. Other particles like clay, silicates, alumina, ferric hydroxide, *etc.*, and oily particles, paraffin, oil, surfactants, and grease, *etc.*, act as hydrophobic substances. Hydrophobic particles show a tendency to cluster or group and form a colloidal state of a material. This state lowers the interfacial free energy, *i.e.*, surface energy due to the surface area exposed. This tendency results in the formation of the sphere of particles; the sphere has a minimum surface area, and exposes limited to the hydrophilic (water) environment [46 - 48]. Thus, for the applications related to membranes fouling can be reduced by using a layer or a cell membrane with surface chemistry which has been modified to render them hydrophilicity (Fig. (**5**)).

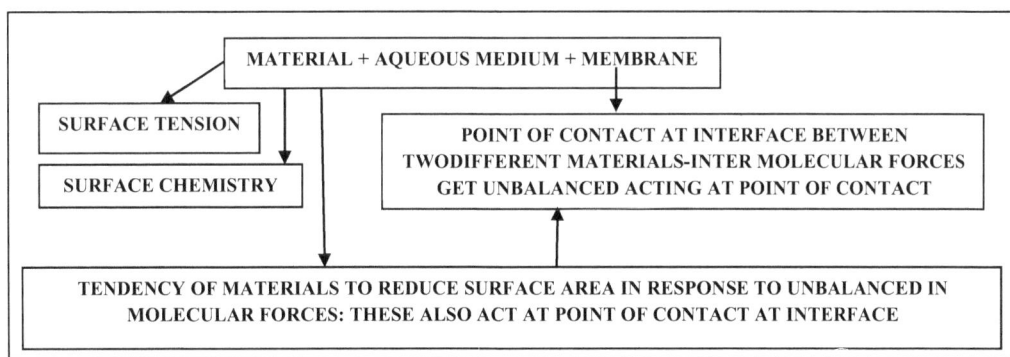

Fig. (5). Probable Behavior of Material in Aqueous Medium and at the Cell Membrane.

INFLUENCE OF BIOCOMPATIBILITY AND BIODISTRIBUTION OF NANOMATERIALS IN BIOSYSTEMS

Biocompatibility refers to the state in which there is no molecular, biochemical, physiological interference by the administered materials in the body of the recipient. It neither creates toxic impact or derogative implications on the structural and functional aspects of the recipient's biosystem. Biosystem has a specific feature that reflects on the accommodating or tolerating the sudden drastic changes caused in the external and internal environment. The administrated nanomaterials do not get influenced by the immune system, reticuloendothelial endoplasmic system, and other components of this system, *etc.*, and reach the target in the biosystem. Since most of the nanomaterials can quickly move across most of the biological barriers. For this reason, nanomaterials can get translocated throughout the body and without getting modified and can deliver its cargo to the set target. During this journey, these nanomaterials do not disturb the body fluids, physiological processes, and cell signaling processes.

PATHWAYS RELATED TO INTERACTION OF NANOMATERIALS AND TISSUES –CLASSICAL PATHWAY; C-REACTIVE PROTEINS; LECTIN PATHWAY; ALTERNATIVE PATHWAY

Whenever there is an interaction between two interacting components, the activation process takes place. During this activation process, the interacting components undergo some structural, conformational changes. The physicochemical and physiological conditions prevailing in the ambient environment influence these changes. These changes accomplish the interaction towards its completion. Nanomaterials, when interact, are likely to induce host defense interaction, host response relation, host immune response, clearing of foreign body from the host. There is a complement system among a vertebrate biological system that plays a vital role in defense of the host. This complement system mostly includes more than thirty plasma protein and membrane proteins. During the process of activation, these members of the complement interact in a synchronized and controlled manner. These interactions bring about the promotion of inflammation, opsonization of pathogen/foreign particles, lytic mechanism, *i.e.*, killing the pathogen. Mostly only antigen surfaces are subjected to complementary activation and involve many soluble and biologically active and influential fragments or components.

The complementary system influences natural, fabricated, specifically engineered nanomaterials with precisely target orientation. The physicochemical features of the surface of nanomaterials induce various pathways to complement cascade activation [32]. There are three specific pathways that such nanomaterials face in a biosystem. These pathways are (i) classical pathway, (ii) Mannan binding lectin pathway, and (iii) alternative pathway. Under regular pathogen-host interaction in classical pathway antigen and antibody complex formation takes place, this, in turn, activates complement activation. In mannan-binding lectin, pathway lectin binds to the surface of a pathogen or foreign particles, and this conjugate followed by complement activation. During the alternative pathway, the surface of the foreign particle activates the complement system [32, 49].

Complement (C) and a single number (C3) represent the components of the classical pathway. The proteolytic activity and proteases activate the elements of this pathway in the host. The deposition of specific protein and other materials like adsorption of protein (C-reactive), non-coated single-walled carbon nanotubes, negatively charged radical/group, *etc.*, also activate this pathway (Zhang ., 2012). The lower case letter, *e.g.*, C3a; C3b, are the products of cleavage and represent the products formed. Mannan binding lectin associated serine protease-1 (MASP-1) represents the components of the Mannan-binding-lectin pathway. If the element is inactive, then this representation is like the lower case, Factor-B, and Factor-D, and represents the components of alternative pathways [49].

Once there is a formation of the immune complex, the classical pathway activates immediately. The C1 component is capable of recognizing the antigen-antibody complex. There are conformational changes in the constant region of the antibody when the antibody and antigen-binding take place. As a result, the possible sites of the Fc region get exposed; this can bound to the first component of the classical pathway (C1). This component is a macromolecule.

The activation of the Mannan-binding-lectin pathway is similar to the classical pathway, but there is one difference in that a protein activates this pathway. This pathway can also be triggered. If the pathogen-associated motifs identify the mannan/mannose-binding-lectin at a surface interface; this also enables the Mannan/mannose-binding-lectin pathway [32]. Mannan-binding-lectin (MBL) is similar to C1q, and it binds with the carbohydrates and mannose related complexes present on the surface of the pathogen. MBL, mannan-binding-lectin, is related to two serine proteases, *i.e.*, MASP- and MASP-2. Once MBL gets associates with the pathogen, these two components activate. C3 convertase interacts with PEG-coated nanomaterials, double-stranded nucleic acids, or the component of the pathogen. These respective conjugates lead to a C5 convertase [32].

The alternative complement pathway is associated with the slow hydrolysis of the C3 component, and this is quite spontaneous in plasma. This hydrolyzed component (C3) cleaves factor B and binds, forming the C3-H2O-Bb complex. This complex is convertase and is capable of forming additional molecules of C3b elements. When there is a preferential deposition of the product of C3b on the surface of the interacting nanomaterials, there is a magnification of the cascade activation. The C3 convertase amplifies involving factor D, properdin, Bb, Ba. All these pathways direct to C5-convertases; this results in the lysis of pathogen and senescent cells by getting attached to the membrane of these cells. During this process, there is a release of proinflammatory mediators, and these show a yellow outburst [32].

It is challenging to balance the functionalization of the surface of nanomaterials and the low rate of immune identification, specifically in the case of PEG-coated nanomaterials. The third component of the complementary element (C3) after opsonization facilitates immune cell identification. C3 complement under the influence of the lectin pathway opsonizes the superparamagnetic nano worms. The degree of opsonization declines in the case of cross-linked dextran shell and its functionalized form with epichlorohydrin is under study. Its macrophagous uptake of the peritoneum of the mouse also declines.

Further, when the modified cross-linked SPION nano worms, using either polyethylene glycol or polyethylene glycol-antibody such as IgG (about 160, molecule or particles); the opsonization does not increase, through C3, even in the case of macrophage internalization in the peritoneum of the mouse. Intra caudal vein administration of plain cross-linked nano worms and PEGylated cross-linked nano worms show a decline in the degree of C3 opsonization and a lower rate of leukocyte uptake in the mouse. Antibody conjugated cross-linked nano worms exhibit a prominent degree of C3 opsonization and increased internalization depending on the complement in the experimental model. This functional feature is in proportion to the number of the conjugated antibody with 46 IgG molecules or particles. If the functionalization of surface or nanomaterials can be finely tuned, then the complement activation and complement related immune internalization are useful to get the desired results [30].

ENDOCYTOTIC MECHANISM OF UPTAKE OF NANOPARTICLES

Nanoparticles interact with the extracellular macromolecules forming a covering around the nanoparticles which further interact with the components of the cellular plasma membranes and are internalized by the process of endocytosis. Depending on the type of extracellular biomolecular covering the nanoparticles and the type of cell interacting through their plasma membrane the endocytic mechanism is classified into 5 different ways as explained below -

Phagocytosis

A procedure mainly used for the defense mechanism by the host organism mainly through the cells called as professional phagocytes such as macrophages, monocytes, neutrophils, and dendritic cells. However, there are other cell types like epithelial, endothelial and fibroblast cells, also perform the phagocytosis. Usually, phagocytosis is initiated when the antibodies, proteins present in the complement system and blood proteins such as laminin and fibronectin are adsorbed on to the surface of the nanoparticles. This process of adsorption is called opsonization and the opsonized nanoparticles interact with the plasma membrane through ligand-receptor interaction. This process activates a series of signaling cascade for actin assembly which helps in the cell surface invaginations; engulf the nanoparticles forming structures called phagosomes. Some of the receptors which help in the formation of phagosomes include Fc receptors, receptors involved in the complement system, scavenger receptors and glucose receptors. As mentioned earlier the biological response towards the engulfed nanoparticles depends on the type of the receptor involved in phagocytosis. The processes where complement receptor-based uptake of the nanoparticles is involved it does not elicit an immune response whereas Fc receptor-dependent uptake of nanoparticles mediates immune response by the generation of proinflammatory cytokines. It is seen in many studies phagocytosis is involved in the uptake of nanoparticles which are of bigger size *i.e.*, 500 nm or more. This can be clearly seen from phagocytotic studies of either human mononuclear cells or mouse peritoneal macrophages where the radiolabeled nanoparticles of albumin in the size range 200 – 1500 nm and polystyrene nanoparticles of sizes between 1000 – 2000 nm are efficiently phagocytosed. The shape of the nanoparticles is another important aspect of the mechanism of phagocytosis of nanoparticles by different cells. Murine macrophages show a higher amount of phagocytosis of PEGylated gold nanospheres as compared to nanorods which explained with other *in vivo* studies that show higher accumulation of nanospheres in the ovarian tumor-bearing mice as compared to nanorods. Similarly, other critical aspects that determine the phagocytic uptake efficiency of nanoparticles include the surface properties which influence the opsonization process and subsequently affect the membrane receptors binding and phagosome formation. Hydrophilic coatings on the nanoparticles usually lower or prevents protein coating formation or the opsonization process. This is possible only when a minimum thickness of the polymers such as PEG cover the surface of the nanoparticles which in turn depends on the molecular weight of polymer, conformation, and extent of branching chains that adsorb on the nanoparticle surface. This can be clearly explained considering the case of Doxil® (FDA approved anticancer doxorubicin-liposomal formulation) where the increased PEGylation decreases the cellular uptake of nanoparticles by phagocytosis and improves the blood half-life by increasing the circulation with reduced uptake by reticuloendothelial cells system (RES) and enhancing the pharmacokinetics of the nanoformulation. An increase in the cellular uptake with decreased blood circulation times or reduced half-life could be alternatively achieved by changing to the hydrophobic surface coating which effectively allows the adsorption of complement proteins and hence increases the greater cellular uptake by phagocytosis [14]. So such strategies could be exploited to help appropriate design and use of nanoparticles either for increased blood circulation/improving half-life or increased reticuloendothelial systems (RES) cellular uptake. Improving RES strategy could be used for the treatment of diseases that affect organs like liver and cases of hepatocellular carcinomas.

Caveolae-mediated Endocytosis

Most of the epithelial and non-epithelial cells have flask-shaped structures within the dense bodies anchoring the cytoskeleton forming membrane invaginations called caveolae. The caveolae form at least 75% of the cell membrane proportion of most of the adipocytes and smooth muscle cells which form the non-epithelial origin. The flask shape is usually attained due to the presence of a protein called caveolin resulting in the overall size of caveolae in the range of 50-80 nm. The structural stabilization of the caveolae is because of another protein called caveolin-2. Other proteins which are involved in the overall endocytosis mediated through caveolae include the membrane stabilization mediated by cavin proteins, budding, and dissociation of vesicles is mediated by dynamin and vesicle fusion is mediated by vesicle-associated membrane protein (VAMP2) and synaptosomal-associated protein (SNAP). This endocytotic mechanism mediated by caveolae has vital roles in numerous biological processes like the cellular signaling, transcytosis, regulation of lipids, fatty acids, membrane proteins, and membrane tension and plays a critical role in some diseases like diabetes, various viral infections, and cancer. Cancer cells have the ability to highly uptake the nanoparticles if they are bound with albumin as it binds to the gp60 and albumin receptor present in the caveolae of the endothelial cells. This will facilitate the transport of nanoparticles through the tumor interstitial spaces and ultimately acts on the cancer cells. This was especially seen in the commercial nanoparticle formulation of Abraxane®. It is also learned from other studies that the caveolae-mediated internalized nanoparticles sometimes escape the lysosomal degradation and such phenomenon finds immense applications in the field of drug delivery especially for carriers such as proteins and genes. In other mechanisms, sometimes, the caveolae-mediated engulfed nanoparticles are acid degraded in the lysosomes and such phenomenon takes the lead in the release of some drugs such as doxorubicin sensitive to the pH variations as seen in the lysosomes of cancer cells.

Clathrin-mediated Endocytosis (CME)

Clathrin is a protein assembly of 3 heavy and 3 light chains in the form of 3-legged structure assembly which plays a vital role in membrane curvature stabilization and vesicle formation. It forms about 2% of the cell surface and is responsible for the transport of nutrients across the membrane such as cholesterol and iron through low-density lipoproteins (LDL) and transferrin carriers. CME can of be two various types either through the non-specific absorption/uptake or through the receptor-specific uptake. In the case of non-specific uptake usually, the hydrophobic or electrostatic interactions dictate the uptake mechanism. The presence of adaptor proteins, on the docking site of the cytoplasmic side of the plasma membrane, helps in the recognition of various cargoes and signals responsible for sorting. Accessory proteins (epsin, amphiphysin, SNX9) along with the adaptor proteins help in the nucleation of clathrin on the side of the membrane internalization process. During the nucleation process, pentagonal and hexagonal curved structures of lattices are formed in which the clathrin assemble into these pits into which the membrane invaginates and helps in the formation of endocytic vesicles of size in the range 100–150 nm. During this process other proteins that help in the formation of vesicles are accessory proteins helps in the formation and stabilization of curvature of the membrane, a membrane scission protein called dynamin helps in the releasing of vesicles as they are recruited to the neck of the budding vesicles. Usually, the contents of the vesicles during the CME are frequently routed to the lysosomes enzymatic action for the degradation mechanism. Hence this might not be the appropriate path for the nanoparticles. It is seen in many studies that the nanoparticles with a positive charge are usually up taken by CME mechanism in the HeLa cervical cancer cells. However, it is seen that anionic particles are uptaken by both the CME and the caveolae method. Similarly, positively charge silica nanoparticles are uptaken by the breast cancer MDA-MB-231 cells by the CME mechanism.

Other Mechanisms of Cellular Uptake

In a mechanism called as macropinocytosis, large regions of the membranes called 'ruffles' are formed with the help of cytoskeleton arrangements, and does not involve the membrane rafts or proteins that help in the formation of pits. Therefore, these large invaginations of the membrane thus formed, usually of the size 0.2 - 5µm, collect large volumes of extracellular fluid as well as dissolved particles in the absence of any specific receptors. This is an important aspect of the cellular uptake mechanism where the nanoparticles are of the order of micrometer size and not nanometer. Unlike the micropinocytosis mechanism, there are other mechanisms that are independent of clathrin or caveolae endocytosis process of cells that lack the presence of clathrin or caveolae and mediate the nanoparticles uptake by the presence of specific lipid components such as cholesterol. Nanoparticles modified with folic acid are the best example exhibiting such a phenomenon and the advantage being that such mechanism escapes the lysosomal enzymatic degradation process and are retained in the endocytic vesicles or directly released into the cytoplasm].

NOTE: Overall representation of uptake of nanomaterials is represented in Figs. (**6** and **7**).

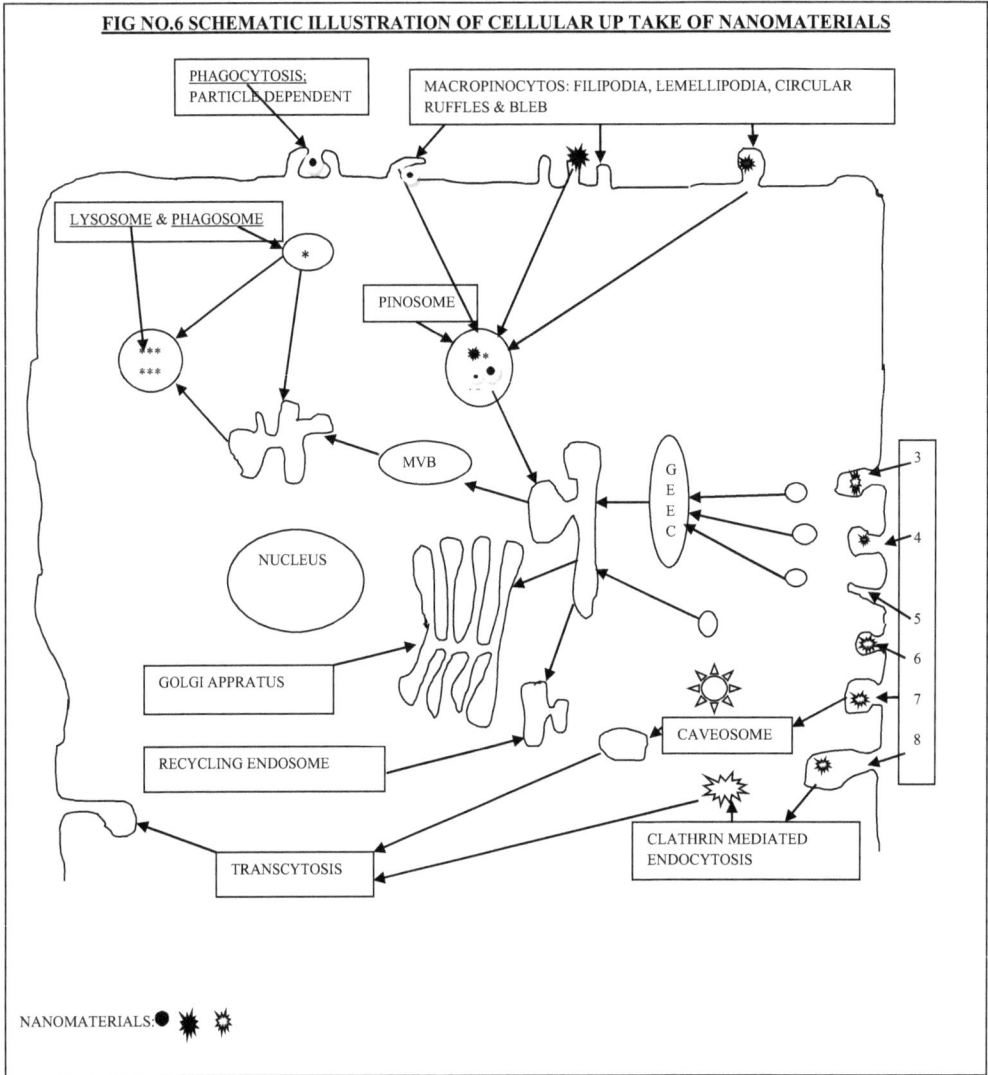

Fig. (6). Phagocytosis (particle dependent); Macropinocytosis (<2μm) Filipodia, Lamellipodia, Circular ruffles, Bleb); 3-Flotilin mediated (<100 μm); 4-ARF-6 (Alternative reading frame) mediated, >100 nm); 5-CDC42 (a type of Rheokinase) mediated, <50 nm); 6-Rho A- mediated, related to Rheokinase, <200 nm); 7-Caveolae mediated <8 nm); 8-Clathrin mediated (<300 nm); MVB- Multi vascular body; GEEC-GP-1 early endosome component.

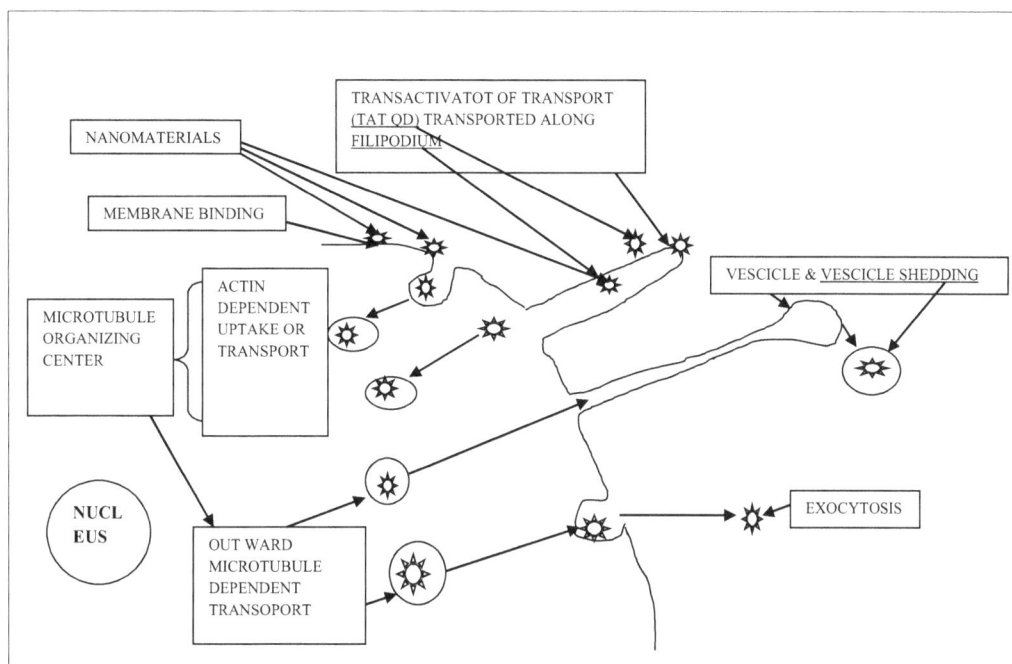

Fig. (7). TAT-functionalized QDs enters the cell through micro-pinocytosis- fluid phase endocytosis triggered by TAT QD binding to negatively charged cell membrane. In addition to TAT QDs RGD (arginyl glycocylate aspartic acid) is another peptide motifs, allostatin-I, arginine rich peptides may also be involved.

THE BIOCHEMICAL, BIOPHYSICAL AND FUNCTIONAL ASPECTS OF GLYCOCALYX AND ITS INFLUENCE ON NANOMATERIALS

The glycocalyx or pericellular matrix is a specific layer present between the extracellular matrix and the cell membrane. The glycocalyx differentiates into the attached layer and unattached layer. The connected region is relatively firm, and it includes an inherent component that belongs to the cell membrane and is challenging to separate mechanically. If forcefully separated, then the cell membrane also gets damaged [50 - 54]. Biochemically glycocalyx consists of glycoproteins and oligosaccharides or glycolipids. The glycocalyx is also present in some of the bacteria. A fuzz coating is present on the external aspects of the cell membrane representing glycocalyx. The glycocalyx is a prominent feature of freely moving cells in blood circulation and also of the composite cells located as lining in the case of blood vessels, alimentary canal, urinary bladder, ureter, urethra, *etc*. Biochemically glycocalyx is made of glycoproteins and oligosaccharides. Proteins like selectin-E, selectin-P, and selectin-L are the transmembrane proteins present in the cell membrane. These intermingle with freely moving cells like white blood cells. The transmembrane proteins help to recognize specific oligosaccharides present on the interacting cells. The oligosaccharides that are present in glycocalyx act as cell markers and facilitate cell identification [51].

Endothelial cells have carbohydrate-rich glycocalyx. It acts like a mesh-like structure and has soluble molecules [55]. The glycocalyx is a dynamic structure having equilibrium between the blood flowing along and its soluble components. This equilibrium depends on its composition. Proteoglycans, glycosaminoglycan, and glycoproteins are the main constituents of the glycocalyx. Proteoglycans have syndecans and glypicans, mimecan, perlecan, biglycan, and present in very less amount. These reside in glycocalyx but diffuse in blood circulation [56, 57]. The major glycosaminoglycan like heparan sulfate, chondroitin sulfate, dermatan sulfate, keratan sulfate, and hyaluronan are also present [58, 59]. Glycocalyx in endothelial cells is like a brush-like and appears on their luminal side. The glycocalyx in this region possesses anti-thrombogenic characters. The presence of hyaluronic acid and heparan sulfate in glycocalyx renders it richly hydrophilic substance. A hydrophilic surface in nature cannot adsorb nonspecific proteins in comparison to hydrophobic surfaces. Water molecules bind the hydrophilic surface strongly and securing it firmly and it is difficult to replace the binding efficiency of protein [54].

The glycocalyx is responsible for maintaining a healthy functional state, functional connection between the cell membrane and itself. It also helps to and fro inter and intracellular signaling. It associates with selectin, integrins, and immunoglobulins to ensure appropriate cellular signaling. It restores and maintains vascular permeability even under stressed conditions. This stress may due to colloids, fluids moving along the endothelium. It provides a check on the molecules that approach endothelial cells or interact during the interactions between the wall of vessels, blood cells, platelets, and other circulating cells healthy or sick cells. It acts as a regulatory factor for the rheological fluctuations due to the flow of blood and other body fluids like urine, cerebrospinal fluid, and lymph, *etc*. It regulates the interplay during stimulation to the cell that includes enzymes, cytokines, and reperfusion on the onset of adhesion of molecules. It functions as a mechanotransducer and facilitates cell proliferation during transplant. It acts as a potential barrier, but if damaged, then it leads to pathological conditions. It plays a significant role during cellular adhesion, cellular recognition, and signaling [53, 55, 60 - 70]. The glycocalyx, if damaged, the risk of pro-atherosclerotic, ischemia, a higher concentration of low-density lipids, adhesion of leucocytes, hyperglycemia, mismanagement of intravenous fluids, and renal disorders highly increase. Damage to glycocalyx aggravates unwanted permeability through inter-endothelial pores (fenestrated zone), leading to the independent functioning of growth factors, cytokines, and elevated cell-matrix [54, 71 - 73].

Endothelins are potential reactive peptides and are the enzymes with 21 amino acids. These function as regulators in the secretory pathway in maintaining the cell surface and as endothelium-derived vasoconstrictors. There are three significant endothelins (ET) found in mammals, ET-1, ET-2, and ET-3. These produced in various mammalian tissues. Endothelins are the product of around 200 residue prepropeptides; subtilisin–a prohormone present in these endothelins. Phosphoramidon inhibits the vasopressor effect of endothelin. Endothelial converting enzymes regulate the structural and functional aspects of the endothelium [74]. Endothelin plays a significant role in regulatory processes in the mammalian cardiovascular system. It is a potential vasoconstrictor and can increase blood pressure by activating constriction of blood vessels (pressor action). It also accomplishes other biological functional aspects like regional vasodilatory impact, activation of proliferative activity of vascular smooth muscle cells, contraction of airways' and intestinal smooth muscles', the release of endothelial-derived relaxing factors from the vascular bed, stimulation of arterial natriuretic peptides secretion from atrial cardiocytes, inhibition of rennin release from glomeruli, modulation of norepinephrine release from sympathetic terminals, *etc*. Endothelin is also related to cardiac functioning. There are three significant aspects of cardiac functioning, namely inotropic, chronotropic, and dromotropic impacts. Inotropic impact deals with the energy/force of muscle contraction; these are either negative or positive. Negative impact reduced the energy of contraction while positive elevates this energy.

The chronotropic impact affects the heart rate, while dromotropic impact affects the velocity of conduction through the conducting tissue of the heart. Endothelin affects the inotropic and chronotropic effects of the myocardium [75]. Bax and Saxena suggested that endothelins are involved in many pathological conditions [76], the findings of Inoue support these conditions [75]. Endothelial shedding enzymes (endothelial converting enzymes (ECE) are the members of the metalloprotease family. These can shed the ectodomain of the endothelium. For experimental purposes endothelium of from human umbilical vein is considered. These enzymes play a crucial role in either regulating or shedding glycocalyx present in the luminal region [77].

Interactions Between Nanomaterials and Glycocalyx

The endothelial glycocalyx is present on the luminal side of the endothelial cells of blood vessels and as a peripheral layer around the freely moving macrophages and other cells. The glycocalyx is a dense network or meshwork which consists of glycoproteins and polysaccharides. It is a protective coat, and that acts as structural and functional aspects to regulate the interaction and translocation of materials that come in contact with endothelial cells. Glycocalyx adheres to the luminal side of the endothelial cells. It covers adhesion and signaling molecules present on the respective loci on the cell membrane; as a result, it prevents attachment, uptake, and translocation of the nanomaterials through endothelial cells. But once the glycocalyx gets degraded, these activities do take place [78]. Uptake of nanoparticles or metabolites takes place through endothelial cells only at those places where glycocalyx is absent due to its erosion. This erosion is the result of enzymatic activity of glycocalyx shedding enzymes. The human endothelial cells (from human umbilical vein) pickup polystyrene nanospheres (50 nm) readily during cell culture experiments, and immunostaining of heparan sulfate technique is suitable for this type of study [79]. Endothelial cells produce angiopoietin-2. It acts antagonist to the endothelial stabilizing receptor called Tie-2. The angiopoietin-regulates the vascular permeability during contraction and disintegration of cell junctions. Angiopoietin can shed or reduce the layer of the glycocalyx; the enzyme heparanase regulates this shedding and process and removes heparan sulfate. Thus there is increased heparanase enzyme activity and the amount of heparan sulfate as a result of the angiopoietin-2 impact. The leakage of plasma was noticed once glycocalyx was removed from endothelial cells [80].

There is a possibility to adopt a strategy in which glycol nanoparticles can mimic particles as a part of glycocalyx and drug, or the modified nanoparticles that can move across glycocalyx and endothelial cells. Self-assembled glycol nanoparticles could move across macrophages. The induced polarization of primary peritoneal macrophages can facilitate the uptake of glycol nanoparticles and converts immunosuppression phenotype M-2 into inflammatory M-1. Due to polarization, the cell surface signaling molecules like CD206 and CD23 decrease, but the expressions of CD23 get elevated. There was the secretion of cytokines, thereby supporting this polarization. The same technique applies to the glycocalyx of endothelial cells [81].

The glycocalyx is permeable with varying degrees. The studies related to permeable efficacy are time-bound and initiate within 2 minutes, but the profile of glycocalyx keeps on fluctuating even after this duration. Thus, permeability refers to the pattern of glycocalyx [82]. Endothelium glycocalyx is a membrane-bound layer consisting of carbohydrates and absorbed plasma proteins, and exhibits variations in its thickness (50 nm to 500 nm). The hydrodynamics of blood and body fluids influence the conformation and functionality of glycocalyx. The culture conditions and pathogenic conditions also affect its structure and functionality, likewise its chemical aspects. These features may not be useful *in vitro*, as found *in vivo* conditions [83]. Gold nanoparticles (less than 5 nm) coated with PEG amine or galactose translocate *via* endothelial glycocalyx in renal and brain endothelial glycocalyx. Renal endothelial glycocalyx exhibits a higher degree of uptake (4 times) in comparison to that of brain endothelial glycocalyx. In both, these endothelial glycocalyx is different, as confirmed by lectin binding. Reduction in glycocalyx reduces the internalization of coated gold nanoparticles in renal tissue, but this is not so in the case of brain tissue. The glycocalyx of endothelium exhibits tissue specificity [84].

Nanosized glycomaterials can mimic different components of the glycocalyx and can influence its biological functions. Glycoconjugates reflect on the possibility of glycomaterials like glycol-clusters, glycodendrimers, glycopolymers, *etc.*, which can help to understand the mechanism of its functionality. Glycocalyx affects processes like cell signaling, binding of immunoglobulins, activation of B-cell antigen receptors, initiation of a signaling cascade concerning the regulation of B-cell proliferation, and their differentiation, and senescence. Glycosylated nanomaterials having lipid anchors interplay with cellular glycocalyx and involve extracellular leaflet that belongs to the plasma membrane. Investigation in this direction can lead to understanding the role of glycocalyx during the translocation of nanomaterials through it [85]. The glycocalyx envelopes the apical or luminal sides of endothelial cells and performs endocytosis; it acts as a selective biological barrier. This barrier functions concerning specific materials with surface molecules. The glycocalyx can restrict the translocation of nanomaterials depending on its chemical composition, density, thickness, ability to renew, and negative charge on its surface [86]. Glycopeptide dendrimers inhibit adhesion of bacteria to glycocalyx or any material that mimics glycocalyx, and this process follows the effects of macromolecules [87].

CELL MEMBRANE: OVERVIEW

The cell membrane or plasma membrane or cytoplasmic or plasmalemma is a cellular component that binds the contents of and demarcates the limits between the cell and its ambient extracellular matrix. Structurally it is lipid bilayer with some specific proteins either embedded in it or linked on the outer and inner phases. These proteins are transmembrane proteins. Some of them act as ion channels and also water channels. Functionally it maintains the structural and functional integrity of a cell and on the whole architectural and functional unit of a biosystem. Cell membrane helps in cellular functions like cellular adhesion, the conductivity of ions and selective biomolecules, inter and inter and intracellular signaling, *etc.* It is in communication with glycocalyx on the outer side and cytoskeleton internally. Biochemically glycocalyx is glycoproteins and oligosaccharides or glycolipids, while cytoskeleton has protein components. Under the field of synthetic biology, the cell membrane reassembles artificially [88, 89].

Some of the specific biomolecules associated with cell membranes receive stimuli, particular molecules that bring some information to the cell, and vice versa. These are structurally and functionally specific and bring about inter and intracellular communication and accomplish cellular working, maintains homeostatic equilibrium, and facilitate the suitable cellular response to adverse as well as normal stimuli. Most of them are proteins or conjugated proteins and act as integral membrane proteins. The molecules move as extracellular entities are hormone, neurotransmitter, cytokines, growth factors, cyto-adhesive molecules, nutrients, and potential activators of receptors located on the cell membrane. Mostly their interactions involve various signal transductions, binding with ligands, and by inducing chemical change cascades. The transmembrane proteins, *i.e.*, receptors, are generally glycoproteins and lipoproteins in nature and exhibit a wide variation in their structure and functions. Their classification depends on their three-dimensional structure, a tertiary structure. Those receptors are three-dimensional aspects and related to the topology of the membrane. The most common receptors are in the form of polypeptide chains cross lipid bilayer only on time. The complex receptors, such as G-protein coupled receptors, cross the lipid bi membrane many times. The membrane receptors are distributed on cell membranes depending on their structure and functions thus do not exhibit uniformity. But generally, these are located as clusters on the membrane [90 - 93].

There are two prime mechanisms that participate during the functioning of transmembrane receptors, (i) dimerization and (ii) rotation. During dimerization, the monomeric form of receptor changes into active dimmer form before it binds with a specific ligand. Under normal conditions, the transmembrane receptors are monomeric forms. During the process of rotation or some conformational changes take place in the helical region of transmembrane receptors. These conformational changes get exposed to the intracellular side of the cell membrane. Thus these changes facilitate suitable interaction with protein located inside the cell [94].

The most accepted concept of the model of the cell membrane is that of the fluid mosaic model is made up of lipid bilayers studded with proteins forming a protein-protein complex. Cytoskeleton binds the inner phase of the cell membrane, while the outer aspect is by glycocalyx. There are specialized ordered micro-domains and compactly packed components consisting of glycosphingolipids and protein receptors. These are the lipid rafts and are capable of moving within the membrane without any hindrance. These act as prime centers for trafficking of membrane proteins, affecting the fluidity of membrane and assembly centers for molecules related to cell signaling, synchronized neurotransmission, and trafficking of receptors. These lipids rafts are present in the membranes of cell organelles like lysosomes and the Golgi apparatus [95 - 97]. The compositional aspect of the cell membrane is a very dynamic and specific lipid, and protein molecules generally keep on fluctuating. As a result, there are changes in the fluidity of the membrane and its ambient environment. Such changes are of frequent occurrence during the developmental stages of the cells [98]. The components of a stable membrane are likely to have low solubility in aqueous conditions. Probably there is an exchange of lipid and protein molecules during the aqueous phase. Endocytosis, exocytosis, and intracellular vesicles can selectively either incorporate or eliminate molecules from the cell [99, 100].

The lipid bilayer present in the cell membrane is the product of self-assembly of amphipathic phospholipids. This biomolecule consists of the hydrophobic tail and a hydrophilic head. During this self-assembly head region arranges towards the outer (extracellular phase) and inner phases (intracellular phase) of the cell membrane while the hydrophobic region (tail) gets sandwiched in between. As a result, the cell membrane exhibit hydrophilic and hydrophobic behavior. These two features concerning cell membranes act as the major causes of the lipid bilayer nature, reactivity with hydrophilic and hydrophobic molecules, and change in entropy. Other interacting forces, noncovalent forces like van der Waals, electrostatic, and hydrogen, also participate in cellular functions. Further, this arrangement of head and tail of amphipathic phospholipids is capable of preventing the movement of polar ions or polar molecules like amino acids, nucleic acids, carbohydrates proteins, *etc*. Thus cell membrane acts as a selectively permeable structure. The cell membrane, along with the cell, can control the movement of the various molecules involving complexes of transmembrane proteins, pores, channels, and gates. Besides, other compounds facilitate the translocation of some of the specific molecules through the lipid bilayer that carries a negative charge. Among these, the most common ones are lipases (an ATP dependent on a transmembrane lipid transporter protein), scramblase (a protein that concentrates phosphatidylserine with a negative charge and facilitates the translocation of phospholipids through the lipid bilayer of the cell membrane). There is a complex molecule N-acetyl neuraminic acid (NANA – a derivative of neuraminic acid- a monosaccharide having 9-carbon backbone). These complex molecules function as additional barriers for the charged moieties that need to move across the cell membrane [101, 102]. Accurately, in the case of epithelial cells, protein complexes like ion channels, pumps can shift from the basal surface to the lateral surface of these cells and back, and facilities because of the fluid mosaic nature of the cell membrane. In the case of epithelial cells, the tight junctions are placed closer to the apical surface (towards the lumen) and prevent movement of the ion channels and pump to the apical surface of the epithelial cells [101].

Interaction Between Nanomaterials and Cell Membrane

Nanoscaled materials are natural products and synthetic products under the guidelines of nanotechnology and are tissue targets involving materials, explicitly science, and their biomedical applications. These nanomaterials induce physical, chemical, and biological impacts on biosystems. The cell membrane is one of the prime sites of such interactions. Nature of the cell membrane and physicochemical features of the nanomaterials like size, shape, quality, topography, charge, functional groups, *etc.*, present on the surface of nanomaterials influence their interactions. Artificial, engineered, and natural nanomaterials like metal, metal oxide, semiconductors, and polymeric nanomaterials exhibit varied impacts on the cell membrane. The understanding of such interactions is of great help in deciding the rationale applications in the materials sciences, pharmacology, and therapeutic drugs and as nano-drug carriers to secure the safety of human beings and the environment [103]. Nanomaterials get internalized directly (inhalation), *via* the dermal route (application of cosmetics), along with food products (oral-ingestion), as a contrast agent during biomedical diagnosis (as a contrast agent for magnetic resonance imaging), *etc.* The cell membrane exhibits specific features like surface charge, composition, surface tension, elasticity, and physicochemical characteristics. The nanomaterials possess features like thickness, physicochemical features, size, shape thickness, surface features, *etc.* The interplay between the cell membrane and nanomaterials depends on their characteristics.

Further, the parameters of the ambient environment like pH of the solvent, polarity, temperature, viscosity, ionic strength, and biological substrates also have their impacts on the interaction between nanomaterials and cell membranes [104]. When nanomaterials come in contact with the cell membranes, the nanoparticles may remain freely moving in the ambient environment or adhere to the cell membrane or wrapped in the cell membrane and become internalized. During adhesion, there can be partial envelopment of nanoparticles. When nanoparticle undergoes passive engulfment, three parameters provide energy to accomplish the process. These are (i) energy of adhesion at the site of the area of contact between the nanoparticle and the membrane, (ii) bending module of the membrane (k), and (iii) surface tension of the cell membrane (σ). Nanoparticles are brought in to membrane by adhesion energy; the resistance of the cell membrane facilitates its deformation and bending takes place concerning the elastic module. In this case, two features, *i.e.*, tensionless cell membrane and non-specific adhesion, play a significant role specifically during engulfment of nanoparticles. In such cases, the adhesion energy must be either equal or enough to overcome the binding energy of the membrane that is around the nanoparticle [104 - 106].

The interaction between metal and metal oxide nanoparticles can be toxic and non-toxic, it depends on the dose exposure, physicochemical properties of nanoparticles, as mentioned earlier [107 - 109]. These nanomaterials act as useful tools or agents for diagnostics, imaging, biosensing if their features undergo suitable modification [110]. Three conventional approaches are available for understanding the mechanism of interactions between cell membranes and nanoparticles: direct diffusion, degree of disruption to the plasma membrane, and endocytosis. These are likely to involve ion-channels, transporting proteins, and endocytic pathways. Interaction between gold nanoparticles with a model as cell membrane depends on the surface charge on the gold nanoparticles, integrity of cell membrane, size, surface modification charge, and state of the phase of the lipid bilayer (it can be gel or liquid phase). The interaction between cation ammonium gold nanoparticles and anionic and carboxyl gold nanoparticles with the lipid layer results in structural and organizational changes in the lipid bilayer (floating lipid bilayer as model membrane and solid substrate carrying bilayer on it). The arrangement of the lipid bilayer is such that one lipid layer floats around 2-3 nm away from the adsorbed layer on the solid substrate. This lipid bilayer is highly hydrated and exhibits fluctuations in its dynamic properties resembling natural cell membranes. This technique represents biomimetic samples and is useful to examine the structural integrity of the cell membrane. This technique also enables us to study the fate of gold nanoparticles using neutron reflectometry. The concentration of gold nanoparticles is an essential parameter because its higher concentration results in the disruption of the membrane, while low concentration causes the incorporation of gold nanoparticles in the hydrophobic moiety of the cell membrane. The functionalized gold nanoparticles with cationic head groups penetrate the center of the lipid bilayer and disrupt the cell membrane. In contrast, gold nanoparticles functionalized with anionic head groups do not permeate through the membrane and prevent the disruption of the membrane.

Polycationic organic nanoparticles generally disrupt the model biological membrane and living cell membrane at nano-molar concentration. The degree of disruption of cell membrane relates to the size, shape, charge on the nanoparticles, fluid or gel, or crystalline state of biological membrane. The internalization of nanoparticles also depends on the shape of nanoparticles. Spherical nanoparticles of the same size internalize 500% more than rod-shaped nanoparticles. The elongated nanoparticles take more time to form membrane wrapping in comparison to the spherical nanoparticles. The size of nanoparticles strongly affects the binding and activation of membrane receptors and subsequent protein expression. There is a possibility of inducement of toxicity because of their interaction with the components of the membrane. Scanning probe microscopy helps to study disorder or disruption on the living membranes; fluorescence microscopy and diffusion assay are beneficial to investigate cellular leakage [111 - 113].

Nanomaterials internalize typically *via* bounded endosomes and do not gain access to the cellular machinery. Some biomacromolecules can penetrate or fuse with the cell membrane without disrupting it. In most cases, synthetic materials of comparable size do not exhibit such behavior. Hydrophobicity facilitates the internalization of cationic nanoscaled substance only to some extent. These also gain entry by generating transient openings/holes associating with cytotoxicity [114]. When the cell membrane comes in contact with Al_2O_3 nanoparticles (13 and 2nm; 10 and 25 µg ml⁻1) for 24 hours, these nanoparticles significantly declined the cell viability. On comparing the impact of three metal oxide nanoparticles, *i.e.*, TiO_2, Al_2O_3, and CeO_2, the degree of decline in cell viability was in the order TiO_2< Al_2O_3 < CeO_2. A relatively higher degree of depolarization results with smaller nanoparticles (13 nm Al_2O_3) in comparison to larger nanoparticles of Al_2O_3 (30 nm). Maximum toxicity is in the case of CeO_2 (20 nm) but results in significantly very less depolarization and this reflects on the effectiveness of other factors like exposure duration, surface chemistry degree of the depolarization. These metal oxides nanoparticles exhibit cytotoxicity concerning size as CeO_2 (20 nm)>Al_2O_3 (22 nm)>TiO_2 (40nm) [115]. Different factors may contribute to the degree of toxicity due to nanoparticles. First, maybe the exposure time consumed during the imaging in addition to the experimental exposure time. Thus, any change during imaging experimental represents only the initial effect of nanoparticles on the cell; the cellular environment limits the duration involved during the imaging study. During this time, it is challenging to control temperature and gas concentration. Secondly, it is entirely reasonable to propose that Al_2O_3 nanoparticles can exhibit faster kinetics than CeO_2 while moving through the solution and interacting with the cell due to the lower density of Al_2O_3 nanoparticles. Thirdly, CeO_2 nanoparticles can induce toxicity through mechanisms that do not induce disruption of the cell membrane, but other modes like oxidative stress or lipid peroxidation or both can cause their impacts. Fourth, since the surface chemistry of CeO_2 and Al_2O_3 is different, their interaction with the membrane will also be altered. The metal atoms on the surface of the two different metal oxide nanoparticles are different in size, charge-charge density, valance electron, and valence electron orbital arrangements. These are some of the factors that are likely to result in various interactions with cell and cell membrane [114, 115].

CYTOSKELETON

The cell is the structural and functional unit of the biosystem, may it be prokaryote or eukaryote. Operational aspects of a given cell relate to the behavior of cells and the cell organelles. Cellular behavior includes cellular communication, elasticity, mobility, adhesion, invasion, proliferation, differentiation, phagocytosis, endocytosis, and exocytosis, and responding to various stimuli all depend on the cytoskeleton. The cytoskeleton retains the specific temporal and spatial orientation of prokaryotic and eukaryotic cells. Cells are the ultimate targets that face sudden physical, biochemical, physiological, molecular, and physicochemical changes in their ambient environment. These changes bring about a derogative and stressful influence on cells [50, 54]. The specific physicochemical features, like the ability for surface modifications, charge modification, stealth features like size, shape, area to volume ratio, can induce morphological, molecular, biochemical fluctuations in a cell.

The cytoskeleton is primarily responsible for the cellular orientation suitable to its morphological, histological, and functional aspects, and ensures its physiological homeostatic state. It acts as a framework for the cell to provide distinct cellular shape, size, and highly intricate intracellular organization [54]. The cytoskeleton consists of wiry and tubular structures along with a dynamic matrix. The dynamism of the cytoskeleton helps to restore the fundamental structural and functional aspects. This cellular component also concerned with the functioning of cilia, flagella, pseudopodia, and the intracellular process like chromosomal movements during cell replication, muscular contraction, and relaxation, the orientation of cell organelles [52, 54, 116]. Biochemical and molecular interactions related to tyrosine kinase, lipid kinase, phospholipids, serine/heroine kinase, *etc.*, for example, poisonings trigger the process of phagocytosis also involve cytoskeleton [54]. Although signaling pathways and receptors like Fc have their respective parts to play, cytoskeleton also has a dynamic role in these functions [117]. During the activation of signaling molecules, some of those molecules are like GTPase of Rho family, guanine-nucleotide exchange factors concerned to regulate GTPase, and the factors that promote nucleation and molecular motors of myosin family, *etc.*, and remodeling of components of the cytoskeleton. All these factors are dedicated to the synchronized and chronological performance of the cytoskeleton to facilitate smooth cellular functions [54, 118].

The behavior of the three major types of components of the cytoskeleton, namely microtubules, intermediate filaments, and actin or microfilaments, facilitate the execution of various cellular functions. Intermediate filaments provide mechanical strength and enable the cell to bear sheer/stress imposed. Microtubules help to ensure the respective positioning of cell organelles and during intracellular transportation. Actin or microfilaments are responsible for the maintenance of the shape of the cell surface and the cellular movements. The functionality of the components of the cytoskeleton is due to the accessory proteins. These proteins get linked with each filament and also with the cellular components as a result, ensuring controlled and coordinated assembly of these components concerning specific cellular locations [53, 117]. Motor proteins are responsible for the regulatory and coordinated roles in the efficient functioning of components of the cytoskeleton as one unit. The further cytoskeleton also facilitates in regulating the protein-protein interaction during the signaling pathway and ligand-dependent receptor signaling [53, 117].

The cytoskeleton is present on the inner phase of the cell membrane, and it acts as scaffolding to provide anchorage to the membrane proteins and even structuring cell organelles extending from the cell. The components of the cytoskeleton have intimate structural and functional interaction with the cell membrane. The cytoskeleton also participates in appendages like organelle cilia and microtubule-based extensions based on actin [119]. Permeability is one of the essential functional aspects of the cell membrane, and this process is responsible for passive diffusion molecules, and these are permanent molecules. This property is dependent on an electric charge, the polarity of the molecules and, to some extent, molar mass also affects this process. Hydrophobicity of the cell membrane plays a crucial role during the translocation of small electrically neutral molecules in comparison to the charged ions. When charged particles fail to move across the cell membrane, it results in the partition of pH throughout that specific fluid compartment [120].

The Interactions of Nanomaterials with Cytoskeleton

Natural and man-made nanomaterials are suitable materials for the applications in various fields like as mentioned earlier. Nanomaterials can cause of derogative and beneficial impacts depending on the nature of nanomaterials, type of the cell, dose, and ambient physiological conditions. These changes can influence cellular elasticity, mobility, adhesion, invasion, transport of cell organelles within the cell, *etc*. Nanomaterials like carbon-based nanomaterials, metal oxide; quantum dots, *etc*., induce biochemical, molecular, and biophysical impacts at the interface formed at the site of interaction [121]. Cytoskeleton dynamic and its components undergo either polymerization or depolymerization. This dynamism plays a significant role in cell differentiation, cell morphology, and proliferation, maintenance of cells as a structural and functional entity. Interactions between nanoparticles and cytoskeleton cause dysfunction due to reactive oxygen species formation, direct physical interaction; imbalance in the intracellular energetics, and changes in electrophysiological aspects of cytoskeleton specifically in the case of nerve cells [122].

Supra paramagnetic iron oxide nanoparticles cause stress in the dermal fibroblast cells and deceive the signaling pathway, cytoskeleton, and receptors dependent endocytosis, and aggregate along the cell membrane [123]. Silicon dioxide nanoparticles (5 nm to 150 nm) induced the process of formation of the actin filament [124]. The polymerization of cortical F-actin filaments present in a microvascular endothelial cell is affected by the uncoated magnetic nanoparticles (50µg/ml, duration 10 min to 5 hours). These nanoparticles elevate the degree of cellular permeability [125]. Weakly protected gold nanoparticles on interacting with microtubules inhibit their polymerization, and these nanoparticles get aggregated within the free cell system. Single citrate capped gold nanoparticles resulted in the aggregation of tubulin heterodimers. The conformational changes are caused in and act as the leading cause of this interaction. These gold nanoparticles got internalized in A549 cells (used as test model) and also caused the arrest of the cell cycle, at GO or G1 phase and also apoptosis, these gold nanoparticles induce mitochondrial apoptosis; these gold nanoparticles influence the proteins concerning apoptosis-like Bax, p53 (up-regulating proteins) and Bcl-2 and cleavage of poly (ADP-ribose polymerase (downregulating proteins). The degree of aggregation of insoluble intracellular tubulin elevates in the case of the experimental cell but not in a control set of the cells. The extent of inhibition of polymerization is in congruence with size, concentration, and the duration of exposure to the gold nanoparticles. Cellular microtubules play a significant role during the interaction of anti-cancer drugs and when it is affected adversely during the interaction with gold nanoparticles as a result of conformational changes, the tubulin and microtubule system becomes dysfunctional [126].

Titanium-di-oxide (100µg/ml) can change the cell morphology of human neuroblastoma cell line SH-SY5Y (used as test model) but not the cellular viability in these cells. There is a disorder, disruption, retraction, and a distinct declined intensity among microtubules in these cells resulted due to the interactions between TiO_2 nanoparticles. Further, the percentage of soluble tubulin elevates, and the expression of α and β tubulin did not fluctuate. These dynamic changes in microtubule reflect on the declined growth rate and increased rate of reduction in microtubule and its lifetime. There are possibilities that the interaction between TiO_2 and microtubules results in the involvement of heterodimers, microtubules, and tau-proteins. This interaction plays a significant role in stabilizing of microtubules [127]. Single-walled carbon nanotubes (SWCNTs) and actin component of the cytoskeleton at the molecular level reflect on its binding with SWCNT. This binding is stable and involves hydrophobic interactions. This type of binding permits the sliding of nanotubes and rotation on the surface of actin. There is the formation of actin-SWCNT complexes that may result in the induction of the reorganization of filaments of actin [128].

Quantum dots are among one of the favored options to probe the active and passive transport across the cell membrane. Functionalized quantum dots exhibit a double population depending on their different constants and the degree of depolymerization of actin. Functionalized quantum dots act as tools to target various kinesin motors. There are chances that during this study, some indicative refractory as stable actin induced transverse displacement, may not be observed. Further, non-functional quantum dots diffuse in a cell heterogeneously, and it relates to the presence of F-actin. Microtubules associated quantum dots internalize without the involvement of kinesin and these cause changes in the transverse process and actin.Kinesin-1 bound quantum dots to exhibit direction-dependent fluctuations, possibly because of the modification in microtubules involving motors [129]. Quantum dots also internalize *via* microinjection and cause the spatial coordinate ion can result. The fluorescent technique is useful to study such internalization. Fibroblast cells undergo due to the destabilization of actin and lead to the disruption of a network of actin in the experimental cells [130]. Quantum dots conjugated with epidermal growth factor undergo endocytosis in a living cell using Arpe-19 cells, (human epithelium that has rich actin cortex and polarized microtubules). There are heterozygous movements of individual endosomes due to the quantum dots. Most of the endosomes moved towards the perinuclear zone. This movement involves microtubule and dynein dependent diffusive movements, and actin does not participate. Maneuvers of endosome change with dense microtubule region, interaction with other endosomes, and endoplasmic reticulum. During this movement, merging and splitting of endosomes, changes in the endoplasmic reticulum, and the changes in directions occur. There is high tension on the endosomal membrane, specifically during directional oriented movements. Molecular motors play a significant role during intracellular endosomal trafficking along with components of the local cytoskeleton and endoplasmic reticulum [131].

The actin network acts as one of the biological barriers for the diffusion of nanomaterials using self-fluorescent cationic dendrimers (6 nm as diameter). This nanomaterial interacts with actin filaments. This interaction is reversible and is possibly electrostatic. The rate of diffusion of dendrimers retards and the polymerization of actin also gets affected. When dendrimers are in low concentration, these act like G-binding actin protein and result in declined polymerization of actin. When dendrimers are in higher concentration, these behave as nucleating protein, and polymerization rate got elevated. Dendrimers can induce physical and chemical impacts. When the degree of polymerization is low, its diffusion is more, and when the polymerization rate is higher, its rate of diffusion is less. Thus cationic dendrimers interact with the biochemical aspects of the actin component of the cytoskeleton. Despite its effect on the polymerization of actin still, it can be an option to act as a drug carrier [132]. Functionalized polyamidoamine dendrimers conjugated with photo-physical dye

Cy5 and Cy3 as fluorescent dendritic nanoprobes are useful to study and to image the cytoskeletal network in mammalian cells. The conjugated dendritic nanoprobes interact with immunolabeled microtubules using fluorescence microscopy, the microtubules under 33kW/cm2; in the presence of an oxygen scavenger system, *i.e.*, glucose oxidase or catalase shows the role of dendritic nanomaterials. The ratio of concentrations of the secondary antibody and dendritic probe as 0.4:1 and 0.9:1 for Cy5 and Cy3 probes could give appropriate imaging of the cytoskeletal network [133].

Cylindrical nanoparticles penetrate the cell membrane and cause deformation in the cytoskeleton. During penetration of cylindrical nanoparticles, there is resistance due to the cytoskeleton, and it distorts the cell membrane, conformational fluctuations in the ligand-receptor bonds, and free receptors. These factors influence the engulfment of such nanoparticles. Most of the cylindrical nanoparticles interact with the cell membrane vertically during the process of engulfment cytoskeleton impede their entry. This process depends on stiffness-dependent optimal radiuses of the cylindrical nanoparticles that enter the cell. The stiffness of the cytoskeleton is more active during vertical entry in comparison to horizontal engulfment [134].

F-actin component of the cytoskeleton is primarily responsible for force transduction, transport maintenance of cell structure, and cytokinesis. Highly purified single-walled carbon nanotubes cause changes in F-actin but no interaction and change in G-actin after intracellular biodistribution. Even myosin–II located sub-cellularly is also not affected. Further, the traction force was substantially declined in cells treated with single-walled carbon nanotubes [135].

CONCLUSION

Biosystems are acellular, cellular, unicellular, or multicellular structures. The cell is a well organized structural and functional unit of the biosystem. It exhibits a wide variety, nature, and functions that ensure its survival. In the recent past, there have been tremendous developments in nanoscience, nanotechnology, and nanomaterials; as a result, these products find applications in industries, food technology, agriculture, pharmaceutics, cosmetics, clothing/garment, military ware fare, chemical technology, *etc.* Despite enormous advancements in this field, there have been more reports on their utility in the various area rather than clinical toxicity, indicating the probability of adverse biological interactions. Nanomaterials such as metal and a metal oxide, quantum dots, fullerene, and fibrous nanomaterials are capable of causing chromosomal fragmentation, DNA strand breaks, point mutation, oxidative DNA adduct formation and changes in expression of genes, *etc.* There is a great need to develop and scale up to meet the commercialization and industrial applications (Wolfgang., 2015). Understanding the concepts related to the interactions between cell, biosystem, and different nanomaterials helps in developing suitable nanodevices for future use. Such processes that accomplish this interaction will furnish more relevant information about the intricacies involved. The role of biomolecules and other molecules of the particular size typically liposomes, nanomaterials, and macromolecular exhibit behavior that guide in formulating, functionalizing a specific nanomaterial that can achieve the set target in the biomolecular field and biomedical sciences. One of the significant challenges during designing, fabricating, or formulating specific nanomaterials should ensure biocompatibility, biodistribution, and biodegradation.

REFERENCES

[1]   Lahir YK. Impacts of metal and metal oxide nanoparticles on reproductive tissues and spermatogenesis. J Exp Zool India 2018; 18(2): 594-608.

[2]   Maynard A. Scientists set five grand challenges for nanotechnology risk research 2006.https://phys.org/news/2006-11-scientis-grand-nanotechnology.html

[3]   Parak WJ, Nel AE, Weiss PS. Grand challenges for nanoscience and nanotechnology. ACS Nano 2015; 9(7): 6637-40.
[http://dx.doi.org/10.1021/acsnano.5b04386] [PMID: 26192457]

[4]   Zhang F. Grand challenges for nanoscience and nanotechnology in energy and health. Front Chem 2017; 5: 80.
[http://dx.doi.org/10.3389/fchem.2017.00080] [PMID: 29164100]

[5]   Saier MH Jr, Reddy VS, Tamang DG, Västermark A. The transporter classification database. Nucleic Acids Res 2014; 42(Database issue): D251-8.
[http://dx.doi.org/10.1093/nar/gkt1097] [PMID: 24225317]

[6]   Lau WL, Ege DS, Lear JD, Hammer DA, DeGrado WF. Oligomerization of fusogenic peptides promotes membrane fusion by enhancing membrane destabilization. Biophys J 2004; 86(1 Pt 1): 272-84.
[http://dx.doi.org/10.1016/S0006-3495(04)74103-X] [PMID: 14695269]

[7]   Cheng LC, Jiang X, Wang J, Chen C, Liu RS. Nano-bio effects: interaction of nanomaterials with cells. Nanoscale 2013; 5(9): 3547-69.
[http://dx.doi.org/10.1039/c3nr34276j] [PMID: 23532468]

[8]   Oliverira S, van Roog I, Kranenburg O, Storm G, Schiffeless RM. Fusogenic peptide enhance endosomal escape in improving siRNA – induced silencing of an oncogene. Int J Pharmacol 2007; 331(2): 211-4.
[http://dx.doi.org/10.1016/j.ijpharm.2006.11.050]

[9]   Nakase I, Futaki S. Combined treatment with a pH-sensitive fusogenic peptide and cationic lipids achieve enhanced cytosolic delivery of exosomes. Sci Rep 2015; S10112.
[http://dx.doi.org/10.1038/srep10112]

[10] Milletti F. Cell-penetrating peptides: classes, origin, and current landscape. Drug Discov Today 2012; 17(15-16): 850-60.
[http://dx.doi.org/10.1016/j.drudis.2012.03.002] [PMID: 22465171]

[11] Opalinska JB, Gewirtz AM. Nucleic-acid therapeutics: basic principles and recent applications. Nat Rev Drug Discov 2002; 1(7): 503-14.

[http://dx.doi.org/10.1038/nrd837] [PMID: 12120257]

[12]   Eskstein F. The versatility of oligonucleotides as potential therapeutics. Exp Opin Biolog Ther 2007; 7(7): 1021-34.
[http://dx.doi.org/https://doi;10.1517/14712598.7.1021]

[13]   Stewart KM, Horton KLO, Kelley SO. Cell-penetrating peptides as delivery vehicles for biology and medicine. Org Biomol Chem 2008; 6(13): 2242-55.
[http://dx.doi.org/10.1039/b719950c] [PMID: 18563254]

[14]   Copolovici DM, Langel K, Eriste E, Langel Ü. Cell-penetrating peptides: design, synthesis, and applications. ACS Nano 2014; 8(3): 1972-94.
[http://dx.doi.org/10.1021/nn4057269] [PMID: 24559246]

[15]   Rejinold NS, Han Y, Yoo J, Seok HY, Park JH, Kim YC. Evaluation of cell penetrating peptide coated Mn:ZnS nanoparticles for paclitaxel delivery to cancer cells. Sci Rep 2018; 8: 1.
[http://dx.doi.org/10.1038/s41598-018-20255-x] [PMID: 1899]

[16]   Freeman EC, Weiland LM, Meng WS. Modeling the proton sponge hypothesis: examining proton sponge effectiveness for enhancing intracellular gene delivery through multiscale modeling. J Biomat Sci Polymer Edition. 2013; 24(4): 398-416.
[http://dx.doi.org/ 10.80/09205063-2012.690282]

[17]   Gatenby RA, Gillies RJ. Why do cancers have high aerobic glycolysis? Nat Rev Cancer 2004; 4(11): 891-9.
[http://dx.doi.org/10.1038/nrc1478] [PMID: 15516961]

[18]   Popat A, Liu J, Lu GQMS. Qia O S Z, A pH-responsive drug delivery system based on chitosan-coated mesoporous silica nanoparticles. J Mater Chem 2012; 22: 11173-8.
[http://dx.doi.org/10.1039/c2jm30501a]

[19]   Mekaru H, Lu J, Tamanoi F. Development of mesoporous silica-based nanoparticles with controlled release capability for cancer therapy. Adv Drug Deliv Rev 2015; 95: 40-9.
[http://dx.doi.org/10.1016/j.addr.2015.09.009] [PMID: 26434537]

[20]   Liu J, Luo Z, Zhuang J, *et al.* Hollow mesoporous silica nanoparticles facilitate drug delivery *via* cascade pH stimuli in the tumor microenvironment for tumor therapy. Biomaterials 2016; 83: 51-65.
[http://dx.doi.org/10.1016/j.biomaterials.2016.01.008]

[21]   Meng H, Xue M, Xia T, *et al.* Autonomous *in vitro* anticancer drug release from mesoporous silica nanoparticles by pH-sensitive nanovalves. J Am Chem Soc 2010; 132(36): 12690-7.
[http://dx.doi.org/10.1021/ja104501a] [PMID: 20718462]

[22]   Möller K, Müller K, Engelke H, Bräuchle C, Wagner E, Bein T. Highly efficient siRNA delivery from core-shell mesoporous silica nanoparticles with multifunctional polymer caps. Nanoscale 2016; 8(7): 4007-19.
[http://dx.doi.org/10.1039/C5NR06246B] [PMID: 26819069]

[23]   Xiong L, Bi J, Tang Y, Qia OSZ. Magnetic core-shell silica nanoparticles with large radial mesopores of siRNA deliverysmall 2016; 12(34): 4735-42.
[http://dx.doi.org/https:doi:10.1002/smll.201600531]

[24]   Watermann A, Brieger J. Mesoporous silica nanoparticles as drug delivery vehicles in cancer, Nanomaterials 2017.
[http://dx.doi.org/https://10.0.13.62/nano7070189]

[25]   Durfee PN, Lin YS, Dunphy DR, *et al.* Mesoporous silica nanoparticles supported lipid bilayers (Protocells) for active targeting and delivery to individual leukemic cells. ACS Nano 2016; 10(9): 8325-45.
[http://dx.doi.org/10.1021/acsnano.6b02819] [PMID: 27419663]

[26]   Maeda H. Macromolecular therapeutics in cancer treatment: the EPR effect and beyond. J Control Release 2012; 164(2): 138-44.

[http://dx.doi.org/10.1016/j.jconrel.2012.04.038] [PMID: 22595146]

[27]   Yu M, Wu J, Shi J, Farokhzad OC. Nanotechnology for protein delivery: Overview and perspectives. J Control Release 2016; 240: 24-37.
[http://dx.doi.org/10.1016/j.jconrel.2015.10.012] [PMID: 26458789]

[28]   Nichols JW, Bae YH. Evidence and fallacy. J Cont Rel 2014; 190: 451-64.
[http://dx.doi.org/10.1016/j.jconrel.2014.03.057]

[29]   Nakamura Y, Mochida A, Choyke P, Kabayashi H. Nanodrug delivery: Is the EPR effect sufficient for curing cancer? Bioconjugate Chemistry 2016; 27(10): 2225-38.
[http://dx.doi.org/10.1021/acs.bioconjuchem] [PMID: 6b00437]

[30]   Wang J, Huang J, Chen H, *et al.* Exporting EPR the effect is driven by ultrafine iron oxide nanoparticles with T-1 and T-2 switchable magnetic resonance imaging contrast. ACS Nano 2017; 11(5): 4582-92.
[http://dx.doi.org/10.1021/acsnano.7b00038] [PMID: 28426929]

[31]   Narayanan E, Wakaskar R. Utilization of nanoparticulate therapy in cancer targeting. Cogent Med 2018; 5(1): 15045004.
[http://dx.doi.org/10.1080/2331205X.2018.1504504]

[32]   Zhang X-Q, Xu X, Bertrand N, Pridgen E, Swami A, Farokhzad OC. Interactions of nanomaterials and biological systems: Implications to personalized nanomedicine. Adv Drug Deliv Rev 2012; 64(13): 1363-84.
[http://dx.doi.org/10.1016/j.addr.2012.08.005] [PMID: 22917779]

[33]   Torchilin VP. Recent advances with liposomes as pharmaceutical carriers. Nat Rev Drug Discov 2005; 4(2): 145-60.
[http://dx.doi.org/10.1038/nrd1632] [PMID: 15688077]

[34]   Wang EC, Wang AZ. Nanoparticles and their applications in cell and molecular biology. Integr Biol 2014; 6(1): 9-26.
[http://dx.doi.org/10.1039/c3ib40165k] [PMID: 24104563]

[35]   Gradishar WJ, Tjulandins S, Davidson N, *et al.* Phase III trials of nanoparticles albumin-bound paclitaxel compared to with polyethylene castor oil-based paclitaxel in women with breast cancer. J Clin Oncol 2005; 23(31): 7768-81.
[http://dx.doi.org/10.1200/JCO.2005.04.937] [PMID: 16204007]

[36]   Huang X, Jain PK, El-Sayed IH, El-Sayed MA. Gold nanoparticles: interesting optical properties and recent applications in cancer diagnostics and therapy. Nanomedicine (Lond) 2007; 2(5): 681-93.
[http://dx.doi.org/10.2217/17435889.2.5.681] [PMID: 17976030]

[37]   Chan J, Valencia P, Zhang L, Langer R, Farokhzad O. Cancer nanotechnology. Humana Press 2010; 624.

[38]   Mintzer MA, Grinstaff MW. Biomedical applications of dendrimers: a tutorial. Chem Soc Rev 2011; 40(1): 173-90.
[http://dx.doi.org/10.1039/B901839P] [PMID: 20877875]

[39]   Mahmoudi M, Sant S, Wang B, Laurent S, Sen T. Superparamagnetic iron oxide (SPION): Nanoparticles development, surface modification, and application in therapy. Adv Drug Deliv Rev 2011; 63(1-2): 24-46.
[http://dx.doi.org/10.1016/j.addr.2010.05.006] [PMID: 20685224]

[40]   Chang YP, Pinaud F, Antelman J, Weiss S. Tracking bio-molecules in live cells using quantum dots. J Biophotonics 2008; 1(4): 287-98.
[http://dx.doi.org/10.1002/jbio.200810029] [PMID: 19343652]

[41]   Vlasova II, Kapralov AA, Michael ZP, *et al.* Enzymatic oxidative biodegradation of nanoparticles: Mechanisms, significance and applications. Toxicol Appl Pharmacol 2016; 299: 58-69.
[http://dx.doi.org/10.1016/j.taap.2016.01.002] [PMID: 26768553]

[42] Kamaly N, Xiao Z, Valencia PM, Radovic-Moreno AF, Farokhzad OC. Targeted polymeric therapeutic nanoparticles: design, development and clinical translation. Chem Soc Rev 2012; 41(7): 2971-3010.
[http://dx.doi.org/10.1039/c2cs15344k] [PMID: 22388185]

[43] Perez-Herreo E, Fernandez-Medarde A. Advanced targeted therapies in cancer: drug nanocarrier, the future of chemotherapy, European. J Pharm Biomorph 2015; 93: 52-79.
[http://dx.doi.org/10.1016/j.ejph.2015.03.018]

[44] Sureshbabu AR, Kurapati R, Russier J, *et al.* Degradation-by-design: Surface modification with functional substrates that enhance the enzymatic degradation of carbon nanotubes. Biomaterials 2015; 72: 20-8.
[http://dx.doi.org/10.1016/j.biomaterials.2015.08.046] [PMID: 26342557]

[45] Francois M, Espinosa A, Luciant N, *et al.* Massive intercellular biodegradation of iron oxide nanoparticles evidenced magnetically single endosome and tissue level.. ACS Nano 2016; 10: 7626-38.
[http://dx.doi.org/10.1021/acsnano.6b02876]

[46] Robert JG. Contact angle, wetting, and adhesion: a critical review. J Adhes Sci Technol 1992; 6(12): 1269-302.
[http://dx.doi.org/10.1163/156856192X00629]

[47] Lee KS, Ivanova N, Starov VM, Hilal N, Dutschk V. Kinetics of wetting and spreading by aqueous surfactant solutions. Adv Colloid Interface Sci 2008; 144(1-2): 54-65.
[http://dx.doi.org/10.1016/j.cis.2008.08.005] [PMID: 18834966]

[48] Clegg C. Contact Angle Spreading Coefficient 2016.www.ramehart.com

[49] Mathews H l . Host defense Complement 2011.www.lumen.luc.edu/Lumen/meded/hostdef/week5.pdf

[50] Becker WM, Kleinsmith LJ, Hardin J. The world of the Cell. 5th Ed., Philadelphia, Delhi: Pearson Education (Singapore) Pvt Ltd 2004.

[51] Cooper GM. The Molecular Approach. 2ND Ed., Sunderland (MA): Sinauer Associates 2015 .

[52] Lahir YK. A dynamic component of a tissue-Extracellular matrix: structural and adaptive approach. Biochem Cell Arch 2015; 15: 331-47.

[53] Lahir YH. Understanding the basic role of glycocalyx during cancer. Journal of Radiation and Cancer Research 2016; 7: 79-84.
[http://dx.doi.org/10.4103/0973-0168.197974]

[54] Lahir YK. Some aspects of interactions between Nanomaterials and the Cytoskeleton of Eukaryotic cells. Ad Clin Toxi 2016; 1(2): 000111.

[55] Reitsma S, Slaaf DW, Vink H, van Zandvoort MA, oude Egbrink MG. The endothelial glycocalyx: composition, functions, and visualization. Pflugers Arch 2007; 454(3): 345-59.
[http://dx.doi.org/10.1007/s00424-007-0212-8] [PMID: 17256154]

[56] Carey DJ. Multifunctional cell surface coreceptors. Biochem J 1997; 327(Pt-1): 1-16.

[57] Fransson LA, Belting M, Cheng F, Jönsson M, Mani K, Sandgren S. Novel aspects of glypican glycobiology. Cell Mol Life Sci 2004; 61(9): 1016-24.
[http://dx.doi.org/10.1007/s00018-004-3445-0] [PMID: 15112050]

[58] Esko JD, Selleck SB. Order out of chaos: assembly of ligand binding sites in heparan sulfate. Annu Rev Biochem 2002; 71: 435-71.
[http://dx.doi.org/10.1146/annurev.biochem.71.110601.135458] [PMID: 12045103]

[59] Sugahara K, Mikami T, Uyama T, Mizuguchi S, Nomura K, Kitagawa H. Recent advances in the structural biology of chondroitin sulfate and dermatan sulfate. Curr Opin Struct Biol 2003; 13(5): 612-20.

[http://dx.doi.org/10.1016/j.sbi.2003.09.011] [PMID: 14568617]

[60]   van Haaren PM, VanBavel E, Vink H, Spaan JA. Charge modification of the endothelial surface layer modulates the permeability barrier of isolated rat mesenteric small arteries. Am J Physiol Heart Circ Physiol 2005; 289(6): H2503-7.
        [http://dx.doi.org/10.1152/ajpheart.00587.2005] [PMID: 16100247]

[61]   Gouverneur M, Spaan JA, Pannekoek H, Fontijn RD, Vink H. Fluid shear stress stimulates incorporation of hyaluronan into endothelial cell glycocalyx. Am J Physiol Heart Circ Physiol 2006; 290(1): H458-2.
        [http://dx.doi.org/10.1152/ajpheart.00592.2005] [PMID: 16126814]

[62]   Jacob M, Bruegger D, Rehm M, Welsch U, Conzen P, Becker BF. Contrasting effects of colloid and crystalloid resuscitation fluids on cardiac vascular permeability. Anesthesiology 2006; 104(6): 1223-31.
        [http://dx.doi.org/10.1097/00000542-200606000-00018] [PMID: 16732094]

[63]   Constantinescu AA, Vink H, Spaan JA. Endothelial cell glycocalyx modulates immobilization of leukocytes at the endothelial surface. Arterioscler Thromb Vasc Biol 2003; 23(9): 1541-7.
        [http://dx.doi.org/10.1161/01.ATV.0000085630.24353.3D] [PMID: 12855481]

[64]   Pries AR, Secomb TW. Rheology of the microcirculation. Clin Hemorheol Microcirc 2003; 29(3-4): 143-8.
        [PMID: 14724335]

[65]   Lipowsky HH. Microvascular rheology and hemodynamics. Microcirculation 2005; 12(1): 5-15.
        [http://dx.doi.org/10.1080/10739680590894966] [PMID: 15804970]

[66]   Henry CB, Duling BR. TNF-alpha increases entry of macromolecules into luminal endothelial cell glycocalyx. Am J Physiol Heart Circ Physiol 2000; 279(6): H2815-23.
        [http://dx.doi.org/10.1152/ajpheart.2000.279.6.H2815] [PMID: 11087236]

[67]   Florian JA, Kosky JR, Ainslie K, Pang Z, Dull RO, Tarbell JM. Heparan sulfate proteoglycan is a mechanosensor on endothelial cells. Circ Res 2003; 93(10): e136-42.
        [http://dx.doi.org/10.1161/01.RES.0000101744.47866.D5] [PMID: 14563712]

[68]   Lanctot PM, Gage FH, Varki AP. The glycans of stem cells. Curr Opin Chem Biol 2007; 11(4): 373-80.
        [http://dx.doi.org/10.1016/j.cbpa.2007.05.032] [PMID: 17681848]

[69]   Xu GK, Qian J, Hu J. The glycocalyx promotes cooperative binding and clustering of adhesion receptors. Soft Matter 2016; 12(20): 4572-83.
        [http://dx.doi.org/10.1039/C5SM03139G] [PMID: 27102288]

[70]   Alphonsus CS, Rodseth RN. The endothelial glycocalyx: a review of the vascular barrier. Anaesthesia 2014; 69(7): 777-84.
        [http://dx.doi.org/10.1111/anae.12661] [PMID: 24773303]

[71]   Devuyst O. Glycocalyx: the fuzzy coat now regulates cell signaling. Perit Dial Int 2014; 34(6): 574-5.
        [http://dx.doi.org/10.3747/pdi.2014.00221] [PMID: 25228208]

[72]   Biddle C. Like a slippery fish, a little slime is a good thing: the glycocalyx revealed. AANA J 2013; 81(6): 473-80.
        [PMID: 24597010]

[73]   Luciana T, Salgado C, Valdez C, *et al.* Protective function of endothelial glycocalyx during hemorrhage shock in skeleton muscle: integration of systemic and the local parameter *in vivo*. FASEB Journal 2015; 29(Supply): 800-9.

[74]   Emoto N, Yanagisawa M. Endothelin-converting enzyme-2 is a membrane-bound, phosphoramidon-sensitive metalloprotease with acidic pH optimum. J Biol Chem 1995; 270(25): 15262-8.
        [http://dx.doi.org/10.1074/jbc.270.25.15262] [PMID: 7797512]

[75]   Inoue A, Yanagisawa M, Kimura S, *et al.* The human endothelin family: three structurally and pharmacologically distinct isopeptides predicted by three separate genes. Proc Natl Acad Sci USA 1989; 86(8): 2863-7.
[http://dx.doi.org/10.1073/pnas.86.8.2863] [PMID: 2649896]

[76]   Bax WA, Saxena PR. The current endothelin receptor classification: time for reconsideration? Trends Pharmacol Sci 1994; 15(10): 379-86.
[http://dx.doi.org/10.1016/0165-6147(94)90159-7] [PMID: 7809954]

[77]   Kuruppu S, Reeve S, Ian Smith A. Characterisation of endothelin converting enzyme-1 shedding from endothelial cells. FEBS Lett 2007; 581(23): 4501-6.
[http://dx.doi.org/10.1016/j.febslet.2007.08.028] [PMID: 17761169]

[78]   Uhl B, Hirn S, Immler R, *et al.* The endothelial glycocalyx control interaction between quantum dots with endothelium and their translocation across the blood tissue barrier. ACS Nano 2017; 11(2): 1498-508.
[http://dx.doi.org/10.1021/acsnano.6b06812] [PMID: 28135073]

[79]   Möckl L, Hirn S, Torrano AA, *et al.* The glycocalyx regulates the uptake of nanoparticles by human endothelial cells in vitro, Nanomedicine (London). 2017; 12: pp. (3)207-17.
[http://dx.doi.org/10.2217/nnm.2016-0332]

[80]   Lukasz A, Hillgruber C, Oberleithner H, *et al.* Endothelial glycocalyx breakdown is mediated by angiopoietin-2. Cardiovasc Res 2017; 113(6): 671-80.
[http://dx.doi.org/10.1093/cvr/cvx023] [PMID: 28453727]

[81]   Su L, Zhang W, Wu X, *et al.* Glycocalyx mimicking nanoparticles for stimulation and polarization of macrophages via specific interaction Small 2015; 11(33): 4191-200.

[82]   van Haaren PM, VanBavel E, Vink H, Spaan JA. Localization of the permeability barrier to solutes in isolated arteries by confocal microscopy. Am J Physiol Heart Circ Physiol 2003; 285(6): H2848-56.
[http://dx.doi.org/10.1152/ajpheart.00117.2003] [PMID: 12907418]

[83]   Potter DR, Damiano ER. The hydrodynamically relevant endothelial cell glycocalyx observed *in vivo* is absent *in vitro*. Circul Res 2008; 102: 770-6.

[84]   Gromnicova R, Kaya M, Romero IA, *et al.* Transport of gold nanoparticles by vascular endothelium from different human tissues. PLoS One 2016; 11(8): e0161610.
[http://dx.doi.org/10.1371/journal.pone.0161610] [PMID: 27560685]

[85]   Huang ML, Godula K. Nanoscale materials for probing the biological functions of the glycocalyx. Glycobiology 2016; 26(8): 797-803.https://doi.orh/10.1093/gycoob/cww022
[http://dx.doi.org/10.1093/glycob/cww022] [PMID: 26916883]

[86]   Bouwmeeter H, van der Zande M, Jepson MA. Effect of food born nanomaterials on gastrointestinal tissues and microbial. WIRES Nanomedicine and Nanobiotechnology 2018; 2018: 10.e1481.
[http://dx.doi.org/10.1002/wnan.1481]

[87]   Sebestik J, Reins M, Jezek J. Biomedical applications of peptide glycol and glycopeptide dendrimers and analogous dendrimeric structure. Springer Verlag Wien 2012.
[http://dx.doi.org/10.1007/978-3-7091-1206-9]

[88]   Albert B, Johnson A, Lewis J, *et al.* Molecular Biology of the Cell. 4th ed., New York: Garland Sciences 2002.

[89]   Budin I, Devaraj NK. Membrane assembly driven by a biomimetic coupling reaction. J Am Chem Soc 2012; 134(2): 751-3.
[http://dx.doi.org/10.1021/ja2076873] [PMID: 22239722]

[90]   Cuatrecasas P. Membrane receptors. Annu Rev Biochem 1974; 43(0): 169-214.
[http://dx.doi.org/10.1146/annurev.bi.43.070174.001125] [PMID: 4368906]

[91]   Jacobson C, Côté PD, Rossi SG, Rotundo RL, Carbonetto S. The dystroglycan complex is necessary

for stabilization of acetylcholine receptor clusters at neuromuscular junctions and formation of the synaptic basement membrane. J Cell Biol 2001; 152(3): 435-50.
[http://dx.doi.org/10.1083/jcb.152.3.435] [PMID: 11157973]

[92]   Dautzenberg FM, Hauger RL. The CRF peptide family and their receptors: yet more partners discovered. Trends Pharmacol Sci 2002; 23(2): 71-7.
[http://dx.doi.org/10.1016/S0165-6147(02)01946-6] [PMID: 11830263]

[93]   Rivière S, Challet L, Fluegge D, Spehr M, Rodriguez I. Formyl peptide receptor-like proteins are a novel family of vomeronasal chemosensors. Nature 2009; 459(7246): 574-7.
[http://dx.doi.org/10.1038/nature08029] [PMID: 19387439]

[94]   Maruyama IN. Activation of transmembrane cell-surface receptors via a common mechanism? The "rotation model". BioEssays 2015; 37(9): 959-67.
[http://dx.doi.org/10.1002/bies.201500041] [PMID: 26241732]

[95]   Singer SJ, Nicolson GL. The fluid mosaic model of the structure of cell membranes. Science 1972; 175(4023): 720-31.
[http://dx.doi.org/10.1126/science.175.4023.720] [PMID: 4333397]

[96]   Simons K, Ehehalt R. Cholesterol, lipid rafts, and disease. J Clin Invest 2002; 110(5): 597-603.
[http://dx.doi.org/10.1172/JCI0216390] [PMID: 12208858]

[97]   Pike LJ. The challenge of lipid rafts. J Lipid Res 2009; 50 (Suppl.): S323-8.
[http://dx.doi.org/10.1194/jlr.R800040-JLR200] [PMID: 18955730]

[98]   Noutsi P, Gratton E, Chaieb S. Assessment of membrane fluidity fluctuations during cellular the development reveals time and cell-type specificity. PLoS One 2016; 11(6): e0158313.
[http://dx.doi.org/10.1371/journal.pone.0158313] [PMID: 27362860]

[99]   Mitra K, Ubarretxena-Belandia I, Taguchi T, Warren G, Engelman DM. Modulation of the bilayer thickness of exocytic pathway membranes by membrane proteins rather than cholesterol. Proc Natl Acad Sci USA 2004; 101(12): 4083-8.
[http://dx.doi.org/10.1073/pnas.0307332101] [PMID: 15016920]

[100]   Wessel GM, Wong JL. Cell surface changes in the egg at fertilization. Mol Reprod Dev 2009; 76(10): 942-53.
[http://dx.doi.org/10.1002/mrd.21090] [PMID: 19658159]

[101]   Lodish H, Berk A, Matsudaira P, *et al.* Molecular Cell Biology. 5 th Ed., New York : W H Freeman New York .

[102]   Hankins HM, Baldridge RD, Xu P, Graham TR. Role of flippases, scramblases and transfer proteins in phosphatidylserine subcellular distribution. Traffic 2015; 16(1): 35-47.
[http://dx.doi.org/10.1111/tra.12233] [PMID: 25284293]

[103]   Nabika H, Unouya K. Interaction between nanoparticles and cell membrane, In Surface Chemistry of nanobiomaterials 2016.

[104]   Claudia C, Schneemilch M, Gaisford S, Quirke N. Nanoparticles-membrane interactions. J Exp Nanosci 2018; 13(1): 62-81.
[http://dx.doi.org/10.1080/17458080.2017.1413253]

[105]   Helfrich W. Elastic properties of lipid bilayers: theory and possible experiments Z Nat forsch Teil-C Biochem Biophys Biol Virol. 1972; 28: pp. (11)693-703.

[106]   Deserno M. Elastic deformation of a fluid membrane upon colloid binding. Phys Rev E Stat Nonlin Soft Matter Phys 2004; 69(3 Pt 1): 031903.
[http://dx.doi.org/10.1103/PhysRevE.69.031903] [PMID: 15089318]

[107]   Pardeshi P, Nawale AB, Mathe VL, Lahir YK, Dongre PM. Effects of Zinc Oxide nanoparticles on the hepatic tissue of chicken embryo: A histopathological Approach Bionano Frontier 2014; 7(2): 176-80.
online: 2320-9593. www.bionanofrontier.org

[108] Shirsekar PP, Mathe VL, Lahir YK, Dongre PM. Interaction of ZnO NPs on human red blood cell Bionano Frontier 2016; 9(1): 99-104. www.bionanofrontier.org

[109] Jackson TC, Patani BO, Israel MB. Nanomaterials and cell interaction: A review. J Biomat Nanobiotech 2017; 8: 220-8.
[http://dx.doi.org/https://doi,org/10.4236/jbnb.2017.84015]

[110] Lahir YK, Samant M, Dongre PM. The role of nanomaterials in the development of biosensors. Global Journal of Bioscience and Biotechnology 2016; 5(2): 146-63.

[111] Giovanna F, Thierry C, Daillant J. Floating lipid bilayer: models for physics and biology. Eur Biophys J 2012; 41(10): 863-874 .

[112] Sabina T, Marco M. Understanding nanoparticles interaction with the cell membrane Biophysics nanotechnology 2013; 6www.2physics.com/2013/07/understanding-nanoparticles-interaction.html

[113] Wahid F, Khan T, Shehzad A, Ui-Islam M, Kim YY. Interaction of nanomaterials with cells and their medical applications. J Nanosci Nanotechnol 2014; 14(1): 744-54.
[http://dx.doi.org/10.1166/jnn.2014.9016] [PMID: 24730294]

[114] Verma A, Uzun O, Hu Y, *et al.* Surface-structure-regulated cell-membrane penetration by monolayer-protected nanoparticles. Nat Mater 2008; 7(7): 588-95.
[http://dx.doi.org/10.1038/nmat2202] [PMID: 18500347]

[115] Lin W, Stayton I, Huang Y, Zhou X-D, Ma Y. Cytotoxicity and cell membrane depolarization induced by aluminum oxide nanoparticles in human lung epithelial cells A540. Toxicol Environ Chem 2008; 90(5-6): 983-96.
[http://dx.doi.org/10.1080/02772240701802559]

[116] Herrmann H, Aebi U. Intermediate filaments: molecular structure, assembly mechanism, and integration into functionally distinct intracellular Scaffolds. Annu Rev Biochem 2004; 73: 749-89.
[http://dx.doi.org/10.1146/annurev.biochem.73.011303.073823] [PMID: 15189158]

[117] Mattila PK, Batista FD, Treanor B. Dynamics of the actin cytoskeleton mediates receptor cross talk: An emerging concept in tuning receptor signaling. J Cell Biol 2016; 212(3): 267-80.
[http://dx.doi.org/10.1083/jcb.201504137] [PMID: 26833785]

[118] Wickstead B, Gull K. The evolution of the cytoskeleton. J Cell Biol 2011; 194(4): 513-25.
[http://dx.doi.org/10.1083/jcb.201102065] [PMID: 21859859]

[119] Doherty GJ, McMahon HT. Mediation, modulation, and consequences of membrane-cytoskeleton interactions. Annu Rev Biophys 2008; 37: 65-95.
[http://dx.doi.org/10.1146/annurev.biophys.37.032807.125912] [PMID: 18573073]

[120] Rapin JR, Wiernsperger N. Possible links between intestinal permeability and food processing: A potential therapeutic niche for glutamine. Clinics (São Paulo) 2010; 65(6): 635-43. [Review].
[http://dx.doi.org/10.1590/S1807-59322010000600012] [PMID: 20613941]

[121] Wu YL, Putcha N, Ng KW, *et al.* Biophysical responses upon the interaction of nanomaterials with cellular interfaces. Acc Chem Res 2013; 46(3): 782-91.
[http://dx.doi.org/10.1021/ar300046u] [PMID: 23194178]

[122] Kang Y, Liu J, Song B, *et al.* Potential links between cytoskeleton distribution and electroneurohysiological dysfunctions induced in the central nervous system by inorganic nanoparticles. Cell Physiol Biochem 2016; 40(6): 1487-505.
[http://dx.doi.org/10.1159/000453200] [PMID: 27997890]

[123] Berry CC, Charles S, Wells S, Dalby MJ, Curtis AS. The influence of transferrin stabilised magnetic nanoparticles on human dermal fibroblasts in culture. Int J Pharm 2004; 269(1): 211-25.
[http://dx.doi.org/10.1016/j.ijpharm.2003.09.042] [PMID: 14698593]

[124] Lee J, Chu BH, Chen KH, Ren F, Lele TP. Randomly oriented upright SiO2 coated nanorods for reduced adhesion of mammalian cells. Biomaterials 2009; 30(27): 4480-93.

[http://dx.doi.org/10.1016/j.biomaterials.2009.05.028]

[125] Apopa PL, Qian Y, Shao R, Guo NL. Schwegler, Iron oxide nanoparticles induce human microvascular endothelial cell permeability through reactive species of oxygen production and microtubule remodeling. Part Fibre Toxicol 2009; 6: 1.
[http://dx.doi.org/10.1186/1743-8977-6-1] [PMID: 19134195]

[126] Chaudhury D, Xavier PL, Chaudhari K, *et al.* Unprecedented inhibition of tubulin polymerization directed by gold nanoparticles inducing cell cycle arrest and apoptosis. Nanoscale 2013; 5(10): 4476-89.
[http://dx.doi.org/10.1039/c3nr33891f]

[127] Mao Z, Xu B, Ji X, *et al.* Titanium dioxide nanoparticles alter cellular morphology *via* disturbing the dynamics of the microtubule. Nanoscale 2015; 7(18): 8465-75.
[http://dx.doi.org/10.1039/C5NR01448D] [PMID: 25891938]

[128] Shams H, Holt BD, Mahboobi SH, *et al.* Actin reorganization through dynamic interactions with single-wall carbon nanotubes. ACS Nano 2014; 8(1): 188-97.
[http://dx.doi.org/10.1021/nn402865e] [PMID: 24351114]

[129] Katusha EA, Mikhayalova M, Hugo X, *et al.* Probing cytoskeleton modulations of passive and active intracellular dynamics using nanobody functionalized quantum dots. Journal Nature Communications PY, 2017/03/21 online VL -8 SP-14772EP-PB;.
[http://dx.doi.org/10.1038/ncomms14772L3]

[130] Grady ME, Parrish E, Caporizzo MA, Seeger SC, Composto RJ, Eckmann DM. Intracellular nanoparticle dynamics affected by cytoskeletal integrity. Soft Matter 2017; 13(9): 1873-80.
[http://dx.doi.org/10.1039/C6SM02464E] [PMID: 28177340]

[131] Zajac AL, Goldman YE, Holzbaur ELF, Ostap EM. Local cytoskeletal and organelle interactions impact molecular-motor- driven early endosomal trafficking. Curr Biol 2013; 23(13): 1173-80.
[http://dx.doi.org/10.1016/j.cub.2013.05.015] [PMID: 23770188]

[132] Ruenraroengsak P, Florence AT. Biphasic interactions between a cationic dendrimer and actin. J Drug Target 2010; 18(10): 803-11.
[http://dx.doi.org/10.3109/1061186X.2010.521159] [PMID: 20932230]

[133] Kim Y, Kim SH, Tanyeri M, Katzenellenbogen JA, Schroeder CM. Dendrimer probes for enhanced photostability and localization in fluorescence imaging. Biophys J 2013; 104(7): 1566-75.
[http://dx.doi.org/10.1016/j.bpj.2013.01.052] [PMID: 23561533]

[134] Wang J, Li L. Coupled elasticity-diffusion model for the effects of cytoskeleton deformation on cellular uptake of cylindrical nanoparticles. J R Soc Interface 2015; 12(102): 20141023.
[http://dx.doi.org/10.1098/rsif.2014.1023] [PMID: 25411410]

[135] Holt BD, Shams H, Horst TA, *et al.* Altered cell mechanics from the inside: dispersed single wall carbon nanotubes integrate with and restructure actin. J Funct Biomater 2012; 3(2): 398-417.
[http://dx.doi.org/10.3390/jfb3020398] [PMID: 24955540]

CHAPTER 6

Interactions Between Proteins and Nanomaterials

Abstract: Proteins are among the significant biomolecular constituents in a biosystem. The structure of proteins and the nanomaterials, intracellular interactions, type of the cell, cell organelles, cell signaling and sensation, *etc.*, affect the interactions between proteins and the nanomaterials. The interface formed between proteins and nanomaterials is the original site of contact and the interplay. The behavior of the interacting components reflects on the regulatory aspects, assembly of biomolecules, and various applications in the normal functioning of a biosystem. The fundamentals related to the tendency of biomolecules and nanomaterials help to retain their stable physicochemical conformations. The interactions involving proteins and nanomaterials bring changes in both. It is essential and beneficial to understand the mechanism of these interactions and their impacts on each other. This chapter deals with the nature, structure, and behavior of protein in general and nanomaterials, their stability, the significance of zeta potential, opsonins and their role, protein corona, and the factors influencing their dynamics.

Keywords: Biodistribution, Bionanointerface, Nanomaterials, Opsonization, Protein Structure, Protein Corona, Protein Chip, Zeta Potential.

OVERVIEW OF STRUCTURAL ASPECTS OF PROTEINS

In a biosystem, the prime share of the biomolecules is proteins and protein-related molecules. The internalized nanomaterials encounter proteins and interact with them within a biosystem. These interactions involve the molecular assembly of specific proteins, inter and intracellular communication, and sensation related to the cell and cell organelles. The physicochemical features of nanomaterials, proteins, and adducts formed, affect the beneficial or derogative interplay [1]. A protein molecule is a polymer made up of a specific monomer called amino acids. Generally, a protein molecule is a peptide having a minimum of 40 amino acid residues arranged in an unpredictable linear pattern. A molecule of an amino acid is basic and acidic because it has a primary group $-NH_2$ at one end and acidic group $-COOH^-$ at the other end. Thus, each amino acid exhibits a double function, *i.e.*, bifunctionality. This feature of amino acids tends to link linearly forming peptide bonds involving an amine group of one amino acid and the carboxyl group of another amino acid. The structural, physical, chemical, bio-

Yogendrakumar H. Lahir & Pramod Avti

chemical properties are related to the substituent present on the side chain of the amino acid. Further, the side-chain substituent plays a vital role in the functionality of the amino acids as acidic, basic, and neutral molecules. This feature also ensures the functions of amino acids as essential amino acids, and these are essential for an individual [2].

A protein may not be a polypeptide as it is a structurally very long polypeptide. Generally, there can be about 40 or 100 amino acids residues to 10,000 amino acids in various proteins. Monomeric and the multimeric proteins are the two categories of protein molecules. Monomeric proteins and multimeric protein classes depend on the type of nature and the number of peptides present in it. A protein molecule having one peptide chain is monomeric protein, while protein molecules having more than one peptide chain is multimeric protein. The peptide chains of multimeric protein are the protein subunits. These sub-units are either identical to each other or different. Insulin is a multimeric protein and consists of two sub-units. Of the two sub-units, one consists of 21 amino acid residues, and the second sub-unit consists of 30 amino acid residues [3, 4].

Proteins are also grouped based on their chemical composition. Simple proteins consist of only amino acid residues and can have more than one sub-unit. All these sub-units are amino acids. The conjugated proteins have one or more non-amino acid entities as their structural components. These components may be either inorganic or organic. These components act as additional fundamental aspects of the protein. Such parts are the prosthetic groups. Proteins are also classified based on the types of prosthetic groups present in their molecules. If a prosthetic group present in a protein molecule is a lipid, the protein is called lipoprotein; when a prosthetic group present in protein is a carbohydrate, the protein is a glycoprotein. Some of the proteins have metal-ion; such proteins are the metalloproteins. Prosthetic groups are a very significant component of a conjugated protein [4].

THE STRUCTURAL ASPECTS OF PROTEIN

The structural aspects of proteins include primary, secondary, tertiary, and quaternary structural components.

The Primary Structure

The primary structure of protein consists of amino acids organized in a linear sequence and constitutes peptide or protein. This sequential arrangement starts with amino-terminal N- and ends at the C-carboxyl terminal of amino acid. There are two terminals on an amino acid residue, one terminal residue shows a free amino group, while the other terminal amino acid has a free carboxyl group.

According to the peptide theory, there are three types of peptides. These are open peptides, cyclic peptides, and branched peptides [5]. The open peptide includes amino acids linked by a peptide bond to form a polypeptide chain. Each amino acid as a unit is residue. A polypeptide chain formed by the same sub-units is an open peptide. Cyclic peptides are composed of mixtures of amino acids containing L and D oliguria residues and glycine. The backbone is composed of H-bonds. These are also cyclic compounds having a peptide bond along with primarily L-amino acids and non-protein amino acids. The cyclic peptides also referred to as cyclotides, are disulfide-rich sub-macro cyclic proteins having around 28 to 37 amino acids. It contains an amide head, and the tail is cyclized peptide as a backbone having cyclic cystine knot. Structurally these sub-macromolecules are chains of polypeptides containing a circular sequence of bonds, *e.g.* cyclosporine [6, 7]. The branched peptides represent a non-continuous chain of carbon bonds and compulsorily have a carbon atom; the carbon atoms show a linear pattern, and it acts as a branching point or branching site. All branches have one or more aspects of the non-continuous link. Such cases are most common in plants [8].

Secondary Structure of Protein

The secondary structure of the protein includes α-helix and β-sheet (β-pleated sheets). These two components interact with each other, involving a hydrogen atom of an amino group and oxygen atom of the carboxyl group resulting in the formation of peptide linkage. This type of linkage ensures a strong structural and functional backbone of the protein. Alpha (α) helix, a component of the secondary structure of protein adopts a coiled spring-like shape; it has an established coiled structure based on hydrogen bonds. Hydrogen bonds are present between the =N-H of amide group and =O of the carboxyl group. This helix is right-handed spirally coiled clockwise. The hydrogen bonds formed are oriented parallel to the axis of the helix formed. One turn of this coiled structure or spiral consists of 3.6 amino acid residues. The H-bond is present between the carboxyl group of one amino acid and the amino group of other amino acid oriented with four amino acid residues. In a given spiral of the helix group, R- is always on the outer side, not within the spiral [4, 9].

The (β) beta-pleated sheets constitute the second component of the secondary aspect of protein. Two β-pleated sheets are either the same or different in a protein molecule. The two β-pleated sheets are bonded with each other by H-bonds. The H-bonds in peptide linkage and atoms involved are of different regions of a single chain that folds or bends on itself, forming an intrachain bond or comprise atoms that belong to different peptides chain in a given protein molecule. Proteins have more than one intra-chain. A single protein molecule containing a β-sheet should have many u-turns to retain its specific structure.

In most cases, the β-sheets and the u-turns present provide fundamental conformational aspects to a given protein molecule. Two aspects of β-sheets are essential, and the H-bonds formed are present within the plan of the β-sheet. Group R- is positioned below and above the plan of sheet and away from the axis of the helix [4, 9].

Tertiary Structure of Protein

The polypeptide chain represents the tertiary structure of a protein as a backbone having either one or more secondary structures of a protein in a specific protein domain. Its atomic coordinates characterize the tertiary structure. These coordinates also reflect on the entire protein structure or the respective protein domain [10]. The tertiary structure of protein brings about the structural functionality of the polypeptide chain. At the tertiary fundamental structural level, the protein molecule under consideration attains three-dimensional aspects, and its functional groups are present at its surface, and that ensures appropriate interaction with other molecules in a unique mode. This conformational state also ensures the lowest energy state of the resultant molecule. The bonding interactions between the groups of amino acids of the side chain provide random and irregular three-dimensional shapes to the protein molecule at a tertiary structural level. The hydrophobic side chain of protein has neutral, non-polar amino acids like phenylalanine or isoleucine get buried within this 3-dimensional structure on the surface of a protein molecule. This arrangement of amino acids protects this molecule from the adverse effects of aqueous physiological conditions existing in a biosystem. The amino acids, like alanine, valine, leucine, and isoleucine, have an alkyl group. This group of the amino acids facilitates the establishment of hydrophobic interaction amongst them. At the same time, aromatic groups and amino acids like phenylalanine and tyrosine stack together. All these arrangements result in exposing acidic or basic amino acids at the surface of a protein molecule and are responsible for its hydrophilic nature. The tertiary structure gets stabilized by the formation of disulfide bridges between cysteine and as a result of oxidation of sulfhydryl groups present or formed on cysteine. These disulfide bridges, along with the H-bonds together, bring the two parts of the chain closer, giving them a firm configuration. Thus, the tertiary protein structure stabilizes because of the salt bridges; ionic interactions involving site positive and negative charge sites are present on the amino acid side chain [4, 9].

Quaternary Structure of Protein

The quaternary structure of a protein is an organized product made of simple dimmers and complex components- homo-oligomers having specific, well-defined sub-units of variables. Proteins consist of sub-units. These sub-units are multiple polypeptide chains or protein subunits. These are either the same or different; when these subunits are the same, the resultant protein is homodimer protein. When the subunits are different, then the resultant protein is called heterodimer protein. A quaternary aspect of protein structure reflects on the arrangement of the protein subunits resulting in the formation of complex aggregate protein. Interactions involving H-bonding, disulfide bridges, salt bridges, play an essential role in the formulation of complex and stable protein structure [2, 11].

PROTEIN STABILITY

The structural aspects of protein reflect on the weak interactions that regulate three-dimensional formulations of protein molecules. This feature also indicates the sensitive nature of the protein molecules. Protein, in its natural form, is a representation of functionally stable natural conformation in a biosystem. The physiological parameters like pH, temperature, elimination of water, exposure to hydrophobic surfaces, metal ions can distort the native protein structure, and elevated shear force [2].

The disulfide bridges or bonds disrupt during the denaturation of proteins. These disulfide bonds are present between two molecules of cysteine amino acid; any suitable reducing agent (2H-) can break these disulfide bridges causing denaturation of a protein molecule. This process ensures the unfolding of protein and causes either random or wrong folding. This change alters the overall shape of the protein molecules. Denaturation of protein exhibits a wide range of physiological processes like loss of biological activity of proteins, aggregation under stressed conditions, disturb immune responses, *etc*. The structural aspects, chemical degradation, oxidation, deamidation, hydrolysis based on the peptide-bond, reshuffling or breaking of a disulfide bond, *etc*., take place during denaturation [2, 12].

ZETA POTENTIAL (Z-POTENTIAL)

The zeta potential represents the electro-kinetic potential existing on the protein molecule in the dispersion state. It also refers to the potential difference between the dispersed medium and the stationary layer of fluid attached to the dispersed particles. It is an electrical potential present in the interfacial double layer. Ionic concentration and potential difference both act as a function of distance from the charged surface of a particle suspended in a given dispersion. Zeta potential (ζ) is an essential parameter for evaluating the stability of the given colloidal dispersion. The multitude of zeta potential reflects on the electrostatic repulsion between adjacent, similarly charged particles in the distribution. Molecules and particles (small enough) exhibit high zeta potential reflecting stability. Colloids that show higher zeta potential (positive or negative) indicate their electrical stability [13].

Zeta potential measuring is one of the most straightforward modes that facilitate direct characterization of the surface of the charged colloidal particles. This mode helps conclude concentration, distribution, adsorption, ionization, and exposure or the protection or shielding of specifically charged moieties [13, 14], and expressed in millivolt. Zeta potential is an empirical indication representing reactivity and degree of coagulation of nanoparticles in a given aqueous medium. Variations in pH and ionic strength reflect on the zeta potential of the nanoparticles in dispersion. It also reflects on the measure of the electrostatic double layer. The usual range of zeta potential varies from 0 to ±60mV. If the value is above ±30mV, it represents the high stability of the nanoparticles. In this state, the electrostatic repulsion is stable, while lower estimates than ±30vM are the indication of coagulation. The electrostatic forces are most economical and offer a higher possibility of coagulation that occurs at the isoelectric point. If ionic strength declines, the zeta potential elevates. The nanoparticles disperse because the electrostatic double-layer expands to equilibrate electrostatic repulsive and attractive forces under this condition. If the ionic strength of the solution containing nanoparticles increases, the zeta potential declines, and nanoparticles exhibit aggregation and precipitation. Thus, zeta potential depends on the ionic strength of the solution containing nanoparticles [15].

OPSONIZATION

Opsonization is related to the immune system, and it identifies an invader (foreign particle, organism to a biosystem). This recognition induces the process of the formation of the antibody for the specific invader. After that, the invader undergoes phagocytosis. Opsonins bind to the binding site present on the surface of the invader (it corresponds to an epitope of antigen) and prepare it for opsonization. This conjugated product exhibits affinity for phagocytes shows a chemical relationship, and finally gets engulfed by them [16]. The opsonins like IgG and IgE immunoglobulins and complement proteins like C3b and C4b, *etc.*, participate in the process of opsonization. The Fab component of the antibody binds to the epitope of an antigen. The opsonization and phagocytosis processes proceed in a stepwise manner and involve chemotaxis, attachment of opsonin to the invader, ingestion (engulfment), and digestion. The opsonins interact with the invader and render them susceptible to absorption. This process is essential because the organisms having similar charges (both have negatively charged cell membranes) on their surfaces are made to interact, *i.e.*, immune cell and bacterial cell both have a negative charge and interact. The receptors participate by binding with each other. The opsonization process is an innate molecular response that involves molecules, microbes, or apoptotic cells, and get modified biochemically that induce a strong affinity between opsonized particles and the receptors present on the cell surface of phagocytes, natural killer cells, and dendritic cells [17]. Generally, the physiological fluids of biosystems are the sites of opsonization. The immune system involves non-specific (innate) response and specific (adaptive) response. The natural response gets strongly induced whenever there is an activation of the complement cascade.

OPSONINS

Opsonins are mostly proteins and grouped as antibodies; complement proteins, and circulating proteins. Antibodies come under adaptive immune response, and these are the products of B-cells concerning the specific antigen. Immunoglobulins (Ig) participate as opsonins, specifically, IgG and IgE. IgG and IgM act as activating complement, hence, considered as an opsonin [18]. Complement proteins are the components of innate immune response or system. Some of these, like C3b, C4b, and Clq, are the prime complements and referred to as opsonins. When complement cascade gets activated, C3b changes to C4b and binds with the receptor on the antigen. The complement proteins mediate antibody-independent opsonization, phagocyte involvement leading to the lysis of microbial organisms [19]. Circulating proteins are the one that moves in the physiological fluid of a biosystem. These act as potential sources of biomarkers. Circulating biomarkers are not directly associated with tumor cell properties. Pentraxin, collectins, and ficolins are common circulating proteins [20].

Pentraxins

Pentraxins or pentaxins get their names from five berries like structures arranged in a radial pattern and contain five protein domains. These relate to the pattern recognition receptors (PPRs) functionally and are classical Acute Phase Proteins (APPs). The five monomers are arranged in a ringed manner (95Å x 35Å). There are three principal members namely serum amyloid P component, and C-reactive proteins [21 - 24].

Collectins

Biochemically and functionally collectins are trimers made of (i) a domain rich in cysteine at their N-terminal, (ii) collagen-like area, (iii) neck like coiled domain, and (iv) carbohydrate recognizing domain (CRD). It is a C-type lectin domain. Carbohydrate recognizing domain CRD acknowledges some of the specific parts of the microbe. This recognition is related to the parameters like the degree of oligomerization of collectin, Ca++ and ligand present on the bacteria. The collectins perform many functions like inhibition of growth of microbes, apoptosis, and modulation of adaptive response, immune and inflammatory responses, *etc.*, and cause the opsonization and activation of phagocytosis [25 - 27].

Ficolins

Biochemically, the ficolins are oligomeric lectins (carbohydrate-binding proteins). These consist of fibrinogen, collagen, and lectin. Fibrinogen is a globular domain; collagen is long thin extensions. The ficolins identify the receptor [28].

The opsonins, like complement components (C1q) along with other opsonins, are capable of restricting the autoimmunity and sustaining self–tolerance and ensure appropriate enhancement in the clearance of apoptotic substances and the activation of phagocytic functionality. The process of removal of foreign or unwanted molecules depends on the involvement of many receptors, opsonins, and the degree of their binding with ligands. The opsonins involved belong to apoptotic cells, complement factors (C1q), activation of members of the pentraxin family, and these play a significant role in the clearance of debris from the biosystem. In this process, the phagocytic system, systemic susceptibility to autoimmunity, and complement proteins also actively help the operation of the clearance [29].

Opsonins are present in the physiological fluid of biosystems and show a high affinity for any foreign bodies, microbes, antigens, *etc.* The opsonins bind to these foreign bodies and activate them for phagocytosis by multinuclear macrophages (a component of the mononuclear phagocytic system) to ensure clearance from the biosystem. The conventional nanomaterials are useful as drug carriers, undergo opsonization, and undergo phagocytosis. This interaction hampers the designated function of the delivery of the drug. Thus, the process of opsonization functions as one of the biological barriers specifically for the delivery of drugs, and by using their polyethylene glycol (PEG) grafted from, help this process. It acts as a hydrophilic protective layer. Thereby, these opsonins prevent the adsorption, involvement of steric repulsive forces that avoids opsonization [30].

PROTEIN CORONA

The formation of protein corona is a common phenomenon depicting the interaction between nanomaterials and proteins. The structure of protein corona reflects the cellular uptake of nanoparticles, predictive biological responses, and expression of receptors concerning the target cell. The proteins adsorb on the surface of the internalized nanomaterials and the soluble proteins adsorb on metal nanoparticle resulting in the formation of protein corona with a specific composition. This composition depends on the physicochemical properties of nanoparticles. It differentiates into the inner hard or permanent corona and outer soft or distal corona. There exists a state of equilibrium of corona formation. The protein corona imparts some beneficial impacts like reducing attractive aggregative forces, inflammatory effects due to nanoparticles, and their toxic influence [15, 31 - 34].

Appropriate interactions involving nanomaterials and proteins provide biological structural and functional status related to the interacting components. Adsorption of protein takes place following surface properties and physicochemical properties like nature, size, shape, surface morphology, chemistry, and modification of the nanomaterials. There are specific conditions in the ambient biological media that guide the administration of nanomaterials systemically. These conditions include temperature, pH, dynamic shear stress, duration of exposure, components present in the medium, and the nature of the medium used [35]. During protein corona formation, proteins behave differently depending on their amount and molecular affinity. If, in a given medium, two or more proteins are present having different amounts and relationships, then during the initial phase of the interaction with nanomaterials, the protein having more concentration will get absorbed. During the later stage of this interaction, the adsorbed protein gets replaced by the protein having a higher degree of affinity, although its amount is less in comparison to the protein adsorbed earlier. This effect is called the Vroman effect or competitive protein adsorption on a surface of nanomaterials [35, 36].

Impact of Size of Nanoparticle on Protein Corona

Size is one of the physical parameters that can make a massive difference in the formation of the protein corona. Nanoparticles with larger sizes provide more coverage for proteins. Nanoparticles with a small scale cannot do so, but the thickness of corona increases on the small-sized nanoparticles. There are relatively less conformational variations in protein corona. The size of nanoparticles permits a relatively higher amount of adsorbed protein and the evolution of composition [35]. Different proteins may exhibit different binding affinity towards different sizes of nanomaterials and their nature. The prothrombin and gelsolin show a more binding kinship for silica nanoparticles having a larger size (110 nm), while the clusterin protein shows a higher affinity for smaller silica nanoparticles (20 nm). There are proteins like some immunoglobulin, *i.e.*, IgG, do not display any sized dependent kinship towards nanoparticles [37]. Thus, this physical parameter changes in nature and the composition of protein corona [35].

Impact of Morphological Aspects of Nanoparticles on Protein Corona

The shape of nanoparticles or the morphology of nanomaterials induces the process of adsorption of proteins on them. The morphology of nanomaterials can also affect the structural aspects of the interacting proteins. Gold nanospheres (10.6±1 nm) and gold nanorods (10.3±2 X 36.4±9) induce 10% and 15% loss of secondary structure α-chymotrypsin and lysozyme, respectively. This behavior of gold nanoparticles reflects on the degree of conjugation or aggregation of the protein under study and also the corresponding decline in their enzyme activities. The protein loading aspects are a monolayer formation, a 40% loss of secondary structure, and 86% loss of enzyme activity. These also show more adsorption of protein on the nanorods rather than on nanospheres [35, 38]. There is a need to continue such studies with morphologically different nanomaterials.

Impact of Surface Charge Present on Nanomaterials During the Formation of Protein Corona

During the formation of the protein corona, the surface charge present on the nanomaterial plays a significant role. This charge may alter the composition or cause conformational changes in the corona formed. A higher electrostatic charge on the surface of nanoparticles induces the degree of conformational changes in proteins. The conformational variations are maximum in the presence of the positively charged nanomaterials, a medium degree of difference occurs with negatively charged nanomaterials, and the minimal changes when occurring when neutrally charged nanoparticles are in contact with the surface of nanomaterials. The charged nanoparticles internalize at a higher rate, and it is relatively higher in comparison to the electrically neutral nanoparticles. Among positively and negatively charged nanoparticles, the positively charged nanoparticles get incorporated intracellularly in large numbers and at a faster rate in comparison to those that have negative charges [35, 39]. This aspect is helpful in pharmacological formulations and therapeutic aspects. The side effects of therapeutic drugs are restricted or minimized, and when these products load or conjugate with nanomaterials, the degree of bioavailability increases. The nanoparticulate systems provide a better therapeutic functional option for the targeted sites. The surface charge can be either modulated or mediated according to the set requirements to attain maximum benefit and minimum derogative effect [40].

Influence of Hydrophobicity on Protein Corona

The hydrophobic nanomaterials have a higher affinity for biomolecules and bring about more conformational changes in proteins. The process of opsonization is convenient in the case of hydrophobic nanomaterials in comparison to the hydrophilic nanomaterials [35]. Hydrophobic nanomaterials increase the amount of adsorbed protein. The copolymer nanoparticles interact with protein and adsorbed more amount of protein. This adsorption is either absent or least at high curvature sits [41]. The hydrophobic nanoparticles also exhibit qualitative adsorption depending on the degree of hydrophobicity. Latex particles get modified with surface modulation with different levels of hydrophobicity readily. These nanoparticles show qualitative adsorption of plasma proteins following the degree of hydrophobicity [42]. Fabricated poly [acrylonitrile-co-(N-vinyl pyrrolidone)] nanoparticles (133-181 nm) are hydrophobic, and this nature enhances the amount of adsorbed protein in comparison to its hydrophilic form [43].

Patterns of Protein Corona

The design of protein corona and nature, both are related to the different properties of nanomaterials, the composition of protein, and the ambient biological environment or interacting media. Parameters such as binding sites or active sites of proteins and enzymes affect the protein corona pattern due to those few amino acids that are responsible for the action of the protein. Further, the prime factors that regulate the adsorption of protein and its pattern are (i) interaction between proteins and nanomaterials, (ii) protein-protein interaction. Factors like composition and the relative amount of protein present in the medium to which nanomaterials are exposed play a lesser role in the pattern of protein corona formation. In blood, the amount of albumin is higher in comparison to fibrinogen, complement proteins, and immunoglobulins. The adsorption of a type and the nature of protein influence the opsonization of nanomaterials; it also affects the allocation, and their efficacy towards the set target [32, 44]. Corona formed is differentiated into the hard corona and the soft corona. Hard corona is composed of proteins that have a comparatively higher affinity in comparison to other proteins. The proteins present in the fuzzy corona have a lower kinship and bind or adsorb reversibly on the nanomaterials, and get exchange quickly. Overall protein corona is a dynamic conjugate formation, and the participating proteins are affected concerning the biological and physicochemical properties or parameters [32, 35, 44].

Influence of Composition of Medium on the Formation of Protein Corona

The composition of interacting media is an important feature that plays a significant role in the formation of protein corona and the behavior of such nanomaterials concerning biosystems. When pristine nanomaterials are incubated in the medium having no serum, then protein present on the cell membrane and cytosol adsorb on them. Generally, nanomaterials exhibit quite effective interaction with the structural, functional aspects of cell membranes and cells. The type of protein coat formed on nanomaterials affects their behavior within body fluids, the cells, and the cytosol. The pristine nanomaterials tend to adhere to the cell membrane and show a higher degree of internalization, but this is not so in the case of nanomaterials having protein corona. This aspect of their behavior reflects on the degree of uptake, distribution, and the fate of nanomaterials in a given biosystem [45]. Amount of protein that binds with nanoparticles as protein corona is dependent on size, mass, surface chemistry, *etc*. This binding of protein also interferes with their interaction with cells, cell organelles, and the cytosol [46]. The protein corona formed on the nanomaterials regulates the intracellular translocation, pathways, and uptake, *etc*., of nanomaterials in a biosystem. The nanoparticles with protein corona move from one body fluid to another. Possibly different compositions of the respective body fluids affect the structure of the protein corona. These body fluids do not exhibit quantitative and qualitative similarities of proteins. These altered proteins induce changes in the proteins of the corona of the nanomaterials. Nanoparticles show different protein corona formation when incubated with cytosolic fluid and plasma, while the hard corona is mostly undisturbed [47].

There exists diversity in the degree of adsorption of proteins on to nanoparticles, including human plasma proteins. The surface functional groups present on the nanomaterials affect the adsorption of proteins on the nanomaterials. Proteins, like apolipoprotein B-100, component factor H, fibronectin, complement-C3, gelsolin, complement C4b, apolipoprotein-A, complement-C1r-subcomponent, thrombospondin-1, prothrombin, *etc*., readily adsorb on the pristine silica nanoparticles. The NH_2 functional groups present on the silica nanoparticles influence the adsorption of proteins, specifically apolipoprotein B-100, complement-factor-H, complement-C3, gelsolin, apolipoprotein-A, thrombospondin, complement-C1r subcomponent, inter-α-trypsin inhibitor heavy chain-H4, coagulation factor-V. The functional group –COOH affects the adsorption of apolipoprotein B-100, complement factor H, complement C3, fibronectin, gelsolin, coagulation factor V, thrombospondin-1, complement-C1r subcomponent, serum albumin and myosin on to the silica nanoparticles [48]. The albumin, IgG, IgA, IgM, α-1-antitrypsin, transferrin, fibrinogen, α-2 macroglobulin, complement C3, and haptoglobin (haptoglobin plasma protein) present in human plasma bind with free hemoglobin (Hb). The erythrocytes release these proteins and help to prevent oxidative interaction [49].

Influence of Protein Conformation on Protein Corona Formation

Protein conformation reflects on the three-dimensional characteristic shape or structure formed due to the arrangement of amino acids constituting polypeptide in a given protein. This specific arrangement of amino acids renders a characteristics shape or structure to the primary, secondary, super secondary, tertiary, and quaternary structure of a polypeptide that ensures the structure of the protein. The overall structure or shape is specific, and some aspects of it are reversible. Super-secondary structural characteristics are responsible for the particular motif, while the tertiary structure is for the domain of a given protein. The structural changes, *i.e.*, conformational changes, take place during the protein interactions. These changes are reversible or alternative and provide some functional variations in protein interaction. These changes are related to the secondary and tertiary structure of the protein. These variations are dynamic and depend on the conditions of the ambient environment of interaction. Such modifications are structural changes and are quite common during the interaction between protein and nanomaterials [35, 50]. There are changes in the conformation of protein during its adsorption on to the surface of nanoparticles. The secondary structure of BSA changes due to the cationic polystyrene nanoparticles, but not in the presence of anionic polystyrene nanoparticles; instead, a binding establishes with the albumin receptors [50].

Influence of Protein Concentration in Interacting Medium on Protein Corona

During adsorption of protein, its concentration plays an important role and also influences the motif. A protein having lower binding affinity but is present in higher quantity replaces the protein in less amount but having a higher kinship. Silica nanoparticles, when incubated using 3%, 20%, and 80% plasma; the proteins present in less concentration are replaced by those present in a higher amount (0.5 – 50wt/v %). Apolipoprotein-E gradually replaces albumin and transferrin at 50% serum concentration. There is a possibility that proteins coated nanoparticles undergo phagocytosis. Generally, dendritic cells engulf such nanoparticles having different amounts of protein [35, 46, 51, 52]. This behavior is due to the reorganization of protein components rather than on concentration [53].

Influence of Exposure Duration on Protein Corona Formation

Interaction between protein and nanomaterials is significantly dependent on the exposure duration. The development and formation of protein corona are a biologically relevant and time-bound process. The body fluids in a biosystem are in the physiological dynamic state, and this energetic state influences the protein corona formation quantitatively, qualitatively concerning the exposure time [35]. The processes like hemolysis, activation of thrombocytes, cellular uptake of nanomaterials, and endothelial cell death, affect the formation of the corona. All these activities are also related to the time of exposure. The effective duration of exposure may be less than 30 seconds [48]. As the exposure time increases, that rate of adsorption of proteins also increases. The pattern of protein corona formed is likely to change concerning time provided for exchanging protein to be adsorbed in the corona. This process is based on the binding affinity of the participating proteins [35, 54].

Role of Static and Dynamic States of Body Fluids on Protein Corona

Body fluids are in constant motion or dynamic state in biosystems. Animals and plants have a specific dynamically equilibrated state of their body fluid. These body fluids move within a particular direction along with its contents with physiologically suited velocity. In the case of human, the average speed of blood flow is around 0.03cm/sec in most of the capillaries, the average rate of flow of blood in superior and inferior vena cava the rate of flow is approximately 15cm/sec, and around 40cm/sec in the aorta [55]. These velocities are suggestive of the existence of sufficient shear stress on the biomolecules, the cells moving in the respective fluid circulation, and the luminal epithelial cells of the blood vessels, *etc*. After the systemic administration of nanomaterials or fabricated nanomaterials experiences, the shear stress under hydrodynamic conditions existing in the body fluid of the biosystem. Nanofabrication involves explicitly designing for carrying drugs, and that includes cross-linking, non-cross-linking, and stimuli-responsive polymers. These techniques facilitate drug delivery. The formations of the protein corona, its nature, pattern, *etc.*, influence the bio-nan--interface. The protein coronas significantly affect the mode of release of drug from either protein conjugated nanocarriers or surface loaded drug nanocarriers (Kang *et al*., 2016; 2016a; 2016b). Polymer nanocapsules protein corona changes the profile of drug release slightly. Other factors like buffer and salt concentration in the body fluid or salt solution also influence the drug release profile *in vivo* conditions [35, 56, 57].

Role of Static and Dynamic States of Body Fluids on Protein Corona

Body fluids are in constant motion or dynamic state in biosystems. Animals and plants have a specific dynamically equilibrated state of their body fluid. These body fluids move within a particular direction along with its contents with physiologically suited velocity. In humans, the average speed of blood flow is around 0.03cm/sec in most of the capillaries, the average rate of flow of blood in superior and inferior vena cava is approximately 15cm/sec and about 40cm/sec in the aorta [55]. These velocities are suggestive of the existence of sufficient shear stress on the biomolecules, the cells moving in the respective fluid circulation, and the luminal epithelial cells of the blood vessels, *etc.* The administered nanomaterials or fabricated nanomaterials experience the shear stress under hydrodynamic conditions present in the body fluids. Nanofabrication involves explicitly designing for carrying drugs, and that includes cross-linking, non-cros--linking, and stimuli-responsive polymers. These techniques facilitate drug delivery. The formations of the protein corona, its nature, pattern, *etc.*, influence the bio-nano-interface. Protein corona significantly affects the release of drugs from protein conjugated nanocarriers or surface loaded drug nanocarriers [58 - 60]. Polymer nanocapsules protein corona changes the profile of drug release slightly. Other factors like buffer and salt concentration in the body fluid or salt solution also influence the drug release profile *in vivo* conditions [35, 56, 57].

Influence of Temperature on Protein Corona Formation

Temperature is another parameter that affects protein corona formation. Endothermic animals have specific body temperature. Most of the experimental models used *in vitro* belong to this group. Each such model has different thermal zones like inner and peripheral thermal zones, *etc.* Protein also exhibits specific a process referred melting of proteins. During this process, a given protein loses its three-dimensional conformation. The parameters like pH and salt concentrations also facilitate this process. Melting temperature or point is one of the variables that help in the characterization of a particular protein. This feature is more suitable in the case of the formulation of pharmaceutical products involving nanomaterials [61]. Protein corona formed on gold nanoparticles gets modified because of the electrostatic interactions and H-binding. This change concerns thermodynamic principles. The negative values of entropy of binding reflect on the limitations during the adsorption of protein. Adsorbed protein molecules on to the surface of nanoparticles and free protein molecules exhibit thermal related behavior, and these are affected by the factor melting point [62]. Temperature regulates the extent of the surface of the nanoparticles covered during protein adsorption and the composition of the protein adsorbed during the formation of the protein corona. Even the rate of uptake of nanoparticles having protein corona is different. Temperature significantly changes the pattern of protein corona within the temperature range 37°C to 41°C when nanoparticles incubated with human serum albumin. Even the quality of adsorbed protein on copper nanoparticles increases with the rise in temperature from 15°C to 27°C and 37°C to 42°C. There can be some significant changes in the interactions involving protein and nanoparticles and after that, their biodistribution and bioavailability in a biosystem at body temperature [35]. There is a need to investigate the role of temperature during the evaluation of bio-nano-interactions [63].

Impact of pH on Protein Corona Formation

The pH condition of the interacting medium influences the interaction between nanoparticles and proteins. The biological fluids like blood, intracellular fluid, and lysosomes, exhibit neutral, 6.8, and 4.5-5.0 pHs, respectively. This different pH of the body fluids affects the interactions between nanomaterials and protein, protein corona formation, biodistribution, and the profile of drug delivery of nano-drug carriers. This aspect is essential to evaluate the fate of nanoparticles or materials, bioavailability, and therapeutic significance [35]. The protein corona formation and its composition can be changed if the pH of the interacting medium is changed suitably [64]. The physicochemical properties like surface charge isoelectric point, pH, and point of zero electrostatic charges (isoionic point) play significant roles in the stability of nanomaterials. When the pH value is lower than the isoelectric point (electrically, no net electrostatic or electrically neutral), the equilibrium of nanomaterials declines [65]. When nanoparticles are present in a medium that has a pH lower than the isoelectric point, these nanoparticles have a positive charge on their surface. If the pH of the medium is higher than an isoelectric point, the nanoparticles have a negative charge on their surface, but this condition exists only if potential determining ions are absent in the medium. These conditions facilitate aggregation due to a change in the pH of the medium. Similar principles apply to the process of protein corona formation [66]. Most of the nanomaterials and proteins have a specific point of zero charges or isoionic point. The isoionic point is a pH value at which an electrolyte shows zero net electrostatic charges on the surface). These pH-dependent parameters and surface charge play a significant role in regulating nanoparticles and protein interactions. Doctor et al. suggested that the pH of medium affects the corona formation only if the surface of nanoparticles or their chargeable groups or sites, and the presence of hydroxyl group, are present on the surface of metal or metal oxide nanoparticles. The pH causes a decline in the reversal of the charge within the limits of the environment or medium. The isoelectric point is within a pH range of 5.0 to 9.0 [66].

Humic acid increases the negative surface charge on nano TiO_2 at pH 5.7, and in this condition, the humic acid-induced transport of titanium dioxide nanoparticles at low adsorption densities. The natural organic matters (NOM) change the physicochemical properties of an aqueous C60 (fullerene). These properties include size, shape, and surface charge, *etc*. These changes induce translocation and toxicity of aqueous fullerene in a natural aqueous medium. Natural organic matter and pH of solution regulate the stability and transportation of nanoparticles like TiO_2 in the natural aquatic environment. There is a decline in the degree of adsorption of bacterial exo-polysaccharides on silver nanoparticles and TiO_2 nanoparticles when pH is higher than the isoelectric point [67, 68]. Most of the natural organic matter and cells of biosystems (both prokaryotic and eukaryotic cells) have a negative charge on their surfaces. These negatively charged cells of biosystems and natural organic matter should be positively charged to ensure their electrostatic interaction with nanomaterials [68]. The stronger electrostatic attraction exists between negatively charged nanoparticles and the humic acid (one of the components of natural organic matter) within acidic pH; the surface charge on the iron hydroxide gets reversed under alkaline range [69]. The intensity of this interaction declined as the pH range changes to the alkaline range (8.0-10.5). There are some cases when it is difficult to reverse the charge of the surface of silica nanoparticles. In such cases, the impact of pH does not appear to be effective. This interplay indicates that there exists a mechanism other than electrostatic attraction during the protein corona formation. The mechanisms of an exchange of ligand or hydrophobic interaction are also involved during the corona formation [70].

Impact of Colloidal Stability on Protein Corona Formation

The colloidal stability helps corona formation. Repulsive barriers add to the solution to keep nanoparticles in a stabilized state, and this acts as a barrier, and it is the result of either electrostatic repulsion or steric repulsion. Electrostatic repulsion is related to the charge on nanoparticles, and this can be minimized or nullified by the addition of electrolytes. This effect follows the principle of the Derjaguin-LandauVerway Overbeek (DLVO) theory. Steric repulsion is not related to the charge on nanoparticles. It is the result of the ability of some additives that prevent the aggregation or coagulation of the suspension. These additives may be hydrophilic polymers or surfactants. These additives bind with the system in such a manner so that loops and tails extend in the solution. The chains of additives are hydrophilic and prefer to remain in contact with the water solution [71, 72]. Agglomeration involves a binary collision between the particles. The collision rate follows the principle of the second-order rate process. The overall numbers of strikes divided by numbers of bangs that form agglomerate with increasing concentration of protein cause a decline in the effectiveness of collision. The surface coverage against the log of the amount of protein appears to be the binding curve that is evaluated based on the Hill Equation. The binding affinity [KD] represents the amount of protein adsorbed. The desorbed on to the nanoparticles is an important feature that directs the nanoparticle-protein interaction. There is a need for being cautious during such type of investigations. Nanoparticle formulation is another feature that affects the stability of the system, and some proteins increase the phenomenon of consistency, while some destabilize the system [37, 73]

Applications and Significance of Protein Corona

In most cases, when pristine nanoparticles expose to physiological, components of the environment, and biomolecules like protein or lipids, get adsorbed on to their surface and envelop or cover the chemical elements or reactive groups present on their surface. All these interplays affect the size, surface chemistry, charge, solubility, dispersibility, degree of aggregation, or agglomeration of nanomaterials are greatly affected. These features further influence the biodistribution, engulfment by macrophages and uptake of nanoparticles by a cell of the systems [74]. The opsonization and the dysopsonization affect the fate of protein-coated nanoparticles. During the opsonization process, the phagocytes identify the receptors on the antigen, dead cells, dying cells. This selection helps in clearing these structures from the body or the physiological system. Opsonization involves opsonins like antibodies complement protein (part of immune system-components), circulating proteins (pentraxins), *etc.*; prepare the protein-coated nanoparticles for phagocytosis by macrophages. This process elevates the level of phagocytosis and the role of the liver and spleen in clearing the dead or dying cells, cell debris, *etc* [75]. Dysopsonization dysopsonin, like albumin, elevates the shelf life or extends the duration of the protein-coated nanoparticles in the physiological medium. PEG provided steric stability to multiblock copolymers in serum *in vivo*. Dysopsonization facilitates this stability because serum proteins have enveloped the reactive complex. As a result, the active compound, *i.e.*, the multiblock copolymer, can deceive the components of the reticuloendothelial system [76]. The changes in protein corona formation represent the biological interaction between nanomaterials and biomolecules and cellular components. These changes are the predictive fingerprints, and these explain the said associative interactions at least 50 to 60%. The formation of protein corona is related to size, state of aggregation and surface charge, mode of binding, *etc.*, of the nanoparticles, and cellular interaction. The comprehensive database on the protein corona is available [77].

Most of the nanoparticles and protein interactions are taking place *in vivo* after being administered in the physiological medium of a biosystem and overall interplay between nanoparticles and proteins grouped in the following modes.

(i) in the first mode, the interactions between nanoparticles are instant. These interactions may involve cells and platelets. (ii) The second mode of interplay is time-related and changes in the adsorbed protein and cellular processes. (iii) The third model of this interaction involves the inducement of the protein cascades. (iv) in the following way of interactivity, components of the immune system, and reticulate endoplasmic network because of the high degree of phagocytosis, the protein-coated nanoparticles, and the identification of their elimination, (v) This mode of interaction involves the deposition of nanoparticles in non-macr--phagocytes, (vi) In this type of reaction between nanoparticles and nanoparticles, the influences of the overall interactivity within the cell. This mode includes inducing signaling pathways that lead to either survival or death of the cell. Most probably, the affected cell is under stress, and responses are as per the related cellular interplay [74, 78, 79] (Fig. **1**).

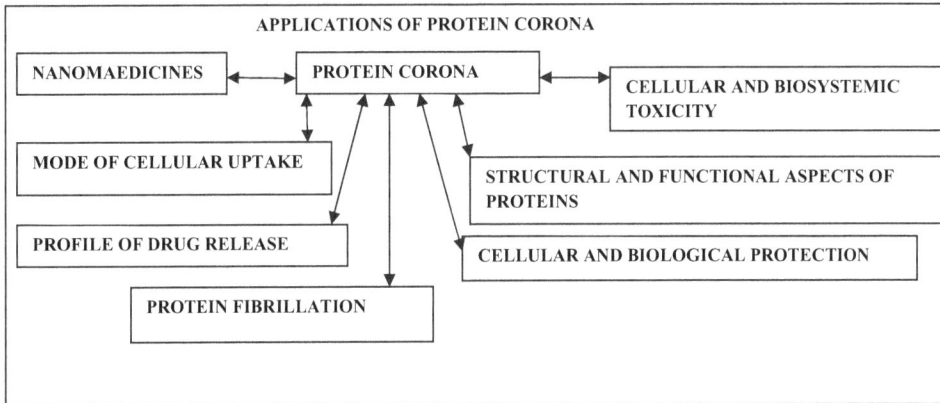

Fig. (1). Applications of protein corona.

Nanoparticles and protein interaction are related to the properties of the surface of nanoparticles, type, and quantity of protein adsorbed reflects on this interaction in the biosystem. This interaction depicts the degree of toxicity, biological fate, and therapeutic efficacy. Protein corona formation offers both advantages and challenges [80].

Protein corona provides information about the functional aspects of protein. During the corona formation, the binding sites on protein corona indicate the possibility of identifying functional motifs, binding sites, and their position on protein. This interplay also helps to identify the biological identity of nanoparticles. The functional protein motifs interact at the nano-bio-interface. Transferrin coated polystyrene nanoparticles show functional motifs, but the spatial organization is not distinctly indicated. This adsorption process seems to be stochastic and irreversible [81]. Nanoparticles selectively interact with protein during the formation of corona involving adsorption. Proteins that participate in corona formation maintain their functionality. These proteins interconnect along with different proteins in given immobilized human proteins. This technique is deployed to design a target-specific system. This process is related to the concentration and amount of protein-protein interaction [82]. Chemical functionality of nanoparticles and their biological effects differentiate with the help of free particles of the corona. The concept that zwitterionic nanoparticles can inhibit adsorption of protein is the basis of this feature of the free particles of the corona. Further, investigation in this direction will help to design the nanoparticles and their ability to regulate nanomaterials and biosystems without the involvement of protein corona [83]. It is possible to modulate the composition of protein corona to use it as a targeting agent for specific types of cells. The 'opsonin-induced-phagocytosis utilizes an agent to corona-mediated-nanoparticle to reach a particular target. Immuno-labeling reflects on the possibility of binding of opsonin or any other specific protein to interact or bind to a particular receptor. This interaction is very selective and specific [84]. Protein corona is a useful agent to reach a specific target. During the protein corona formation, the adsorbing protein/s and chemical aspects of nanomaterials modify adsorption of a specific protein on nanoparticles. These protein-coated nanoparticles attract other interacting proteins. If the surface coated with silica nanoparticles and gamma globulin, the resultant product involves immunoglobulins and complement proteins present in the physiological fluids of the biosystem, and protein-protein interaction attains it. The coating of gamma globulin and complementary protein can ensure targeting macrophages. Sometime this targeting may fail due to the inability of the adsorbed proteins to bind with the specific receptor present on the surface of the target cell surface. One may improvise functional protein corona to enhance the accessibility to the target receptor. Specific protein to a particular position that corresponds to the binding motif is accessible to the receptor present on the surface of target cells or tissue [80].

The nanoparticles have enormous applications in industries. In biosystems, these minute particles behave in a toxic manner and indicate that some of the parameters overlooked in the investigation related to nanoparticles, biological systems, and industrial, specifically interaction occurring at the nano-bi--interface. This aspect appears to be the prime cause of conflicting reports on the use of nanomaterials in context with the biological system. The parameters of protein corona possibly help to find appropriate solutions to the investigatory problems [80]. The nature of nanoparticles is quite reactive in comparison to their respective bulk forms. Some of the physicochemical features of nanomaterials exhibit high surface area to volume ratio possess high reactive energy. The nanomaterials may change the pH of the nutrient of culture medium and interact with biomolecules, and tendency makes the nutrient-deficient for the organisms. The observations of Mahmoudi et al. support this observation. Superparamagnetic iron oxide nanoparticles cause changes in the culture medium. Cells treated with these nanoparticles show gas vesicles, change in pH, and the components of culture medium. These nanomaterials also change the dye used in the MTT technique, tetrazolium dye to formazan because of these changes, there are negative impacts on the observations obtained [85].

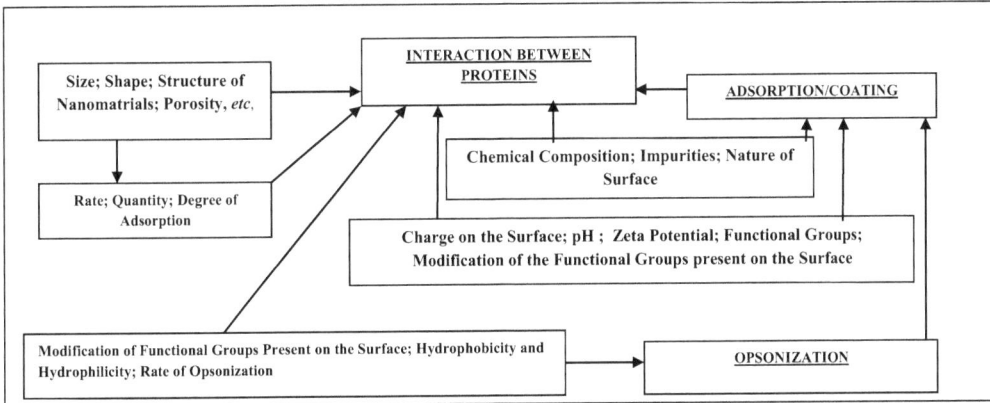

Fig. (2). Physicochemical features of nanomaterials influncing protein binding.

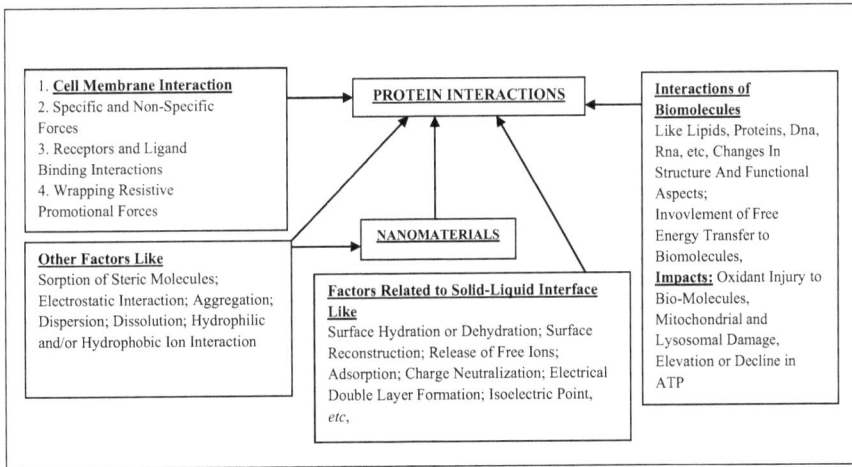

Fig. (3). Role of nano-bio-interface during the interaction between nanomaterials and proteins.

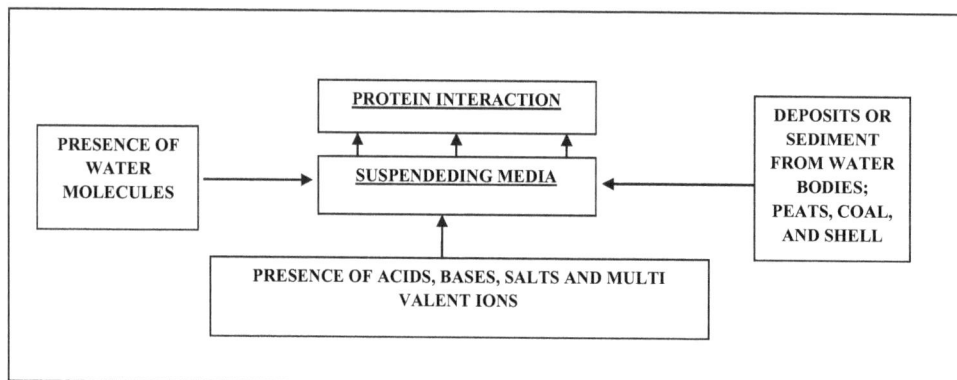

Fig. (4). Factors related to the suspending media affecting interaction between nanomaterials and protein.

The carbon nanotubes, cerium oxide, and polyethylene-coated nanoparticles, *etc.*, have the potential to cause derogative structural changes in proteins. Amyloidogenic proteins can form amyloids. Amyloids are the fibrils like structures that are the product of the misfolding process. The misfolded protein becomes the component of plaques. Mostly, it happens when such proteins get precipitated and cause neurodegenerative diseases and dialysis amyloidosis. Nanomaterials like carbon nanotubes, cerium oxide, and polyethylene glycol-coated quantum dots induce such conditions. This type of induction reflects on the ability of nanomaterials to exhibit aggregation, precipitation, and misfolding of amyloidogenic proteins, and shows the strength of such nanoparticles to affect the motifs and moieties of the proteins. Hydrated fullerene (C60) reduces the process of fibrillation by inducing ant-amyloidogenic impacts on the amyloid β 25-35 peptides. Slight thermal fluctuations within the physiological limits influence the process of fibrillation. Amyloid-β monomer that has amino acid sequence 17-24 is hydrophobic, and it is considered to be the prime support. It plays a significant role during fibrillation. Hydrophilic silica nanoparticles elevate the degree of fibrillation [86 - 89] (Figs. **2-4**) deposits or sediments from water bodies, peats, coal, shells deposits or sediments form water bodies, peats, coal, and shells

SIGNIFICANCE OF INTERFACE DURING INTERACTIONS BETWEEN NANOMATERIALS AND PROTEINS

The interface is not only a site for interactions between two reactants, but it provides appropriate mechanical suitable support and surface that influences the interaction. There is every possibility of structural changes either in one of the reactants or both. The interface is sophisticated architectural aspects that provide useful and appropriate mechanical features and mechanical amplificative impacts on the related interacting components. The success of an interface depends on the mechanical improvement of non-biological and biological materials. This success leads to similar search features that are meeting in synthesized composites expressly varied industrial applications so that the anatomical aspects preserve for better options. A suitable perfect interface should exhibit a hierarchical structure; it should be strong enough to provide required mechanical stress, cohesion within the units, or building blocks to ensure the structural integrity of the constitutional materials. These features offer correct stiffness, tensility, toughness, while the interaction completes. The interface should also be delicate enough to sense and gaps in the resultant configuration or fabricated materials that provide suitable toughness, stiffness required for the interface. The biological interfaces offer much-needed information to fabricate interface from non-biological materials [89]. When gold nanoparticles and bovine serum albumin conjugate, the BSA attains more flexible conformational aspects, at least along the boundary surface of gold nanoparticles, *i.e.*, bio-nano-interface. Higher concentrations of gold nanoparticles induce conformational changes. Further, investigations based on the circular dichroism technique, it is readily noticed that α-helical contents get reduced as a result of bioconjugation [90]. During the interaction between nanomaterials and proteins, interfacial interplay occurs. These interactions are complicated and have the potential to endure changes in the structural and functional aspects of protein. For example, a conformational transformation of protein takes place during the modulation of aggregation. The specific pattern formed due to this assemblage influences the functions of a molecule of protein, and alter the morphological features like size, surface changes, density, hydrophobicity, hydrophilicity influence the activity of nanomaterials [91]. The physicochemical parameters of nanomaterials like size, shape, area, chemistry, charge, roughness, porosity, valance conducting state, *etc.*, of the surface, along with the functional groups or ligands present also have their impacts of interfacial interactions. Even the constituents of the suspending media such as water molecules, acids, bases, salts, and multivalent ions influence interaction at the interface. The products like natural organic matter, humic acid, proteins, lipids, surfactant polymers, and polyelectrolytes affect the interactions at the interface. The hydration and dehydration of the surface, surface construction, the release of free surface energy, ion adsorption, and charge neutralization, *etc.*, affect the solid-liquid interface play an essential role during the interactions. Factors like aggregation, dispersion, dissolution, hydrophobic and hydrophilic reactions, electrical double layer formation, and zeta potential, and isoelectric point, sorption of steric and electrostatic interactions interfere with the interactions at the interface [92].

Whenever nanomaterials interact with biological materials like a membrane, cell organelles, a specific nano-bio-interface is the result. Specific interactions occur at nano-bio-interface on the membranes involving non-specific forces, receptor-ligand binding interactions, and membrane wrapping resistive and promotional forces. The biomolecules interactions involving lipids, proteins, DNA, RNA, bring change the structure and functions of the membranes and the biomolecules. These changes are for a short duration, temporary, permanent, or long-lasting. Free energy transfer takes place during the interface interactions that involve biomolecules, and it causes oxidant injury to the biomolecules and cell organelles, like mitochondria, the lysosome. There is a decline or increase in ATP. Formation of non-specific binding takes place during nano-bio-interface interactions. These can be the primary concerns because these may lead to false-positive signals and low-signal-to-noise ratio during such estimations. One can block the possible sites of non-specific binding and treat nanomaterials with specific surface coatings that combine an ultra-low fouling background that has abundant biorecognition elements using non-fouling coated materials like Zwitterionic polymers, polyethylene glycol and its derivative [93 - 96].

INTERACTIONS BETWEEN NANOMATERIALS AND PROTEINS

The nanomaterials come in contact with the contents of body fluids. The substances like proteins, lipids, adsorb on their surface after their internalization. These biomolecules are present either as free or in conjugated form. The extent and manner of adsorption depend on the physicochemical features of nanomaterials and the surface features of the biomolecules. These interactions influence the structural and functional aspects of biomolecules. Such interplay has the ability to influences bio-dispersion of the nanomaterials in the biosystem. The dendrimers are helpful in biomedical, biomolecular, pharmaceuticals, and biotechnology fields. Their use in the biomedical field is quite challenging because their behavior *in vivo* is ambiguous. These challenges include rapid clearance from the blood vascular system, increased chances of development of blockages in capillaries, diversion or decline in targetability and toxic or derogative impacts, *etc*. The specific physicochemical properties of nanomaterials and reactivity of protein molecules present in the physiological fluids, cytosol, interstitial fluids of the biosystem, and the formation of protein corona are affected. All these add to the practical challenges when concerned nanomaterials are *in vivo* use [97]. Since proteins are relatively more qualitatively and quantitatively in body fluid, hence the related discussion is dedicated to proteins.

The nanomaterial-protein complex formed, and the biodistribution of nanomaterials within the biosystem depends on the response of the biosystem. This biological response also affects the movements across physiological barriers and access to the target cell or tissues. Surface bounded proteins promote cell-specific uptake, activation of inter, and intracellular signaling pathways. A higher abundance of proteins may dominate the surface of nanomaterials during the initial phase, but later these are displaced by the proteins having lesser quantity depending on the higher affinity and slower kinetics under the principles of interaction is dynamics of interaction [92].

Adsorption of proteins on a surface of materials at the nanoscale dimension (substrate) is a complex phenomenon. Interactions between nanomaterials, proteins, and materials with the nano, meso, and macro scale are quantitatively and qualitatively different. The protein-coated nanomaterials exhibit change in their surface charge and properties of like hydrodynamic size, aggregation, *etc.* These two changes play a relatively prominent role during the kinetics related to protein and nanomaterial interaction. The proteins like albumin, fibrinogen, apolipoproteins, and immunoglobulins bind strongly to nanomaterials. The aggregates of nanomaterials are retained in blood circulation and result in embolism (a process of lodging; any detached solid, liquid, or gaseous intravascular mass and moves within the blood vessels. It is capable of clogging blood vessels). The coated and non-coated nanomaterials exhibit short circulation half-life, low efficacy in a biosystem, can cause toxicity. This behavior mostly happens when these get accumulated in the liver and spleen because these bring about primary limitation for the systemic administration of the nanomaterials. Biological interactions between functionalized nanoparticles and proteins affect the adsorption of proteins on the surface of the interacting nanoparticles. This adsorption is dependent on the size, shape, and surface charge of the nanoparticles. The Fluorescence quenching technique and affinity capillary electrophoresis techniques are suitable to investigate the process of adsorption. When electrostatic conditions are favorable, the comparative binding of bovine serum albumin (BSA) is faster, specifically towards the cationic surface of gold nanoparticles. The amount of charge does not influence this process. When surfaces of gold nanoparticles become electrically neutral or as PEGylated form, the adsorption of BSA declines. During fluorescence quenching titration, the amount of light absorbed due to gold nanoparticles interferes with the process of adsorption. The values of adsorbed protein differ when the affinity capillary electrophoresis technique is used to study the adsorption. The rate of adsorption is about 105 times less in comparison to that during the spectroscopic titrations. The optical properties of gold nanoparticles interfere with the determination of adsorption constant (Ka). The compatibility aspects of the materials like positively charged gold nanoparticle induce fluctuations during affinity capillary electrophoresis because the positive particles adhere to the wall of the capillary [98].

The degree of nonspecific protein interaction varies with different proteins and nanomaterials. There is an extensive change in surfaces during the adsorption of protein on nanomaterials and affects their distribution. The binding of immunoglobulins with metal nanoparticles induces and promotes the opsonization. The mononuclear phagocyte system identifies the immunoglobulin coated nanoparticles, removes these particles rapidly from the blood circulation. The macrophage, phagocytes capture the coated nanoparticles, and this is more prevalent in hepatic sinusoids and spleen filtration [98, 99].

The size and shape of nanomaterials are the variables that have their own structural and functional importance during nonspecific and specific protein interactions. The hydrodynamic diameter (scattered-light-intensity-weighted average diameter) of gold nanoparticles with size 30nm and 50 nm (as the gold colloidal form) increases up to 76.1 and 100.0 nm (when incubated with human plasma). The treatment with trypsin, the coating of plasma proteins, reduces the hydrodynamic diameter, but gold nanoparticles help to regain their original hydrodynamic diameter. The uncoated gold nano colloidal particles after being washed with PBS/saline change the color of the solution containing, it turns black. After cleaning the colloidal nanoparticles, the Coulombic Repulsion declines, and protein coating inhibits aggregation of gold nanoparticles. Studies related to adsorption of plasma protein on these gold nanoparticles reveal that in all 69 proteins are identified; the adherence of plasma proteins to gold nanoparticles of sizes, 30 nm, and 50nm exhibits different degrees. In the case of gold nanoparticles having 30 nm sizes, 48 proteins got adhered to while 21 proteins adhered to the gold nanoparticle with 50 nm. Around 14 proteins are common that adhered to both the dimensions of gold nanoparticles. The elevation of particle diameter and decline of absolute zeta potential after incubation with plasma proteins reflect that positively or neutrally charged proteins adsorb to gold nanoparticles when in the bloodstream. This type of protein binding may influence the immune cells by involving receptor-mediated endocytosis or due to a reduction in Coulombic repulsion between the nanoparticles and negatively charged cell membranes [99]. Bovine serum albumin gets adsorbed on colloidal Al_2O_3 nanoparticles in an aqueous medium. The degree of adsorption is proportional to the charge of the isoelectric point of a colloidal mixture of Al_2O_3 and protein. Single-layer is needed to mask the charge on Al_2O_3 nanoparticles and charge on Al_2O_3 compromises it. About 30 to 36% of the negatively charged groups of a protein molecule bind to the protonated and charged surface of the metal oxide under study. There is a possibility that BSA exhibits various modes of adsorption [100]. When BSA and gold nanoparticles interact, flexible conformational changes take place in the structure of bovine serum albumin. These changes are the result of conjugation between BSA at the surface of Au nanoparticles. When conjugation completes, there is a decline in the α-helical contents. These conformational changes are under the concentration of gold nanoparticles [90]. The cellular uptake of TiO$_2$ nanoparticles is size depended, internalize readily, and reach cell organelles, *i.e.*, these react with proteins of cytoskeleton like microtubules, *etc.* TiO$_2$ nanoparticles are the risk of specific neural and other organ systems. These nanoparticles cause adverse impacts on microtubules. These nanoparticles result in significant conformational changes in tubulin protein. The polymerization of tubulin protein is affected, and the degree of polymerization also declines [101]. Interaction between nanomaterials and proteins bring about aggregation and folding of proteins. These dysfunctional aspects can lead to some clinical conditions. Under normal circumstances, there are intracellular chaperons that accomplish standard protein folding. Nanoparticles can be fabricated using gelatin, chitosan, and inorganic molecules like iron, gold, and silver. Such fabricated nanoparticles form dynamic-nanoparticle-protein-corona. This resultant product induces conformational changes in the adsorbed proteins. This interaction is potent enough to change the bio interaction of such fabricated nanoparticles that results in the wrong or an abnormal type of protein folding. Correct or suitable protein aggregate plays a decisive role in curing neurodegenerative disorders. Aggregations of wrong folded proteins affect such corrective processes. Investigations related to the interactions between nanoparticles and proteins provide a better understanding of the related mechanism. In a biosystem, the protein misfolding or aggregation causes neurodegenerative diseases, accumulation of amyloid aggregates, dominant-negative mutations, elevate the degree of cytotoxicity, and defective degradation of protein, and results in subcellular localization. The nanoparticles may cause or act as artificial chaperons and induce fibrillogenesis or detect intermediate folding. The nanoparticle-protein-corona induces conformational changes in protein and also affects cellular interaction [102]. Zinc oxide (ZnO) Nanoparticles are useful as an agent for molecular docking involving proteins, chemokines, cytological proteins, and disrupt the cell physiology. The interaction between these two is intense and affects the active sites by blocking the activation of stimuli that help the functioning of chemokines, and cytological proteins. Such interactions result in the change of the nature of some of the critical residual amino acids like lysine, valine, threonine, tyrosine, tryptophan, leucine, *etc*. Thus, ZnO nanoparticles interrupt the cell signaling pathway, blocks active sites, and renders the affected cell dysfunctioning [103].

The nanomaterials are small enough to enter readily in most of the cells, cell-organelles, and access any physiological fluids of a biosystem. This reactive nature further complicates the interactive phenomena. The existence of proteins in higher concentration provokes the avidity related to the close spatial repetition of the same protein. This interaction also induces and involves cooperative effects like either promotion or restriction of self-assembly, fibrillation utilizing macromolecules that function as mold [104].

The interactions between nanomaterials and proteins result in derogative impacts in a biological system. The nanohydroxyapatite like nano-HA, nano-SiO_2, and nano-TiO_2 nanomaterials cause inflammation, fibrosis, pulmonary, and DNA damage. These particles elude the phagocytic pathway and get access to the nucleus through nuclear pores. These nanoparticles form H-bonds and result in the formation of positive and negative charged particles. This attraction results in peptide chain binding on the particle surface hydrogen bonds (N-H-----O- and O-H-----O-). This binding results in the twisting and transmutation of the peptide chain. There is a charged attraction between cytochrome-C and hemoglobin (Hb). HA particles get aggregated and form a colloidal formulation at the sites where a change of the secondary structure takes place. Nanohydroxyapatite, nano-SiO_2, and nano-TiO_2 get adsorbed on the surface of the cell membrane of the embryo of zebrafish. Specifically, SiO_2 blocks the passages on the cell membrane, delays hatching time, and results in axial malformation, while TiO_2 causes a deformity in the cardiovascular system [105].

THE INFLUENCE OF THE INTERACTIONS BETWEEN NANOMATERIALS AND PROTEINS ON THEIR BIODISTRIBUTION

The interactions between nanomaterials and protein affect their biodistribution and cellular uptake. This cellular uptake is carried out by regular cells of the tissues or by the circulating macrophages or resident phagocytes of the bio tissues like Kupffer's cells, dendritic cells, and natural killer cells. The disturbed biodistribution is the cause of missing the target cells or the issue. Such unwanted biodistribution of nanomaterials brings toxicity, a decline in therapeutic efficacy, or other unwarranted physiological activities [106]. The protein binding profile relates to the physicochemical properties of nanomaterials. The disoriented binding pattern is the cause of disturbance in biodistribution. The cause and effect' correlation of protein binding also influences distribution. The type of protein binding alternates the size and surface charge of nanomaterials; as a result, biodistribution, and biocompatibility become unsuitable [75]. Amorphous silica nanoparticles coated with Pluronic F-127 show reduced adsorption of serum protein and the degree of toxicity declines. Albumin (human serum albumin and fetal bovine serum) adsorbs on to the surface of single-walled carbon nanotubes (SWCNTs) and albumin forms the Nagase. The albumin from Nagase analbuminemic rates also adsorbs on SWCNTs. The profiles are different; the pattern of albumin from the Nagase analbuminemic rat is damaged or structured. Such nanoparticles get readily cleared from the circulation, and scavenger receptors played a significant role in this clearing. The profile pattern of albumin from human serum or fetal bovine serum) remains in circulation for the designated time. Lipopolysaccharide coated SWCNTs restrict the cyclooxygenase-2 enzyme activities. The scavenger receptors, like Fucoidan, limit the anti-inflammatory state. It also declines the cellular uptake of fluorescent SWCNTs. These observations show that the adsorbed protein pattern influences the behavior of SWCNTs and even their toxic behavior. The response of amorphous silica nanoparticles is also gets affected [107]. Fetuin (one of the blood proteins) plays an active role in mediating the uptake of polystyrene nanoparticles. The polystyrene nanoparticles (size 50 nm; negative zeta potential -21.8±2.3v M) coated with BSA, and fetuin proteins get internalized involving scavenger receptors (SR-A). The polyinosinic acid, an inhibitor of scavenger receptor-A inhibits the uptake of plain polystyrene nanoparticles. This behavior of the nanomaterials indicates that SR-A is not affected by BSA and fetuin proteins [108]. Human serum albumin, nanoparticles coated with apolipoproteins E-3, A-1, and B-10 involving NHS-PEG-Mai 3400 conjugate with apolipoprotein E-3, A-1, and B-100 within 15 minutes of their intravenous administration. These show stronger antinociceptive impact (impact of the action or process of blocking the detection of a painful or injurious stimulus by sensory neurons) for the one-hour duration of administration. The nanoparticles loaded with drug Loperamide did not show antinociceptive impact (the action or process of blocking the detection of a painful or injurious stimulus by sensory neurons) after their intravenous administration. This observation indicates that nanoparticles carrying Loperamide drug could move across the blood-brain barrier, endothelial cells, and these nanoparticles show access to the set target [109]. Grafting of PEGylated polymers and their architecture are essential parameters that facilitate cellular uptake and biodistribution. There are alterations in the surface properties, protein binding profile of such grafted nanoparticles. The phagocytes decline the cellular uptake of the grafted copolymer nanoparticles. When the nanoparticles deceive the phagocytic recognition (mononuclear), these nanoparticles reach the designated target cell or tissue without any interruption [110].

The physicochemical features of nanomaterials like size, surface morphology, chemistry, and molecular weight, nature, and affinity of protein to be adsorbed, play a significant role in the formation of the protein corona. The size, quality, and charge of protein corona (zeta potential) are also crucial for their distribution in the biosystem.

PROTEIN CHIP OR PROTEIN MICROARRAY AND NANOMATERIALS

It is a micro-device fabricated based on DNA microassay. These devices are used to follow the protein interaction and to determine the functionality of proteins. This device helps study and track many proteins at the same time [111]. A protein chip consists of support, nitrocellulose membrane, and bead or microtitre plate. Glass slide acts as a support, and on this support, nitrocellulose spreads as a membrane. The study sample consists of a collected or captured array of proteins and is applied to this membrane. A probe molecule is loaded with a fluorescent dye and incorporates a mixture of a variety of proteins under investigation. If there is an interaction between the protein present in a sample mixture and the probe, then there is a signal in the form of fluorescent emission. That is read using a laser scanner [112]. Protein chip or protein microarray technique is fast, automated, cost-effective, and highly sensitive. This technique involves small quantities of samples and reactants. This technique works on the basic principles of antibody microarray and DNA microarray [113]. Generally, there are two types of microchips or microarrays, namely analytical microchip and functional protein chip. In the analytical protein chip technology, the antibody, and a molecule mimicking antibody (biomolecules having a similar antigen-binding domain but do not act as an actual antibody), using other immobilized proteins are involved. This device is useful for measuring the presence and concentration of protein in a given sample. The second type of protein chip includes a set of proteins, or the total protein proteome present in a given sample to examine the biochemical interaction. Microarray technique or protein chip is among the powerful, sensitive, well predictive diagnostic and research tool in the fields of biological, biochemical, biomedical, and pharmaceutical sciences. This tool is versatile and handy to the microscopic test samples investigations concerning genomics, proteomics, transcriptomics, glycomics, metabolomics, and colonics aspects of biomolecular and biochemical fields, *etc* [114 - 120]. Nanotechnology-on-a-chip provides a step better in the field of microfluidics. Lab-on-a-chip technology offers a quick, sensitive, and broader scope for manipulability involving nanoparticles as tags or labels. For example, magnetic nanoparticles act as a label to detect specific antigens during the magnetic immunoassay technique. The magnetically labeled target generates a magnetic field, and the magnetometer represents the observations [121].

There are two prime objects of using protein chip technology: (i) identification and quantification of protein applied for diagnosis, screening, searching biomarkers and protein drugs; and (ii) Investigating the function of a protein, biomolecule/s that involves intermolecular interactions, characterization of binding parameters, enzyme functions, and interactions of novel biomolecules. The parameters like (i) production of capture agent, (ii) chemistry of the surface, and (iii) development of high throughput detection mode support the protein chip technology [122]. Lab-on-chip is an essential platform for the advantage of a wide variety of nanomaterials like nanoparticles, quantum dots, nanowires, graphene, *etc*. Nanomaterials and Lab-on-chip technique both have applications reciprocating each other and together act as a versatile technology that facilitates diagnostics, drug delivery, and nanotoxicology [123]. The protein enzyme and transmembrane properties help in mobilizing biomolecules using substrates like micro/nano-printing, nanovessels, or nanoliposomes. The nanotechnology facilitates real-time detection during the miniaturized assay and advanced detection using labeled markers like quantum dots, gold nanoparticles, carbon nanotubes (CNTs), *etc* [124]. Quantum dots nanomaterials are semiconductors and possess electrochemiluminescent properties that give an excellent yield of fluorescent emission. This feature makes them a useful agent for diagnostics. For instance, quantum dots coated with lectin (energy donor) gets immobilized on a glass slide. The carbohydrate coated gold nanoparticles quench the photoluminescence of quantum dots covered with lectin (energy donor). In the presence of glycoproteins, photoluminescence of quantum dots increases. The intensity of quantum dots is dependent on the concentration of glycoprotein. The carbon nanotubes (CNTs) exhibit suitable electrical, mechanical, higher conductance, and chemical stability, and electrocatalytic activity, and these features make such nanomaterials advantageous agents as electrochemical sensors for biomolecular sensing, specifically for glucose and neurotransmitters, proteins, DNA. The peptide-coated carbon nanotubes are useful tools as immunosensors to detect the serological autoantibodies [125].

There are varied and multiple interactions of nanomaterials within the biosystem depending on the factors that represent the characteristic ambient environment. Nanomaterials exhibit interplay within proximal fluids, outside and inside a cell, within culture media, and with cell organelles. This reflects a variety of microenvironments, nanobiointerface, and challenges. Since nanomaterials show multiple interplays, their designing, development, heterogeneity, biocompatibility, and biodistribution are significant during their applications [126, 127].

CONCLUSION

A significant share of the biomolecules in a biosystem is proteins, and these play structural and functional roles. The physicochemical features of nanomaterials influence their cellular uptake and access to the cell organelles. Nanomaterials and proteins have a specific reactive nature. The relative higher concentration of proteins plays a prime role in this interactive phenomenon. Proteins or other biomolecules are present either as free or as a conjugated form in the body fluids. The process of coat formation depends on the physicochemical features of the nanomaterials and the biomolecule. The interactions between nanomaterials and protein affect their biodistribution and cellular uptake. This cellular uptake may be by regular cells of the tissues or by the circulating macrophages, resident phagocytes of the biological tissues like Kupffer's cells, dendritic cells, natural killer cells, or diseased cells (cancer cells). The protein chip technology is versatile and is the most suitable for studying many samples for genomics, proteomics, transcriptomics, and glycomics at the same time.

REFERENCES

[1]   Kane RS, Stroock AD. Nanobiotechnology: protein-nanomaterial interactions. Biotechnol Prog 2007; 23(2): 316-9.
[http://dx.doi.org/10.1021/bp060388n] [PMID: 17335286]

[2]   Science P. https://www.particlescience.com/echnical-brief/2009/Proteinstructura.html2009.

[3]   Karp G. Cell biology. 7th Ed International Student Version. Wiley 2013.

[4]   Stephen Stokes H. Organic and biological chemistry. 7th Ed., Boston, MA, USA: Cengage Learning 2015. ISBN: 13:978-1-305-08107-9.

[5]   Sanger F. The arrangement of amino acids in proteins Advances in Protein Chemistry 1952; 7: 1-67.

[6]   Xu Y, Tang J, Jiu J-C, He WJ, Tan NH. Application of a TLC chemical method to detection of cyclotides   in   plants.   Chin   Sci   Bull   2008;   53(11): 1671-4.www.scichina.com/cab.scichina.com/www.springerlink.com
[http://dx.doi.org/10.1007/s11430-010-4046-4]

[7]   Bhardwaj G, Mulligan VK, Bahl CD, *et al.* Accurate de novo design of hyperstable constrained peptides. Nature 2016; 538(7625): 329-35.
[http://dx.doi.org/10.1038/nature19791] [PMID: 27626386]

[8]   Starkie Sowers CN. A primer on branched-chain amino acids Huntington College of Health Sciences 2009. www.hchs.edu.

[9]   Cox MM, Nelson DL. Lehninger Principles of Biochemistry. NY: W H Freeman and Co 2011.

[10]   Branden C, Tooze J. Introduction to protein structure. New York: Garland Publishing 1990; 1991.

[11]   Rehman I, Botelho S. Biochemistry, secondary protein structure. Treasure Islands (FL), USA: Start Pearls Publishing LLC 2018.

[12]   Dunbar J, Yennawar HP, Banerjee S, Luo J, Farber GK. The effect of denaturants on protein structure. Protein Sci 1997; 6(8): 1727-33.
[http://dx.doi.org/10.1002/pro.5560060813] [PMID: 9260285]

[13]   David F. An overview of Zeta potential-part-2: Measurement Monday O', American Pharmaceutical Review 2013. www.americanpharmaceuticalreview.com/1429-AuthorProfile.3030- David-Fairhurs--PhD

[14]   Honary S, Zahir F. Effects of zeta potential on the properties of nanodrug delivery system-A review part-1. Trop J Pharm Res 2013; 12(2): 255-64.
[http://dx.doi.org/10.4314/jtpr.v12i2.19]

[15]   Teske SS, Detweiler CS. The biomechanisms of metal and metal-oxide nanoparticles' interactions with cells. Int J Environ Res Public Health 2015; 12(2): 1112-34.
[http://dx.doi.org/10.3390/ijerph120201112] [PMID: 25648173]

[16]   Barret J. Basic Immunology and its medical application. 2nd ed., St Louis: The C V Moby Co 1980.

[17]   Zhang Y, Hoppe AD, Swanson JA. Coordination of Fc receptor signaling regulates cellular commitment to phagocytosis. Proc Natl Acad Sci USA 2010; 107(45): 19332-7.
[http://dx.doi.org/10.1073/pnas.1008248107] [PMID: 20974965]

[18]   Williams MR, Hill AW. A role for IgM in the *in vitro* opsonisation of Staphylococcus aureus and Escherichia coli by bovine polymorphonuclear leucocytes. Res Vet Sci 1982; 33(1): 47-53.
[http://dx.doi.org/10.1016/S0034-5288(18)32358-0] [PMID: 6753075]

[19]   Tosi MF. Innate immune responses to infection. J Allergy Clin Immunol 2005; 116(2): 241-9.
[http://dx.doi.org/10.1016/j.jaci.2005.05.036] [PMID: 16083775]

[20]   Boldrup L, Troiano G, Gu X, *et al.* Evidence that circulating proteins are more promising than miRNAs for identification of patients with squamous cell carcinoma of the tongue. Oncotarget 2017;

8(61): 103437-48.
[http://dx.doi.org/10.18632/oncotarget.21402] [PMID: 29262574]

[21]   Emsley J, White HE, O'Hara BP, *et al.* Structure of pentameric human serum amyloid P component. Nature 1994; 367(6461): 338-45.
[http://dx.doi.org/10.1038/367338a0] [PMID: 8114934]

[22]   Romero IR, Morris C, Rodriguez M, Du Clos TW, Mold C. Inflammatory potential of C-reactive protein complexes compared to immune complexes. Clin Immunol Immunopathol 1998; 87(2): 155-62.
[http://dx.doi.org/10.1006/clin.1997.4516] [PMID: 9614930]

[23]   Li XA, Yutani C, Shimokado K. Serum amyloid P component associates with high density lipoprotein as well as very low density lipoprotein but not with low density lipoprotein. Biochem Biophys Res Commun 1998; 244(1): 249-52.
[http://dx.doi.org/10.1006/bbrc.1998.8248] [PMID: 9514915]

[24]   Coe JE, Rose MJ. Electrophoretic polymorphisms of hamster pentraxin, female protein (Amyloid P-Component). Scand Journal of Immunology 1997; 46(2): 180-2.
[PMID: 11090]

[25]   Weis WI, Crichlow GV, Murthy HM, Hendrickson WA, Drickamer K. Physical characterization and crystallization of the carbohydrate-recognition domain of a mannose-binding protein from rat. J Biol Chem 1991; 266(31): 20678-86.
[PMID: 1939118]

[26]   Weis WI, Drickamer K, Hendrickson WA. Structure of a C-type mannose-binding protein complexed with an oligosaccharide. Nature 1992; 360(6400): 127-34.
[http://dx.doi.org/10.1038/360127a0] [PMID: 1436090]

[27]   O'Riordan DM, Standing JE, Kwon KY, Chang D, Crouch EC, Limper AH. Surfactant protein D interacts with Pneumocystis carinii and mediates organism adherence to alveolar macrophages. J Clin Invest 1995; 95(6): 2699-710.
[http://dx.doi.org/10.1172/JCI117972] [PMID: 7769109]

[28]   Matsushita M. Ficolins: complement-activating lectins involved in innate immunity. J Innate Immun 2010; 2(1): 24-32.
[http://dx.doi.org/10.1159/000228160] [PMID: 20375620]

[29]   Roos A, Xu W, Castellano G, *et al.* Mini-review: A pivotal role for innate immunity in the clearance of apoptotic cells. Eur J Immunol 2004; 34(4): 921-9.
[http://dx.doi.org/10.1002/eji.200424904] [PMID: 15048702]

[30]   Owens DE III, Peppas NA. Opsonization, biodistribution, and pharmacokinetics of polymeric nanoparticles. Int J Pharm 2006; 307(1): 93-102.
[http://dx.doi.org/10.1016/j.ijpharm.2005.10.010] [PMID: 16303268]

[31]   Viota JL, González-Caballero F, Durán JD, Delgado AV. Study of the colloidal stability of concentrated bimodal magnetic fluids. J Colloid Interface Sci 2007; 309(1): 135-9.
[http://dx.doi.org/10.1016/j.jcis.2007.01.066] [PMID: 17346730]

[32]   Lundqvist M, Stigler J, Elia G, Lynch I, Cedervall T, Dawson KA. Nanoparticle size and surface properties determine the protein corona with possible implications for biological impacts. Proc Natl Acad Sci USA 2008; 105(38): 14265-70.
[http://dx.doi.org/10.1073/pnas.0805135105] [PMID: 18809927]

[33]   Lynch I, Dawson KA. Protein nanoparticle interaction. Nano Today 2008; 3: 40-7.
[http://dx.doi.org/10.1016/S1748-0132(08)70014-8]

[34]   Hotze EM, Phenrat T, Lowry GV. Nanoparticle aggregation: challenges to understanding transport and reactivity in the environment. J Environ Qual 2010; 39(6): 1909-24.
[http://dx.doi.org/10.2134/jeq2009.0462] [PMID: 21284288]

[35]   Nguyen VH, Lee BJ. Protein corona: a new approach for nanomedicine design. Int J Nanomedicine 2017; 12: 3137-51.
[http://dx.doi.org/10.2147/IJN.S129300] [PMID: 28458536]

[36]   Vroman L, Adams AL, Fischer GC, Munoz PC. Interaction of high molecular weight kininogen, factor XII, and fibrinogen in plasma at interfaces. Blood 1980; 55(1): 156-9.
[http://dx.doi.org/10.1182/blood.V55.1.156.156] [PMID: 7350935]

[37]   Treuel L, Docter D, Maskos M, Stauber RH. Protein corona - from molecular adsorption to physiological complexity. Beilstein J Nanotechnol 2015; 6: 857-73.
[http://dx.doi.org/10.3762/bjnano.6.88] [PMID: 25977856]

[38]   Gagner JE, Lopez MD, Dordick JS, Siegel RW. Effect of gold nanoparticle morphology on adsorbed protein structure and function. Biomaterials 2011; 32(29): 7241-52.
[http://dx.doi.org/10.1016/j.biomaterials.2011.05.091] [PMID: 21705074]

[39]   Hühn D, Kantner K, Geidel C, *et al.* Polymer-coated nanoparticles interacting with proteins and cells: focusing on the sign of the net charge. ACS Nano 2013; 7(4): 3253-63.
[http://dx.doi.org/10.1021/nn3059295] [PMID: 23566380]

[40]   Chellat F, Merhi Y, Moreau A, Yahia L. Therapeutic potential of nanoparticulate systems for macrophage targeting. Biomaterials 2005; 26(35): 7260-75.
[http://dx.doi.org/10.1016/j.biomaterials.2005.05.044] [PMID: 16023200]

[41]   Lindman S, Lynch I, Thunil E, *et al.* Systemic investigation of thermodynamics of human plasma protein adsorption to N-isopropylacrylamide/N-tert-butyl acrylamide copolymer nanoparticles, effects of particles size and hydrophobicity. Nano Lett 2007; 7(4): 914-20.
[http://dx.doi.org/10.1021/nl062743+] [PMID: 17335269]

[42]   Gessner A, Waicz R, Lieske A, Paulke B, Mäder K, Müller RH. Nanoparticles with decreasing surface hydrophobicities: influence on plasma protein adsorption. Int J Pharm 2000; 196(2): 245-9.
[http://dx.doi.org/10.1016/S0378-5173(99)00432-9] [PMID: 10699728]

[43]   Staufenbiel S, Marino M, Li W, *et al.* Surface characterization and protein interaction of a series model of poly [acrylonitrile-co-(N-vinylpyrrolidone) nanocarriers for drug targeting International Journal of Pharma 2015; 485(1-2): 87-96.

[44]   Sacchetti C, Motamedchaboki K, Magrini A, *et al.* Surface polyethylene glycol conformation influences the protein corona of polyethylene glycol-modified single-walled carbon nanotubes: potential implications on biological performance. ACS Nano 2013; 7(3): 1974-89.
[http://dx.doi.org/10.1021/nn400409h] [PMID: 23413928]

[45]   Lesniak A, Fenaroli F, Monopoli MP, Åberg C, Dawson KA, Salvati A. Effects of the presence or absence of a protein corona on silica nanoparticle uptake and impact on cells. ACS Nano 2012; 6(7): 5845-57.
[http://dx.doi.org/10.1021/nn300223w] [PMID: 22721453]

[46]   Gräfe C, Weidner A, Lühe MV, *et al.* Intentional formation of a protein corona on nanoparticles: Serum concentration affects protein corona mass, surface charge, and nanoparticle-cell interaction. Int J Biochem Cell Biol 2016; 75: 196-202.
[http://dx.doi.org/10.1016/j.biocel.2015.11.005] [PMID: 26556312]

[47]   Lundqvist M, Stigler J, Cedervall T, *et al.* The evolution of the protein corona around nanoparticles: a test study. ACS Nano 2011; 5(9): 7503-9.
[http://dx.doi.org/10.1021/nn202458g] [PMID: 21861491]

[48]   Tenzer S, Docter D, Kuhar J, *et al.* Rapid formation of plasma protein corona critically affects nanoparticle pathophysiology International Nanotechnology 2013; 8(10): 722-81.

[49]   Schuchard M, Melm C, Crawford A, *et al.* NCI proteomic technologies reagents resources workshop 2005. citeseerx.ist.psu.edu/viewdoc/download?doi=10.1.1590.7879&rep=rep1&type=pdf

[50]   Fleischer CC, Payne CK. Nanoparticle-cell interactions: molecular structure of the protein corona and cellular outcomes. Acc Chem Res 2014; 47(8): 2651-9.
[http://dx.doi.org/10.1021/ar500190q] [PMID: 25014679]

[51]   Monopoli MP, Walczyk D, Campbell A, *et al.* Physical-chemical aspects of protein corona: relevance to *in vitro* and *in vivo* biological impacts of nanoparticles. J Am Chem Soc 2011; 133(8): 2525-34.
[http://dx.doi.org/10.1021/ja107583h] [PMID: 21288025]

[52]   Yallapu MM, Chauhan N, Othman SF, *et al.* Implications of protein corona on physico-chemical and biological properties of magnetic nanoparticles. Biomaterials 2015; 46: 1-12.
[http://dx.doi.org/10.1016/j.biomaterials.2014.12.045] [PMID: 25678111]

[53]   Thiele L, Diederichs JE, Reszka R, Merkle HP, Walter E. Competitive adsorption of serum proteins at microparticles affects phagocytosis by dendritic cells. Biomaterials 2003; 24(8): 1409-18.
[http://dx.doi.org/10.1016/S0142-9612(02)00525-2] [PMID: 12527282]

[54]   Maiorano G, Sabella S, Sorce B, *et al.* Effects of cell culture media on the dynamic formation of protein-nanoparticle complexes and influence on the cellular response. ACS Nano 2010; 4(12): 7481-91.
[http://dx.doi.org/10.1021/nn101557e] [PMID: 21082814]

[55]   Marieb EN, Hoehn K. The Cardiovascular System: Blood Vessels. 9th ed., Upper Saddle River, NJ: Pearson Education, Inc 2013.

[56]   Behzadi S, Serpooshan V, Sakhtianchi R, *et al.* Protein corona change the drug release profile of nanocarriers: the "overlooked" factor at the nanobio interface. Colloids Surf B Biointerfaces 2014; 123: 143-9.
[http://dx.doi.org/10.1016/j.colsurfb.2014.09.009] [PMID: 25262409]

[57]   Palchetti S, Colapicchioni V, Digiacomo L, *et al.* The protein corona of circulating PEGylated liposomes. Biochim Biophys Acta 2016; 1858(2): 189-96.
[http://dx.doi.org/10.1016/j.bbamem.2015.11.012] [PMID: 26607013]

[58]   Kang T, Park C, Lee BJ. Investigation of biomimetic shear stress on cellular uptake and mechanism of polystyrene nanoparticles in various cancer cell lines Archives of Pharmacal Research 2016; 39(12): 9.

[59]   Kang T, Cho Y, Park C, *et al.* Effect of biomimetic shear stress on intracellular uptake and cell-killing efficiency of doxorubicin in a free and liposomal formulation. Int J Pharm 2016; 510(1): 42-7.
[http://dx.doi.org/10.1016/j.ijpharm.2016.06.017] [PMID: 27286636]

[60]   Kang T, Park C, Cui JS, Cui JH, Lee BJ. Effects of shear stress on the cellular distribution of polystyrene nanoparticles in a biomimetic microfluid system. J Drug Deliv Sci Technol 2016; 31: 130-6. b
[http://dx.doi.org/10.1016/j.jddst.2015.12.001]

[61]   Panalytical M. Protein melting point characterization using Zetasizer nanosystem 2005. www.azonano.com/article.aspx?ArticleID=1224

[62]   Goy-Lopez S, Juarez J, Alatorre-Meda M, *et al.* Physicochemical characteristics of protein nanoparticles Bioconjugate: the role of particles curvature and solution conditions on human serum albumin conformation and fibrillogenesis. Langmuir 2012; 28(24): 9113-26.
[http://dx.doi.org/10.1021/la300402w] [PMID: 22439664]

[63]   Mahmoudi M, Abdelmonem AM, Behzadi S, *et al.* Temperature: the "ignored" factor at the NanoBio interface. ACS Nano 2013; 7(8): 6555-62.
[http://dx.doi.org/10.1021/nn305337c] [PMID: 23808533]

[64]   O'Brien J, Shea KJ. Tuning the protein corona of hydrogel nanoparticles the synthesis of abiotic protein and peptide affinity reagents. Acc Chem Res 2016; 49(6): 1200-10.
[http://dx.doi.org/10.1021/acs.accounts.6b00125] [PMID: 27254382]

[65]   Adamson AW, Gast AP. Physical Chemistry of Surfaces. John Wiley and Sons 1997.

[66]   Docter D, Westmeier D, Markiewicz M, Stolte S, Knauer SK, Stauber RH. The nanoparticle biomolecule corona: lessons learned - challenge accepted? Chem Soc Rev 2015; 44(17): 6094-121.
[http://dx.doi.org/10.1039/C5CS00217F] [PMID: 26065524]

[67]   Khan S, Mukherjee A, Chandrasekaran N. Silver nanoparticles tolerant bacteria from sewage environment. J Environ Sci (China) 2011; 23(2): 346-52. [CODEN: JENSEN].
[http://dx.doi.org/10.1016/S1001-0742(10)60412-3] [PMID: 21517011]

[68]   Chen G, Liu X, Su C. Distinct effects of humic acid on transport and retention of TiO_2 rutile nanoparticles in saturated sand columns. Environ Sci Technol 2012; 46(13): 7142-50.
[http://dx.doi.org/10.1021/es204010g] [PMID: 22681399]

[69]   Wang D, Bradford SA, Harvey RW, Gao B, Cang L, Zhou D. Humic acid facilitates the transport of ARS-labeled hydroxyapatite nanoparticles in iron oxyhydroxide-coated sand. Environ Sci Technol 2012; 46(5): 2738-45.
[http://dx.doi.org/10.1021/es203784u] [PMID: 22316080]

[70]   Pakarinen K, Petersen EJ, Alvila L, *et al.* A screening study on the fate of fullerenes (nC60) and their toxic implications in natural freshwaters. Environ Toxicol Chem 2013; 32(6): 1224-32.
[http://dx.doi.org/10.1002/etc.2175] [PMID: 23404765]

[71]   Kittler S, Greulich C, Gebauer JS, *et al.* The influence of protein on the dispersibility and cell Biology activity of silver nanoparticles. J Mater Chem 2010; 20: 512-8.
[http://dx.doi.org/10.1039/B914875B]

[72]   Segets D, Marczak R, Schafer S, *et al.* Experimental and theoretical studies of the colloidal stability of nanoparticles- a general interpretation based stability maps, American Chemical Society. Nano 2011; 5(6): 4659-69.
[http://dx.doi.org/10.1021/NN200465]

[73]   Gebauer JS, Malissek M, Simon S, *et al.* Impact of the nanoparticle-protein corona on colloidal stability and protein structure. Langmuir 2012; 28(25): 9673-9.
[http://dx.doi.org/10.1021/la301104a] [PMID: 22524519]

[74]   Rahman M, Laurent S, Tawil N, Yahia L-H, Mohamoudi M. Protein corona: Applications and challenges.Protein-nanoparticles interaction, Springer Series in Biophysics, 15. Berlin, Heidelberg: Springer-Verlag 2013.
[http://dx.doi.org/10.1007/978-3-642-37555-2_3]

[75]   Aggarwal P, Hall JB, McLeland CB, Dobrovolskaia MA, McNeil SE. Nanoparticle interaction with plasma proteins as it relates to particle biodistribution, biocompatibility and therapeutic efficacy. Adv Drug Deliv Rev 2009; 61(6): 428-37.
[http://dx.doi.org/10.1016/j.addr.2009.03.009] [PMID: 19376175]

[76]   Bikram M, Lee M, Chang CW, Janát-Amsbury MM, Kern SE, Kim SW. Long-circulating DNA-complexed biodegradable multiblock copolymers for gene delivery: degradation profiles and evidence of dysopsonization. J Control Release 2005; 103(1): 221-33.
[http://dx.doi.org/10.1016/j.jconrel.2004.11.011] [PMID: 15710513]

[77]   Walkey CD, Olsen JB, Song F, *et al.* Protein corona fingerprinting predicts the cellular interaction of gold and silver nanoparticles ACS Nano 2014; 8(30): 2439-55.

[78]   Moghimi SM, Hunter AC, Murray JC. Long-circulating and target-specific nanoparticles: theory to practice. Pharmacol Rev 2001; 53(2): 283-318.
[PMID: 11356986]

[79]   Karmali PP, Simberg D. Interactions of nanoparticles with plasma proteins: implication on clearance and toxicity of drug delivery systems. Expert Opin Drug Deliv 2011; 8(3): 343-57.
[http://dx.doi.org/10.1517/17425247.2011.554818] [PMID: 21291354]

[80]   Zanganeh S, Spitler R, Erfanzadeh M, Alkilany AM, Mahmoudi M. Protein corona: Opportunities and challenges. Int J Biochem Cell Biol 2016; 75: 143-7.

[http://dx.doi.org/10.1016/j.biocel.2016.01.005] [PMID: 26783938]

[81]   Kelly PM, Åberg C, Polo E, *et al.* Mapping protein binding sites on the biomolecular corona of nanoparticles. Nat Nanotechnol 2015; 10(5): 472-9.
[http://dx.doi.org/10.1038/nnano.2015.47] [PMID: 25822932]

[82]   O'Connell DJ, Bombelli FB, Pitek AS, Monopoli MP, Cahill DJ, Dawson KA. Characterization of the bionano interface and mapping extrinsic interactions of the corona of nanomaterials. Nanoscale 2015; 7(37): 15268-76.
[http://dx.doi.org/10.1039/C5NR01970B] [PMID: 26324751]

[83]   Moyano DF, Saha K, Prakash G, *et al.* Fabrication of corona-free nanoparticles with tunable hydrophobicity. ACS Nano 2014; 8(7): 6748-55.
[http://dx.doi.org/10.1021/nn5006478] [PMID: 24971670]

[84]   Mirshafiee V, Kim R, Parks S, Mahmoudi M, Kraft ML. Impact of protein pre-coating on the protein corona composition and nanoparticles cellular uptake, Biomaterials 2016; 75: 295-304.
[http://dx.doi.org/http://dx/doi.org/10.1016/j.biomaterials. 2015.10.019]

[85]   Mahmoudi M, Simchi A, Imani M, *et al.* A new approach for the *in vitro* identification of the cytotoxicity of superparamagnetic iron oxide nanoparticles. Colloids Surf B Biointerfaces 2010; 75(1): 300-9.
[http://dx.doi.org/10.1016/j.colsurfb.2009.08.044] [PMID: 19781921]

[86]   Podolski IY, Podlubnaya ZA, Kosenko EA, *et al.* Effects of hydrated forms of C60 fullerene on amyloid 1-peptide fibrillization *in vitro* and performance of the cognitive task. J Nanosci Nanotechnol 2007; 7(4-5): 1479-85.
[http://dx.doi.org/10.1166/jnn.2007.330] [PMID: 17450915]

[87]   Laurent S, Ejtehadi MR, Rezaei M. Kehoe, Mahmoudi M, Interdisciplinary challenges and promising theranostics effects of nanoscience in Alzheimer's disease. RSC Advances 2012; 2: 5008-33.
[http://dx.doi.org/10.1039/c2ra01374f]

[88]   Ghavami M, Rezaei M, Ejtehadi R, *et al.* Physiological temperature has a crucial role in amyloid β in the absence and presence of hydrophobic and hydrophilic nanoparticles. ACS Chem Neurosci 2013; 4(3): 375-8.
[http://dx.doi.org/10.1021/cn300205g] [PMID: 23509973]

[89]   Wan S, Cheng Q. Role of interface interaction in the construction of GO-based artificial Nacres. Adv Mater Interfaces 2018; 5(12): 1800107.
[http://dx.doi.org/10.1002/admi.201800107]

[90]   Wangoo N, Suri R, Shekhawat G. Interaction of gold nanoparticles with protein: A spectroscopic study to monitor protein conformational changes Applied Physics Letters 2008; 92>133104.

[91]   Hou J-F, Yang Y-L, Wang C. Molecular mechanism of interface interactions between Nanomaterials and proteins. Wuli Huaxue Xuebao 2017; 33(1): 63-79.
[http://dx.doi.org/10.3866/PKU.WHXB201608233]

[92]   Nel AE, Mädler L, Velegol D, *et al.* Understanding biophysicochemical interactions at the nano-bio interface. Nat Mater 2009; 8(7): 543-57.
[http://dx.doi.org/10.1038/nmat2442] [PMID: 19525947]

[93]   Dalsin JL, Hu BH, Lee BP, Messerssmith PB. Mussel adhesive protein mimetic polymer for the preparation of non- surfaces. J Am Chem Soc 2003; 125: 4233-58.
[http://dx.doi.org/10.1021/ja0284963]

[94]   Spisak S, Tulassay Z, Molnar B, Guttman A. Protein microchips in biomedicine and biomarker discovery. Electrophoresis 2007; 28(23): 4261-73.
[http://dx.doi.org/10.1002/elps.200700539] [PMID: 17979160]

[95]   Johnson CJ, Zhukovsky N, Cass AE, Nagy JM. Proteomics, nanotechnology and molecular diagnostics. Proteomics 2008; 8(4): 715-30.

[http://dx.doi.org/10.1002/pmic.200700665] [PMID: 18297650]

[96]   Vaisocherová H, Yang W, Zhang Z, *et al.* Ultralow fouling and functionalizable surface chemistry based on a zwitterionic polymer enabling sensitive and specific protein detection in undiluted blood plasma. Anal Chem 2008; 80(20): 7894-901.
[http://dx.doi.org/10.1021/ac8015888] [PMID: 18808152]

[97]   Peng Q, Mu H. The potential of protein-nanomaterial interaction for advanced drug delivery. J Control Release 2016; 225: 121-32.
[http://dx.doi.org/10.1016/j.jconrel.2016.01.041] [PMID: 26812004]

[98]   Boulos SP, Davis TA, Yang JA, *et al.* Nanoparticle-protein interactions: a thermodynamic and kinetic study of the adsorption of bovine serum albumin to gold nanoparticle surfaces. Langmuir 2013; 29(48): 14984-96.
[http://dx.doi.org/10.1021/la402920f] [PMID: 24215427]

[99]   Dobrowolski MA, Patr AK, Zheng J, *et al.* the interaction of colloidal gold nanoparticles with human blood: effects on particle size and analysis of plasma protein binding profile. Nanomedicine 2009; 5(2): 106-17.

[100] Rezwan K, Meier LP, Rezwan M, Vörös J, Textor M, Gauckler LJ. Bovine serum albumin adsorption onto colloidal Al2O3 particles: a new model based on zeta potential and UV-*vis* measurements. Langmuir 2004; 20(23): 10055-61.
[http://dx.doi.org/10.1021/la048459k] [PMID: 15518493]

[101] Gheshlaghi ZN, Riazi GH, Ahmadian S, Ghafari M, Mahinpour R. Toxicity and interaction of titanium dioxide nanoparticles with microtubule protein. Acta Biochim Biophys Sin (Shanghai) 2008; 40(9): 777-82.
[http://dx.doi.org/10.1093/abbs/40.9.777] [PMID: 18776989]

[102] Romana P, Shamsi TN, Fatima S. Nanoparticles and protein interaction: Role in protein aggregation and clinical implications International Journal of Biological Macromolecules, 94-Part A2017 2017; 386-95.2017;

[103] Singh K P, Dhasmana A, Rahman Q. Elucidation the toxicity mechanism of ZnO nanoparticles using molecular docking approach with proteins Asian Journal of Pharmaceutical and clinical research 2018; 11(3): 441-6.

[104] Shemetov AA, Nabiev I, Sukhanova A. Molecular interaction of proteins and peptides with nanoparticles. ACS Nano 2012; 6(6): 4585-602.
[http://dx.doi.org/10.1021/nn300415x] [PMID: 22621430]

[105] Zhang XQ, Xu X, Bertrand N, Pridgen E, Swami A, Farokhzad OC. Interactions of nanomaterials and biological systems: Implications to personalized nanomedicine. Adv Drug Deliv Rev 2012; 64(13): 1363-84.
[http://dx.doi.org/10.1016/j.addr.2012.08.005] [PMID: 22917779]

[106] Ilinskaya AN, Dobrovlskaia MA. Interaction between nanoparticles and plasma proteins: Effects on nanoparticle distribution and toxicity.Polymer nanoparticles for nanomedicine, Springer Chem. 2016.https://doi.org
[http://dx.doi.org/10.1007/978-3-319-41421-8_15]

[107] Dutta D, Sundaram SK, Teequarden J, *et al.* Adsorbed proteins influence the biological activity and molecular targeting of nanomaterials. 2007; 100: pp. (1)303-15.
[http://dx.doi.org/10.1093/toxsci/kfm217]

[108] Nagayama S, Ogawara K, Minato K, *et al.* Fetuin mediates hepatic uptake of negatively charged nanoparticles via scavenger receptor. Int J Pharm 2007; 329(1-2): 192-8.
[http://dx.doi.org/10.1016/j.ijpharm.2006.08.025] [PMID: 17005341]

[109] Kreuter J, Hekmatara T, Dreis S, Vogel T, Gelperina S, Langer K. Covalent attachment of apolipoprotein A-I and apolipoprotein B-100 to albumin nanoparticles enables drug transport into the

brain. J Control Release 2007; 118(1): 54-8.
[http://dx.doi.org/10.1016/j.jconrel.2006.12.012] [PMID: 17250920]

[110] Sant S, Poulin S, Hildgen P. Effect of polymer architecture on surface properties, plasma protein adsorption, and cellular interactions of pegylated nanoparticles. J Biomed Mater Res A 2008; 87(4): 885-95.
[http://dx.doi.org/10.1002/jbm.a.31800] [PMID: 18228249]

[111] Melton L. Protein arrays: proteomics in multiplex. Nature 2004; 429(6987): 101-7.
[http://dx.doi.org/10.1038/429101a] [PMID: 15129287]

[112] Mark S. Protein microarrays. Jones and Bartlett Learning 2005. ISBN 978-0-7637-3127-4.

[113] Mitchell P. A perspective on protein microarrays. Nat Biotechnol 2002; 20(3): 225-9.
[http://dx.doi.org/10.1038/nbt0302-225] [PMID: 11875416]

[114] DeRisi JL, Iyer VR, Brown PO. Exploring the metabolic and genetic control of gene expression on a genomic scale. Science 1997; 278(5338): 680-6.
[http://dx.doi.org/10.1126/science.278.5338.680] [PMID: 9381177]

[115] MacBeath G, Schreiber SL. Printing proteins as microarrays for high-throughput function determination. Science 2000; 289(5485): 1760-3.
[PMID: 10976071]

[116] Zhu H, Bilgin M, Bangham R, et al. Global analysis of protein activities using proteome chips. Science 2001; 293(5537): 2101-5.
[http://dx.doi.org/10.1126/science.1062191] [PMID: 11474067]

[117] Park ML, Hur J, Cho HS, et al. High throughput single-cell, cell imaging. Lab Chip 2011; 11: 79-86.
[http://dx.doi.org/10.1039/C0LC00114G] [PMID: 20957290]

[118] Phelps TJ, Palumbo AV, Beliaev AS. Metabolomics and microarrays for improved understanding of phenotypic characteristics controlled by both genomics and environmental constraints. Curr Opin Biotechnol 2002; 13(1): 20-4.
[http://dx.doi.org/10.1016/S0958-1669(02)00279-3] [PMID: 11849953]

[119] Feizi T, Fazio F, Chai W, Wong CH. Carbohydrate microarrays - a new set of technologies at the frontiers of glycomics. Curr Opin Struct Biol 2003; 13(5): 637-45.
[http://dx.doi.org/10.1016/j.sbi.2003.09.002] [PMID: 14568620]

[120] Gobert GN, McInnes R, Moertel L, et al. Transcriptomics tool for the human Schistosoma blood flukes using microarray gene expression profiling. Exp Parasitol 2006; 114(3): 160-72.
[http://dx.doi.org/10.1016/j.exppara.2006.03.003] [PMID: 16631746]

[121] Abraham AM, Kannangai R, Sridharan G. Nanotechnology: a new frontier in virus detection in clinical practice. Indian J Med Microbiol 2008; 26(4): 297-301. www.ijmm.org.
[http://dx.doi.org/10.4103/0255-0857.43551] [PMID: 18974480]

[122] Syahir A, Usui K, Tomizaki K-Y, Kajikawa K, Mihara H. Label and label-free detection technique for protein microassay. Microarrays (Basel) 2015; 4(2): 228-44.
[http://dx.doi.org/10.3390/microarrays4020228] [PMID: 27600222]

[123] Medina-Sánchez M, Miserere S, Merkoçi A. Nanomaterials and lab-on-a-chip technologies. Lab Chip 2012; 12(11): 1932-43.
[http://dx.doi.org/10.1039/c2lc40063d] [PMID: 22517169]

[124] Sanfins E, Dairon J, Fernando R-L, Dupret J-M. Nanoparticles-protein interactions: from crucial plasma protein to key enzymes. J Phys Conf Ser 2011; 304: 012039.
[http://dx.doi.org/10.1088/1742-6596/304/1/012039]

[125] Krizokova S, Heger Z, Zalewoka M, et al. Nanotechnologies in protein microassay Nanomedicine (London) 2015; 10(17): 2743-55.

[126] Carlota A-S, Tabata N, Pabio J-V, et al. Interactions of nanoparticles and biosystems: The

microenvironment of nanomaterials and biomolecules. Nanomaterials (Basel) 2019; 9: 1365.
[http://dx.doi.org/10.3390/nano9101365]

[127] Sharma VK, Sayes CM, Guo B, *et al.* Interactions between silver nanoparticles and other metal nanoparticles under environmentally relevant conditions: A review. Sci Total Environ 2019; 653: 1042-51.
[http://dx.doi.org/10.1016/j.scitotenv.2018.10.411] [PMID: 30759545]

CHAPTER 7

Interactions Between Nanomaterials and Genetic Material (DNA and RNA)

Abstract: Genetic material is a stable biomolecule in an organism. The intact and integrated transfer of genetic information from the parental generation to the offspring (daughter cells) is essential. This transfer acts as a basis and ensures the conveyance of somatic and sex-linked traits from generation to generation. The DNA contains genetic information and is present in eukaryotes and prokaryotes, while viruses have genetic information either in DNA or RNA. The genetic information plays a prime role in maintaining structural, physiological originality and modifications by retaining the specific pattern of transcription, translation, and replication of genetic material during cell proliferation, cell cycle, cell differentiation, *etc.* Cellular behavior reflects on the structural, functional, and genetic health of a cell, tissues, and an organism. The formulations of nanomaterials are in concern with the targeted moieties. The nanomaterials have spread their tentacles in most of the fields following the functional and procedural aspects of the biological and non-biological sciences. Different types of nanomaterials are produced in order to meet the demands of various domains like biotechnology, biomedical sciences, industrial, material sciences, *etc.* Nanomaterials cause either beneficial or harmful effects in a biosystem and the environment. The disoriented biochemical, biophysical, and biomolecular impacts are due to the adverse effects of nanomaterials on genetic contents. This condition brings disorganized functionality of the genetic information and the cell. The evaluation of their implications on biomolecules like DNA and RNA is essential to understand the mechanism involved. This chapter deals with the overall biochemical, physiological, and biophysical aspects of genetic contents, along with the impacts of various types of nanomaterials

Keywords: Annealing, Bionanointerface, Carbon Nanomaterials, Dendrimers, DNA, Histone- Proteins, Hydrophobicity And Hydrophilicity, Quantum Dots.

INTRODUCTION

Genetic integrity is the foundation of morphological, physiological, and genetic functionality of all the organisms. This genetic information also maintains the phylogenetic status of a species. Studies related to genetics help in understanding the number of malfunctioning in a life form. Genetic integrity appears to be a significant aspect related to the synthesis of proteins that are needed from time to time in the life of a creature. The genetic materials play a prime role in maint-

Yogendrakumar H. Lahir & Pramod Avti

aining the structural, physiological originality during the transcription, translation, and replication of genetic elements, and the normal cellular functions like cell proliferation, cell cycle, and cell differentiation [1]. There are natural processes like repairing DNA damages; this process retains and restores the originality of genetic materials in species. The rate of repair or restoration of genetic information varies in different cells, tissues, and organisms. Parameters like age, physiological, and pathological status of a species are also related to genetic integrity. A biological cell with damaged genetic material is likely to undergo irreparable dormancy state, senescence, apoptosis, and uncontrolled defective cell division, *etc* [1].

Nanotechnology and nanoscience have made exceptional advancements, and their products have found suitable applications in most of the fields. The nanomaterials are useful and beneficial because of their specific features like small size, the higher surface to volume ratio, ease of modifications of surface chemistry, and their ability to conjugate with multivalent ligands. These physicochemical features enhance the degree of avidity of the fabricated nanomaterials for targets like biological tissues, cells, biomolecules, and cell organelles [2]. DNA forms several alternative structures like non-B forms of DNA. These forms are more in numbers and detected in a genome. These non-B forms of DNA play an active role in the varied cellular processes and cause instability in the genetic information of an organism. These non-B Forms of DNA influence gene functions, regulation of immune response, telomere maintenance, and recombination in a cell. The antigen variations in human concerning pathogens and developmental conditions result in the diversity in the genome of a biosystem. These non-B forms of DNA are concerned with transcription and translation also [3].

The polymorphic form-DNA exhibits different assemblies and conformational forms like right-handed A form, left-handed Z form, triplex, G-quadruplex forms, i-motif forms under different physiological conditions [4]. Sathees and Leiber mentioned the existence of cruciform-DNA, Z- DNA, sticky DNA, slipped DNA structure (RNA–DNA hybrid), E DNA (e-motif) [5]. Different conformational and assembled forms of DNA are potential agents for human diseases. The cruciform (hairpin) form of DNA concerns with genetic instability, male infertility, recurrent abortions, Emanuel syndrome, and polycystic kidney disease. The triplet form causes hereditary neurological disorders, follicular lymphoma, and other types of cancers [5].

AN OVERVIEW OF BIOCHEMICAL AND BIOPHYSICAL ASPECTS OF DNA

A genetic material, *i.e.*, DNA, is bestowed with storing information related to the originality of the cells, species, and also for the functionality of specific cells in an organism. Generally, the genetic material in a life form is stable and intact and is essential for the DNA to maintain its structural, functional, and phylogenetic integrity and identity. This fundamental and technical originality of the genetic information in totality is an essential aspect during normal processes, such as transcription, translation, replication, even during the hereditary transfer from one generation to another. This feature is functionally significant to avoid erroneous genetic configuration. DNA is among one of the macro biomolecules correctly attributed to genetic aspects of an organism and protein synthesis. DNA is a double-stranded bipolar helical structure. The two strands are polymer composed of monomer units referred to as the nucleotide. The monomer unit consists of one of the nitrogen nucleobases among cytosine, adenine, guanine, thymine, pentose sugar (deoxyribose and ribose), and phosphate groups. Phosphate groups and sugar are bounded, and nitrogen nucleobases involving covalent bond and hydrogen bonds. These nitrogen bases are pyrimidines (thymine and cytosine) and purines (adenine and guanine).

The backbone of DNA resists cleavage. Each strand is anti-parallel and coiled around the same axis having pitch 34 Å (3.4 nm) and radius 10 Å (1.0 nm). Mendelkerm and co-workers reported the width of DNA between 22 to 26 Å (2.2 to 2.6 nm) and the size of one monomer unit 3.3Å (0.33nm) [6 - 8]. Primarily two forces maintain the stability of the DNA molecule. The hydrogen bonds are present between nucleotides, and base stacking interactions involving aromatic nucleobases are also responsible. Nucleotide bases align at a right angle to the axis of the DNA, forming π-bonds, thereby reducing the interactions between them [9].

PHYSICOCHEMICAL   FEATURES   OF   DNA - IMPACT   OF TEMPERATURE

Melting temperature is an essential parameter during its interaction with nanomaterials [10]. DNA is prone to fluctuations in temperature. DNA gets denatured when its double-stranded structure is disturbed or the double-stranded changes into two single-stranded conformations. The denatured DNA elevates the degree of absorption of UV radiation. It is a potential parameter to denature DNA molecules. As temperature increases, the frequency of breaking of Hydrogen bonds between the two strands also increases. A temperature at which 50% of DNA gets denatured, *i.e.*, the double-stranded DNA sample becomes 50% single-stranded form, is called its melting temperature (Tm). Tm-50- represents the annealing temperature of DNA. The annealing temperature is related to the length of DNA, the compositional sequence of nucleotides, strand concentration, and ionic concentration of the salt added. When the sample of DNA is having more contents of G-C, the value of Tm is higher. Thus, during the polymerase chain reaction (PCR), the Tm technique acts as an essential parameter and also while investigating the C-G contents of DNA. The enthalpy ($\Delta H°$) and entropy ($\Delta S°$) for the melting of each nearest neighbor doublet base pairs in the DNA molecule is essential to predict the stability of the DNA molecule. In terms of changes in free energy concept, the biochemical secureness of DNA is free energy equals to enthalpy -entropy ($\Delta G° = \Delta H° - \tau \Delta S°$). Regular B-DNA conformation of the helix with Watson and Crick's base pairs stabilizes with the nearest neighbor stacking has ten possible kinds of nearest-neighbor doublet [11].

The interaction between the fixed phosphate charges influences the alteration of Tm because of the salt concentration. There is a relationship between the electrostatic free energy changes in the helical coil to a specific potential. This potential exists because of the repulsion between the phosphate charges; functionally, it depends on the distance between the two interacting strands [12]. The defective, mismatched, and deleted sequences of base pairs influence the Tm melting temperature [13].

OPTICAL PROPERTIES OF DNA DURING RADIATION ABSORPTION

DNA concentration is measured at 260 nm, using spectrophotometer and quartz cuvette. The buffer used should be checked as it interferes with the absorption of radiation at 280 nm. A pure DNA solution should have a 260/280 ratio in the range of >1.8. Contaminants at 280 may be proteins that lower this ratio. A-260 nm reading of 1 in a 1mL cuvette with 1 cm path length indicates the concentration of DNA to be as 50µg/ml while the same optical density reading for single-stranded DNA solution indicates the amount to be as 30µg/ml [14]. Absorbance measurement is a standard option to detect DNA and RNA if the interference of contaminants and buffer are important parameters during their estimation. The relatively narrow range of wavelengths is suitable to detect purity and concentration of DNA. Commonly wavelength A-260 is used to identify the purity of nucleic acids up to microgram quantity. At A-260, there is no distinction between DNA and RNA at A-260 [15]. The estimation of DNA depends on the absorbance on the mode of detection concerning the contents of phosphorous [16] (Holden *et al.*, 2009). The nitrogen bases in DNA absorb UV light at wavelength 260 nm, and this is easily monitored using spectrophotometry. Free bases absorb 1.60 units at 260 nm. Single-stranded DNA absorbs 1.37 units at 260 nm while double-stranded DNA absorbs 1.00 units at 260 nm [17 - 19, 29].

RELATIONSHIP BETWEEN DENSITY AND DNA

The buoyant density centrifugation technique is suitable to separate and find the density of the desired molecule present in a sample solution. The centrifugation technique helps to isolate genetic materials and the isotopic forms of biomolecules from the test sample. Under the standard experimental practice, cesium or cesium chloride solution is used to separate the desired molecule from the sample solution. The density of the desired molecule equals the mass of the cesium chloride at specific concentrations that settles at a particular level of denseness, forming a distinct band. This technique is suitable to isolate DNA molecules using a particular amount of cesium chloride using buoyancy [18 - 20]. The cesium chloride acts as a density gradient centrifugation. The sample solution of DNA is mixed with cesium chloride (CsCl), and centrifuged at 50000rpm for long hours. The opposing forces of sedimentation and diffusion force produce a stable and linear gradient of CsCl with the lightest density at the top and the heaviest at the bottom. If DNA in a sample solution is having one single, then density band is the result. Thus, DNA with different densities forms of a different number of optical groups equal to the amounts of DNA present in the sample. Macromolecules, like DNA, concentrate in the zone of CsCl that has the same mass. Thus, denser DNA migrates downward, and the less dense DNA is upward, forming respective bands [17 - 20].

The parameter of density is conveniently used to investigate the G-C contents of DNA. Typically, G-C base pairs are denser than the A-T base pairs. Thus, DNA having more of G-C base pairs will form band lower than the sample having A-T Base pairs. DNA density study also helps in the investigations about the satellite DNA. If chromosomal DNA is cut into equal-sized pieces and subjected to CsCl-density-ultracentrifugation, two bands are the result. The first band contains most of the DNA from the genome, and the second band consists of about 5% of DNA from the genome and has a highly repetitive sequence. This technique is dependable to establish the phylogenetic relationship among the various groups of organisms. In addition to biochemical, immunological, and biophysical methods, the buoyant density technique is also applicable to establish a taxonomical relationship. This technique is suitable to affirm taxonomical inter-relationship among prokaryotes [21]. The observations obtained from the density detection technique and the corresponding optical constants related to DNA show a relationship. This relationship plays a significant role in the characterization of interactions within DNA [22].

INFLUENCE OF HYDROPHOBICITY AND HYDROPHILICITY CONCERNING DNA AND NANOMATERIALS

Hydrophobic molecules tend to decline the Tm of DNA. The hydrophobic molecules dissolve the bases present in DNA, thereby, making it easy to disrupt the H-bonds between nitrogen bases and DNA strands. Hydrophilic surfaces exhibit competitive interactions between terminal hydrogen bonds and central base stacking. These surfaces stabilize the central base stacking.

On the other hand, hydrophobic surfaces make the H-bonds stronger but disrupt the stacking of central bases. Whatever the surface chemistry of hydrophobic surfaces may be concerning terminal nitrogen bases, the DNA shows the melting process by breaking H-bonds. Unsettling of the ambient nitrogen bases follow the melting of the DNA molecule. The central nitrogen bases present in the vicinity of hydrophobic surface the H-bonds disturbs the resulting melting process, and finally disrupts the stacking. There is a faster melting rate of central nitrogen bases in the region of hydrophobic surfaces during hybridization, and such setup results in the conformational changes in the biomolecule under consideration. This feature is helpful to get the desired conformation as per the need [17 - 20, 23].

THE IMPACT OF pH

Acids or acidic molecules with pH less than 1.0 break the phosphodiesters bonds between nucleotides. These molecules also break the N-glycosidic bonds between the sugar and purine bases. When the pH is around 4.0, there is a selective breaking of N-glycosidic bonds between sugar and purine, and DNA after this treatment, the resultant DNA is apurinic. Bases when interacting with DNA cause change in the polarity of the groups involved with H-bonds. At pH 11.3, all H-bonds disrupt, and the DNA becomes denatured. Typically, the DNA molecules resist hydrolysis at pH 13.0, but apurinic DNA hydrolyzes [18 - 20].

IMPACT OF IONIC STRENGTH

The backbone of DNA, *i.e.*, sugar and phosphate, are negatively charged. After treatment with distilled water, the DNA spontaneously denatures and becomes single-stranded DNA. Salts dissociate into ions neutralize the charge of phosphate groups. Thus, salts can stabilize DNA and elevate the Tm of DNA [17 - 20].

A-DNA, B-DNA, and Z-DNA

In the biosystem, three conformations of DNA molecules are present, namely A-DNA, B- DNA, and Z-DNA. All these conformations exhibit changes in the twist or helical pitch. This form of DNA depends on the stacking force exerted by each base on the neighboring base present in the chain. These conformations are also related to the absolute configuration of the nitrogen bases, which directs the helical curve [24].

A-DNA: This form of DNA is one of the many possible double helical structures of DNA along with B-DNA and Z-DNA. It is a popular form and is similar to the B-DNA conformation. This form shows the elaboration of geometrical attributes of various kinds of DNA. A-DNA is a right-handed double helix but shorter and compact helical structure. This form is found mostly in dehydrated samples of DNA and is useful in crystallographic experiments. A-DNA has right-handed helical orientation exhibiting 1bp repeating unit, 33.0°A rotating, 11.0 mean bp/turn, 19° inclination of bp to the axis, 0.24 nm rise/turn of helix 18p mean propeller twist and 2.6 nm diameter [17, 20, 24].

Biological Significance of A-DNA

The confirmation of DNA is an essential factor in the biochemical aspects of cell functionality. Any transition in the forms of DNA within the biosystem is quite tedious. There is a biological relevance of these different DNA conformations [25]. The transformation from B-DNA conformation to A-DNA conformity is one of the first reversible structural transitions noticed among biomolecules. Although B-DNA conformation recognizes the standard form of the double helix, the interchange between A-DNA and B-DNA conformity is of biological, biochemical, and biomedical significance [17, 26]. The understanding of this inter-transition is of relevance because it explains the structure and functionality of DNA in the cellular physiology. This transition is an appropriate target related to anticancer drug cisplatin form a biomedical point of view. The interaction between transcriptional promoter among eukaryotes-TATA-binding protein and A-DNA conformation induces exaggerated base pairs, thereby influencing the transcriptional process [26]. The growing numbers of DNA crystal structures exhibit properties of A- and B-DNA conformations. When methylated and brominated variants of the sequenced $(GGCGCC)_2$ crystallized as standard A-DNA, this serves as intermediate conformation. The composite structural strand formed has half A-DNA and half B-DNA and the A-nucleotides paired with the complementary B-nucleotides across duplex. These intermediate conformations have C3'-endo-sugars, therefore, are the allomorphic forms or conformities of A-DNA rather than true transition intermediates [26].

The conformation of A-DNA resembles the native structure of B-DNA. There is some difference between the conformity of A-DNA and that of B-DNA. A-DNA is smaller, relatively with more compact helical in comparison. The base pairs are not aligned at a right angle to the helical axis in contrast to B-DNA. Rosalind Franklin, its discoverer, suggested that the native DNA transforms into A-DNA conformity when it undergoes dehydration. Whenever the DNA molecule crystallizes, this condition is suitably applicable. A crystalline form of native DNA resembles A-DNA conformation, and also in the DNA-RNA hybrid doubles helices [27]. A-DNA conformation is the commonly identified conformation. This conformation of DNA is related to the defense mechanism of the cell. The structure of A-DNA attains the stability of the conformity of B-DNA. Studies related to the functionality of A-DNA indicate that it enhances the understanding of the process of evolution of nucleic acids, and the cellular biochemical processes [25]. Prokaryotes like a rod-like virus, *Sulfolobus islandicus,* thrive at extreme temperatures (around 80°C) and high acidic pH (3.0). In this case, the DNA exists as the conformity of A-DNA and possibly helps to withstand the highly adverse survival conditions. The capsid protein of the virus forms an α-helical extension that encases the viral DNA. Probably, it also stabilizes the conformation of A-DNA. This mechanism is similar to that occurring in spores of bacteria to protect bacterial DNA under highly unfavorable environmental conditions [17 - 20, 28].

Biological Significance of B-DNA

B-DNA is the most common conformational form present in organisms. This form is a right-handed double-helical structure. It has one bp repeating units, 35.9°A rotating sub-unit, 10.5 mean bp/turn, +16° inclination of bp to axis, 0.34nm rise/bp along the axis, 3.32 nm rise /turn of the helix, +16° mean propeller twist and diameter is 2.0 nm. The anionic and sodium counter ion phosphate helps the stacked base orientation. The conformity of the charge-neutralized counter ion (ion that maintains neutrality) phosphate model is a very similar model to that present in the conformation of B-DNA [17, 29]. The computational studies and dispersion-correlated density functional theory evaluates the structure and stability of B-DNA (natural DNA). These techniques affirm the different combinations of four DNA bases A, T, G, and C in a di-2'-deoxyribo-e-nucleoside, the monophosphates model DNA strand complex [30]. Other techniques and components like optimizing geometries, the inclusion of implicit water solvent, sodium counter ions, and neutralization of negative charge of phosphate, help the conformation of B-DNA. The results confirm the relative stability of isomeric single and double strands of B-DNA. These observations also reflect the involvement of energy needed with the number of H-bonds per base pair, *i.e.,* two H-bonds between the adenine and thymine, three H-bonds between guanine and cytosine. The confirmation of the arrangement of bases is in 5' to 3' end strand pattern and it after the establishment of superimposition state [17 - 20, 30].

Biological Significance of Z-DNA

Z-DNA is one of the three active conformational forms of DNA. It is a left-handed double-helical biomolecule. This conformity of DNA is confirmed while investigating the repeating polymer of inosine cytosine [31]. Their observations are based on the dichroism phenomenon, for such DNA, and these are in a reverse circular pattern. The interpreted comments show that the strands wrap around one another in a left-handed manner, and this is confirmed crystallographic studies. The DNA has a self- complementary DNA hexamer [32]. It is challenging to study the double-helical conformation of Z-DNA because it is a structurally unstable molecule. This form of DNA is a transient structure that induces during some intracellular biological activities (related to the concentration of ions or salts); after that, it disappears. The Z-DNA biomolecule is narrow and more prolonged in comparison to A-DNA and B-DNA. The major grooves are not like real grooves, and minor grooves are smaller. This conformation appears under when higher salt concentration. This conformation shows alternating purine and pyrimidines sequence with some base substitute. The backbone of this conformation is in zigzag formation because the C-sugar (cytosine) conformation compensates for the G-glycoside bond conformation [33].

This form of DNA can be formed *in vivo* if the proper sequence and the superhelical tension exist. Its functions remain relatively elusive. The formation of Z-DNA is mostly unfavorable, but some of the conditions like alternating purine-pyrimidine sequence, especially poly $(dGC)_2$, negative DNA supercoiling or high salt conditions, and some cations, *etc.*, at physiological temperature 37°C, pH 7.3-7.4. Z-DNA can form a B- T0 Z- junction box with B-DNA, and as a result, some of the base pairs are extruded from the main structure [18 - 20, 34].

The biological significance of Z-DNA is an open field for the investigators. The Z-DNA is quite an unstable molecule and stabilizes; during the process of transcription, the negative supercoiling generates. Its instability is associated with transcription. It is possible to convert B-DNA into Z-DNA when B-DNA exposes to 4M sodium chloride under UV circular dichroism of poly (dG-dC). The B-DNA inverts into Z-DNA. The base pairs are flipped over and become upside down in their orientation, and form syn-configuration concerning B-DNA, during this inversion, and the phosphate groups get closer [18 - 20, 34].

Z-DNA is associated with immune reactions involving monoclonal and polyclonal antibodies. There are specific antibodies related to Z-DNA. Thus, there is a possibility that Z-DNA plays a significant role in human autoimmune diseases. Those proteins that bind to Z-DNA are RNA editing enzymes. These are the double-stranded RNA adenosine deaminase [35]. There is some specific correlation between Z-DNA and chromosome 22. The Z-DNA exhibits some tendency to participate in the process of gene transcription (Champ *et al.*, 2004). It is commonly understood that Z-DNA is associated with providing torsional relief (supercoiling) during DNA transcription [32]. The Z-DNA either directly or indirectly inhibits DNA replication. Ethidium bromide results in shifting the kinetoplastid. The kinetoplastid is an organelle containing a large quantity of DNA and is present in parasitic flagellated protists. The ethidium bromide intercalates during the shifting of kinetoplastid, and this step loosens the DNA molecules. This loosening of DNA causes unwinding [36].

BASE PAIR GEOMETRY IN DNA

The base-pair geometry is related to the precise and well-defined location and orientation in the space of every base or base pair in a given nucleic acid molecule. This orientation relates to its predecessor along the axis of the helix. If any distortion or disturbance or structural damage exists, there will be a change in the concern coordinates. This feature elucidates deformities in DNA. Some coordinates help in the characterization of base-pair geometry or base-pair step. These include a shift, slide, rise, tilt, roll, and twist, *etc.*, as mentioned below:

Shear: It is the strain produced by the pressure in the structure.

Stretch: It occurs when strain exerts to the utmost or legitimate limit.

Stagger: It is due to base stacking forces.

Buckle: It is related or linked with the type and force between the components.

Propeller twist: It is due to the rotation of one base concerning others in the same nitrogen base pair. Opening: It is a condition in which a site or position where strand or bases get separated corresponding to the breaking of H- bonds.

Shift: It is the displacement along the axis in the base-pair directed from the minor to the more extensive grove.

Slide: It is the displacement along the axis in the plane of the base-pair directed from one strand to the other.

Rise: It is the displacement along the helical axis.

Tilt: It is the rotation change around the shaft axis.

Roll: It is the rotation along the slide axis.

Twist: It is a rotation around the rise axis.

VX displacement: It is about the circularization;

Y displacement: It is concerning the small DNA;

Inclination: It is the angle at which DNA nitrogen bases locate concerning the helical axis (helix based coordinates).

Tip: It is a position that suggests the probable first contact with the host membrane concerning bacteriophages.

Pitch: It is the number of base pairs per complete turn of the helix [37 - 39].

MAJOR AND MINOR GROOVES OF DNA

DNA molecule is rigid, the twisted double helix. It is a viscous molecule with enormous length and has a minute width. The twin helical strands structurally and functionally act as the backbone of this molecule. These macromolecules exhibit major and minor grooves. Since the two strands are not symmetrical and are positioned asymmetrically related to each other, the grooves formed are of unequal or non-uniform patterns. Hence, the prominent grooves are significant, and relatively less noticeable groves are the minor grooves. The major grooves are more in-depth and broader while the minor grooves are narrower and shallower. The major grooves are present where the backbone is away from each other while minor grooves are nearer to each other. The grooves twist around the helical axis opposite side to each other. These grooves or voids are adjacent to the base pairs and provide binding sites. Based on crystallographic observations, major groove and minor groove are 22Å and 12Å wide, respectively [17, 40].

DNA AND ITS HELIX OR HELICAL STRUCTURE

This helical nature is due to the dual nature of DNA. The phosphate and sugar both are hydrophilic while the nitrogen bases which are stacked on each other and are hydrophobic. These bases articulate between the two backbones of the DNA. Structurally, each nitrogen base is a configuration flat but possesses three-dimensional orientations. These are stacked one on others. Thus, the zone within the two backbones is a hydrophobic mini-environment. The nitrogen bases twist to accommodate each other and to maximize their hydrophobic interactions with each other. This twisting is due to the hydrophobic nitrogen bases; these, in turn, act as the bases of the helical configuration of the DNA molecule. The structure of most of the forms of DNA, *i.e.*, A-DNA, B-DNA, and Z-DNA are firmly dependant on the sequences of nitrogen bases. The series that is oriented strategically to open the DNA helix for initiation of transcription. The other forms render it either more rigid or flexible and the potential sites for protein binding and formation of the complex [41]. The overall structure of DNA is quite different from the oligonucleotide and non-histone-protein-DNA complexes. The geometrical aspects of DNA base-pair-step also add to the superhelical structure of DNA. The unusual conformation of DNA also induces the binding between double-stranded DNA helical structure and histone protein. It also influences the size and mobility of nucleosome depending on the sequence and recognition of proteins [17, 24]. The unusual conformation also concerns with the hydrophobic nature and the interaction between nitrogen bases, and there is another feature that influences the helical structure of DNA is the arrangement and the orientation of the nitrogen bases on each DNA strand. It also exhibits the interaction between the nitrogen bases causing the stack forces. This force affects the helical nature of the DNA. The stack force is the cause of attractive forces between the nitrogen bases present below and above on the strand of DNA [42].

PROTEIN AND DNA BINDING INTERACTIONS

Major and minor grooves play their specific roles during binding interactions between proteins and DNA. Certain proteins bind to DNA and result in the modification of DNA or help in regulating transcription or replication, *etc*. DNA binding proteins interact with the bases that constitute the internal parts or regions of DNA at the site of the major groove. During these interactions, the backbone of DNA does not interfere. Transcription factors that bind to specific sequences in double-stranded DNA make contact with the sides of the bases exposed in the major groove. The situation varies in unusual conformations of DNA within a cell [43]. DNA-protein interactions are essential aspects of cell functions like transcription, activation or repression, DNA replication, and repair, *etc*. Proteins bind at the floor of the DNA grooves using specific binding, van der Walls interactions, generalized electrostatic interactions, *etc*. Proteins identify H-bonds donors, H-bond acceptors, hydrophobic groups like the methyl group, *etc*. Sometimes proteins bind with DNA in the region of the significant groove while some binds in the zone of the minor groove. Some proteins bind with DNA in both the grooves [44]. The basic concept of protein-DNA interaction is related to the energetic profile of the interacting components that exhibit comparable Gibbs free energy depends on temperature. The Gibbs free energy is the available thermodynamic potential, *i.e.*, capacity to do work. This potential becomes minimized as the system reaches equilibrium at constant temperature and pressure. There are quantitative differences in the energetic parameters related to the binding to minor and a major groove of DNA and reflect on the intrinsic differences existing between these groves. There is a comparable Gibbs free energy that is associated with the major groove, and the enthalpy process affects it. The binding to the minor groove is characterized by the unfavorable enthalpy that is compensated by the favorable entropic contributions. These distinct energetic factors depend on the binding with the major versus the minor grooves. These are irrespective of the magnitude of DNA binding and the extent of binding induced by the protein refolding. The primary determinants of their energetic profiles appear to be distinct hydration properties of major and minor grooves. The minor groove is rich in A-T and water, and this site is in a highly ordered state, but after removal of water, substantial positive contribution to binding entropy is the result. Since the entropic forces drive protein binding into a minor groove and depend on the displacement of water by the regular arrangement of polar contents, they behave like hydrophilic components [44].

EFFECTS OF ENTROPY ON DNA

The concept and principles of thermodynamics apply to live and non-living components of the environment, may it be physical, chemical, or biological. Thermodynamics may not express the functional aspects of a process as a function of time, but its second law depicts the direction of time as arrow reflects the trend of the increase or decrease of entropy in a system [45]. Temperature is one such parameter that facilitates the structural and functional aspects of almost all physical, chemical, biochemical, and biomolecular interactions. This parameter does not permit the destruction by the chaotic thermal changes or motions. During the impact of temperature, a delicate and subtle line of balance exists [45]. With the help of optical tweezers and atomic force microscopy (AFM), it is possible to hold a macromolecule of DNA and stretch it mechanically. Completely stretched molecule exhibits zero entropy when it is free from the mechanical stretch; it gains near-zero entropy. During the early phase of relaxing it may fold randomly, and in the later stage, it regains its natural bends or folded structure and maintains the standard entropy [45].

Entropy is a physical feature, and it plays an essential role in maintaining the stability of biomolecules, macromolecules, and DNA. The intracellular environment is suitable for the existence of DNA as a double-stranded conformation. The intracellular biochemical conditions, primarily like pH, salt concentration, and temperature, *etc.*, favor the formation of the double-helical structure. These factors contribute to the net elevation of entropy that ensures the double-stranded conformation even if the two strands of DNA get separated. A-, B- and Z-conformations of DNA exhibit varying degrees of compression, stretching limits, but of these three conformations, the degrees of compression and stretching are moderate in the conformity of B-DNA. Hence, the B-DNA conformation is relatively more stable. There may be some evolutionary significance in the existence of such structural forms of DNA [46]. Effects of entropy on conformational aspects express as micro-cantilever deflections cannot be ignored, especially at the highly packed density conditions or long-chain systems. The expressed variations are observable because the proposed analytical model can record these fluctuations qualitatively and quantitatively. The deflections of the cantilever obtained for the ssDNA molecule authenticates the principle of minimal energy. The information theory suitably explains the thermodynamic, physical, and functional relationship of the entropy in the case of molecular biology, and Shannon proposed this theory in 1940 [47].

ELASTICITY OF DNA

The DNA also acts as an indicator of the intracellular biochemical, biophysical, and physiological changes. The molecule is one of the most suitable examples that support the theorist's notion of a polymer. DNA is a linear chain of nucleotides having nitrogen bases arranged precisely in a controllable manner and these bases can interchange. The long linear DNA is a macromolecule, and it acts like a flexible entity, self-avoiding polymer, while short length scale of DNA behaves like stiff polymer that exhibits resistance to bending and twisting [48]. The DNA macromolecule regulates the various types of proteins that play vital roles in mechanical, biochemical, and enzymatic activities. N-association with lipids or fatty acids stores energy that participates in the structural and functional aspects of the membrane, specifically during sealing inter and intracellular media [49].

DNA is a long, thin elongated structure and has a 2nm diameter. It exhibits properties of elasticity that facilitates its functionality. One end of a single molecule of DNA can be attached to a fixed point and the other end to a large bead. Now, if the small force is applied using hydrodynamic drag, optical tweezers, or low magnetic field, a displacement under different strengths is readily observed. Since the individual molecule is highly fragile, the power is almost unimaginably small to the range of 10−12 Newton. This force is corresponding to the width of the single bacterium. The DNA exhibits behavior concern is with its total length v/s applied force. The difference arises from the entropic effect, and this facilitates the understanding of the elasticity of DNA. The entropic effects include freely jointed chain resists elongation; this effect is entropic elasticity. Simple polymers do not exhibit any appreciable twist-resistance, but complicated polymers like DNA can display, and even actin molecule also shows a similar tendency. This effect is measured by stretching long DNA strand keeping in mind that the ends are not allowed to rotate. Then, the extension of the chain depends on the applied tension and the amount of extra twist or stress imposed on the molecule beyond its natural twist. These studies provide the values for bonding and the persistent twisting length. Bending and twisting are the two independent parameters that control the behavior of DNA *in vivo* [49]. The non-stretched state exhibits higher entropy. In the light of statistical physics, in this state, the object has lower free energy in comparison to the stretched state of the object even though no power is involved when each link bends. The impact of this effect is proportional to the Boltzmann's constant and temperature. This impact is not noticeable as it is very less (4×10^{-21} J). If the object is a polymer, then this aspect is of great significance. Even freely jointed chain opposes resistance to elongation. This impact is entropic elasticity. The effect of temperature on entropic elasticity varies with the change in temperature; at the higher temperature, the entropic elasticity is lower [49].

The utility of studying the elasticity of DNA enhances the understanding of binding of DNA with drugs, its role in genetic regulations, binding to NH_3, and during the compact packing within a cell. DNA responds to the forces applied as an elastic body. Elasticity is a complicated and unintuitive that obeys quantum laws.

OVERVIEW OF THE HARMFUL IMPACTS OF NANOMATERIALS CONCERNING DNA

The unique physicochemical properties of nanomaterials bring the changes in the biological, biochemical, and biophysical activities including biochemical, conformational stress and oxidative stress, *etc*. These stresses depend on the size, concentration, duration of exposure, composition, types, and nature of nanomaterials, *etc*., act as acellular factors. The cellular parameters include mitochondrial respiration, chloroplast functioning, immunological response, cytological interaction with nanomaterials, intracellular transportation, and cell signaling, *etc*. There can be physiopathological impacts like inflammation, fibrosis, disruption, necrosis, senescence, and disordered inter and intracellular communication. There can be direct or indirect genotoxic consequences because of secondary damage resulting from the induction of ROS and RNS formation. Nanoparticles of transition metals cause chromosomal aberration, DNA strand breaks, oxidative damage to DNA, mutation, (OH·) is one of the significant potential radical formed during such interactions. It forms 8-hydroxyl--'-deoxyguanosine adduct, and this adduct acts as a biomarker of OH·- mediated DNA lesion. Nanoparticles of copper, iron, titanium, and silver are potential agents for damaging micronuclei and DNA. Pro-oxidant metals like copper and iron react with lipid hydroperoxides to induce DNA damage, and malondialdehyde and 4-hydroxynonenal are the end products of this interaction. These interactions also mediate inflammations and are the potential risk factors for carcinogenesis. Carbon nanotubes cause genotoxicity either directly or through inducing oxidative stress and the inflammatory response; these impacts cause chromosomal aberrations, the DNA strand breaks, the micronuclei induction, the formation of $\gamma H_2 A X$, *etc*. H_2A is histone family member, $\gamma H2 A X$, which is the product of phosphorylation on serine 139 and causes the DNA double-strand breaks. Multi-Walled and Single-Walled Carbon Nanotubes damage the genetic materials involving oxidative dependant DNA breakage, repair, and activation of signaling pathways affecting poly-ADP-ribose polymerase, AP-1, NF-kβ, p38 and protein kinase B pathway (Akt), *etc*. As a result of this signaling interruption, structural and numerical chromosomal damages are the result. The specific mechanism involving the transmission of purine nucleotides, ATP, and intercellular signaling within a barrier *via* connexins gap junction or hemichannels, and pannexin channels, and these mediate such damages. Fullerene and its derivatives form a complex with DNA and induce DNA strand breaks, mutagenicity, and damage to chromosomes [50 - 54].

INTERACTIONS BETWEEN NANOMATERIALS AND DNA

Nanomaterials interact with genetic materials, and the fundamental intermolecular forces in these interactions are similar to those in the interaction between nanomaterials and proteins. These interactions depend on the physicochemical features of nanomaterials, and the nature of genetic material. Changes in the conformity of DNA molecules influence the impacts of these interactions. These interactions involve binding inducing either change in DNA and degradation of DNA or RNA. Dynamic light scattering, measuring ζ (zeta) potential, and the changes in the resultant complex formed, are the results of different types of binding between nanomaterials and DNA and RNA. The electron absorption spectroscopic pattern based on UV-VIS spectrometry helps in detecting the changes in the resultant complex. Near IR fluorescence technique is suitable to evaluate the binding ability and degree of coupling between DNA with nanomaterials. Non-canonical structures of DNA include (i) A-DNA with right-handed twist conformity, (ii) Z-DNA with left twist handed conformation, (iii) the triplex form, (iv) the G-quadruplex form, and (v) i-motif. These different conformations of DNA exist within the cellular environment, and these forms interact with the internalized nanomaterials.

Metal and metal oxide nanoparticles are in fields like human health care, veterinary, biological sciences, biomedical, various industrial aspects, and the formulations of pharmaceuticals. Metal and metal oxide nanoparticles have either direct or indirect access to the components of the environment either in their pristine or in their combined form and act as the potential agents for derogative impacts on the biota too [54]. The specific physicochemical features like size, shape, surface properties, *etc*., and behavioral features like agglomeration or ion releasing features help to characterize the nanomaterials. These properties make them prone to biochemical interaction within cells and biosystems [54]. The nonspecific interaction appears to be capable of disrupting the pre-existing H-bonds in a given short double-stranded DNA. The salts of small metal nanoparticles get absorbed strongly nonspecifically during the DNA melting of oligonucleotides. These inhibit the hydration of complementary DNA sequences in a standard buffered solution. If the particle size of metal nanoparticles increases, then the intensity of nonspecific interaction gets weakened [55].

Titanium dioxide is chemically inert but has the potential to cause cancer in the respiratory tract and affects genetic materials due to oxidation and inflammation in rats. The biophysical techniques, like the comet assay, micronuclei assay, *yH2HX immuno-staining assay, and the estimation of 8-hydroxy-'-deoxyguanosine level, are suitable methods to evaluate DNA damage. Metal nanomaterials cause genetic instability. This feature acts as the endpoint for determining the related derogative effects. DNA deletion and checking the m-RNA level of inflammatory cytokines are related to the peripheral blood. The levels of 8-hydroxy-2'-deoxyguanosine, micronuclei, and DNA deletion get elevated as a result of exposure to TiO_2 [51].

In general, biomolecules behave as a suitable template attaching ligand with nanoparticles. The resultant product acts as a catalyst to accomplish the desired interaction. For example, DNA nicking is brought about by using nanoparticles [56]. Arylhydrozone bound onto the surface of gold nanoparticle (13 nm), and the resultant product is useful for the nicking of DNA under UV radiation (312nm). Arylhydrozone binds on the surface of gold nanoparticles, and polyethylene glycol acts as a spacer. The resultant product acts as photo-induced-DNA cleaving nanomaterials [57]. Zinc complex having auxiliary H-bond donor binds with gold nanoparticles (2nm). The resultant product is capable of hydrolyzing diester-bond as the need and cause cleavage in double-stranded DNA [58].

There is a formation of a complex between capped gold nanoparticles with N-(--mercaptopropionyl) glycine, DNA, and salt solution. The rate of formation of this complex depends on the reciprocal fast and slow relaxation duration depending on the DNA concentration. The kinetic curves indicate biexponential, three dynamic stages of the interaction. The first stage shows rapid diffusion controlled the formation of the complex between gold nanoparticles and the DNA. The second stage reflects on the creation of complex compounds because of the binding affinity between hydrophilic groups of tiopronin and DNA groove. Tiopronin, a drug under the trade name Thiola prescribed to control the rate of cystine precipitation. These stages interpret the consequence of conformational changes in the complex formed, *i.e.*, DNA-AuNP. In the first stage, this complex is not compact [DNA-AuNP (I)], while in the second stage, it becomes more compact [DNA-AuNP (II)]. The value of rate constant relates to steps also to the increase or decrease of concentration of NaCl and the viscosity of the medium [52].

Soluble metal and metal oxide nanoparticles access lungs, brain, liver, spleen, and bones of the test animals causing toxic impact depending on their surface properties in comparison to the insoluble nanoparticles. These nanomaterials induce changes in redox-potential that cause harmful implications on genetic materials [59]. Titanium dioxide (TiO_2) nanoparticles are a potential carcinogenic and genotoxic agent at the micro-scale. When genetically engineered nanoparticles gavage in the case of mice (pink-eyed variety having the dilute gene (C-57 BL/6) during gestation to test mutagen-induced instability, different levels of DNA deletion and increased oxidative damage in the off-springs. The interaction between TiO_2 nanoparticles and DNA is uncertain, reflecting on its potential genotoxic nature in the fetal population, specifically during organogenesis [60].

Different forms of zinc oxide nanoparticles like nanowires, nanotubes, *etc.*, interact with the different nucleotides adenine, cytosine, guanine thiamine, and uracil. The theoretical studies show restricted binding energy order and strength of nucleobases to the interaction; these reflect on the nature of the surface of nanoparticles. Zinc oxide nanoparticles show preference to bind with the (i) top site of nucleobases, (ii) the ring nitrogen atom that has a lone pair related to a binding site on the nitrogen base [55] (Figs. **1**, **2**).

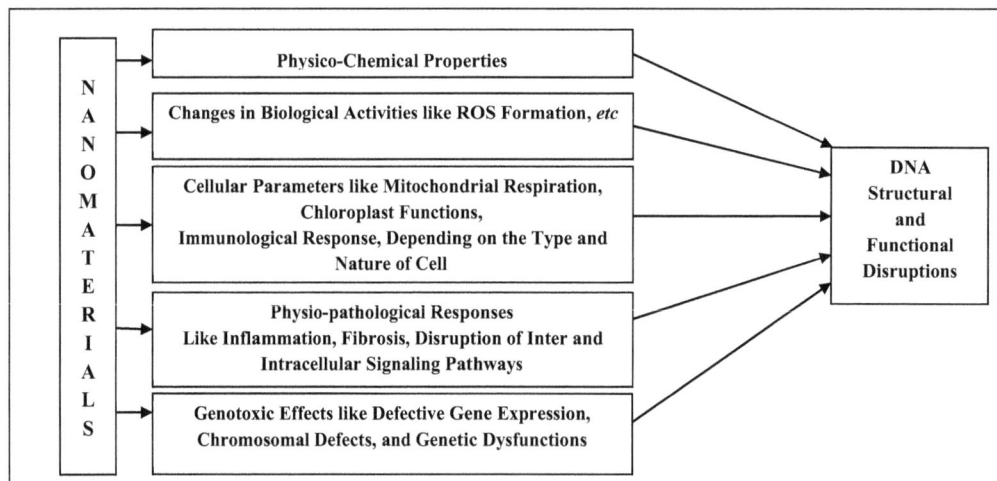

Fig. (1). Overall interactions between nanomaterials and DNA.

INTERACTION BETWEEN CARBON NANOPARTICLES AND GENETIC MATERIALS

The carbon-based nanomaterials are very reactive and applicable in the fields like nanotechnology, biochemical, biomedical, tissue engineering. They have specific physicochemical properties like better conductivity of heat and electricity, high mechanical strength, optical features, *etc*. The most common carbon-based nanomaterials are carbon nanotubes, graphene, fullerene, and nanodiamonds [61].

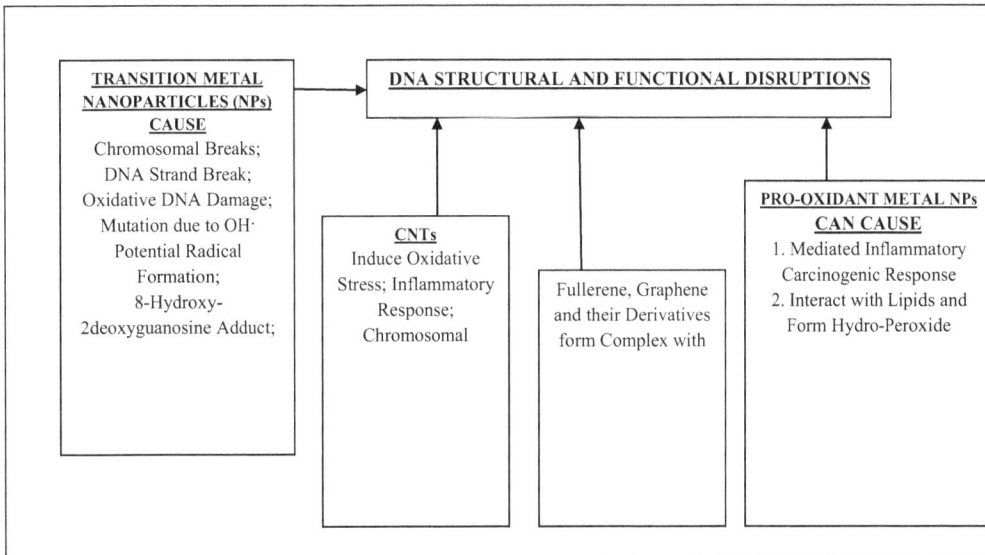

Fig. (2). Interactions between some nanomaterials and DNA.

A complex of single-stranded DNA and single-walled carbon nanotubes (SWCNTs) involves pi-stacking (π-stacking), and the hypochromic absorbance shows that this arrangement follows the pi stacked structure concept. This mode of stacking is involved in DNA dipole transition along the axis of optically anisotropic single-walled carbon nanotubes. Binding of DNA with nanotube is dependent on the orientation of the four adjacent nucleotides concerning the long axis of a nanotube [62]. Single-walled carbon nanotubes (SWCNT) induce destabilization and conformational transition in the DNA strand, and it depends on the sequence present in the DNA strand, and it lowers the (Tm) melting time of DNA, causes the conformational shift of B- DNA to A-DNA conformation, or G-C base transition. These nanomaterials cannot change the A-T base related sequence. Competitive binding and destabilization of triplex DNA confirm these transitional changes [63].

DNA labeled with platinum (DNA-Pt) binds to the external surface of carbon nanotubes (CNTs). This bond is strong and gets separated only during electrophoresis. Some percentage of the DNA-Pt enters the cavity of CNT and gets enclosed in it. Individually, single-stranded DNA (ssDNA) (oligonucleotide) inserts within CNTs and involves van der Waal interactional forces [130]. Single-stranded DNA (ssDNA) gets adsorbed around the surface of carbon nanotubes (CNTs), and most of the bases of ssDNA interact with CNTs involving the van der Waal dispersive forces. One can compute the binding affinity and the conformational fluctuations in ssDNA and double-stranded DNA (dsDNA). CNTs and ssDNA exhibit stable binding configuration. CNTs and dsDNA get adsorbed partially with longer simulation time. Binding in the case of CNTs and small interacting DNA (siDNA) depends on the strong base-pair interacting energy of A-T base pair in comparison to the interacting energy level of A-U base pair. The structural deformations minimize in the case of dsDNA in comparison to ssDNA. About half of the nucleobases of ssDNA reach the hydrophobic interior of CNTs within 20 ns and do not relocate for quite some time as the nucleobases stuck in a metastable (minimum) energy. The dispersion is related to the interaction and aromatic ring present on CNTs, and regulate the translocation of nucleobases. The other half of the ssDNA remains outside CNTs and gets wrapped around the outer surface of CNTs. The dsDNA does not reach the interior of CNT. These CNTs have the same diameter as that used in the case of ssDNA. The more substantial energy barrier may be present in the cavity of CNTs and act as an energy barrier in this case. This behavior does not allow dsDNA to enter the interior of CNTs [64].

The hydroxylated carbon nanotubes and carboxylated carbon nanotubes exhibit different binding energies towards nucleobases. This tendency is of potential use in many fields, specifically for drug delivery and investigations concerning DNA sequences. The hydroxylated carbon nanotubes show binding preference in the decreasing order cytosine, guanine, thymine, adenine, and uracil. The carboxylated carbon nanotubes exhibit different binding capacity with nitrogen bases preference in decreasing order as cytosine, guanine, adenine, thymine, and uracil. These interactions involve the hydroxylated and carboxylated carbon nanotubes and nucleobases the H-bonding is the main participating force [65].

Carbon nanotubes exhibit polydispersibility and poor solubility in aqueous and non-aqueous media. This feature of carbon nanotubes causes problems during their studies. Carbon nanotubes exhibit distinct dispersion. When agglomerated, single-walled carbon nanotubes undergo sonication in water, the single-stranded DNA disperses distinctly. Techniques like optical absorption, atomic force microscopy, and fluorescent spectroscopy can readily evaluate this type of dispersion. Pi-stacking (π-stacking) is involved in the binding of carbon nanotubes. This behavior results in the wrapping of ssDNA around carbon nanotube in a helical manner. The binding free energy between ssDNA and CNTs is the leading cause that prevents the adherence of other CNTs [50].

Graphene is the thinnest, the strongest material, and is almost transparent, exhibiting large surface area. It has the highest intrinsic mobility more than silicon and the ability to conduct electricity within the limits of numbers of electrons available at the site [66, 67]. The intrinsic and extrinsic features of graphene have transformed it from a laboratory research product to the real world product stage. The graphene is a preferred option for the different applications in biological, biomedical, nanotechnological aspects like nanomedicine, nanobiosensing, suitable electronic tools to interact with a biomolecule [68, 69]. There is a need to understand the underlying mechanisms of biomolecule interaction, and to a greater extent, graphene can help in this aspect. The interaction between graphene and genetic material involves van der Waals forces under the density functional theory [69]. There are preferential interactions between graphene and nucleobases in decreasing order; guanine > adenine > thymine > cytosine > uracil. During this interaction, the electronic and structural aspects of both get to participate. The degree of binding between graphene and nucleobases is regulated if one can add or remove electrons from the system during the interaction [69].

Graphene oxide can conveniently identify the sequence of DNA during PCR amplification. The *csp* gene from E.coli can be a suitable DNA template, and amplify without labeling. The graphene oxide binds to the single-stranded DNA but not to the double-stranded DNA. Temperature denatures the DNA and hybridizes with single-strand DNA; the oligonucleotides serve as a probe. These probes identify a specific region of the target DNA that shows the changed sequence. This interaction is specific and exhibits a higher degree of effectiveness, specifically when target and probe show near-perfect matching. By the addition of adding graphene oxide helps to record the free fraction and using the fluorescence technique. Further, graphene oxide detects the fluorescence quenching [70].

Graphene shows good dispersal interaction with stacked aromatic compounds that include the interaction with amino acids and nucleobases. The quantum chemical equation and atomic force microscopic techniques show that in this interaction, the binding energy is an important feature. The natural bond charge establishes the involvement of weak H-bonds and π-bonds during efficient dispersions [71]. Graphene also exhibits physisorption; (-a process in which hardly any disturbance in the electronic structure of atom or molecule takes place when other molecules or atoms are attached). The nitrogen bases, guanine, adenine, thymine, and cytosine physisorbed on graphene as a result of the induction of interfacial dipole provides energy shift during interaction with nucleobases [72]. Fullerene is a natural carbon nanomaterial and is a product of combustion. This carbon nanomaterial exhibits chemical reactions in which there is either donation or acceptance of the electron [73]. The interaction between fullerene and genetic materials follows the basics of molecular dynamics simulation. The fullerene binds with nucleotides resulting in the formation of hybrids forms. These hybrid forms are energetically stable in an aqueous medium. In such binding energy includes free energy and entropic energy. There is an overall change in entropy, and it overpowers the hydrophobic interactions when non-polar molecules are present in an aqueous medium [74]. The reason for such impact is that when non-polar molecule tries to disperse in an aqueous medium, there is a rearrangement of the water molecules or in the aqueous medium, and during this rearrangement higher level of entropy declines. Fullerene binds to the free ends and the minor grooves of double-stranded B-DNA and causes no change in its shape. When fullerene interacts with A-DNA and attaches to the open ends of DNA strands, it breaks the H-bonds between the end base pairs. When fullerene binds to minor grooves, it changes the stacking angles of the bases. Fullerene tends to bind strongly to single-stranded DNA resulting in a higher degree of deformation of the affected nucleotides [74].

Quite often, fullerene binds to defective site present in double-stranded DNA and forms a stable adduct. These interactions are likely to interrupt the standard functionality of the DNA [74]. Fullerene can access the bacterial DNA and bind with it. The fullerenols can enter the bacterial cells and cause changes in the structure and biochemical functionality of its DNA molecule. Fullerenols enhance the thermal stability of DNA 70-85°C, depending on the dose. When the DNA interplay with higher doses of fullerenols, the activities of enzymes like DNase and Type-II site-specific deoxyribonuclease restriction enzymes declines or diminishes [75]. The Fullerene and fullerenol adversely affect the mechanism of antibacterial activity by increasing the degree of cell damage and the rate of mutations [75].

The DNA molecule docked with varied forms of fullerene (with different molecular weights) ranging from C30, C40, and C50 exhibits low binding activity. Fullerene docked DNA molecules are not responsive to the enzyme activities that are involved with the process of replication, and may not ensure the functioning of the primase enzyme [76]. Fullerene docked DNA can interrupt the initiation of DNA polymerase activity [76]. The interaction involving fullerenes with a specific range of molecular weights (C20 to C180) exhibit non-uniform behavior with different forms of DNA. The degree of binding increases with A-

DNA and B-DNA forms, while in the case of the Z-DNA form, this binding activity is irregular and fluctuates. The optimum coupling is with the A-DNA form. The sequence of nucleotides and nuclear bases did not affect the mode and extent of the binding process. The interaction between fullerene and the B-DNA form shows variations in the binding ability between them; small fullerenes < (C48) bind with the minor grooves of the B-DNA while fullerenes with higher molecular weight bind to the major groves only. During the interplay between the A-DNA forms and fullerene with different sizes, the binding limits to only the major grooves of A-DNA. In the case of Z-DNA, the binding pattern is erratic and did not follow any specific arrangement or design [77].

INTERACTION BETWEEN QUANTUM DOTS AND GENETIC MATERIALS

The binding energy and interactivity of nanomaterials explicitly depend on the physicochemical properties of the interacting components. Quantum dots and DNA interact and result in the formation of a conjugate. These conjugates have the combined features of both these interacting molecules. Such interactions are dependent on the surface entities of the reactants. Quantum dots coupled with DNA, and form conjugate, these conjugates are sensitive specific in function. These products are better options for versatile biomedical and biochemical probes. Such probes are suitable in the case of human and other mammalian metaphase chromosomes, single nucleotide polymorphism, detecting multi-allele DNA, finding the code, and sorting out crystals. These DNA-quantum dots conjugates show convenient binding with their complements *in vivo* and fixed cells [78]. The orders of binding and interactive ability of quantum dots with nucleobases are directly related to nature, surface properties of ZnO-quantum dots. These quantum dots exhibit preferences to those binding sites where a lone pair is available, may it be at the top location of nucleobases, or the nitrogen ring atom [53]. For biomedical imaging and sensing nucleotides concerning the genetic investigation, it is essential to design a suitable nano complex [79]. The concentration of DNA is a necessary parameter while studying the interactions between quantum dots and DNA. The amounts of DNA within range (2.5 and 0.25 μg/ml) are suitable for imaging, while concentrations with range 0.5 to 0.25 μg/ml are suitable for the binding of DNA with quantum dots. These concentrations are favorable to study and to image a single molecule of DNA [80]. Quantum dots conjugate effectively with oligonucleotides. These conjugated forms act as suitable probes to detect a specific nucleic acid. A technique called quantum dots electrophoretic mobility assay concern with the affinity between quantum dots and DNA. This technique is suitably applied to study the electro-hydrodynamic property of a target DNA. Streptavadin tagged with biotin and quantum dot (CdSe-QD) to reach the target DNA. These undergo electrophoresis along with the respective standard. The degree of conjugation between DNA and QD is an essential aspect of electrophoretic mobility. Relative migration distance is recorded and is quantified depending on the fluorescent intensity [81].

INTERACTION BETWEEN DENDRIMERS AND GENETIC MATERIALS

The cellular processes like cellular repairs, molecular recognition, biological transport, and other biochemical and biomolecular issues associate with macromolecules and polycationic biomolecules a cell. The polycationic molecules and other, electrostatic interaction between DNA, histone, DNA, and spermidines (polyamine present in the ribosome and living tissues from semen) lodge readily within dendrimers. These macromolecules play significant roles in the stabilization of DNA; regulate transcription and drug delivery, *etc*. Interactions between DNA and amine-terminated ethylenediamine core polyamidoamine dendrimer, DNA wraps around G-7 dendrimer but not around G-2 and G-4. There is a tendency for a specific preference for binding with nucleotides or nucleobases present in DNA. The ionic strength and pH affect the electrostatic interactions that act as a major binding force between the two [81]. Binding between DNA and dendrimer reveals that there are two zones, namely tight bound DNA zone and linker bound DNA zone. In the region of tightly bound DNA zone, there is no space for the ethidium bromide binding. This behavior relates to the least accessibility of DNA to ethidium bromide and also to the concentration of dendrimer (saturation threshold). G-7 of the dendrimer is the zone of higher binding force; hence, this is the functional zone of tight DNA binding zone [82].

The dendrimers are systematically intensely branched, reactive nanomaterials, and are suitable for the different applications in nanotechnology, biomedical and molecular fields. Dendrimers cause hemolysis and other hemocytotoxic impacts. Their toxicity depends on the chemical core of the respective dendrimers, surface end product, and cationic features. The positively charged polymer and strong chemical response to the negatively charged cell membrane also add to their toxic impacts. These charges are the features that result in cell destabilization, cell leakage, and finally, cells undergo lysis [83]. Parameters like pH of solvent, size of the dendrimer, length of DNA, the ratio of charge between DNA and dendrimer regulate the interaction between DNA and dendrimers [64]. Binding energy is concerned with electrostatic energy. If there is a higher positive charge on the dendrimer, then the binding interaction is equally stronger. When poly-amid--amine (PAMAM) dendrimer G4 generation and single-stranded DNA interact and a complex is formed. During this interaction, electrostatic attraction is involved between positive charge present on the dendrimer and negative charge on the backbone of DNA [64]. Dendrimer and the DNA wrap around each other during their interaction, forming a complex. The charge ratio between positively charged dendrimer and the negative charge on the DNA play an essential role in the formation of this complex. Generally, lower generations of the dendrimers do not get wrapped around the DNA. The wrapping of DNA around the nucleosome is similar to the wrapping of DNA around the higher structural generations of dendrimers [64]. During the formation of complex, the numbers of contacts increase within nanoseconds. During the next few nanoseconds, there exists optimizing of the affinity with DNA. There is a higher degree of deformation during the said interaction where lower or smaller generations are involved. This structural deformation of the complex formed, and asphericity factor is useful to evaluate the deformities in the complex built. On the bases of the observations using this technique, changes in the conformation of G-3 dendrimer and G-4, G-5, can be investigated [64]. This aspect is beneficial in understanding and the mechanism involved in the applications of dendrimers in the fields of biochemistry, molecular biology, biomedical, and DNA technology.

INTERACTION BETWEEN HISTONE PROTEINS AND NANOMATERIALS

Histone proteins are rich in lysine and arginine amino acids. DNA wraps around the Histone proteins. One of the significant interactions between them is the formation of salt bridge scaffolds for DNA. Histone proteins are the highly alkaline proteins and are present in the nuclei of eukaryotic cells. These proteins regulate packaging and orderly arrangement of DNA in the form of structural units called nucleosomes. These are the chief proteins components of chromatin and act as spools around which DNA winds. These also play a functional role during gene regulation [84].

During the interaction between histone and DNA with caped cadmium nanoparticles, there is an intense activity at around an intensity peak of 328 nm, indicating the formation of a complex involving cadmium. In this interaction, DNA is optional to detect histone proteins [85]. The modified dot blot technique fulfills the interaction between DNA, RNA, core histone, bovine serum albumin, and nuclear lysate, *etc.*, by using quantum dots. An increased luminescence represents the endpoint of this reaction. This technique involves a stabilized state of quantum dots. The parameter represents the end product, and the binding of quantum dots with core histone and nuclear lysate but not with DNA, RNA, and bovine serum protein [86]. The binding between quantum dots and histone proteins attribute to the negative charge of quantum dots to core histones. Histones have around 30 to 40% positive charge. Their positive charge is because of the presence of amino acids like lysine and arginine on their N-terminals, while DNA, RNA, is negatively charged components. As a consequence, quantum dots have a negative charge, possessing negative forces [86].

THE RNA EXHIBITS THE PHYSICOCHEMICAL AND BIOPHYSICAL FEATURES

RNA biomolecule formulates and disintegrates as per the need of the cellular activities. It is a single-stranded poly nucleic acid and exhibits different forms. The interaction between polymeric molecules and other molecules involving RNA is an exciting topic for the researchers. Biotechnological aspects of RNA are the attraction of researchers in recent years [87]. The RNA molecule participates in the fabrication of nanoparticles, polygons, arrays, bundles, membranes, and microsponge, *etc.*, which are the essential component of nanotechnology [88]. These resultant products are potential agents that have applications in the biomedical, biomolecular and material science fields, *etc.* RNA exhibits some specific physicochemical features and as a polymer shows multiple uses in the field of nanobiotechnology. It is a polymer of nucleotides like adenosine, cytosine, guanosine, and uridine. RNA is available in the homopolymeric form and the heteropolymeric configuration. The nucleotides in the case of RNA link covalently with ribose pentose sugar, nucleobases, and a phosphate group. Between two nucleotides, there is a phosphodiesters bond constituting its negatively charged backbone. The RNA exhibits various conformations like loops, hairpins, bulges, stems, pseudoknots or junction, *etc.*, because of the orientation of glycoside bonds and H-bonds (both the cis and the transforms of RNA) [17 - 20, 87]. Generally, RNA exhibits solid-phase chemistry and the process of transcription. These modes involve stepwise reactions during the formation of RNA. The resultant RNA is a polymer and shows the well-defined sequence of nucleobases, the specific structure, and molecular weight. These products exhibit a particular index of poly-dispersibility, which is physiologically suitable. RNA, as a polymer, shows thermoplastic behavior. The mechanical strength of RNA is related to the temperature. As the temperature increases, its molecular power also increases, and that affects its integrity. A decline in temperature reduces its mechanical integrity of RNA. This feature of RNA indicates that it is influenced directly with thermal impact. Higher temperature causes structural deformation, and the deformed structure regains the original construction when the temperature lowers. Therefore, RNA molecules are thermodynamically stable at a lower temperature but not at higher temperatures [89 - 92].

ROLE OF RNA AS POLYMER

Any biochemical polymer should be non-toxic, non-immunogenic, with suitable mechanical features, metabolizable, sterilizable, thermostable, and scalable, and with a reasonably good degree of biodegradability and biocompatibility, *etc*. These properties of the polymer are useful for its applications in industries and biomedical fields [88]. RNA is quite different from polymer biomolecules than DNA because of the presence of 2'-OH group. RNA is a multifunctional biomolecule and is capable of increasing the rate of reaction like acid-base catalysis, electrostatic effects, influencing the proximity and orientation of the substrate, *etc*. Among many such interactions, the most prominent is the one that involves the interplay between RNA nucleobases and bound metal ions. In the case of RNA, the 2'-OH group is active (intramolecular of 2'-OH on the phosphorus of the phosphate group). It interacts with the adjacent 3'-phosphodiesters bond. This interaction causes cleavages in RNA sugar-phosphate affecting the backbone of RNA [93]. There is a possibility that 2'-OH group is an easy target for nuclease digestion. This behavior is because the RNase cnzyme identifies this specific site, and this can be of disadvantage. This disadvantage conveniently overcomes if 2'-OH group is modified [87]. There is a new 2'-OH group in RNA. Its presence in RNA is responsible for biogenesis and storing of RNA oligonucleotides. The melting temperature of RNA elevates when 2'-OH group adds to RNA biomolecule. RNA stabilizes if RNA duplex locks into a compact A-form helix; this helical conformation is stable in comparison to the double-stranded DNA helix [17 - 20, 94].

RNA IS A THERMODYNAMICALLY STABLE BIOMOLECULE

RNA is a highly flexible biomolecule. This feature facilitates and improvises its functional diversity. RNA can change into different conformations because of its folding behavior [95]. The thermodynamic stability of RNA involves Gibbs free energy ($\Delta G°$), and this free energy plays a significant role during the formation of the double-helical structure. It can also be evaluated by Gibbs free energy for unwinding double helix. The difference between the values of double helix winding and unwinding also reflects on its thermodynamic stability. The double helix of RNA is more stable than double-stranded DNA. Average $\Delta G°$ ranges −3.6 to −8.5 kJ/mol per base pair stacked while $\Delta G°$ for DNA double helix is around −1.4kJ mole per nitrogen base-pair stacked [88]. Proteins play an essential role in the modification of the conformation of RNA. The chaperone process influences the high structural stability or folding of the RNA. Some of the proteins, like RNA-chaperone proteins and metals like magnesium (II) coordinate with RNA and facilitate thermodynamic stability [17 - 20, 96].

STABILITY OF RNA CONCERNING ENZYMES

The m-RNA form of RNA is a potential biomolecule and plays a significant role during gene expression, regulation of biogenesis, and defenses against viral infection. The RNase enzymes modulate the RNA related activities. These enzymes have different specificities that indicate distinct intracellular interactions and various mRNA interacting proteins. There is a need to regulate the overall RNA mechanism so that RNA is available as per the cellular requirements. These are of two types, exoribonucleases, and endoribonucleases. Endoribonucleases group of enzymes act on the phosphodiesters bonds of the backbone of RNA while exoribonucleases group of enzymes act on the phosphodiesters bond at 3' or 5' end of RNA [88, 97 - 99]. The RNA biomolecule is of limited application concerning pharmacokinetics because of its nature, and it tends to interact with the RNase enzyme. The RNA must be RNase resistant for the successful applications using its modification form the 2'-OH group using 2'-F (2'-Fluro group). This change is applicable to develop ribosomes, to enhance Tm, selection of aptamer, and resistance to nuclease enzyme. The RNA with 2'-F (2'-Fluoro) group is stable, and because of folding, it preserves its nature for correct dimer formation, procapsid binding conformation [100, 101]. The double-stranded small interfering RNA (siRNA) helps to harmonize the biological processes. Thus, RNA is made suitable for therapeutic uses. The specific RNA sequence and degradation of both short and long RNA duplex play significant roles is stabilizing siRNA because it gets degraded at UA/UA or CA/UG sites. The cleavage site is RNA sequence-dependent [17 - 19, 88, 100 - 102].

THE RNA EXHIBITS ADAPTABILITY AND PLASTICITY

RNA is very adaptable and exhibits structural and functional diversities under natural technical conditions. The RNA shows a wide range of structural and functional aspects varying from simple helical structured single-stranded hairpin model to complicated structure with multi-framed junctions like pseudoknots. A pseudoknot is a double helical structure that is sensitive to salt concentration, metallic ions, and the sequence of nucleobases [87]. The simulated molecules indicate various fluctuations within the range of common biomolecules. Many variations resemble those fluctuations that occur during stochastic thermal changes. During protein binding with rRNA, its stability increases [103]. A polymer shows basic mechanical features related to flexibility and stiffness. If the polymer has a short length, then it will be relatively rigid. The double-stranded RNA (dsRNA) is more reactive like proteins like helicases, polymerases nuclease, *etc.*, and interacts with biomolecule having a helical structure. These interactions involve electrostatic, biochemical, and mechanical features of dsRNA. Single molecular techniques like magnetic tweezers, atomic force microscopic techniques help in analyzing these interactions. These analyses reflect on the related real-time and qualitative analysis [104]. The RNA assumes double-helical conformation and the base pairing in a complementary pattern. This helical conformation is related to the functional aspects of RNA. The mechanical and structural changes in RNA are essential for the activities of biosynthesis. The behavior of RNA depends on the forces, torques, elastic nature, bending, stretching, and stiffness due to the string of dsRNA [105]. The persistence length in a polymer is an indication of its mechanical properties. These properties quantify the stiffness or plasticity of a given polymer. The piece of polymer shorter then persistence length behaves like a flexible and elastic object. Mechanical features and dynamics of polymer, in this case, RNA, include features like its folding involving flexibility of the nucleic acid chain, its dependence on base contents, arrangement, conditions of the solution and the molecular status, *etc* [106].

The plastic nature of RNA is expressed as a disturbance in the stiffness in its molecule. Disruption in the stiffness and plastic nature of RNA, are related to the position and sequence of bases. These features are either softened or stiffened [88]. The dynamic role of RNA during numerous cellular functionalities is of frequent occurrence, but the general biochemical and biophysical mechanism is not understood. The structural conformation takes place within a time scale ranging from picoseconds to a few seconds. This change concerns with transitions, secondary structural modifications, base-pair/tertiary dynamics, stacking dynamics with a time scale of a nanosecond to microseconds. All these biochemical transitions are concerned with (i) regulatory functions, (ii) aspects related to coupling conjugation, and (iii) achievement of greater functional complexity [107].

The interactions between base pairs add to the rigidity of RNA molecules, while the non-base paired part of this molecule causes changes in the conformation of RNA. These changes are essential to attain the different functionality of RNA [108]. The mechanical abilities of the RNA play vital functions and modulatory roles during an immune response, specifically, cancer immunotherapy, development of vaccine adjuvant, and inflammation, *etc*. The changes in the strand of RNA in the different modes of shape-change like triangle, square, pentagon, have their specific functionality. When adjuvant of immunological aspects and modulations occur, there is an induction of cytokine TNF-α and interleukin-6 (IL-6) up to the limit of 100 times. At the same time, RNA polygon regulates the unnoticed impacts of immune responses. The structural changes are the reflection of the flexibility, plasticity, and versatility of RNA biomolecule [109].

INTERACTIONS BETWEEN NANOMATERIALS AND RNA

There are practical and functional interactions between nanomaterials and RNA. The RNA, in its various conformational forms, is useful in different fields, including nanosized imaging, therapeutics, diagnostics, delivery system, and RNA technology, *etc*. The formation of conjugates or complexes between nanomaterials and RNA depends on the charge, type, and surface chemistry of the nanomaterials and the sequences of the bases in various conformations of this nucleic acid. Cationic nanoparticles involve electrostatic interactions between the backbone of the nucleic acid and nanomaterials. The interactions between the charged ligand take place regardless of the charge on nanoparticles. Single-stranded nucleic acids become compact when these encounter highly charged nanoparticles. Pi-pi stacking (π-π stacking) within the strand and H-bonds are the site of interaction [84]. Parameters such as binding energy, the force of interaction of nucleobases, and the nature of the nanoparticles play significant roles in these interactions. Metal oxide like ZnO nanoparticles binds *via* upper nucleobases or with N-atom of the ring that is having a lone pair (single) only [53].

Mostly at lower temperature (below 100°C) salts gets converted into solution (liquid forms). These fluid forms behave non-molecular with ionic characters. These liquids and mixtures of organic products both produce a biphasic system. These solutions (liquids) provide favorable media for multiple and multiphase reactions. Homogeneous catalysts are conveniently separated. Further, there may be changes or switches over between the organic solvent and ionic liquid. This change results in a novel and unusual chemical interactions. Thus, this model offers an open field to investigate various interactions and catalytic processes [110]. The binding between metal ions and RNA influence the structure and functions of the ribozyme, riboswitches. A riboswitch acts as a molecular controlling zone of messenger RNA. It binds with a small molecule causing a change in the formation of a protein that is encoded by mRNA), hammer-headed mRNA, and proteins. Mg^{2+} causes change in the original cleavage site of iron-responsive element-mRNA and repressor complex in response to Fe^{2+}. The study of cocrystals of iron-responsive element-mRNA and the iron repressor protein shows the formation of respective complexes. Iron repressive element-mRNA is sensitive to Fe^{2+} --EDTA, ascorbate, H_2O_2 compound, *etc*. It reflects on the selective interactions between RNA, metal, and a solvent [111]. Interaction between RNA and metal and metal oxide nanoparticles is related to 3dimensional folding of RNA. Thus, this site serves as an active site for the enzymatic action of RNA. Metal ions interact differently with RNA having different conformations. Such interactions concern with randomness. Brownian motion occurring within the three-dimensional region of the electrostatic field formed within the folded zone of RNA. These interactions are predictable [112].

ZnO and MgO nanoparticles show protective tendencies. ZnO protects RNA while MgO protects DNA. MgO nanoparticles caused a decline in the activity of DNase at a higher temperature (65°C). Nucleic acids possess some basic features that reflect on their suitable applications in biochemical, biomedical, pharmacological fields [53, 107].

EFFECTIVE ROLE OF RNA IN THE DEVELOPMENT OF THREE DIMENSIONAL FORMULATIONS OF NANOSCALE MATERIALS

In most of the biological processes like cellular and gene expression, transcription, translation and protein synthesis, *etc.*, DNA participates as double-stranded and RNA as single-stranded functional biomolecules. These interactions or processes are dependent on the base sequence, orientation of helix, and stacking of nucleobases in RNA. RNA biomolecule exhibits various conformations that are most suitable in accomplishing the aforementioned cellular functions. Specifically, RNA shows the ability to get folded as per the functional requirement. This folding of RNA is related to Watson-Crick base pairing, non-Watson-Crick base pairing, and orientation of helix and stacking. These specific biochemical and biophysical features bring about the structural bases for the different functionalities of RNA, and also to formulate nanostructures to be used in varied fields [88]. The double-stranded region of RNA expresses Watson-Crick base pairing in principle, and non-Watson-Crick base pairs interlink this zone. The natural form and folded form of RNA molecule display complex and closely-knit nucleotides. These nucleotides form non-bonded links with ambient nucleotides or involve in folded conformations. These interactions require (i) phosphate to phosphate link using H_2O molecule or positively charged ions, (ii) phosphate-sugar H-bonding involving 2'-OH group, (iii) sugar-sugar H-bonding, (iv) phosphate to base or sugar H-bonds with base, (v) phosphate to base or base to sugar stacking interactions, (vi) base-H-bonding-base, (vii) base to base stacking action. Most of these interactions involve water molecule or ion or both at the same time. The folded conformation of RNA shows properties that can recognize the base sequence. The molecular interaction adds to the stability of the folded structure in totality. The studies based on crystals of RNA prove the significant functional role of stacking [95, 113, 114]. The specific tendency of folding of RNA is the bases of structural aspects and also for its various biochemical and biophysical functionalities. Biomolecules like proteins, small molecular ligands, mono, and divalent metal ions act as mediators in the process of RNA folding, and at the same time, complicate the formulation of RNA nanostructures [95]. The interference pathway plays a specific role during the therapeutic applications of the native ribose phosphate backbone of small interfering RNA. The RNA nucleobases may show fewer modifications but these modifications help to get the insight of gene silencing process [115]. The specifically engineered siRNA molecule exhibits the potential for stability and immunogenicity in a biosystem [116]. The 2'-amino-LNA scaffolds act as sensitive analytical probes for the detection of DNA and RNA in therapeutics while 2'-amino and isomeric 2'-α-l-amino-LNA (the 2′-amino derivative of locked nucleic acid) scaffolds stabilize and detect nanostructures. The molecular movements of nanodevices and chemically programmed LNA/DNA are traced using a bright fluorescence signaling technique [117]. The RNA nanoparticles get diluted after their administration in the circulatory system. These nanoparticles in low concentrations do not dissociate and act as therapeutic agents [118].

STABILITY OF RNA CONCERNING CHEMICAL AND BIOCHEMICAL ASPECTS

RNA is a prime biomolecule. It is essential for processes like the expression of the gene, protein synthesis, and RNA technology. It also exhibits the ability to neutralize mRNA. This functional aspect of RNA is useful in increasing its chemical and biochemical stability [86]. RNA interference represents the ability of RNA biomolecule to restrict the expression of the gene. During this process, it neutralizes the set or targeted mRNA and results in its functional inhibition along with the co-suppressors and post-transcriptional silencing processes. This ability of RNA interference is a good potential option for optimizing its capabilities related to therapeutics and to increase the utility of modified RNA. Its stability *in vitro* and *in vivo* conditions, its role in cellular delivery, biodistribution, potency, and specificity related to pharmacokinetics are biochemically essential aspects [86] (Fig. **3**).

(1) ALTERATION OF INTERNUCLEOTIDE PHOPHODIESTER BACKBONE:
Replacing non-bridging oxygen atom on the phosphate backbone with sulfur atom and creating phosphorothioate (ref), boranophosphate (ref), phosphoramidate/methylphosphonate

(2) SUBSTITUTION OF 2'-OH GROUP BY INCORPORATING 2'-FLUORO:
Substitution by 2'-O-methyl, 2'-fluoro-β-D-ribonucleotide; IMPACT: Cytotolerant, increased resistance, and immunogenicity, either non-toxic or least toxic, Tm increased, applicable in cancer therapy, resistant to I-124; Cs-131radiation and other clinical doses, gene silencing application

(3) CHANGE IN LINKAGES BETWEEN LOCKED AND UNLOCKED NUCLEOTIDES: Locked nucleotides are of type 2'modification; -2'-O and 4'C are bonded through a methyl bridge. This modification constrains the ribose ring into C3'-endo conformation and it increases thermostability and resistance to nuclease
IMPACT/USES: good option as biosensor, aptamer, and other nanomaterials possibility of Hepatotoxicity

(4) MODIFICATION OF RIBONUCLEOTIDE BASES:
RNA nanotubes can be formed using RNA scaffoldings 5-biotinylated modification; 5-aminoallyl modified with urea (139), even 2-thio, 4-thio, 5-iodo-5-bromo, dihydro, pseudo-uracil, and diamino-purine can be used. IMPACT: increased stability and specificity of base-pair interaction

(5) MODIFICATION IN RIBOSE MOIETY: This mode id specific for siRNA using altritol nucleic acid and hexitol nucleic acid, RNAi induction can be controlled by photo-capping IMPACT: increased potency and nuclease resistance.

STABILITY OF RNA

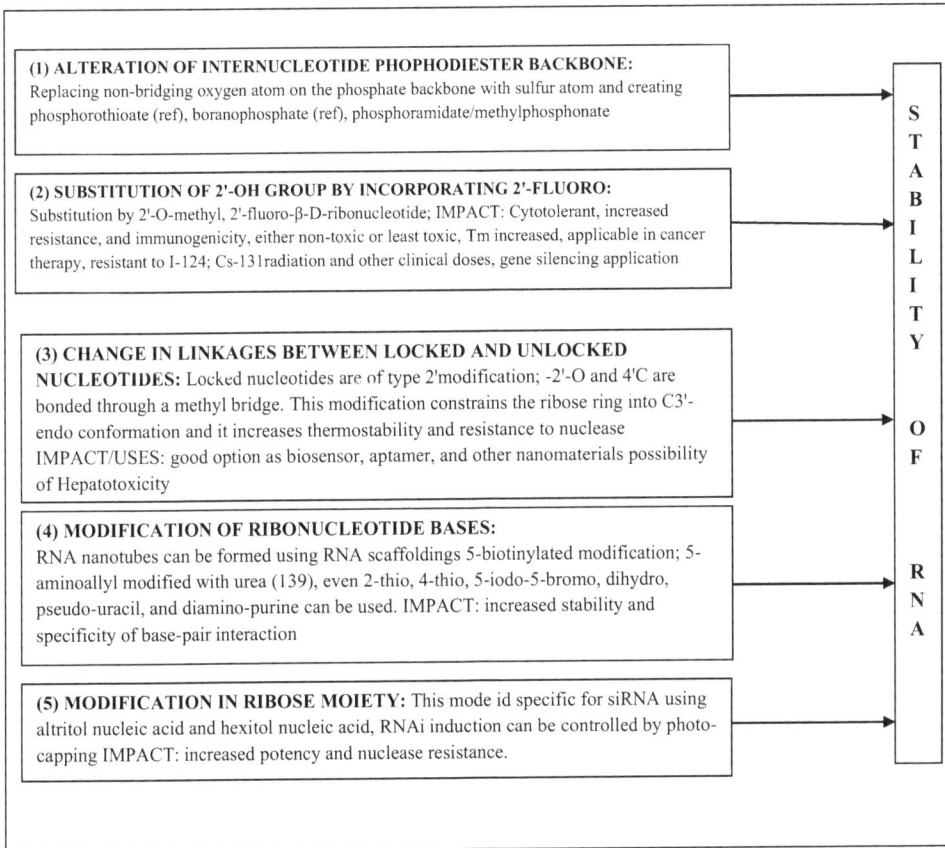

Fig. (3). Modes of enhancing chemical and biochemical stability of RNA.

CONCLUSION

Genetic material is one of the prime hubs of cellular activities. The DNA and RNA represent the genetic material in a biosystem. The DNA and the related molecules-histones, and non-histone proteins exhibit specific biochemical, biomolecular, and biophysical structure and functionalities. Their interactions with nanomaterials are of significance. These aspects are affected when various nanomaterials interact with genetic materials. In pursuing investigations related to the mechanism involved in such interplays, the role of nanomaterials in the fields of genetic engineering, DNA technology, expression of the gene, RNA technology, *etc.* is of great significance. These help to improve modes and the types of nanomaterials and guide in the designing, formulating, fabricating the nanomaterials and biochemical molecules for their suitable applications. Thus, the future line of action envisages promotions, the enhancement in fields like gene expression, DNA and RNA technologies and related investigations, transporting drugs and nuclear medicines, *etc*. The products of the interactions between nanomaterials and the nucleic acids are the potential materials for DNA and RNA technologies. Their clinical and diagnostic applications, such as imaging, and sensing tools, cargo delivery agents, and as components in therapeutics, *etc*., are improved and open a new horizon in the field of health and safety of man and the environment

REFERENCES

[1]   Lahir YK. Natural cure for genetic integrity. Biochem Cell Arch 2013; 13(1): 1-17. www.connectjournals.com/bca

[2]   Zhang X-Q, Xu X, Bertrand N, Pridgen E, Swami A, Farokhzad OC. Interactions of nanomaterials and biological systems: Implications to personalized nanomedicine. Adv Drug Deliv Rev 2012; 64(13): 1363-84.
[http://dx.doi.org/10.1016/j.addr.2012.08.005] [PMID: 22917779]

[3]   Sharma Sudha. Non-B DNA secondary structures and their resolution by Rec Q Helicases J Nucleic Acids 2011; Vol: 15. article ID 724215.
[http://dx.doi.org/https://doi.org/10.4061/2011/714215]

[4]   Sun H, Ren J, Qu X. Carbon nanomaterials and DNA: From molecular recognition to Applications Acc hem Res 2016; 49(3): 461-70.
[http://dx.doi.org/https://doi:10.1021/acs.accounts.5b00515]

[5]   Raghavan SC, Lieber MR. DNA structure and human diseases. Front Biosci 2007; 12(12): 4402-8.
[http://dx.doi.org/10.2741/2397] [PMID: 17485384]

[6]   Watson JD, Crick FH. Molecular structure of nucleic acids; a structure for deoxyribose nucleic acid. Nature 1953; 171(4356): 737-8.
[http://dx.doi.org/10.1038/171737a0] [PMID: 13054692]

[7]   Mandelkern M, Elias JG, Eden D, Crothers DM. The dimensions of DNA in solution. J Mol Biol 1981; 152(1): 153-61.
[http://dx.doi.org/10.1016/0022-2836(81)90099-1] [PMID: 7338906]

[8]   Irobalieva RN, Fogg JM, Catanese DJ Jr, *et al.* Structural diversity of supercoiled DNA. Nat Commun 2015; 6: 8440.
[http://dx.doi.org/10.1038/ncomms9440] [PMID: 26455586]

[9]   Yakovchuk P, Protozanova E, Frank-Kamenetskii MD. Base-stacking and base-pairing contributions into thermal stability of the DNA double helix. Nucleic Acids Res 2006; 34(2): 564-74.
[http://dx.doi.org/10.1093/nar/gkj454] [PMID: 16449200]

[10]   Pederson R, Marchi AN, Majikes J, *et al.* Properties of DNA, In book: Handbook of nanomaterials properties. (1st Ed)., Berlin Heidelberg: Springer 2014.

[11]   Petruska J, Goodman MF. Enthalpy-entropy compensation in DNA melting thermodynamics. J Biol Chem 1995; 270(2): 746-50.
[http://dx.doi.org/10.1074/jbc.270.2.746] [PMID: 7822305]

[12]   Schilkraut C, Lifson S. Exploration of metal ion binding sites in RNA folds by Brownian-Dynamics simulations, Structure 1998; 6: 1303-1314. http://biomednet.com/elecref/0969212600601303

[13]   Harris NC, Kiang C-H. Defects can induce melting time of DNA, Nanoassemblies. J Phys Chem B 2006; 110(33): 16393-6.
[http://dx.doi.org/10.1021/jp062287d] [PMID: 16913768]

[14] Mitra S. Sample preparation Wiley-Interscience. Hoboken, New Jersey: A John Wiley & Sons, Inc 2003.

[15] Gallagher SR. Dependence of melting temperature of DNA on salt concentration Biopolymers 1965; 3(3): 195-208.
[http://dx.doi.org/https://doi,org/10.1002/bip.360030207]

[16] Holden MJ, Haynes RJ, Rabb SA, Satija N, Yang K, Blasic JR Jr. Factors affecting quantification of total DNA by UV spectroscopy and PicoGreen fluorescence. J Agric Food Chem 2009; 57(16): 7221-6.
[http://dx.doi.org/10.1021/jf901165h] [PMID: 19627145]

[17] Voet D, Voet TJ, Pratt CW. Fundamentals of Biochemistry. Hoboken, NJ, USA: John Wiley and Sons 2002.

[18] Sherwood L. Human physiology- From cells to systems. California, USA,: Thomson Learning Inc 2004. ISBN 053-39501-5.

[19] Becker WM, Kleinsmith LJ, Hardin J. The world of the cell. USA: Pearson Education 2004. ISBN 81-297-0415-3.

[20] Cox MM, Nelson DL. Lehninger Principles of Biochemistry. 5th ed., New York: W H Freeman and Co 2011.

[21] Arrighi FE, Mandel M, Bergendahl J, Hsu TC. Buoyant densities of DNA of mammals. Biochem Genet 1970; 4(3): 367-76.
[http://dx.doi.org/10.1007/BF00485753] [PMID: 4991030]

[22] Smialek MA, Jones NC, Hoffman SV, Mason NJ. Measuring the density of DNA film using U V –VIS interferometry. Physical Rev J 2013; E87: IS56060701.

[23] Elder RM, Pfaendtner J, Jayaraman A. Effect of hydrophobic and hydrophilic surfaces on the stability of double-stranded DNA. Biomacromolecules 2015; 16(6): 1862-9.
[http://dx.doi.org/10.1021/acs.biomac.5b00469] [PMID: 25961882]

[24] Richmond TJ, Davey CA. The structure of DNA in the nucleosome core. Nature 2003; 423(6936): 145-50.
[http://dx.doi.org/10.1038/nature01595] [PMID: 12736678]

[25] Whelan DR, Hiscox TJ, Rood JI, Bambery KR, McNaughton D, Wood BR. Detection of an en masse and reversible B- to A-DNA conformational transition in prokaryotes in response to desiccation. J R Soc Interface 2014; 11(97): 20140454.
[http://dx.doi.org/10.1098/rsif.2014.0454] [PMID: 24898023]

[26] Vargason JM, Henderson K, Ho PS. A crystallographic map of the transition from B-DNA to A-DNA. Proc Natl Acad Sci USA 2001; 98(13): 7265-70.
[http://dx.doi.org/10.1073/pnas.121176898] [PMID: 11390969]

[27] Franklin R. The structure of sodium thymonucleate fiber-I. Acta Crystallogr 1953; 6: 673-7.
[http://dx.doi.org/10.1107/S0365110X53001939]

[28] DiMaio F, Yu X, Rensen E, Krupovic M, Prangishvili D, Egelman EH. Virology. A virus that infects a hyperthermophile encapsidates A-form DNA. Science 2015; 348(6237): 914-7.
[http://dx.doi.org/10.1126/science.aaa4181] [PMID: 25999507]

[29] Churchill CD, Wetmore SD. Developing a computational model that accurately reproduces the structural features of a dinucleoside monophosphate unit within B-DNA. Phys Chem Chem Phys 2011; 13(36): 16373-83.
[http://dx.doi.org/10.1039/c1cp21689a] [PMID: 21842033]

[30] Barone G, Fonseca Guerra C, Bickelhaupt FM. B-DNA structure and stability as functional nucleic acid composition: Dispersion-correlated DFT study of nucleoside monophosphates single and double strands. ChemistryOpen 2013; 2(5-6): 186-93.

[http://dx.doi.org/10.1002/open.201300019] [PMID: 24551565]

[31]   Mitsui Y, Langridge R, Shortle BE, *et al.* Physical and enzymatic studies on poly d(I-C)-poly d(I-C), an unusual double-helical DNA. Nature 1970; 228(5277): 1166-9.
[http://dx.doi.org/10.1038/2281166a0] [PMID: 4321098]

[32]   Ha SC, Lowenhaupt K, Rich A, Kim YG, Kim KK. Crystal structure of a junction between B-DNA and Z-DNA reveals two extruded bases. Nature 2005; 437(7062): 1183-6.
[http://dx.doi.org/10.1038/nature04088] [PMID: 16237447]

[33]   Zhang H, Yu H, Ren J, Qu X. Reversible B/Z-DNA transition under the low salt condition and non--form polydApolydT selectivity by a cubane-like europium-L-aspartic acid complex. Biophys J 2006; 90(9): 3203-7.
[http://dx.doi.org/10.1529/biophysj.105.078402] [PMID: 16473901]

[34]   de Rosa M, de Sanctis D, Rosario AL, *et al.* Crystal structure of a junction between two Z-DNA helices. Proc Natl Acad Sci USA 2010; 107(20): 9088-92.
[http://dx.doi.org/10.1073/pnas.1003182107] [PMID: 20439751]

[35]   Rich A, Zhang S. Timeline: Z-DNA: the long road to biological function. Nat Rev Genet 2003; 4(7): 566-72.
[http://dx.doi.org/10.1038/nrg1115] [PMID: 12838348]

[36]   Roy Chowdhury A, Bakshi R, Wang J, *et al.* The killing of African trypanosomes by ethidium bromide. PLoS Pathog 2010; 6(12): e1001226.
[http://dx.doi.org/10.1371/journal.ppat.1001226] [PMID: 21187912]

[37]   Dickerson RE. Definitions and nomenclature of nucleic acid structure components. Nucleic Acids Res 1989; 17(5): 1797-803.
[http://dx.doi.org/10.1093/nar/17.5.1797] [PMID: 2928107]

[38]   Lu XJ, Olson WK. Resolving the discrepancies among nucleic acid conformational analyses. J Mol Biol 1999; 285(4): 1563-75.
[http://dx.doi.org/10.1006/jmbi.1998.2390] [PMID: 9917397]

[39]   Olson WK, Bansal M, Burley SK, *et al.* A standard reference frame for the description of nucleic acid base-pair geometry. J Mol Biol 2001; 313(1): 229-37.
[http://dx.doi.org/10.1006/jmbi.2001.4987] [PMID: 11601858]

[40]   Wing R, Drew H, Takano T, *et al.* Crystal structure analysis of a complete turn of B-DNA. Nature 1980; 287(5784): 755-8.
[http://dx.doi.org/10.1038/287755a0] [PMID: 7432492]

[41]   David WU. DNA structure: A-, B- and Z-DNA helix families Encyclopedia of LifeSciences 2002. www.els.net

[42]   Johnson S. What causes the double helix to twist in a DNA molecule? 2017. https://sciencing.com/causes-double-helix-twist-dna-picture-2848.html

[43]   Pabo CO, Sauer RT. Protein-DNA recognition. Annu Rev Biochem 1984; 53: 293-321.
[http://dx.doi.org/10.1146/annurev.bi.53.070184.001453] [PMID: 6236744]

[44]   Peter LP, Anatoly ID, Robinson CC, *et al.* Factors affecting driving protein to the major and minor groove of DNA. J Mol Biol 2007; 365(1): 1-9.
[http://dx.doi.org/10.1016/j.jmb.2006.09.059] [PMID: 17055530]

[45]   Udgaonkar JB. Entropy in Biology, Resonance 2001. https://ncbs.res.in/sitefiles/jayant-reson.pdf

[46]   Brennen J. The structural stability of the DNA double helix 2017. http://sciencing.com/structural-stability--dna-double-helix-2232.html

[47]   Schneider TD. Biologist in information theory used in biology. IEEE Eng Med Biol Mag 2006; 25(1): 30-3.
[http://dx.doi.org/10.1109/MEMB.2006.1578661] [PMID: 16485389]

[48] Marko JF, Siggla ED. Bending and twisting elasticity of DNA. Macromolecules 1994; 27(4): 981-8.
[http://dx.doi.org/10.1021/ma00082a015]

[49] Austin RH, Brody JP, Cox EC, Duke T, Volkmuth W. Stretch gene. Phys Today 1997; 50(2): 32-8.
[http://dx.doi.org/10.1063/1.881674]

[50] Zheng M, Jagota A, Semke ED, *et al.* DNA-assisted dispersion and separation of carbon nanotubes. Nat Mater 2003; 2(5): 338-42.
[http://dx.doi.org/10.1038/nmat877] [PMID: 12692536]

[51] Trouiller B, Reliene R, Westbrook A, Solaimani P, Schiestl RH. Titanium dioxide nanoparticles induce DNA damage and genetic instability *in vivo* in mice. Cancer Res 2009; 69(22): 8784-9.
[http://dx.doi.org/10.1158/0008-5472.CAN-09-2496] [PMID: 19887611]

[52] Prado-Gotor R, Grueso E. A kinetic study of the interaction of DNA with gold nanoparticles: mechanistic aspects of the interaction. Phys Chem Chem Phys 2011; 13(4): 1479-89.
[http://dx.doi.org/10.1039/C0CP00901F] [PMID: 21132199]

[53] Saha S, Sarkar P. Understanding the interaction of DNA-RNA nucleobases with different ZnO nanomaterials. Phys Chem Chem Phys 2014; 16(29): 15355-66.
[http://dx.doi.org/10.1039/c4cp01041h] [PMID: 24942064]

[54] Lahir YK. Impacts of metal and metal oxide nanoparticles on embryos, Austin Endocrinology Diabetes Case Report (Special Issue on climate change) 2017. www.austinpublishiggroup.com

[55] Yang J, Lee JY, Too HP, Chow GM. Inhibition of DNA hybridization by small metal nanoparticles. Biophys Chem 2006; 120(2): 87-95.
[http://dx.doi.org/10.1016/j.bpc.2005.10.011] [PMID: 16303234]

[56] Heddle JC. Gold nanoparticle biological molecule interactions and catalyst. Catalysts 2013; 3: 683-708.
[http://dx.doi.org/10.3390/catal3030683]

[57] Hsu M-H, Josephrajan T, Yeh CS, Shieh DB, Su WC, Hwu JR. Novel arylhydrazone-conjugated gold nanoparticles with DNA-cleaving ability: the first DNA-nicking nanomaterial. Bioconjug Chem 2007; 18(6): 1709-12.
[http://dx.doi.org/10.1021/bc700222n] [PMID: 17953439]

[58] Bonomi R, Selvestrel F, Lombardo V, *et al.* Phosphate diester and DNA hydrolysis by a multivalent, nanoparticle-based catalyst. J Am Chem Soc 2008; 130(47): 15744-5.
[http://dx.doi.org/10.1021/ja801794t] [PMID: 18975902]

[59] Kumar V, Kumari A, Guleria P, Yadav SK. Evaluating the toxicity of selected types of nanochemicals Rev Environ Contam Toxicol 2010; 30(39): 215-21.
[http://dx.doi.org/https://doi:10.1007/978-1-4614-1463-6-2]

[60] Bergin IL, Witzmann FA. Nanoparticle toxicity by the gastrointestinal route: evidence and knowledge gaps. Int J Biomed Nanosci Nanotechnol 2013; 3: 1-2.
[http://dx.doi.org/10.1504/IJBNN.2013.054515] [PMID: 24228068]

[61] Cha C, Shin SR, Annabi N, Dokmeei MR, Khadhemhosseini A. Carbon-based nanomaterials for Biomedical engineering. ACS Nano 2013; 7(4): 2891-7.
[http://dx.doi.org/10.1021/nn401196a] [PMID: 23560817]

[62] Hughes ME, Brandin E, Golovchenko JA. Optical absorption of DNA-carbon nanotube structures. Nano Lett 2007; 7(5): 1191-4.
[http://dx.doi.org/10.1021/nl062906u] [PMID: 17419658]

[63] Li X, Peng Y, Qu X. Carbon nanotubes selective destabilization of duplex and triplex DNA and inducing B-A transition in solution. Nucleic Acids Res 2006; 34(13): 3670-6.
[http://dx.doi.org/10.1093/nar/gkl513] [PMID: 16885240]

[64] Nandy B, Santosh M, Maiti PK. Interaction of nucleic acids with carbon nanotubes and dendrimers. J

Biosci 2012; 37(3): 457-74.
[http://dx.doi.org/10.1007/s12038-012-9220-8] [PMID: 22750983]

[65]   Sun W, Zhao J, Du Z. Density functional theory-based study of the interaction of DNA/RNA nucleobases with hydroxyl functionalized armchair carbon nanotube. Comput Theor Chem 2017; 1102: 60-8.
[http://dx.doi.org/10.1016/j.comptc.2017.01.001]

[66]   Dubey N, Bentim R, Islam I, Can T, Antonio HCN, Rosa V. Graphene. A versatile carbon-based materials for bone tissue engineering. Stem Cells International. 2015; 2015: p. 12.
[PMID: 804213]

[67]   Lahir YK. Graphene and graphene-based nano-drug delivery: a critical review, Chapter, Applications of Targeted Nano-Drugs and Delivery Systems, 50 Hampshire, Street, 5th Floor, Cambridge, MA 02139; Elsevier Publication.

[68]   Zou X, Wei S, Jasensky J, *et al.* Molecular interaction between graphene and biomolecules. J Am Chem Soc 2017; 139(5): 1928-36.
[http://dx.doi.org/10.1021/jacs.6b11226] [PMID: 28092440]

[69]   Gurel HH, Salmankurt B. Binding mechanisms DNA/RNA nucleobases adsorbed on graphene undercharging: first principles van der Waals study. Mater Res Express 2017; 4(6): 065401.
[http://dx.doi.org/10.1088/2053-1591/aa6e67]

[70]   Giuliodori AM, Brandi A, Kotla S, *et al.* Development of a graphene oxide-based assay for the sequence-specific detection of double-stranded DNA molecules. PLoS One 2017; 12(8): e0183952.
[http://dx.doi.org/10.1371/journal.pone.0183952] [PMID: 28850626]

[71]   Panigrahi S, Bhattacharya A, Banerjee S, Bhattacharya D. Interaction of nucleobases with wrinkled graphene surface: dispersion correlated DFT and AFM studies. J Phys Chem C 2012; 116(7): 4374-9.
[http://dx.doi.org/10.1021/jp207588s]

[72]   Lee J-H, Choi Y-K, Kim H-J, Scheicher RH, Cho J-H. Physisorption of DNA nucleobases on h-BN and graphene, vdW correlated DFT calculations. J Phys Chem C 2013; 117(26): 13435-4.
[http://dx.doi.org/10.1021/jp402403f]

[73]   Lahir YK. Impacts of fullerene on the biological system, Clinical Immunology Endocrinology and Metabolic Drugs 2017; 4(1): 48-57.
[http://dx.doi.org/https://doi:10.2174/2212707004666171111351624]

[74]   Zhao X, Striolo A, Cummings PT. C60 binds to and deforms nucleotides. Biophys J 2005; 89(6): 3856-62.
[http://dx.doi.org/10.1529/biophysj.105.064410] [PMID: 16183879]

[75]   An H, Jin B. DNA exposure to buckminsterfullerene (C60): toward DNA stability, reactivity, and replication. Environ Sci Technol 2011; 45(15): 6608-16.
[http://dx.doi.org/10.1021/es2012319] [PMID: 21718073]

[76]   Firdaus S, Dhasmana A, Srivastava V, *et al.* Interaction pattern for the complex of B-DNA Fullerene compounds with a set of known replication proteins using docking study. Bioinformation 2015; 11(3): 122-6.
[http://dx.doi.org/10.6026/97320630011122] [PMID: 25987764]

[77]   Firdaus S, Dhasmana A, Haque S, *et al.* Interaction pattern of fullerene (C20 to C180) and carbon nanotubes with different forms of DNA: A computational biological approach, Theoretical Biological Forum ISSN 2282-2593. 2015; 41-56.

[78]   Medintz IL, Uyeda HT, Goldman ER, Mattoussi H. Quantum dot bioconjugates for imaging, labelling and sensing. Nat Mater 2005; 4(6): 435-46.
[http://dx.doi.org/10.1038/nmat1390] [PMID: 15928695]

[79]   Wang T-H. Discerning single-molecule interactions of DNA and quantum dots, Biotechnology Journal (special issue: methods and advances) 2013. https://doi.org/1002/biot.201200309

[80] Li K, Zhang W, Chen Y. Quantum dot binding to DNA: single-molecule imaging with atomic force microscopy. Biotechnol J 2013; 8(1): 110-6.
[http://dx.doi.org/10.1002/biot.201200155] [PMID: 22899656]

[81] Zhang Y, Wang T-H. Quantum dot enabled molecular sensing and diagnostics. Theranostics 2012; 2(7): 631-54.
[http://dx.doi.org/10.7150/thno.4308] [PMID: 22916072]

[82] Chen W, Turro NJ, Tomalia DA. Using ethidium bromide to probe the interaction between DNA and dendrimers. Langmuir 2000; 16(1): 15-9.
[http://dx.doi.org/10.1021/la981429v]

[83] Jain K, Kesharwani P, Gupta U, Jain NK. Dendrimer toxicity: Let's meet the challenge. Int J Pharm 2010; 394(1-2): 122-42.
[http://dx.doi.org/10.1016/j.ijpharm.2010.04.027] [PMID: 20433913]

[84] Nash JA, Singh A, Li NK, Yingling YG. Characterization of nucleic acid compaction with histone-mimic-nanoparticles through all-atom molecular dynamics. ACS Nano 2015; 9(12): 12374-82.
[http://dx.doi.org/10.1021/acsnano.5b05684] [PMID: 26522008]

[85] Kulkarni SK, Ethiraj AS, Kharrazi S, Deobagkar DN, Deobagkar DD. Synthesis and spectral properties of DNA capped CdS nanoparticles in aqueous and non-aqueous media. Biosens Bioelectron 2005; 21(1): 95-102.
[http://dx.doi.org/10.1016/j.bios.2004.09.004] [PMID: 15967356]

[86] Couroy J, Byrne SJ, Gunko YK, et al. Cd/Te QD display tropism to core histones and histone rich cell organelles, Small 2008.
[http://dx.doi.org/https://doi:10.1002/smll.200800088]

[87] Li H, Lee T, Dziubla T, et al. RNA as a stable polymer to build controllable and defined nanostructures for material and biomedical applications. Nano Today 2015; 10(5): 631-55.
[http://dx.doi.org/10.1016/j.nantod.2015.09.003] [PMID: 26770259]

[88] Li H, Rychahou PG, Cui Z, et al. RNA nanoparticles derived from 3WJunctions of Phi29 motor pRNA are resistant to I-125 and Cs 131 radiation, Nucleic Acid Therapeutics. 2015; 25: pp. (4)188-97.
[http://dx.doi.org/https://doi:1010.89/nat.2014.0525]

[89] Yang H, Jossinet F, Leontis N, et al. Tools for the automatic identification and classification of RNA base pairs. Nucleic Acids Res 2003; 31(13): 3450-60.
[http://dx.doi.org/10.1093/nar/gkg529] [PMID: 12824344]

[90] Laing C, Wen D, Wang JT, Schlick T. Predicting coaxial helical stacking in RNA junctions. Nucleic Acids Res 2012; 40(2): 487-98.
[http://dx.doi.org/10.1093/nar/gkr629] [PMID: 21917853]

[91] Laing C, Jung S, Kim N, Elmetwaly S, Zahran M, Schlick T. Predicting helical topologies in RNA junctions as tree graphs. PLoS One 2013; 8(8): e71947.
[http://dx.doi.org/10.1371/journal.pone.0071947] [PMID: 23991010]

[92] Khisamutdinov EF, Jasinski DL, Guo P. RNA as a boiling-resistant anionic polymer material to build robust structures with defined shape and stoichiometry. ACS Nano 2014; 8(5): 4771-81.
[http://dx.doi.org/10.1021/nn5006254] [PMID: 24694194]

[93] Lilley DMJ. The origins of RNA catalysis in ribozymes. Trends Biochem Sci 2003; 28(9): 495-501.
[http://dx.doi.org/10.1016/S0968-0004(03)00191-9] [PMID: 13678961]

[94] Buck HM. Modified RNA with a phosphate-methylated backbone: A serious omission in our (retarded) study at HIV-1 RNA loops and integrated DNA, specific properties of the (modified) RNA and dimmers Journal of Biophysical Chemistry 2016; 7(1): 30-41.
[http://dx.doi.org/https://doi:10.4236/jbp.2016.71003] [PMID: 63521]

[95] Schroeder R, Barta A, Semrad K. Strategies for RNA folding and assembly. Nat Rev Mol Cell Biol

2004; 5(11): 908-19.
[http://dx.doi.org/10.1038/nrm1497] [PMID: 15520810]

[96]   Zhang H, Endrizzi JA, Shu Y, *et al.* Crystal structure of 3WJ core revealing divalent ion-promoted thermostability and assembly of the Phi29 hexameric motor pRNA. RNA 2013; 19(9): 1226-37.
[http://dx.doi.org/10.1261/rna.037077.112] [PMID: 23884902]

[97]   Zuo Y, Deutcher MP. Exoribonuclease superfamilies: structural analysis and phylogenetic distribution, Nucleic Acid Research 2001; 29(5): 1017-26.: 11222749.

[98]   Parker R, Song H. The enzymes and control of eukaryotic mRNA turnover. Nat Struct Mol Biol 2004; 11(2): 121-7.
[http://dx.doi.org/10.1038/nsmb724] [PMID: 14749774]

[99]   Li WM, Barnes T, Lee CH. Endoribonucleases--enzymes gaining spotlight in mRNA metabolism. FEBS J 2010; 277(3): 627-41.
[http://dx.doi.org/10.1111/j.1742-4658.2009.07488.x] [PMID: 19968858]

[100]   Liu J, Guo S, Cinier M, *et al.* Fabrication of stable and RNase-resistant RNA nanoparticles active in gearing the nanomotors for viral DNA packaging. ACS Nano 2011; 5(1): 237-46.
[http://dx.doi.org/10.1021/nn1024658] [PMID: 21155596]

[101]   Hernandez FJ, Stockdale KR, Huang L, Horswill AR, Behlke MA, McNamara JO II. Degradation of nuclease-stabilized RNA oligonucleotides in Mycoplasma-contaminated cell culture media. Nucleic Acid Ther 2012; 22(1): 58-68.
[http://dx.doi.org/10.1089/nat.2011.0316] [PMID: 22229275]

[102]   Hong J, Huang Y, Li J, *et al.* Comprehensive analysis of sequence-specific stability of siRNA. FASEB J 2010; 24(12): 4844-55.
[http://dx.doi.org/10.1096/fj.09-142398] [PMID: 20732955]

[103]   Besseová I, Réblová K, Leontis NB, Sponer J. Molecular dynamics simulations suggest that RNA three-way junctions can act as flexible RNA structural elements in the ribosome. Nucleic Acids Res 2010; 38(18): 6247-64.
[http://dx.doi.org/10.1093/nar/gkq414] [PMID: 20507916]

[104]   Abels JA, Moreno-Herrero F, van der Heijden T, Dekker C, Dekker NH. Single-molecule measurements of the persistence length of double-stranded RNA. Biophys J 2005; 88(4): 2737-44.
[http://dx.doi.org/10.1529/biophysj.104.052811] [PMID: 15653727]

[105]   Lipfert J, Skinner GM, Keegstra JM, *et al.* Double-stranded RNA under force and torque: similarities to and striking differences from double-stranded DNA. Proc Natl Acad Sci USA 2014; 111(43): 15408-13.
[http://dx.doi.org/10.1073/pnas.1407197111] [PMID: 25313077]

[106]   Doi M, Edwards SF. Theory of polymer dynamics. Clarendon: Oxford Press 1986; p. 317.

[107]   Mustoe AM, Brooks CL, Al-Hashimi HM. Hierarchy of RNA functional dynamics. Annu Rev Biochem 2014; 83: 441-66.
[http://dx.doi.org/10.1146/annurev-biochem-060713-035524] [PMID: 24606137]

[108]   Chen H, Meisburger SP, Pabit SA, *et al.* Ionic strength the dependent persistence length of ssRNA and DNA, Proceedings of National Academy of Science USA 2012; 109(2): 799-804.
[http://dx.doi.org/https://doi.org/10.1021/acs.jpcb.9b11510]

[109]   Khisamutdinov EF, Li H, Jasinski DL, Chen J, Fu J, Guo P. Enhancing immunomodulation on innate immunity by shape transition among RNA triangle, square and pentagon nanovehicles. Nucleic Acids Res 2014; 42(15): 9996-10004.
[http://dx.doi.org/10.1093/nar/gku516] [PMID: 25092921]

[110]   Wessels CP, Keim W. Ionic liquids new solutions for transition metal catalysis, Angewandte Chemie 2000; 39(21): 3772-89.
[http://dx.doi.org/https://doi.org/10.1002/1521-3773;2000110339.21]

[111] Khan MA, Walden WE, Gross DJ, Theil EC. sensing by iron-responsive messenger RAN/repressor compressor complexes weaken bindings Journal Biological Chemistry (The American Society for Biochemistry and Molecular Biology) 2009. http://www.jbc.org/cgi/doi/10.1074/jbc.M109.041061

[112] Hermann T, Westhof E. Exploration of metal ion binding sites in RNA folds by Brownian-dynamics simulations. Structure 1998; 6(10): 1303-14. http;//biomednet.com/elecref/0969212600601303 [http://dx.doi.org/10.1016/S0969-2126(98)00130-0] [PMID: 9782053]

[113] Nash JA, Tucker TL, Theriault W, Yingling YG. Binding of single-stranded nucleic acid to cationic ligand 2016. https://avs.scitation.org/doi/am.pdf/10.1116/1.4966653

[114] Lescoute A, Westhof E. The interaction networks of structured RNAs, Nucleic Acid Research 2006; 34(22): 6587-604.

[115] Peacock H, Kannan A, Beal PA, Burrows CJ. Chemical modification of siRNA bases to probe and enhance RNA interference. J Org Chem 2011; 76(18): 7295-300.
[http://dx.doi.org/10.1021/jo2012225] [PMID: 21834582]

[116] Bramsen JB, Kjems J. Development of therapeutic-grade small interfering RNAs by chemical engineering. Front Genet 2012; 3: 154.
[http://dx.doi.org/10.3389/fgene.2012.00154] [PMID: 22934103]

[117] Astakhova IK, Wengel J. Scaffolding along nucleic acid duplexes using 2′-amino-locked nucleic acids. Acc Chem Res 2014; 47(6): 1768-77.
[http://dx.doi.org/10.1021/ar500014g] [PMID: 24749544]

[118] Binzel DW, Khisamutdinov EF, Guo P. Entropy-driven one-step formation of Phi29 pRNA 3WJ from three RNA fragments. Biochemistry 2014; 53(14): 2221-31.
[http://dx.doi.org/10.1021/bi4017022] [PMID: 24694349]

CHAPTER 8

Interactions Between Enzymes and Nanomaterials

Abstract: Enzymes are proteins, but all proteins are not enzymes. Enzyme interactions concern with the biochemical and physiological transformations encompassing most of the life activities. Understanding such events will help to predict particular biochemical, biocatalytic, and enzyme reactions involved. These investigations also help to predict clinical and remedial aspects of dysfunctionalities of physiological processes. Chemical enzymes have their impediments that pose difficulties during their industrial applications. Biological enzymes also referred to as biocatalysts, are chemospecific, and applicable conveniently to carry out varied biological activities. This feature is related to the identification and selection of a particular functional group, among others. This selection is physical or chemical but depends on parameters like the nature of the solvent, atomic orbitals, concentration, pH, temperature, *etc.* Their industrial and biological applications increase using the enzyme immobilization technique. Nanomaterials have occupied significant status in the present day scenario. These materials are better options for this technique because these materials offer features like high specific surface area, improved dispersibility, low mass transfer resistance, *etc.* The mechanism of enzyme activity is quite complicated. The necessary steps incriminated are binding of the enzyme with the specific substrate. The complementary shape, size, charge, hydrophobicity, and hydrophilicity, *etc.*, of a substrate, play a significant role in its binding with an enzyme. Nanomaterials are potential components that act as a matrix during the process of enzyme immobilization. These nanostructures elevate the efficacy of biocatalyst, specific surface area, mass transfer resistance, and loading of the capable enzyme, *etc.* The unique physicochemical features like size, surface properties, ease of modulation of nanomaterials, *etc.*, ensure better performance of enzymes and improve their applications in different fields like biomedical, pharmaceuticals, biomolecular, food, and packaging technology, agricultural practice, and biochemical investigations *in vitro* as well as *in vivo*. Some of the fundamental properties of enzymes can be modified to suit the functionality concerning the set targets. This chapter deals with the structure, nature, and regulatory dynamics of the enzyme. The enzyme immobilization technique, its advantages, and interactions with different nanomaterials along with biomimicking agents are also discussed.

Keywords: Active Energy, Biocatalysis, Biomimicking, Enzymes-Action, Enzyme-Immobilization, Free Energy, Nanomaterials, Toxic-Impacts.

BIOCATALYSIS

It is a challenge to know about biological activities. Most of the biological activities are related to or involve proteins, and organic enzymes are proteins. The interactions involving proteins form a considerable umbrella and encompass most of the biological activities and interactions. The catalysts comprise the biocatalysts and regulate such interplays. The biocatalytic events help to predict a particular biochemical, biomolecular reaction that takes place during physiological functionalities. These interactions take place involving either change at the bond level or the reaction center level or similarity at the reaction center. The studies concerning biocatalysts help to understand their potential applications and biomolecular transformations. These transformations assign functions to an enzyme in a sequential biological, biochemical, and chemical reaction [1]. This feature is related to the identification and selection of a particular functional group, among others present on the substrate and enzyme molecule. The choice is physical or chemical but is dependent on parameters like the nature of the solvent, atomic orbitals, concentration, pH, temperature, *etc.* Prediction of such selectivity is quite tricky [2]. The biological catalytic interactions are generally harmless in the sense that they do not form any unwanted by-products. These interactions are occurring in a specific direction; *i.e.*, these prefer to either make or break a chemical bond in one direction. These interactions take place because of the three-dimensional conformation of the reactants. The enzyme and biocatalysts can differentiate between the varied groups present in different zones of the molecule of a substrate. Sometimes two or more compounds have non-identical structural aspects but are not mirror images of each other and exhibit two different conformations [3]. The biocatalysis concerns the enhancement of the rate of the interaction or transformation of biomolecules, organic compounds within the biosystem. The ability of microorganisms to produce enzymes is useful in food technology, beverage, fermentation technology to get the desired commercial products, and to maintain a specific state of the food and other merchandise of economic importance, *etc.* There is a need to produce fine chemicals and other chemicals that are useful in pharmaceutical and related industries, and these based on biocatalytic and enzymatic interactions [1].

There are two types of enzymes namely, endoenzymes and exoenzymes. The secretory cells are the site of the production and location of actions for biocatalyst. Most of the biocatalysts are considered to be endoenzymes. Sometimes an endoenzyme (single molecule) acts as an exoenzyme also. The endoamylase splits larger amylose molecules into smaller chains of dextrin, while exoenzymes operate on the subunits of a polymer at one end of the polymer [4]. The prokaryotic and eukaryotic cells produce exoenzyme. These biological molecules are the product of cells, but these function outside the cell. Exoenzymes are secreted in the cell but act outside the cell. Most of these enzymes break down macromolecules, add a specific group, conjugate temporarily. These act as participants in a multiplex reaction or associated with two or more subunits of a complex molecule or biological macromolecule. Exoenzymes break down the macromolecules and help in the movements of the smaller micromolecules across the membrane. The subunit molecules conveniently move across the cell membrane or biological membrane. Generally, digestive enzymes come under this group [5 - 9]. The flowchart representing the cellular release of protein enzyme is shown in Fig. (**1**).

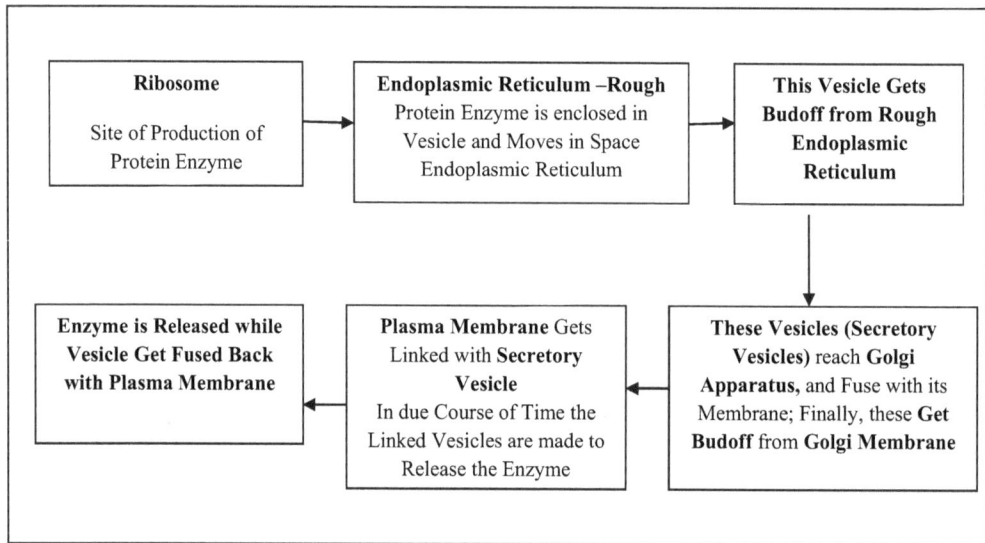

Fig. (1). The flowchart representing the cellular release of protein enzyme.

The biological enzymes are either anabolic or catabolic in action. The anabolic biocatalysts regulate the building up processes leading to the formation of a complex biomolecule from the respective simpler subunits. The resultant complex has a higher molecular weight than the subunits. The photosynthesis, synthesis of proteins, and other macro biomolecules are the products of anabolic enzyme activities. The formation of glucose is one such example. Other examples include synthesis of proteins, nucleic acids, biomolecules, or organic and inorganic molecules having very high molecular weight, complex conformations. During this process, molecular energy in the form of ATP is used up. Catabolic enzymes break the complex molecule with high molecular weight and complex conformation into its sub-units. Such processes release molecular energy in the form of ATP.

The enzymes are macromolecules and regulate specific chemical, biochemical, and biomolecular interactions. These macromolecules act on particular chemicals referred to as the substrate, and the resultant compounds are called the products. Enzymes are suitable agents that play crucial roles in various biomolecular interactions. These are applicable in almost every field, including domestic, industrial, biological science, biotechnological, biomedical sciences, *etc.* Traditional chemical catalysts are generally non-environmentally friendly, costly, have low specificity and specific reaction conditions. The organic enzymes are better options in comparison to these conventional chemical catalysts. The biocatalysts and the technique of enzyme immobilization overcome these pitfalls and provide better prospects to improve their stability and recyclability. The nanomaterials act as suitable support materials during the immobilization of enzyme technique and offer high specific surface area, improved dispersibility, and low mass transfer resistance. These parameters make the support nanomaterials better options in the enzyme technology [10].

Three-dimensional conformations of enzymes play an essential role during their specific interactions. These particular molecules exhibit some of the features like effectiveness in small concentration to accomplish the interplay. These only accelerate the rate of the biochemical interaction by reducing the energy of activation but do not initiate it. At the end of these interactions, these molecules remain unchanged. The properties of the end products do not change because the enzymes do not shift the equilibrium. There is a need for hydration for enzymatic interaction. This efficiency is represented by the numbers of substrate molecules that undergo a change per unit time for each enzyme. The turn over number of an enzyme reaction represents its efficiency, and this number differs from enzyme to enzyme. Lysozymes show relatively lower enzymatic ability. These are highly specific and influenced by parameters such as inhibitors, activators, temperature, and pH, *etc* [11]. Enzymes are sensitive to heat, and its impact influences their activity. Very high and low temperatures decline the enzyme activity and efficiency. The optimum temperature is within 20°C to 35°C. The enzymes with low molecular weight show a relatively higher degree of thermostability, while enzymes with high molecular weight exhibit inhibitory tendency during the activity. In most cases, there is an impact of pH on the enzyme activity, and its range extends from neutral pH to around 7.3. However, some digestive enzymes are the most effective at the acidic range and alkaline pH, at least during digestion in the intestinal region [5 - 8].

Some molecules, like cellular metabolites, drugs, xenobiotics, toxins, *etc.*, influence enzyme activity. These molecules either decline or inhibit enzyme activity ultimately. Molecules that inhibit the enzyme activity are called the inhibitors, and those accelerate are called the promoters. The inhibitor molecules are irreversible and reversible. Irreversible inhibitors or inactivators covalently bind to the amino acid residues present at the catalytic site and cause permanent inactivation of the enzyme activity. This inhibitory situation is overcome by either elevating or diluting the concentration of the substrate or opting for dialysis. The uncompetitive inhibitors get bound to the enzyme-substrate complex. These do not link to the free enzyme. The non-competitive inhibitor or mixed inhibitor gets bound to both the free enzyme and enzyme-substrate complex. In all cases, the feedback inhibitor or allosteric inhibitor becomes the end product, and it is applicable to check the enzyme activity. Hence, it is the feedback inhibitor in a given enzyme interaction.

The allosteric inhibitor is considered as the regulatory component of the enzyme interaction because it binds an effector molecule at the site that is not the active site of the enzyme. The place to which the effector binds is the allosteric site or the regulatory site. There is a change in the conformation of the reactant and transition is in a reversible pattern because of small molecules resulting in modifying the earlier function. Irreversible Inhibitors or inactivators covalently link with the amino acid residue of the catalytic site, and permanently inactivate the enzyme. The penicillin covalently attaches to a serine residue at the active site of the glycopeptide transpeptidase enzyme establishing a cross-linkage in the case of the bacterial cell wall. Phenylmethylsulfonyl fluoride builds a covalent bond with the catalytic site that contains serine residue of proteases like trypsin, chymotrypsin, *etc.*, and influences the enzyme activity. These are useful during the isolation of the enzymes. The malolactic acid is a competitive inhibitor and competes with succinate; it inhibits the activity of succinate dehydrogenase. Likewise, sulpha drugs compete with PABA (Para-aminobenzoic acid, a substrate), which inhibits the synthesis of folic acid (a vitamin) in bacteria. The natural enzymes are exquisite biocatalysts and participate in the biological interactions or interconversions in biological systems. These elevate the rate of biochemical reaction up to 1019 times for specific substrate and reactants in reactions they participate [5 - 8, 12].

Enzymes are globular proteins (spheroproteins). These react either alone or as big complex [13]. The three-dimensional structure of the enzyme is related to the specific sequence of amino acids. The three-dimensional structure and specific sequence of amino acids in an enzyme play a significant role in their catalytic activity, but on the bases of the structure of the catalyst alone, it is difficult to predict the said activity [14]. Temperature influences the structure of the enzymes and can denature it. These molecules have an optimum temperature. Some thermolabile catalysts perform their activity at very high temperatures, like the temperature of hot spring. Generally, the structure of the enzyme is relatively more prominent than their respective substrates. For example, a monomer of 4-oxalocrotonate tautomerase has 62 amino acids residue while there are 2500 residues present in enzyme fatty acid synthase. Fatty acid synthase enzyme is a multifunctional polypeptide, and a multi-enzymatic protein consisting of two similar sub-units with molecular weight 272 kDa, placed in the head to tail position. The catalytic site consists of only 2 to 4 amino acids alone, and these participate in the catalytic activity [15 - 17]. The active site is composed of the catalytic and binding site, while the remaining significant aspect of the enzyme structure helps to retain the precise orientation and dynamics of the active site. In some of the enzymes, more amino acids participate in the catalytic activity. Such catalysts have specific locations to bind and orient the cofactor that takes part in the catalytic activity [17 - 19].

Enzymes are protein macromolecules. Their primary structure consists of amino acids linked with each other by amide bonds or peptide bonds in a linear chain pattern. This chain is the polypeptide chain. The DNA of the corresponding gene regulates and encodes the specific sequence of amino acid. The secondary structure of an enzyme relates to the hydrogen in the amino group (NH_2- group) and oxygen present in the carboxyl group (–COOH) of each amino acid. These form H-bond with each other as a result of interaction between them in the same chain. This interaction also causes the folding of the protein chain on itself and facilitates specific structural conformation of the protein. This folding can be of two modes. In one style of folding the protein, the chain gets wrapped around itself and forms α-helix; other manner is involved in the folding on top of it and forms β-sheet. The β-sheet formed may be antiparallel or parallel in orientation. Folding results in the direction that alternates between each fold. This resultant structure has an antiparallel adjustment. As a result, this type of folding, a 2-dimensional linear chain is the result and represents the secondary structure. The protein in this orientation gets folded up further, and this folding attains a 3-dimensional structure, resulting in a tertiary structure.

In most cases, two or five amino acids participate in catalytic activities at the catalytic site. This site is present close to the binding site. These chemically specific sites bind with corresponding ligand or group having a particular affinity. The strength of the chemical bond formed helps in calculating the power of the chemical bond [17, 18]. In some cases, no amino acids of the enzyme participate in the binding process; instead, there are specific sites that act as a site to get bound with the relative cofactor. The cofactors are either non-protein chemical component or a metallic ion. Some enzymes have allosteric regulation that involves allosteric activation and allosteric inhibition. This interaction incriminates positive and negative allosteric sites and the corresponding binding site or active area. These all components function in coordination to accomplish allosteric regulation. When a molecule binds to the allosteric site, it causes conformational changes. These changes can either increase or decrease the catalytic or binding activity [17, 18]. Ribosomes are few in numbers and are the

RNA based biological catalysts. These act either alone or as complexed with proteins [5 - 8, 17, 18].

The molecular weight of most of the enzymes varies between 10000 to above one million; there are exceptions to this range. Molecular weight is a total of the atomic weight of the atoms of the molecules. Quite a good number of enzymes need a non-protein component called the cofactor. This component is inorganic ions like Fe^{2+}, Mg^{2+}, Mn^{2+}, Zn^{2+}, or this component is an organic or Metallo-organic molecule. This component is the coenzyme [5 - 8].

Pseudoenzymes are a group of molecules that have lost their catalytic activities. This aspect is dependent on the sequence of amino acids. These dead enzymes are being investigated to find their functional and evolutionary diversity. These molecules are non-enzymes, dead enzymes, proenzymes, and zombie proteins. In all probabilities, these might have been associated with biological and cellular and signaling activities [19]. These pseudoemzymes are the protein kinase-like domain that has lost the catalase phosphoryl transferability, but these acts as an essential signaling domain. The thermal-shift assay is a beneficial technique to study these molecules, and these are potential agents to review the nucleotide-binding features related to the kinase-domains. These are also helpful in studying detectable nucleotide binding, cation binding, and nucleotide-binding enhanced by cations [5 - 8, 19]

Directed evolution technique is suitably applicable in optimizing the properties of enzymes. The features of an enzyme are useful criteria to find its economic feasibility despite uncertainty about its molecular basis. Among many properties of a catalyst, its stability has prime significance concerning its functionality. Further, the structural and functional stability of an enzyme is of significant industrial application. The enzymatic stability closely relates to the different issues of protein science, like problems related to protein folding and its concerned physiopathological consequences. The technique of directed evolution facilitates understanding the structural aspects that are concerned with stabilizing a protein. During early investigations, such parameters are not distinct. The optimized surface of protein results in its stability. The thermal stability of an enzyme reestablishes without affecting its catalytic efficacy [5 - 8, 20].

SUBSTRATE

The substrate is a chemical reagent that participates in a chemical reaction and also forms a product. In synthetic and organic chemistry, substrates change. These changes may be structural or conformational. Under the regime of biochemistry, a substrate is a participant of enzyme-substrate interaction, and an enzyme interacts with it. According to the principle of Le Chatelier, the concentration of substrate changes during the interaction between enzyme and substrate. The substrate also acts as a precursor of a product [21]. The product formed as a result of specific enzyme-substrate interaction *in vitro* and *in vivo* may be different. These reactions can be altered in laboratory conditions and within the biosystem. Substrates also exhibit sensitivity, and this feature is a sensitive index substrate. This feature of the substrate molecule is related to pharmacokinetics: drug concentration and drug-drug-interactions, an area under the curve, biochemistry: metabolic pathways [22 - 24]. This parameter is of clinical significance to investigate the physiological and biochemical aspects of metabolisms in a biosystem [22].

The substrate channeling technique involves the intermediate metabolites or the intermediate products formed during the transformation of one enzyme to another enzyme or different active site without releasing this product in the reaction mixture. This process increases the efficiency and reduces the time of a metabolic pathway; it also prevents the random distribution of enzymes in the cytosol and the release of unstable products or intermediates in the cell. These are protected from the effects of other catalysts present in cytosol in the ambient medium [25]. The substrate channeling technique involves any one of the methods: (i) Substrate attaches to one of the domains. This domain is flexible and capable of moving between many active sites, precisely, in the case of multifunctional enzymes, (ii) protein and substrate move through a connecting tunnel present between two active sites, (iii) a charge region develops on the surface of enzyme; it acts as pathway called the electrostatic highway. It directs the substrate that has an opposite charge than the charge present on the active site. The common enzyme complexes that exhibit this phenomenon are pyruvate dehydrogenase complex, tryptophan synthase, dihydrofolate reductase-thymidylate synthase, *etc* [25 - 28].

The substrate channeling technique performs regulatory influences on the multifunctional enzyme action in many ways. It reduces the duration of reaction, checks the process of diffusion to prevent the loss of intermediates of the repercussions that do not allow the release of labile reaction metabolites to go into the reaction mixture, isolates toxic intermediates and restrict them to reach the cell, maintain favorable equilibria of the multiple enzyme reactions, checks the entrance or forestall the entry of competing or interfering metabolites with metabolic pathway, does not allow the excess reaction intermediates accumulate, effectively regulates a block of consecutive reaction, and ensures undisturbed completion of metabolic pathway or multistep enzyme catalytic cycle [29].

IMMOBILIZATION OF ENZYME

An enzyme is said to be immobilized when it links with an inert, insoluble material like calcium alginate. An enzyme under consideration is treated with calcium chloride and is made to interact with sodium alginate solution resulting in the immobilization of the catalyst enzyme. This process entirely or severely restricts the movement of the catalyst, cell organelles, and cells. Quite often, immobilized enzymes are referred to as insoluble, supported, or matrix-lined enzymes. The process of enzyme immobilization deals with either restricting or adsorbing the enzyme or cells understudy on to or on an appropriate inert support. This process ensures the structural and functionality of the immobilized enzyme. The product of this technique is cost-effective and ensures the efficacy of the catalysts. These products are useful for industrial applications. It is essential to protect the immobilized enzyme from harsh and adverse conditions like high temperature, surfactants, oxidizing agents, *etc*. This practice helps the retention of the functionality of the immobilized enzyme and increases the application of these enzymes in industries like food, detergent, textile, bioremediation, and pharmaceutical sectors [30].

The basic concept of the methodology of immobilization involves any of the processes like adsorption, entrapment, cross-linking, and covalent bonding. A physical binding of either enzyme or cell on or with the surface of the inert and inorganic supportive matrix involves adsorption. The support matrix is inorganic like silica gel, calcium phosphate gel, glass, alumina, *etc*. The inert organic matrixes, such as the supporting media, like starch, carboxymethyl cellulose, DEAE-Sephadex, or DEAE-cellulose, are suitable to immobilize the enzyme. The binding between the catalyst and the surface of the inert matrix involves weak forces like hydrogen bonds and van der Waals. This type of binding provides convenient removal of the adsorbed enzyme or cells from the matrix. Immobilization of enzyme is carried out by gently manipulating pH, temperature, ionic strength of the reaction mixtures. This aspect of easy removal of the adsorbed catalyst is not suitable for industrial applications [31, 32].

The entrapment process involves trapping the desired molecule in polymer, a matrix of gel, microcapsule, or fibers. This model consists of the lattice entrapment. One of the essential factors affecting trap is the pore size of the matrix used. The size of the pore should be able to retain the enzyme or cell under consideration and allow the substrate and the product to be drained out. Polymeric network (synthetic or natural) is permeable material; it retains the immobilized enzyme and allows the substrate and product to move out [30, 33]. The incrimination of binding does not cause any conformational and functional changes in the enzyme or cell. The catalysts under consideration bind within the mesh or lattice of fibers or gel to ensure its entrapment. Another mode is to trap the catalyst within the microcapsule. The hydrophilic and hydrophobic forms of the polymers used to make the covering of the microcapsule. In this mode, the trapped enzyme can leak through the microcapsule [30, 33]. The nature of the membrane of the microcapsule is polymeric, based on lipoprotein or non-ionic. This membrane and functionally is semi-permeable and spherical. The formation of the microcapsule is done involving fabricating a particular membrane reaction, the creation of emulsion, and stabilization of the emulsion. The substances like polyacrylamide gel, gelatin, starch, silicon, rubber, or collagen are useful as materials of the membrane of the microcapsule.

Covalent or cross-linking types of bonding take place during enzyme binding. Covalent binding is commonly involved, but it needs the preactivation of the inert matrix before binding. The cyanogen bromide activation occurs when cellulose, Sephadex, having glycol groups, is the choice and ensures the covalent bonding. Diazotisation is the interaction between amino-benzyl cellulose and amino derivatives of polystyrene or amino silanized porous glass help in diazotization. Covalent linkage brings about stable and reliable binding when support materials like polyacrylamide, porous glass, agarose, and porous silica are the choice [30, 34, 35]. Other methods of covalent binding involve peptide bond formation between amino, the carboxyl group of the matrix and carboxyl or amino groups of the enzyme, activation of bi or polyfunctional reagents like glutaraldehyde, albumin, and amino alkylated porous glass. Cross-linking of the catalyst is the preferred option in the absence of a solid inert matrix. In this situation, the catalyst to be immobilized traps involving cross-links between polyfunctional reagents, while interacting between enzyme and the reagents form bridges in-between [36]. Glutaraldehyde, diaminobenzidine, hexamethylene diisocyanate, and toluene di-isothiocyanate incriminate for cross-linking. Glutaraldehyde is a better option because of the cross-links that establish withstands extremes, pH, and temperature. This molecule is an efficient agent and immobilizes enzymes like glucose isomerase and amidase. There are chances of the immobilized catalyst get denatured because of polyfunctional agents used in the process [5 - 8].

The Functional Roles of Nanomaterials in the Process of Immobilization of Enzyme

The conventional methods of enzyme immobilization have industrial applications. This process involves cross-linked enzyme aggregates, microwave-assisted immobilization, click chemistry technology, and mesoporous matrix. The nanomaterials are suitable substances as a matrix and facilitate the immobilization of enzymes. The nanostructures elevate the efficacy of biocatalyst, specific surface area, mass transfer resistance, and loading of the active catalyst. Silica, chitosan, gold, diamond, graphene, and zirconium nanomaterials are the favorite options for the immobilization of enzymes [37].

There is a need to improve the process of enzyme immobilization to meet the set target of various industrial applications. All these efforts are to enhance the degree of catalytic activity and stability of the immobilized enzyme. For example, flower like-organic-inorganic-hybrid nanomaterials can improve stability and enzyme activity. These hybrid nanomaterials are applicable for the immobilized strategy, protein/catalyst, and metal ions that work as organic or inorganic components of hybrid nanoflowers. The resultant immobilized enzyme has a high elevated degree of enzyme activity, stability concerning a relatively more comprehensive range of pH, temperature, and concentration of salt in comparison to the free enzyme [38].

Physicochemical parameters like enzyme density, mass transport, the morphology of surface chemistry of nanoparticles, the orientation of the enzyme participating in the interaction between nano support and the catalysts, *etc.* The surface chemistry of the nanoparticles (as nano support) modifies to regulate the configuration, orientation, and density of the enzyme under consideration. The arrangement and density of nanoparticles and the mobility of catalysts correlate with the degree of target specificity. When multiple catalysts bind to the single nanoparticles among free enzymes, the frequency of the enzyme enhances during the process of immobilization [39, 40]. The morphological variable of nanomaterials influences the process of enhancement of the enzymes. Nanoparticles and nanotubes have functionally larger radii of curvature even though individually they have smaller spaces. Such nanoparticles enhance center-to-center distance between the adjacent immobilized enzymes. These nanomaterials interfere with low interprotein-to-protein interaction because these do not favor the immobilization of enzymes [40]. The interplay between single-walled carbon nanotubes, multiwalled carbon nanotubes, and graphene (as nano support) and the three enzymes, soya bean peroxidase, chloro-peroxidase, and glucose oxidase, depends on their molecular mapping and size of their molecule. Such interplay works in coordination and involves physicochemical features like surface chemistry, charge, and the related ratio. During this interaction, there is an influence on the structure of the enzyme and the curvature of the substrate. The concerned catalytic action at the respective interface formed between the catalyst under study and the carbon nano support. These investigations help to optimize enzyme nano support. This behavior optimizes functionality relates the processes like fermentation, biosensing, generation of biofuel, *etc* [41].

Enzymes are better catalytic agents than their corresponding chemical counterparts. These biomolecules have better functionality. When these associate with chemoselectivity, regioselectivity, and stereoselectivity, these act as active agents to attain the desirable products. These are useful in industries, biomedical, biochemical, biotechnical fields [42]. Immobilization of enzymes using nanomaterials brings about some changes in the resultant products. The immobilized enzymes can be conveniently fabricated or modified. There is no need for surfactants and toxic chemicals to be used in this process. The product formed is homogenous as core-shell nanoparticles. The size of these particles regulates in accordance to the need or set target. As a result of recent advances in technology, even multiple enzymes can be conveniently immobilized. The degree of loading of proteins, enzymes, enzyme activities, and the stability elevates. Above all, these products are cost-effective and applicable in industrial and biotechnological fields. These immobilized enzymes are active within a wide range of pH and temperature. These forms of enzymes exhibit better stability at a relatively higher temperature than their non-immobilized counter forms [43]. Nanoparticles are the novel and use matrices for enzyme immobilization. Nanoparticles possess ideal features to equilibrate fundamental factors that help in determining the biocatalytic efficiency like specific surface area, mass transfer resistance, and active enzyme loading [44]. The interaction between nanomaterials and enzymes is significant. The immobilization of catalysts is one aspect of this interaction and is of immense significance. The immobilized enzymes are potential materials as biosensors, modifications, and degradation of nanomaterials. This interaction functionalizes nanomaterials to mimic enzymes or to act as nanozymes. The specific physicochemical features of nanomaterials serve as general flaws of natural enzymes, and these are applicable to overcome these flaws. A variety of nanomaterials are beneficial options for the matrix that assist the immobilization of the enzymes. Metal nanoparticles and carbon nanomaterials are preferred the better options for this purpose [45]. The enzymes immobilize on nanomaterials exhibit more stability and have better efficacy than their non-immobilized counterparts. The structure of the immobilized enzymes becomes relatively more fixed because of the specific physicochemical features of nanomaterials [46].

In addition to nanosize, a high ratio of area to volume of the nanomaterials also provides enhanced limits of mass transport and better mobility in comparison to corresponding bulk materials. Specifically, the mass transfer limits get many folds increased. The microporous nanomatrices offer a large area for attractive loading space for the enzyme. Thus, an enormous number of molecules of enzyme get adsorbed or bound covalently to such surfaces. The biological activity of the enzyme is directly related to the number of molecules available on a given surface. The value of the operation of the enzyme is immobilization efficiency. This aspect is also influenced by thermodynamics assisting molecules of enzymes to make contact and changes in their conformations. After the optimum value or threshold state, then there is a decline in the enzyme activity, possibly due to the extreme conformational distortion that renders the enzyme deactivated [42, 47]. Mobility is another parameter that elevates when nanomaterials act as immobilization matrices. Wang suggested that the Brownian movements become very useful when nanoparticles or nanostructures are well dispersed or spread in the medium of the said interaction. These molecules, under the influence of the Stokes-Einstein equation, exhibit a correlated degree of diffusibility and mobility of the particles. The rate of movement differs, and it reflects the transition region between homogeneous catalysts with the free enzyme and heterogeneous catalyst with an immobilized enzyme. Biocatalyst systems can be regarded as self-functioning molecular machines that are stable effective, and work as self-targeting units [42, 47].

Quantum dots are luminescent semiconductor nanomaterials. These are applicable in biological, biomedical fields, *etc*. This behavior is because of their ability to form (identifiable) conjugates conveniently with proteins, peptides, genetic materials, drugs, and other biomolecules. Their chemistry concerning conjugation and linkage between the biomolecules is beneficial. Quantum dots exhibit chemoselective ligation either directly or to the biological materials like aniline-catalyzed imine ligation, thiol exchange, thiol-targeting iodoacetate, and Cu (1)-catalyzed azide-alkyne cycloaddition. Bioconjugation depends on the degree of functionalization and solubilizing capacity. The ligands involved are of great significance in this field of enzyme immobilization [48]. Movements of the nanoparticles can influence the enzyme-substrate interaction because of the Brownian actions [49]. During immobilization of enzymes, parameters like minimum diffusion limit, maximum surface area per unit mass, and higher active enzyme loading, *etc*., may cause some confusion. The solutions of nanoparticles are capable of eliminating such confusion. The transitional zone present between the immobilized enzyme (heterogeneous) and soluble free enzyme (homogeneous) influences the catalytic interaction under consideration. The particle mobility depends on the size, viscosity of the solution *via* Stokes and Einstein equation, and collision theory. These features also influence the reaction kinetics of the process. The polystyrene particles (110 to 1000 nm) with α-chymotrypsin (6.6 w %) bind covalently and behave following the basics of collision theory. The catalytic activities are in correlation with the size of the nanoparticles and the mobility of the particles. During this interaction, the intrinsic action because of the nanoparticles attached to the enzyme is affected [50]. Polyvalent and ligand-modified nanoparticles show better binding ability, *i.e*., substrate modified nanoparticles and enzyme-bound nanoparticle. These specifically elevate proteolytic activity [51]. From the viewpoint of pharmaceutical, environmental aspects, the multistep enzymatic cascades play an essential role during the immobilization of enzymes. The multistep coupled enzymatic activities have gained popularity in the industry as they bring economic significance. The pyruvate kinase and lactate dehydrogenase enzymes associate with quantum dots like prototypical nanoparticles. These nanoparticles treated with polyhistidine facilitate conjugation with catalysts. The coupled activity is related to the combination of stabilization and enzyme LDH. The kinetics and intermediate channeling between the quantum dots co-localized catalysts represent the degree of performance of this combination. It also indicates the elevated coupled enzymatic activity and improved application of biocatalysts because of the scaffold used [51].

Carbon nanotubes exhibit specific physicochemical, optical, and magnetic features and are useful materials for immobilizing enzymes. These nanomaterials are applicable in the fields of nanoelectronics, nanocomposites, and nanobiosensors. Single-walled carbon nanotube and multiwalled carbon nanotube offer an enlarged surface area for enzyme loading and exhibit either electrostatic and hydrophobic and hydrophilic interaction. The resultant product is biocompatible and mechanically resistant and is applicable in the biological and nonbiological systems. Multiwalled carbon nanotubes act a suitable surface for hydrophobic and hydrophilic binding. Modified carbon nanotubes concerning ionic properties are useful for the process of tyrosinase. The modified multiwalled carbon nanotubes with the active cationic features act as favorable matrices for the immobilization of enzymes [52]. ZnO Nafion*, *i.e.*, silica-per-fluorosulfonated monomer and ZnO Nafion* multiwalled carbon nanotubes, are applicable for the immobilization of enzymes [53]. [*Nafion is a sulfonated tetrafluoroethylene based fluoropolymer-copolymer]

Inorganic materials like alumina, silica, and zeolites, nanomaterials, and micro and mesoporous silica are the potential options as suitable matrices for immobilization of enzymes. These nanomatrices provide the most appropriate fabricated surface area for the immobilization of enzymes. These meet the requirements and help to attain the set targets. The variables concerning the surface structure, topology, dimensions of pores, and microchannels help to facilitate the most amicable substrate-ligand binding. Besides, the silica gel is mechanically stable, but it is chemically inert even to the solvents used during the process. The surface of the silica gel provides a favorable surface for the wide variety of enzymes [54 - 56]. Magnetic materials exhibit anisotropy and maintain a specific direction. This direction is proportional to the volume of the particle. Magnetic particles like cross-linked-iron-oxide, ultra-small-super-paramagne-ic-iron-oxide, and mono-crystalline-iron-oxide are the favored options for magnetic resonance imaging. Magnetic nanoparticles provide bio element stability, hazard-free isolation for the reaction mixture, and also enhance the sturdiness of the catalyst. Thus, these wonder magnetic nanoparticles are other suitable materials for immobilization of the enzymes [56 - 60].

Nanomaterials and the Process of Immobilization of Enzymes

In comparison to conventional matrices, the nano-based frames and techniques are advantageous in comparison to the traditional framework. These advantages include (1) elimination of the use of toxic and surfactants because nanoenzyme particles are conveniently fabricated or synthesized. (2) The resultant products of the process involving nanomaterials are homogeneous and have described thick core-shell. (3) The size of the particles controllable as per the need or set target. (4) With the advancement of nanotechnology, there are bright chances for co-immobilization of multiple enzymes to attain the cascades of enzyme reactions, at least *in vitro* [43]. Nanomaterials are advantageous as a matrix for the processes of the enzyme immobilization. These materials enhance the suitability of the conditions for the molecule, interacting particles, interface interactions during the immobilization of enzyme, and influence the impact of the Brownian movements. The structure of the enzyme molecule, and the properties, the surface of support play significant roles. These features vary from enzyme to enzyme and matrix to matrix. The immobilized enzymes show a higher degree of enzymatic activity, stability, reusability in comparison to the free catalysts, and it is available for industrial and research applications. All these efforts result in making nanomaterials useful and beneficial options to achieve the desired target.

Free lipase exhibits a much lower degree of activity in comparison to the immobilized enzyme when interacting with the matrix under consideration. Immobilized lipase enzyme shows a higher level of lipase activity when immobilized on sodium bentonite matrix. The cause of the higher lipase activity is the hydrophobic microenvironment that activates the concerned interaction at the interfacial site. Further, the surface of the organo-sodium bentonite provides a much larger surface area. The surface of this matrix is morphologically and structurally suitable for the process of immobilization of enzyme under consideration. The immobilized lipase enzyme on the organo-sodium bentonite matrix and the product obtained is resistant to the changes in the pH, temperature. It also exhibits a better degree of stability and recyclability [61].

Montmorillonite, as 3-aminopropyltriethoxysilane (modified silane), acts as an adsorbent during the immobilization of lipase enzyme extracted from *Candida rugosa*. The immobilized lipase on this matrix shows improved lipase activity, better tolerant for a broader range of pH, temperature, and stability. The immobilized lipase withstands the temperature range (35°C to 55°C) and retains 90% enzyme activity up to 120 min duration. The enzyme in the immobilized form is active for about 40 days. The immobilized lipase enzyme shows a higher level of lipase activity when immobilized on the sodium bentonite matrix

[61]. The lipase enzyme extracted from Thermomyces lanuginosus, get immobilized on different mesoporous organo-silica, having specific treatment with ethane and benzene; this frame has larger pores that help the immobilization of the enzyme. The immobilized lipase enzyme reaction involves hydrolysis and esterification. The surface of the mesoporous matrix exhibits a high specific surface area, the volume of the pore. These features of the support used to enhance the degree of adsorption of the enzyme. The monolayer adsorption gets elevated because of hydrophobic interactions. Immobilized enzymes effectively catalyze the interaction of esterification interactions. The immobilized lipase enzyme in the hydrophobic medium exhibits improved catalytic ability because of the elevated interfacial interaction between the lipase and hydrophobic surface of the matrix. Overall, the elevated interaction of the immobilized enzyme is due to the activation of the catalyst at the interfacial site [62]. The applications of immobilized enzymes are shown in Fig. (2).

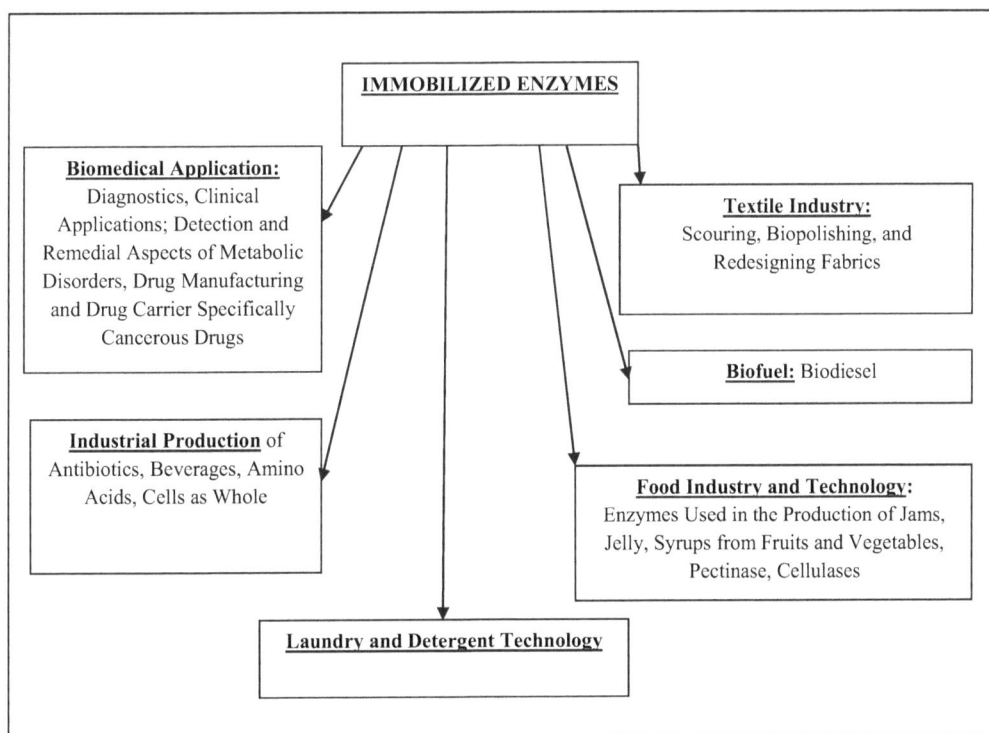

Fig. (2). The overall applications of immobilized enzymes.

APPLICATIONS OF IMMOBILIZED ENZYME

Immobilized enzymes have been in use in various industries since 1916. The charcoal, as a matrix, attained the immobilization of invertase enzyme, and it hydrolyzes sucrose [63, 64]. Since, then the catalysts find multiple applications in fields like cosmetics, paper industry, textile, food, laundry and detergent technologies, pharmaceuticals, *etc.* The extractions, purification, storage of enzymes are tedious, costly, and time-consuming, while the immobilized enzymes can reduce several such hurdles many other hurdles in industrial techniques [64].

Applications of Immobilized Enzymes in Biomedical Field

Immobilized enzymes are one of the most sought after agents used as tools for diagnostics and treating medical pathologies. These inborn deficiencies are curable by the use of concerning metabolic enzymes. Such catalysts are immobilized and encapsulated in red blood cells to carry these enzymes using the electroporation technique. The final products are readily administered in the experimental model. The administered catalysts interact with the related metabolites or biomaterials and rectify the condition, thus restore the normal functioning in the biosystem [64, 65]. Generally, immobilized enzymes are mostly used as biosensors and as bioreactors. The applications of the immobilized enzymes benefit the detection and remedies of clinical pathogenesis. The enzyme penicillinase is a standard requisition for the conversion of the penicillin-G or Penicillin-V into 6-amino penicillanic acid (6-APA), and in turn, converts into the ampicillin [66]. The bacitracin tyrosine, nikkomycin Coenzyme-A, Vit-B5, proinsulin interleukin-II, prostaglandins, and monoclonal antibodies are the biomedical products of the enzyme immobilization technology [67]. L-DOPA drug is useful in Parkinsonism, and it is a product of the immobilized enzyme β-tyrosinase [68]. The immobilized aldolase enzyme helps diagnose muscle disorders.

Application in Textile Technology and Industry

In textile technology, immobilized enzymes play essential roles during processing, texturizing, and other aspects. The transformation of fiber to fabrics and their processing, laundry detergents, and treatment of effluents, *etc.*, involves the immobilized enzymes. The pectinase enzyme is useful in the stages like desizing (removal of starch, by using amylase); bio-sourcing or dissolution of waxes, proteins, pectins, and removal of natural fats from the surface of cotton fibers, lipase and cellulase are used but mostly in solution form [69, 70]. Bleaching, clean up, effluent treatment, are carried out involving catalase enzyme; this treatment removes hydrogen peroxide [71]. Biofinishing deals with the processing of fibers to improve appearance and texture. Biostoning or stone washing processes are the techniques that make new threads appear as fashionably aged looks. Laundering and detergents deal with removing stains from the fibers or tissues [72, 73]. In addition to pectinases enzymes, other enzymes like lipases, proteases, cellulases, xylanases, and cutinases are useful in the scouring process [74].

Application of Immobilization in Food Technology

The immobilized enzymes find enormous applications in the food and dairy technology because these forms of enzymes are economical, stable, and show better enzymatic activity in comparison to the corresponding free types of catalysts. *Aspergillus oryzae* is a good source of immobilized beta-galactosidase, and this extraction involves cellulose acetate polymethylmethacrylate membrane. This enzyme is utilized to hydrolyze lactose in most of the processes of dairy technologies. The immobilized β-galactosidase enzyme has better catalytic activity (83%) in comparison to the free catalyst. It shows better tolerance for pH and temperature, better storage (4°C), and reusability (5th repeated use) as compared to the free catalyst [75, 76]. The immobilized glucose isomerase enzyme helps to obtain high fructose syrup with 42% fructose, 50% glucose, and 8% other sugars. Another enzyme amyloid-glucosidase is the enzyme that will be used to meet the ever-increasing need for high fructose syrup [77, 78].

Application of Immobilization of Enzyme in Biofuel Technology

Biofuels are the products of biochemical and chemical processes. Biochemical processes involve enzymes referred to as a biological catalyst, while chemical processes concern with the use of the pyrolytic process of the raw materials and convert them into fuel. Biomass modifies into biofuel using enzymes or chemical catalysts. The catalysts like lipases and phospholipases play an essential role in the production of biofuel. The lipases are a better option to convert raw material into free acid acids and triacylglycerol into fatty acid methyl ester. The phospholipase enzyme converts phospholipids into diacylglycerol, and this product acts as a substrate for the enzyme lipase. The treatment with methanol and catalysts to eliminate free fatty acids and phospholipids improves the quality of biodiesel. Enzymes use these materials as substrates. The immobilized enzymes are cost-effective and convenient. Hence, it is preferred to use immobilized enzymes in the production of biofuel. Immobilized lipase is involved in the production of biofuel. This enzyme is catalytically active, and stable specifically in a non-aqueous medium. The processes of esterification and trans-esterification take place successfully during the production of biofuel. Even the immobilized lipase enzyme also participates in the adsorption technique [78 - 80].

INTERACTIONS BETWEEN NANOMATERIALS AND ENZYMES

The nanomaterials have unique physicochemical features like size, surface properties, ease of modulation of nanomaterials, *etc.*, make these materials better performers in the enzyme technology, and biosensors *in vitro* and *in vivo* [81]. Some of the fundamental features of enzymes are modified to suit the functionality of enzymes. Enzymes are proteins, the major part of the protein acts as a scaffold to support and position active sites, and their size. The active sites or binding site orients in such a manner that binding between substrate and enzyme is accurate and with highest feasibly. There is a change in the energy of chemical activation at the catalytic site. Generally, this activation energy gets reduced during enzyme-substrate binding or enzyme-complex formation. There is a distinct exhibition of specificity between enzyme and substrate binding. This specificity concerns with the selectivity in complementary shape, region selectivity, and stereospecificity. Further, enzymes exhibit substrate specificity based on the expression of the genome, proofreading regulated by DNA polymerase, *etc.* The binding interactions between catalysts and substrates may involve the following parameters:

i. The specific shape of the active site and the functional group help the reacting molecules during substrate binding *via* noncovalent interaction.
ii. Those amino acids that attract other amino acids at the active site having different charge exhibit electrostatic interactions.
iii. Typical hydrogen bonding between amide and carboxyl groups participates in the binding process.
iv. (Wdf) van der Waal forces present between the concerning amino acids brings about their binding interaction.
v. Those amino acids that repel water exhibit hydrophobic interactions.

The interaction between enzymes and nanomaterials also exhibits some of the specific modes. These methods involve enzymatic modifications, degradation, and enzymatic mimics of nanomaterials. Nanomaterials facilitate the process of immobilization of enzymes and help in developing biosensors [45].

The cationic mixed monolayer protects the nano-clusters of gold (AuMMPCs), and these nanoparticles do not inhibit the activity of chymotrypsin while anionic AuMMPCs can do so, because of the electrostatic complementarity between the carboxyl end group and the location of the whole of the cationic residues around the periphery of the active site. This nanomaterial has little and no ability to inhibit the activity of elastase β-galactosidase or binding of cellular retinoic acid with protein.

Nanoparticles with their surface modified to influence the enzyme function. On comparing the influence nanoparticle and traditional materials, the surface area in the case of nanoparticles is more significant in comparison to the standard conventional materials. Their surface can be reorganized using organic molecules, and this process concerns with covalent and noncovalent bondings, and help to attain the set target. The functionalized enzymes add specific abilities to their surface that allows achieving a particular interaction and prevent undesired interactions like non-specific binding and recognizing specific molecules [81, 82]. The pristine silicon nanowires (SiNW-SiO$_2$), silicon nanowires functionalized with a carboxyl group (SiNW-COOH), and highly reactive forms of (SiNW-H) decline the activities of endonucleases and Taq DNA polymerase enzymes. These three forms exhibit different degrees of inhibitory activity, SiNW-SiO$_2$ the least, and the SiNW-COOH medium and SiNW-H the highest. The size of the nanoparticles influences the activity of adsorption of enzymes. Silica nanoparticles (4 nm 20 nm and 100 nm) show size-dependent interaction. Larger particles cause more protein folding, and medium-sized particles show more adsorption [82]. The single-walled carbon nanotubes also affect enzyme activity. The single-walled carbon nanotubes having alpha-chymotrypsin and soya bean peroxide enzymes influence 1% and 30% enzyme activities. Possibly the secondary structures are either get disturbed or undergo conformational changes. There is hydrophobic interaction between single-walled carbon nanotubes and the enzymes under consideration. Soya bean peroxide had its hydrophobic pocket bound to single-walled carbon nanotube rendering its active site exposed to the solvent to link with the substrate. The denatured chymotrypsin declines its enzymatic efficacy. The surface of the functionalized single-walled carbon nanotube with hydrophilic polymers such as oligomeric PEG moieties elevates the enzyme activity [82]. The iron oxide nanoparticles (30 nm) induce hepatotrophic and damage liver tissue, and increase the activities of alkaline phosphatase, aspartate aminotransferase, and alanine aminotransferases [83].

Silicon, gold, cadmium–selenium nanoparticles reduce or inhibit the activity of purified lactate dehydrogenase (LDH). This inhibitory activity is dependent on the core particle and the composition of the surface functional group. The crude tissue homogenate did not exhibit the inhibitory activity of LDH at the isoelectric point of the protein in the presence of albumin. [Isoelectric point is a pH at which particular molecules do not carry a net electric charge. The pH of the ambient environment influences the total electrostatic charge on the interacting molecules, and these molecules become more positively or negatively charged due to the gain or loss of protons (H$^+$), respectively]. This inhibition of activity reflects on the nanoparticle bonding concerning with abundant proteins *via* charge interaction and conformational changes due to these nanomaterials [83]. Nanoparticles like gold, manganese oxide, and zinc oxide exhibit elevation in the activity of enzyme luciferase. This activity depends on the protein denaturing the protein. The denaturing agents involved are heat, sodium dodecyl sulfate, urea, ethanol, hydrogen peroxide, and pH. These fluctuations in the activity of the luciferase enzyme depend on the interacting nanomaterials and the denaturing agents. The zinc oxide nanoparticles elevate the luciferase activity and the bioluminescence [84].

The size of the nanomaterials influences the activities of the enzymes. The different sizes of silica nanoparticles (4, 20, and 100 nm) affect the structure and enzymatic activity of the chymotrypsin. Based on the studies related to protein structure and circular dichroism showed that the size of these nanomaterials. Further, these influence the specific adsorption pattern and cause a loss in the contents of α-helicity. The degree of loss of α-helicity is directly related to adsorption on to the size of nanomaterials. Larger sized nanoparticles cause significant damage to α-helicity. Even lysozymes exhibit a similar pattern of activity. The activity of adsorbed lysozyme on to silica nanoparticles may be weaker in comparison to the free enzyme. The loss of enzyme activity is proportional to the degree of decline in the contents of α-helix. This behavior reflects that the size of the nanoparticles probably adds to the surface curvature, which affects the structure and efficacy of the enzyme adsorbed [81]. During drug delivery, there is a need to retain the fundamental structural and functional aspects of the catalyst in use. Three functionally different enzymes, and soluble in an aqueous medium, namely horseradish peroxidase, subtilisin-Carlsberg (non-specific protease), and chicken egg white lysozyme interact with single-walled carbon nanotubes using conjugated acids. The kinetic studies indicate that these enzymes retain their high fraction of the fundamental structural and functional aspects as long as these are in the aqueous medium.

Carbon nanomaterials have a role during their interaction with enzymes. Among these carbon nanotubes exhibit an active functional role as there are chances of either promoting or degrading the enzyme activity depending on the conformation of carbon nanotubes. Pristine carbon nanotubes are the ideal option for stabilizing oxidizing conditions in the environment. The carbon nanotubes get incorporated to degrade the waste materials after reaching landfills [85]. The nanofibers, nanoporous media, nanocomposites, and graphene nanomaterials exhibit a substantial-high ratio between size and surface area. This feature enhances the biocatalytic efficacy in the field of industrial applications. The loading of the enzyme and the reaction kinetics get many folds elevated. These nanomaterials are essential for the enzyme biotechnology and improve the production of biofuel, immobilization, and stabilization of enzymes [86]. The conjugates of single-walled carbon nanotubes (SWCNT) and the enzymes are more stable in guanine-hydrochloride (GdHCl) at high temperatures in comparison to their counterparts in an aqueous medium. This behavior indicates the higher stability of their conjugates [87, 88]. The single-walled carbon nanotubes having alpha-chymotrypsin maintain 1% of its native activity while soya bean peroxidase adsorbed single-walled carbon nanotubes retained around 30% of its native enzyme activity. There are substantial secondary structural changes during their binding with single-walled carbon nanotubes as per the FTIR spectroscopic observations [87, 88]. The modified and functionalized surfaces of nanomaterials prevent the non-specific binding and help in recognizing biomicromolecules. Silicon nanowires unmodified (SiNW-SiO$_2$), silicon nanowires modified, functionalized with a carboxyl group (SiNH-COOH), and silicon nanowires modified with highly reactive hydrogen (SiNW-H) interact with each other. These interactions lead to an understanding of the effects on restricted endonucleases and Taq DNA polymerase. The PCR results reflected the inhibition of enzyme activity and concern the effectiveness of the functional groups present on the silicon nanowires (SiNW), and SiNW-H shows the maximum activity, SiNW-COOH shows medium action while SiNW-SiO$_2$ shows the minimum enzymatic activity [89].

Fullerene also referred to as Bucky Ball or Buckminsterfullerene. This nanomaterial is a three-dimensional crystalline allotropic form consisting of carbon atoms only. Its discovery delineates its dimensions at the atomic scale, chemistry, and its derivatives show mechanical, electronic tensility, the conductance of electrons, and heat-related features. These specific features of this allotropic form of carbon make them suitable options for the applications in the electronics, structural materials, biological, biomedical, biomolecular, pharmaceutical fields. This allotropic form of carbon and its derivatives are active with the biomolecules. The fullerene and its derivatives interact with various enzymes, and these interactions are inhibitory and also potential participants in the medical and biomolecular chemistry [90]. Graphene is another form of carbon nanomaterial. It is a very reactive and suitable option for its industrial applications [91].

On the bases of model complex generated using program DOCK-3, show that the derivatives of C60 bind to the active site securely. This binding separates 298A° of primarily non-polar surface and results in the complex formation between ligand and protein. The fullerene occupies the dynamic place and inhibits the interaction with HIV-protease, and there is an interactive competition between HIV-protein and water-soluble C60 derivative–bis (methylamine succinate) during this interaction [92].

The functionalized fullerene and cationic bis-N, N-dimethyl fullero-pyrrolidinium salt, act as non-competitive inhibitors during the hydrolysis of acetylcholine. The molecular modeling technique supports the interaction between the cage of fullerene and amino acids encircling the enzyme. The probable molecular interaction involves ammonium groups and side-chain essential amino acid residue of either principle or external binding sites [93]. The fullerene (C60) influences the restrictive activity of the enzyme, and the derogative impacts [94]. Fullerene, in its pristine form, adsorbs polymerase and inhibits the replication process. There is a competitive interaction for binding [95]. The fullerene derivatives having amino acids influence the transcriptase activity of HIV adversely. Among amino acids, type fullerene derivatives 11 and 12 are stronger inhibitors for protease activity. The position of carboxyl groups at the pyrrolidine ring on fullerene (C60) plays a functional role during the inhibitory activity [96]. Working with DNA or RNA during its synthesis, the enzyme primase is active, and it declines in the presence of C30, C40, and C50.

The fullerene and members of its family interact with various enzymes related to DNA replication. This interaction is in proportion to the molecular weight and the size of the interacting fullerene. Fullerenes get docked with DNA. The DNA anchored with fullerene resists the binding of primase enzyme while its interactions with catalysts like helicases, ligase, topoisomerase, but DNA pol-δ; DNA pol-ϵ and clumping of DNA remain unaffected. The degree of binding of RPA 14 (robotic process automation program) declined non-uniformly, but the interaction elevates as the molecular weight of the fullerene increases [97]. Fullerene (100 µg/L; 99.9% pure) declines the glutathione reductase activity. The intensity of hydrogen peroxide lipid peroxide elevates but acid and alkaline phosphatase activities fall in the presence of fullerene [98].

Fullerene and its derivatives play an influential role in soil ecology. Soil treated with granular fullerene (C60) (1000µg .g−1 soil), nano-C60, tetrahydrofuran (THF-C) (both 1µg g−1) for 30 and 180 days, influence the activities of enzymes like β-glycosidase, urease, soil acid phosphatase, but the effectiveness of dehydrogenase enzyme increases [90, 99]. The polyhydroxylated derivatives [C60 (OH) 24] of fullerene have pro-antioxidant tendencies within biosystems. The activities of enzymes like heme-oxygenase, quinine oxidoreductases-1, and γ-glutamate-cysteine ligase affected among human type-II alveolar epithelial cell line. Fullerene can regulate hydrogen peroxide activity [90, 100].

Graphene and its derivatives are quite active chemically and physically. These nanomaterials are suitable options of applications in the fields of biotechnology, nanotechnology, biomedicine, immobilization of enzymes and drug delivery, *etc*. Graphene oxide is a water-soluble form and has the efficacy to inhibit the enzymatic activity of α-chymotrypsin. This derivative of graphene interacts with the protein actively. The functionalized graphene oxide interacts with well-characterized serine proteases that include trypsin, chymotrypsin, and protein kinase-K concerning varied amine polyethylene glycol located at their terminal region. The activity and the thermal stability of trypsin enhance and thermal stability up to 80°C, but the very slight impact on the action and the functionality exists in the case of chymotrypsin and protein- kinase-K. PEGylated graphene oxide (GO) increases the substrate-dependent activity and its impact is only on the phosphorylated substrates [101]. The molecular dynamics simulation studies reveal that α-chymotrypsin adsorbed on the surface of graphene oxide. The probable mechanism involves the role of α-helix of chymotrypsin in anchoring the interaction between the two. The cationic and hydrophobic residues located on α-chymotrypsin are the bases of this interaction leading to change in the active site of this enzyme along with its inhibition. There is no alteration in the secondary structure of the α-chymotrypsin catalyst. Such studies provide information that helps to design and understand the functionalities of artificial inhibitors. This behavior directs the formulation of a specific inhibitor for the set target within a particular field [102]. The physical and chemical features of water-soluble graphene oxide make it suitable for its application in cancer therapeutics, drug delivery, immobilization of enzymes, and industries. The graphene oxide can prevent structural and conformational changes at the binding sites in flavo adenine dinucleotide (FAD), and thus differentiate the functionalities between flavo and non-flavoenzymes. There is a loss of enzymatic activity in the case of the non-flavoenzyme depending on the concentration; under the similar strength of graphene oxide l-lactate dehydrogenase (LDH) show around 75% of the action in the case of flavoenzyme during glucose oxidase show (GOX) around 45% of activity. The GO results in non-competitive interactions for glucose oxygenase enzymes, and the interplay is competitive in the case of LDH [103].

Dendrimers are highly branched and have a distinct shape and are three-dimensional nanoparticles. These have specific size and structure. The synthesis and fabrication of these nanomaterials are challenging, but these products have multifunctional polymers. Each dendrimer consists of a central functional core and the branches having functional groups at the periphery [104, 105]. The dendrimers influence activities of the aspartate transaminase, alkaline phosphate, and L-lactate dehydrogenase and interact with cationic phosphorous dendrimers G-3 and G-4. During these interactions, mostly, there is a change in the structures of these enzymes. Dendrimers with 3 and 4 generations bind the molecules of these enzymes. As a result, conformational changes and changes in the size of the complex formed also takes place. The surface charge on the phosphorous dendrimers is an important parameter, and it depends on the number and type of groups located at the periphery of the dendrimer under consideration [106]. The polyamidoamine dendrimer G-5 interacts with cytoplasmic protein-L-lactate dehydrogenase and luciferase and affect the integrity of the cell membrane.

Dendrimers have amine groups at their terminals when they come in contact with the cell membrane influences the cellular permeability, causing cell leakage. This impact is not permanent. The cell leakage is concerned with the interaction between dendrimers and the cytosolic enzymes, *i.e.*, L-lactase dehydrogenase and luciferase, and this feature restricts their applications. On the removal of dendrimers, these enzymes regain their functionality resulting in the prevention of cell leakage. The dendrimers interact with L-lactase dehydrogenase and luciferase; during this interaction, holes in the cell membrane develops that permit the intracellular uptake of dendrimers [107]. Synthesized and assembled layer-b--layer dendrimers are an excellent option for immobilizing enzymes. The resultant products are handy materials for biosensors. Such biosensors are highly functional [108]. During the interaction between carbosilane-phosphorous dendrimers with G-2 and 2 types of carbosilane dendrimers, both of these and the three proteins (enzymes) namely alkaline phosphatase (from E. coli), aspartate transaminase (from human) and L-lactase dehydrogenase (from muscles of rabbit), the positively charged hybrid and carbosilane dendrimers interact with negatively charged enzymes (protein) and complexes. The three-dimensional conformations and the degree of flexibility of dendrimers play a significant role in the formation of respective complexes [109].

Metal and metal oxide nanoparticles have their respective impacts while interacting with enzymes or otherwise, (*i.e.*, some of the metal nanoparticles may release the metal ions). Generally, this interaction involves parameters like hydrodynamic radius, mass density, Wigner Radius, and covalent index. The hydrodynamic radius and mass density of the nanomaterials influence their mobility in a medium [110]. The Wigner radius reflects on the mean volume of nanomaterials available for the interaction. Girifalco suggested that the Wigner radius is the radius of a sphere having a size equal to the mean volume per atom of solid based on statistical mechanics. It represents the metal having more mass of an electron, or this radius equals the sphere that has the same size to the size of the free electron [111]. This aspect reflects on the space available for interaction. The covalent index indicates the amount of covalence and the ionicity of the two interacting atoms. It is reflected in the level or degree of relationship between a covalent bond and an ionic bond [112]. The degree of dissolution and characters related to size play major in the activation of the enzyme. Metal oxides like ZnO, CuO, Cr_2O_3, and nickel oxide (NiO) show a relatively higher degree of solubility. Hence, they exhibit a vigorous inhibitory impact on enzyme activity [112]. Metal oxides follow three prime parameters while acting as adsorbent and catalysts. (i) First parameter concerns with their surface atoms, the surface atoms have coordination with the ambient environment. (ii) These have redox properties and (iii) their surface layer exhibits an oxidation state. These metal oxides have only (s) or (p) electron in their valence orbitals that incline effectively towards an acid or base catalysis, whereas those metal oxides that have (f) outer electrons show a tendency for more varied uses [113 - 115].

Intracellular iron oxide nanoparticles participate in pH-dependent peroxide and catalase activities. Iron oxide and γ-iron oxide nanoparticles convert H_2O_2 depending on the electron spin resonance. Hydrogen peroxide present in the cytosol changes into water and oxygen at pH=7.4; this interaction takes place in cytosol involving iron oxide nanoparticles and catalase enzyme. The hydrogen peroxide splits into OH-ions under the influence of peroxidase enzyme using iron oxide nanoparticles bound lysosome at pH=4.8 within the cell. These nanoparticles can convert nascent hydrogen peroxide into water and oxygen, causing a decline in toxicity due to the peroxide enzyme. This aspect is derogative and may harm cells structurally and functionally [116].

THE DEROGATIVE IMPACTS OF NANOMATERIALS ON ENZYMES AND THEIR ACTIVITY

Nanomaterials are beneficial significantly in various fields, but they also show toxic effects. The factors like (i) chemical composition of the reactants, (ii) electrostatic interactions due to the surface charge on the nanomaterials, (iii) hydrophobic nature of nanomaterials and lipophilic groups present on the nanomaterials, (iv) restricted enzyme activity that may be causing either competitive or non-competitive influence on the interaction between the nanomaterials and enzymes [117]. Metal nanoparticles like silver, gold, and copper are the potential inhibitors of the enzymes concerned with an antioxidant defense mechanism. These metal nanoparticles decline the activities of the superoxide dismutase, catalase, and peroxidase enzymes, causing redox imbalance. The prolonged interaction leads to the failure of the antioxidant defense mechanism [118]. Silicon, gold, and cadmium/selenium (CdSe) nanoparticles inhibit the activity of lactate dehydrogenase. This interaction and its impact depend on the nature of the core and the composition of the surface functional groups present [83].

Titanium dioxide nanoparticles cause stress due to ROS in the tissues like gills, brain, and the liver of fish. The enzymes like superoxide dismutase, catalase, peroxidase, and lipid peroxidase cause a decline of oxidative stress due to ROS in the tissues of the fish. The concentration, size, surface area of the TiO_2 [(50, 100, 150 ppm); (50 nm); (30 ± 10 m^2/g)] affect the activities of these enzymes. During the initial stages of exposure, the enzymatic activities increase, but as the oxidative stress increases due to the rise in the concentration of TiO_2 and exposure, these enzymes activities decline. In the case of lipid peroxidase enzyme, the changes are proportional to the formation of the amount of malondialdehyde (MAD) because of oxidative stress in the organs of fish [119]. Silver nanoparticles inhibit enzymatic activities of phosphomonoesterase, arylsulphatase, β-glycosidase, and leucine-aminopeptidase enzymes. The comparative study of impacts of silver chemicals and silver nanoparticles suggests that silver nanoparticles are more derogative to these enzymes in comparison to silver compounds [120].

IS THERE ANY CORRELATION BETWEEN NANOMATERIALS, ENZYMATIC ACTIVITIES, AND BIOMIMETICS?

Biological enzymes and chemical enzymes exhibit their catalytic activities under specific conditions like pH, temperature, substrate concentration, and other physiological parameters. Whenever there is a change or alteration in these functional parameters, the enzymatic activity under consideration gets affected or becomes inactive functionally. This situation may not favor its industrial applications and enzyme technology. The elevated technical and behavioral flexibility of the enzymes can tackle this situation. Scientists at ATEN PROJECT (Europe funded-CORDIS) observed that known catalysts like platinum (Pt) and gold (Au) have common characteristics inter-relation with catalase, superoxide dismutase, and ferroxidase enzymes. These enzymes are useful in cosmetology and redox signaling. These enzymes are also intensely concerned with causing oxidative and inflammatory impacts. These researchers have found inhibitors related to catalase enzyme action. These inhibitors do not affect other interactions co-occurring. Their observations include the characterization of catalase activity of the inorganic nanoparticles precisely when pH changes; there is an enhanced efficiency in the enzyme action. After the completion of the said interaction, there is a recovery of nanoparticles acting as an enzyme (NP-ACTING-AS-ENZYME). This recovery of catalysts is not possible in general cases. Similar behavior occurs when the temperature fluctuates under experimental conditions. This behavior of NP-ACTING-AS-ENZYME is not in congruence with the normal enzyme behavior. In the case of NP-ACTING-AS-ENZYME, the concerned nanoparticle gets engaged within protein-shell under a controlled process of synthesis, resulting in a product, *i.e.*, the protein encapsulated inorganic nanoparticles. It shows better enzymatic activity and greater active surface/area in comparison to traditional enzymes. The protein-shell encloses the nanoparticles and also brings about synergistic action. These products can be good options to deliver drugs, genes, *etc*. These products are structurally and functionally favorable options for applications in biomedical, pharmaceutical, biomolecular fields. This technique is of significance from clinical, technical biotechnology, and therapeutic point of view [121].

The combination of structural and functional aspects of biological phenomena with material science needs a precise technique to design, engineer and regulates interfaces where biological issues intersect material sciences. These functions are essential and implementable in bionanotechnology that helps to visualize the materials at the molecular level and the nanoscale by enhancing the structural, functional information, and the correlation with biomolecules, cells, and tissues. Following the principles of molecular biology, protein sciences, it is possible to develop materials that function under controlled conditions that can exhibit self-assembly. The material-specific-peptide (related to biomolecular aspects) plays a significant role in developing inspired materials for biomolecular engineering [122]. Carboxyl-modified-graphene oxide exhibits peroxidase-like activity and catalyzes the reaction of peroxidase substrate 3.3.5.5-tetramethylbenzidine in the presence of H_2O_2 and forms blue-colored product [123]. Polyoxometalate having Well-Dowson conformation can inhibit the A-β- Heam peroxidase-like activity. This interaction is strong. The mechanism is not understood when these products move across the blood-brain barrier, but it reacts in the brain and restricts the action of A-β-Hem peroxidase [124]. A nanoenzyme is designed peroxidase mimicking nanoparticle grown on porous and thermally stable organic matrix, (AuNPs@MIL 101-nanozymes). These nanozymes behave like peroxidase, mimic and change the leuco-malachite green into malachite green involving H_2O_2 and gave surface-enhanced the Raman scattering signals. When this enzyme interacts with glucose oxidase and lactate oxidase *in vitro*, it detects glucose and lactate using surface-enhanced Raman scattering signals in tissues like brain, tumors, *etc* [125]. The phenomenon of immobilization of enzymes is applicable in the biomedical field. Such applications involve parameters like specific binding between enzymes and substrate, the molecular weight, and hydrophobicity. Such interactions are optimized as per the requirements, [126 - 130].

CONCLUSION

The biocatalysts and conventional enzymes exhibit chemoselectivity. This feature is related to the identification and selection of a particular functional group, among others. These are dedicated to regulate and accomplish chemical, biochemical, and biomolecular interactions. The mechanism of the enzyme appears to be quite complicated. The necessary steps incriminated are binding of the catalyst with the specific substrate. The complex formed between the catalyst and substrate is just before the initiation of interaction. The practice of the immobilization of enzymes helps to retain the functions of the immobilized enzyme. This process also increases the application of these enzymes in industries like food, detergent, textile, bioremediation, and pharmaceuticals. The product formed is homogenous as core-shell nanoparticles, and the size of these particles adjusted according to the requirement. Nanomaterials are of benefit in the various fields but can cause toxic impacts. The biomimetics technique is of significance from clinical, industrial biotechnology, and a therapeutic point of view. Under this technique, a protein-shell encloses the nanoparticles and also brings about synergistic action. One can deliver drugs, genes, using these products. These products are structurally and functionally favorable options for applications in biomedical, pharmaceutical, biomolecular fields. The investigation related to the interaction between nanomaterials, immobilization of enzymes, and kinetics of such interactions opens a vast area to delve and ponder for future applications of nanomaterials concerning enzyme technology.

REFERENCES

[1]   Rahman SA, Cuesta SM, Furnham N, Holliday GL, Thornton JM. EC-BLAST: a tool to automatically search and compare enzyme reactions. Nat Methods 2014; 11(2): 171-4.
[http://dx.doi.org/10.1038/nmeth.2803] [PMID: 24412978]

[2]   Ballard CE. pH-controlled oxidation of aromatic ketone: structure elucidation of the products of two green chemical reactions. J Chem Educ 2010; 87(2): 190-4.
[http://dx.doi.org/10.1021/ed800054s]

[3]   Eliel EC, Wilsen SH. The stereochemistry of organic compounds. New York, 1203, : Wiley Interscience 1994.ISBN: 0-471-01670-5QD481.E52115.

[4]   Pelczar MJ. Microbiology: Application-based approach. Tata McGraw Hill Education 2017.

[5]   Voet D, Voet TJ, Pratt CW. Fundamentals of Biochemistry. Hoboken, NJ, USA : John Wiley and Sons 2002.

[6]   Sherwood L. Human physiology- From cells to systems. California, USA,: Thomson Learning Inc 2004. ISBN 053-39501-5.

[7]   Becker WM, Kleinsmith LJ, Hardin J. The world of the cell. USA: Pearson Education 2004. ISBN 81-297-0415-3.

[8]   Cox MM, Nelson DL. Lehninger Principles of Biochemistry. 5th Ed., New York: W H Freeman and Co 2011. ISBN 978-0-230-22699-9.

[9]   Juturu V, Wu JC. Microbial xylanases: engineering, production and industrial applications. Biotechnol Adv 2012; 30(6): 1219-27.
[http://dx.doi.org/10.1016/j.biotechadv.2011.11.006] [PMID: 22138412]

[10]   Li C, Jiang S, Zhao X, Liang H. Co-immobilization of enzymes and magnetic nanoparticles by metal-nucleotide hydrogel nanofibers for improving stability and recycling. Molecules 2017; 22(1): 179. [www.mdpi.com/journal/molecules].
[http://dx.doi.org/10.3390/molecules22010179] [PMID: 28125003]

[11]   Smith S. The animal fatty acid synthase: one gene, one polypeptide, seven enzymes. FASEB J 1994; 8(15): 1248-59.
[http://dx.doi.org/10.1096/fasebj.8.15.8001737] [PMID: 8001737]

[12]   Lin Y, Ren J, Qu X. Catalytically active nanomaterials: A promising candidates for the artificial enzymes Account for Chemical Research 2014; 47(47): 1097-105.
[http://dx.doi.org/https://doi:10.1021ar400250z]

[13]   Anfinsen C B. Principles that govern the folding of the protein chain, Science. 1973; 181: pp. (4096) 223-30.
[http://dx.doi.org/https://doi:10.1126/science.181.4096.223]

[14]   Dunaway-Mariano D. Enzyme function discovery. Structure 2008; 16(11): 1599-600.
[http://dx.doi.org/10.1016/j.str.2008.10.001] [PMID: 19000810]

[15]   Chen LH, Kenyon GL, Curtin F, *et al.* 4-Oxalocrotonate tautomerase, an enzyme composed of 62 amino acid residues per monomer. J Biol Chem 1992; 267(25): 17716-21.
[PMID: 1339435]

[16]   Smith S, Witkowski A, Joshi AK. Structural and functional organization of the animal fatty acid synthase. Prog Lipid Res 2003; 42(4): 289-317.
[http://dx.doi.org/10.1016/S0163-7827(02)00067-X] [PMID: 12689621]

[17]   Krauss G. The regulation of enzyme activity, Biochemistry of signal transduction and regulation. 3rd ed. Weinheim: Wiley-VCH 2003; pp. 89-114.
[http://dx.doi.org/10.1002/3527601864]

[18]   Suzuki H. Active site structure: how enzymes work.From structure to function. Boca Raton, FL: CRS Press 2015; pp. 117-40.

[19]   Murphy JM, Farhan H, Eyers PA. Bio-Zombie: the rise of pseudoenzymes in biology. Biochem Soc Trans 2017; 45(2): 537-44.
[http://dx.doi.org/10.1042/BST20160400] [PMID: 28408493]

[20]   Eijsink VGH, Gåseidnes S, Borchert TV, van den Burg B. Directed evolution of enzyme stability. Biomol Eng 2005; 22(1-3): 21-30. https://doi.org
[http://dx.doi.org/10.1016/j.bioeng.2004.12.003] [PMID: 15857780]

[21]   Compendium of chemical terminology. 2nd Ed., the Gold Book, substrate. 1997.
[http://dx.doi.org/10.3151/goldbook.S06082]

[22]   Ogu CC, Maxa JL. Drug interactions due to cytochrome P450. Proc Bayl Univ Med Cent 2000; 13(4): 421-3.
[http://dx.doi.org/10.1080/08998280.2000.11927719] [PMID: 16389357]

[23]   Saghatelian A, Trauger SA, Want EJ, Hawkins EG, Siuzdak G, Cravatt BF. Assignment of endogenous substrates to enzymes by global metabolite profiling. Biochemistry 2004; 43(45): 14332-9. [ACS Publication].
[http://dx.doi.org/10.1021/bi0480335] [PMID: 15533037]

[24]   U S. Food and Drug Administration. table of the substrate, Inhibitors, and inducers 2016.

[25]   Huang X, Holden HM, Raushel FM. Channeling of substrates and intermediates in enzyme-catalyzed reactions. Annu Rev Biochem 2001; 70: 149-80.
[http://dx.doi.org/10.1146/annurev.biochem.70.1.149] [PMID: 11395405]

[26]   Miles EW, Rhee S, Davies DR. The molecular basis of substrate channeling. J Biol Chem 1999; 274(18): 12193-6.
[http://dx.doi.org/10.1074/jbc.274.18.12193] [PMID: 10212181]

[27]   Perham RN. Swinging arms and swinging domains in multifunctional enzymes: catalytic machines for multistep reactions. Annu Rev Biochem 2000; 69: 961-1004.
[http://dx.doi.org/10.1146/annurev.biochem.69.1.961] [PMID: 10966480]

[28]   Pravda L, Berka K, Svobodová Vařeková R, *et al.* Anatomy of enzyme channels. BMC Bioinformatics 2014; 15: 379.
[http://dx.doi.org/10.1186/s12859-014-0379-x] [PMID: 25403510]

[29]   Milani M, Pesce A, Bolognesi M, Bocedi A, Asconzi P. Substrate channeling: molecular bases. Biochemistry and Molecular Biology 2006; 31(4): 228-33.
[http://dx.doi.org/10.1002/bmb.2003.494031040239]

[30]   Nisha S, Arun KS, Gobi N. A review on methods, applications, and properties of immobilized enzymes. Chemical Science Review, and Letters 2015; 1(3): 148-55.

[31]   Brady D, Jordaan J. Advances in enzyme immobilisation. Biotechnol Lett 2009; 31(11): 1639-50.
[http://dx.doi.org/10.1007/s10529-009-0076-4] [PMID: 19590826]

[32]   Razi A, Maryam S. Enzyme immobilization: An overview on nanomaterials as immobilization matrix. Biochem Anal Biochem 2015; 4: 178.
[http://dx.doi.org/10.4172.2161-1009.1000178]

[33]   Riaz A, Qadir SAU, Anwar A, Samina T. Immobilization of a thermostable α-amylase on calcium alginate beads from Bacillus subtilis. Aust J Basic Appl Sci 2009; 3(3): 2883-7.

[34]   Ghous T. Analytical application of immobilized enzyme. J Chem Soc Pak 2001; 23(4): 228-34.

[35]   Wong LS, Thirlway J, Micklefield J. Direct site-selective covalent protein immobilization catalyzed by a phosphopantetheinyl transferase Journal of American Chemical Society 2008; 130(37): 12456-12464.
[http://dx.doi.org/https://doi:10.21/ja8030278]

[36]   Yücel Y. Biodiesel production from pomace oil by using lipase immobilized onto olive pomace. Bioresour Technol 2011; 102(4): 3977-80.
[http://dx.doi.org/10.1016/j.biortech.2010.12.001] [PMID: 21190844]

[37]   Cipolatti EP, Josse M, Silva MJA, *et al.* Current status and trends in enzymatic nano immobilization, Journal of Molecular Catalysis B. Enzymatics 2014; 99: 56-67.
[http://dx.doi.org/10.1016/j.molcatb.2013.10.019]

[38]   Altinkayanak C, Tavlasoglu S. Yzdemir, Ocsoy I, A new generation approach in enzyme immobilization: organic-inorganic hybrid nanoflowers with enhanced catalytic activity and stability. Enzyme Microb Technol 2016; (93-94): 105-12.
[http://dx.doi.org/10.1016/j.enzmictec.2016.06.011]

[39]   Carlos Tassa JLD, Lewis TA, Ralph W, *et al.* Binding affinity and kinetic analysis of targeted small molecule-modified nanoparticles, Bioconjugates chemistry 2010; 21: p. (1) 14019.
[http://dx.doi.org/https://doi:10.1021/bc900438a]

[40]   Ding S, Cargill AA, Medintz IL, Claussen JC. Increasing the activity of immobilized enzymes with nanoparticles conjugation Current Opinion in Biotechnology 2015; 34: 242-50. www.sciencedirect.com
[http://dx.doi.org/10.11016/j.copbio.2015.04.005]

[41]   Campbell AS, Dong C, Meng F, *et al.* Enzyme catalytic efficiency: a function of bio-nano interface reactions. ACS Appl Mater Interfaces 2014; 6(8): 5393-403.
[http://dx.doi.org/10.1021/am500773g] [PMID: 24666280]

[42]   Gupta MN, Kaloti M, Kapoor M, Solanki K. Nanomaterials as matrices for enzyme Immobilization, Artificial Cells, Blood Substitutes, and Biotechnology. Ind Health Care 2010; 39: 98-108.
[http://dx.doi.org/10.3100/10731197.2010.5162]

[43]   Ansari SA, Husain Q. Potential applications of enzymes immobilized on/in nano materials: A review. Biotechnol Adv 2012; 30(3): 512-23.
[http://dx.doi.org/10.1016/j.biotechadv.2011.09.005] [PMID: 21963605]

[44]   Ahmad R, Sardar M. Enzyme immobilization: an overview of nanoparticles as immobilization matrix. Biochem Anal Biochem 2015; 4: 178.
[http://dx.doi.org/10.4172/2161-1009.1000178]

[45]   Chen M, Zeng G, Xu P, Lai C, Tang L. How do enzymes meet? Trends Biochem Sci 2017; 42(11): 914-30.
[http://dx.doi.org/10.1016/j.tibs.2017.08.008] [PMID: 28917970]

[46]   Lee C-K, Duong A-N-A. Enzyme immobilization on nanoparticles: Recent applications, in emerging areas in Bioengineering. Wiley-VCH Verlag GmbH and Co KGaA 2018; p. 20180109.

[47]   Wang P. Nanoscale biocatalyst systems. Curr Opin Biotechnol 2006; 17(6): 574-9. https://foi.org/10.1016/j.copbio.2006.10.009
[http://dx.doi.org/10.1016/j.copbio.2006.10.009] [PMID: 17084611]

[48]   Blanco-Canosal JB, Wu M, Gumumu UK, *et al.* Recent progress in the bioconjugation of quantum dots. Coord Chem Rev 2014; 263-264: 101-37.
[http://dx.doi.org/10.1016/j.ccr.2013.08.030]

[49]   Jia H, Zhu G, Wang P. Catalytic behaviors of enzymes attached to nanoparticles: the effect of particle mobility. Biotechnol Bioeng 2003; 84(4): 406-14.
[http://dx.doi.org/10.1002/bit.10781] [PMID: 14574697]

[50]   Algar WR, Malonoski A, Deschamps JR, *et al.* Proteolytic activity at quantum dot-conjugates: kinetic analysis reveals enhanced enzyme activity and localized interfacial "hopping". Nano Lett 2012; 12(7): 3793-802.
[http://dx.doi.org/10.1021/nl301727k] [PMID: 22731798]

[51]   Vranish JN, Ancona MG, Oh E, *et al.* Enhancing coupled enzymatic activity by colocalization on nanoparticles surfaces: kinetic evidence for direct channeling of intermediates, ACS Nano 2018. ASAP. 2018.
[http://dx.doi.org/https://doi:10.1021/acsnano.8b02334]

[52]   Gizik U, Katarzyna HK, Danuta W. Immobilization as a strategy for improving the application to oxidoreductases, Molecules 2014; 19: 8995-9018.
[http://dx.doi.org/https://doi:10.3390/molecules.19078995]

[53]   Lee JM, Xu GR, Kim BK, Choi HN, Lee WY. Amperometric tyrosinase biosensor based carbon nanotubes doped sol-gel derived ZnO-Nafion- composites, film. Electroanalysis 2010; 23: 961-70.

[54]   Blanco R, Terreros P, Fernandez-Perez M, Otero C, Diaz Gonzalez G. Functionalization of mesoporous silica for lipase immobilization-characterization of the support and the catalysts. J Mol Catal, B Enzym 2004; 30: 83-93.
[http://dx.doi.org/10.1016/j.molcatb.2004.03.012]

[55]   Jal PK, Patel S, Mishra BK. Chemical modification of silica surface by immobilization of functional groups for extractive concentration of metal ions. Talanta 2004; 62(5): 1005-28.
[http://dx.doi.org/10.1016/j.talanta.2003.10.028] [PMID: 18969392]

[56]   Homaei A. Enzymatic immobilization and its application in the food technology: In book: Advances food biotechnology. 1st ed., John Wiley and Sons Ltd 2015.

[57]   Kooi ME, Cappendijk VC, Cleutjens KB, *et al.* Accumulation of ultrasmall superparamagnetic particles of iron oxide in human atherosclerotic plaques can be detected by *in vivo* magnetic resonance imaging. Circulation 2003; 107(19): 2453-8.
[http://dx.doi.org/10.1161/01.CIR.0000068315.98705.CC] [PMID: 12719280]

[58]   Krause MH, Kwong KK, Gragoudas ES, Young LH. MRI of blood volume with superparamagnetic iron in choroidal melanoma treated with thermotherapy. Magn Reson Imaging 2004; 22(6): 779-87.
[http://dx.doi.org/10.1016/j.mri.2004.01.052] [PMID: 15234446]

[59]   Tsang SC, Yu CH, Gao X, Tam K. Silica-encapsulated nanomagnetic particle as a new recoverable biocatalyst carrier. J Phys Chem B 2006; 110(34): 16914-22.
[http://dx.doi.org/10.1021/jp062275s] [PMID: 16927981]

[60]   Kluchova K, Zboril R, Tucek J, *et al.* Superparamagnetic maghemite nanoparticles from solid-state synthesis - their functionalization towards peroral MRI contrast agent and magnetic carrier for trypsin immobilization. Biomaterials 2009; 30(15): 2855-63.
[http://dx.doi.org/10.1016/j.biomaterials.2009.02.023] [PMID: 19264355]

[61]   Dong H, Li J, Li Y, Hu L, Luo D. Improvement of catalytic activity and stability of lipase by immobilization on oregano-bentonite Chemical Engineering Journal 2012; 181-182.

[62]   Zhou Z, Inayat A, Schwieger W, Hartmann M. Improved activity and stability of lipase Immobilized in a cage-like large pore mesoporous organo-silicas, Mesoporous and Microporous Materials 2013; 54: 133-41.
[http://dx.doi.org/https://10.1016/j.micromeso]

[63]   Nelson JM, Griffin EG. Adsorption of invertase. Journal of American Society 1916; 38(5): 1105-15.
[http://dx.doi.org/10.1021/ja02262a018]

[64]   Hemlatha V, Kalyani P, Chandana V, Hemlatha KPJ. Methods, applications: A mini-review. International Journal of Engineering Science and Research 2016; 5(11): 523-6.
[http://dx.doi.org/10.5281/zendo.168439]

[65]   Johnson KM, Tao JZ, Kennan RP, Gore JC. Gadolinium-bearing red cells as blood pool MRI contrast agents. Magn Reson Med 1998; 40(1): 133-42.
[http://dx.doi.org/10.1002/mrm.1910400118] [PMID: 9660563]

[66]   Sakaguchi K, Murao S. A preliminary report on a new enzyme, penicillin amidase. Journal of Agriculture Chemical Society, Japan 1950; 23: 411.
[http://dx.doi.org/10.1271/nogeikagaku1924.23.411]

[67]   Liao G, Li J, Li L, Yang H, Tian Y, Tan H. Selectively improving nikkomycin Z production by blocking the imidazolone biosynthetic pathway of nikkomycin X and uracil feeding in Streptomyces ansochromogenes. Microb Cell Fact 2009; 8: 61.
[http://dx.doi.org/10.1186/1475-2859-8-61] [PMID: 19930628]

[68]   Birkmayer W, Hornykiewicz O. [The L-3,4-dioxyphenylalanine (DOPA)-effect in Parkinson-akinesia]. Wien Klin Wochenschr 1961; 73: 787-8.
[PMID: 13869404]

[69]   Qiu GM, Zhu BK, Xu YY. α-Amylase immobilized by Fe_3O_4/poly(styrene-co-maleic anhydride) magnetic composite microspheres: Preparation and characterization. J Appl Polym Sci 2005; 95: 328-35.
[http://dx.doi.org/10.1002/app.21239]

[70]   Pazalioglu NK, Sariisik M, Telefoncu A. Treating denim fabrics with immobilized commercial cellulases. Process Biochem 2005; 40: 767-71.
[http://dx.doi.org/10.1016/j.procbio.2004.02.003]

[71]   Silva CJSM, Prabaharan M, Gübitz G, Cavaco-Paulo A. Treatment of wool fibers with subtilisin and subtilisin-PEG. Enzyme Microb Technol 2005; 36: 917-22.
[http://dx.doi.org/10.1016/j.enzmictec.2005.01.017]

[72]   Yachmenev VG, Bertoniere NR, Blanchard EJ. Intensification of the bioprocessing of cotton textiles by combined enzyme/ultrasound treatment. J Chem Technol Biotechnol 2002; 77: 559-67.
[http://dx.doi.org/10.1002/jctb.579]

[73]   Soares JC, Moreira PR, Queiroga AC, *et al.* Applications of immobilized enzyme technology for the textile industry: A review. Biocatal Biotransform 2011; 29(6): 223-37.
[http://dx.doi.org/10.3109/10242422.2011.635301]

[74]   Wang Q, Fan X, Hua Z, Gao W, Chen J. Degradation kinetics of pectins by an alkaline pectinase in biosourcing of cotton fabrics. Carbohydr Polym 2007; 67: 572-5.
[http://dx.doi.org/10.1016/j.carbpol.2006.06.031]

[75]   Panesar PS, Kumari S, Ahmad A. Potential applications of immobilized β-galactosidase in the food processing industry, Enzyme Research . 2010; 2010.
[http://dx.doi.org/https://104061/2010/473137] [PMID: 473137]

[76]   Ansari SA, Satar R, Zaidi SK, Ahmad A. Immobilization of Aspergillus oryzae β-galactosidase on cellulose acetate polymethylmethacrylate and its applications in the hydrolysisof lactose from milk and whey. Int Sch Res Notices 2014; 2014: 163987.
[http://dx.doi.org/10.1155/2014/163987] [PMID: 27350979]

[77]   Roig MG, Rello JF, Velasco FG, Celis CSD, Cachaza JM. Biotechnology applied biology section applications of immobilized enzymes. Wiley Online Library 2010; 15(4): 198-208.

[78]   Aggarwal S, Sahni S. The commercial exhibition of immobilized enzymes, 2012, International

Conference on Environmental Biomedical and Biotechnology, IPCBEF, Vol (41), 2012, IACSIT PRESS Singapore.

[79]   Yang F. Liquid Biofuels, electronic source 2006. https://english.qibebt.cas.cn/rh/rs/bagcc/libi

[80]   Tan T, Lu J, Nie K, Deng L, Wang F. Biodiesel production with immobilized lipase: A review. Biotechnol Adv 2010; 28(5): 628-34.
[http://dx.doi.org/10.1016/j.biotechadv.2010.05.012] [PMID: 20580809]

[81]   Wu Z, Zhang B, Yan B. Regulation of enzyme activity through interactions with nanoparticles. Int J Mol Sci 2009; 10(10): 4198-209.
[http://dx.doi.org/10.3390/ijms10104198] [PMID: 20057940]

[82]   Badadi VY, Leila N, Najafi A, *et al.* Evaluation of iron oxide nanoparticles effects on tissues and enzymes of the liver in rats. J Pharm Biomed Sci 2012; 23(4): 1-4.

[83]   Maccormack TJ, Clark RJ, Dang MK, *et al.* Inhibition of enzyme activity by nanomaterials: potential mechanisms and implications for nanotoxicity testing. Nanotoxicology 2012; 6(5): 514-25.
[http://dx.doi.org/10.3109/17435390.2011.587904] [PMID: 21639725]

[84]   Barber S, Abdelhakiem M, Ghosh K, *et al.* Effects of nanomaterials on luciferase with significant protection and increased enzyme activity observed for zinc oxide nanomaterials. J Nanosci Nanotechnol 2011; 11(12): 10309-19.
[http://dx.doi.org/10.1166/jnn.2011.5013] [PMID: 22408903]

[85]   Kotchey GP, Hasan SA, Kapralov AA, *et al.* A natural vanishing act: The enzyme-catalyzed degradation of carbon nanomaterials. Account of Chemical Research 2012; 45(10): 1770-181.
[http://dx.doi.org/10.1021/ar300106h]

[86]   Puri M, Barrow CJ, Verma ML. Enzyme immobilization on nanomaterials for biofuel production. Trends Biotechnol 2013; 31(4): 215-6.
[http://dx.doi.org/10.1016/j.tibtech.2013.01.002] [PMID: 23410582]

[87]   Korajenagi SS, Vertegel AA, Kane RS, Dordick IS. Structure and function of enzymes adsorbed on to SWCNT. Langmuir 2004; 20(26): 11594-9.
[http://dx.doi.org/10.1021/la047994h] [PMID: 15595788]

[88]   Lin Y, Allard LF, Sun YP. Protein affinity of SWCNT in water. J Phys Chem 2004; 108: 3760-4.
[http://dx.doi.org/10.1021/jp031248o]

[89]   Yi C, Fong CC, Chen W, Qi S, Lee ST, Yang M. Inhibition of biochemical reactions by silicon nanowires through modulating enzyme activities. ChemBioChem 2007; 8(11): 1225-9.
[http://dx.doi.org/10.1002/cbic.200700136] [PMID: 17566125]

[90]   Lahir YK. Impacts of fullerene on a biological system, Clinical Immunology, Endocrinology and Metabolic Drugs 2017; 4(1): 48-57.
[http://dx.doi.org/https://doi:10.2174/2212707004666171111351624]

[91]   Lahir YK. Graphene and graphene-based nanomaterials are suitable for drug delivery, In Applications of Targeted Nano-Drugs and Delivery Systems. Elsevier Publication 2018; pp. 157-90.ISBN: 978--12-814029.

[92]   Friedman SH, DeCamp DL, Silberman P, *et al.* Inhibition of HIV-1 protease by fullerene derivatives: model-building studies and experimental verification. J Am Chem Soc 1993; 115(15): 6506-9.
[http://dx.doi.org/10.1021/ja00068a005]

[93]   Pastorin G, Marchesan S, Hoebeke J, *et al.* Design and activity of cationic fullerene derivatives as inhibitors of acetylcholinesterase. Org Biomol Chem 2006; 4(13): 2556-62.
[http://dx.doi.org/10.1039/b604361e] [PMID: 16791318]

[94]   Connie AW, Greenwood C, Kang A, Siu D. Investigating possible interactions between nano C60 and Plasmid pBR 322 DNA. Journal of Experimental Microbiology and Immunology 2006; 9: 64-74.

[95]   Maoyong S, Guibin J, Junfa Y, Hailin W. Inhibition of polymerase activity by pristine fullerene

nanoparticles can be mitigated by abundant proteins. Chem Commun (Camb) 2010; 46(9): 1404-6.
[http://dx.doi.org/10.1039/b922711c] [PMID: 20162129]

[96]   Yang X, Ebrahimi A, Li J, Cui Q. Fullerene biomolecule conjugates and their biomedical Applications International Journal of Nanomedicine 2014; 9: 77-92.
[http://dx.doi.org/https://doi:10.2147/JNS52829]

[97]   Firdaus S, Dhasmana A, Srivastava V, *et al.* Interaction pattern for the complex of B-DNA Fullerene compounds with a set of known replication proteins using docking study. Bioinformation 2015; 11(3): 122-6.
[http://dx.doi.org/10.6026/97320630011122] [PMID: 25987764]

[98]   Sumi N, Chitra KC. Effects of fullerene C60 on antioxidant enzymes activities and lipid Peroxidation in gills of cichlid fish, Pseudotrophus maculatus (Bloch 1795). J Zool Stud 2017; 3(6): 31-7. www.journalofzoology.com

[99]   Fortner JD, Lyon DY, Sayes CM, *et al.* C60 in water: nanocrystal formation and microbial response. Environ Sci Technol 2005; 39(11): 4307-16.
[http://dx.doi.org/10.1021/es048099n] [PMID: 15984814]

[100] Qiao R, Roberts AP, Mount AS, Klaine SJ, Ke PC. Translocation of C60 and its derivatives across a lipid bilayer. Nano Lett 2007; 7(3): 614-9.
[http://dx.doi.org/10.1021/nl062515f] [PMID: 17316055]

[101] Jin L, Yang K, Yao K, *et al.* Functionalized graphene oxide in enzyme engineering: a selective modulator for enzyme activity and thermostability. ACS Nano 2012; 6(6): 4864-75.
[http://dx.doi.org/10.1021/nn300217z] [PMID: 22574614]

[102] Sun X, Feng Z, Hou T, Li Y. Mechanism of graphene oxide as an enzyme inhibitor from molecular dynamics simulations. ACS Appl Mater Interfaces 2014; 6(10): 7153-63.
[http://dx.doi.org/10.1021/am500167c] [PMID: 24801143]

[103] Maiti S, Kundu S, Roy CN, Ghosh D, Das TK, Saha A. A comparative evaluation of the activity modulation of flavo and non-flavo enzymes induced by graphene oxide. J Mater Chem B Mater Biol Med 2017; 5(14): 2601-8.
[http://dx.doi.org/10.1039/C7TB00083A] [PMID: 32264038]

[104] Klajnert B. Dendrimers in medicine. Nova Science Pub, Inc 2007.

[105] Menjoge AR, Kannan RM, Tomalia DA. Dendrimer-based drug and imaging conjugates: design considerations for nanomedical applications. Drug Discov Today 2010; 15(5-6): 171-85.
[http://dx.doi.org/10.1016/j.drudis.2010.01.009] [PMID: 20116448]

[106] Leonov M, Ihnatsyen KA, Sylwia M, *et al.* Effect of dendrimers on selected enzymes evaluation of nanocarriers International Journal of Pharmaceutics 2016.
[http://dx.doi.org/https://dx.doi.org/10.1016/j.ijpharm.2015.12.056]

[107] Hong S, Bielinska AU, Mecke A, *et al.* Interaction of poly(amidoamine) dendrimers with supported lipid bilayers and cells: hole formation and the relation to transport. Bioconjug Chem 2004; 15(4): 774-82.
[http://dx.doi.org/10.1021/bc049962b] [PMID: 15264864]

[108] Sato K, Anzai J. Dendrimers in layer-by-layer assemblies: synthesis and applications. Molecules 2013; 18(7): 8440-60.
[http://dx.doi.org/10.3390/molecules18078440] [PMID: 23867653]

[109] Szwed A, Milowaska K, Leonov M, *et al.* Interaction between dendrimers and regulatory proteins: Comparison of effects of carbosilane and viologen-phosphorous dendrimers. Royal Society of Chemistry Advances 2016; 6: 97546-54.
[http://dx.doi.org/10.1039/C6RA16558C]

[110] Zhou D, Bennett SW, Keller AA. Increased mobility of metal oxide nanoparticles due to photo and thermal induced disagglomeration. PLoS One 2012; 7(5): e37363.

[http://dx.doi.org/10.1371/journal.pone.0037363] [PMID: 22624021]

[111] Girifalco LA. Statistical mechanics of solid. ISBN 978-0-19-516717-7. U.K: Oxford University Press 2003.

[112] Sizochenko N, Leszczynska D, Leszczynski J. Modeling of interaction between the Zebrafish hatching enzyme ZHE-1 and A series of metal oxide nanoparticles: Nano QSAR and casual analysis of inactivation mechanism. Nanomaterials (Basel) 2017; 7(10): 330.
[http://dx.doi.org/10.3390/nano7100330] [PMID: 29035311]

[113] Cimino A, Stone FS. Oxide solid solution as a catalyst, Advanced Catalysts 2002; 47: 141-306. https://doi.org

[114] Busca G. The surface acidity and basicity of solid oxides and zeolites.Metal oxides. Boca Raton, Fl: CRC 2006. www.crcpress.com

[115] Fernandez-Garcia M, Roderqus JA. Metal oxide. In Nanomaterials, Inorganic and Bioinorganic perspectives, Brookhaven National Library, Upton NY 2007.www.bnl.gov www.bnl.gov/isd/document/41042.pdf

[116] Chen Z, Yin J-J, Zhou Y-T, *et al.* Dual enzyme-like activities of iron oxide nanoparticles and their implication for diminishing cytotoxicity. ACS Nano 2012; 6(5): 4001-12.
[http://dx.doi.org/10.1021/nn300291r] [PMID: 22533614]

[117] Vineet K. Environmental Toxicity of Nanomaterials 2018. www.crcpress.com

[118] Schrand AM, Rahman MF, Hussain SM, Schlager JJ, Smith DA, Syed AF. Metal-based nanoparticles and their toxicity assessment. Wiley Interdiscip Rev Nanomed Nanobiotechnol 2010; 2(5): 544-68.
[http://dx.doi.org/10.1002/wnan.103] [PMID: 20681021]

[119] Karthigarani M, Navaraj PS. Impact of nanoparticles on enzyme activity in Oreochromis mossambicus. International Journal of Scientific and Technology Research 2012; 1(10): 13-7. www.ijstr.org

[120] Peyrot C, Wilkinson KJ, Desrosiers M. Sauveꞌ S, Effects of silver nanoparticles on soil enzyme activity within and without added organic matter. Environ Toxicol Chem 2014; 33(1): 115-25.
[http://dx.doi.org/10.1002/etc.2398] [PMID: 24115203]

[121] Kenz CORDIS. (Community Research and Development Information Service). Project ID-322158/Funder under FP7-PEOPLE; Record no 214317 .

[122] Tamberler C, Saikaya M. Modular biomimetics: nanotechnology and bionanotechnology using genetically engineering peptides. Philos Trans R Soc Lond, A 1894; 2009(367): 1705-26.
[http://dx.doi.org/10.1098/rsta.2009.0018]

[123] Song Y, Qu K, Zhao C, Ren J, Qu X. Graphene oxide: intrinsic peroxidase catalytic activity and its application to glucose detection. Adv Mater 2010; 22(19): 2206-10.
[http://dx.doi.org/10.1002/adma.200903783] [PMID: 20564257]

[124] Gao N, Sun H, Dong K, *et al.* Transition-metal-substituted polyoxometalate derivatives as functional anti-amyloid agents for Alzheimer's disease. Nat Commun 2014; 5(5): 3422.
[http://dx.doi.org/10.1038/ncomms4422] [PMID: 24595206]

[125] Hu Y, Cheng H, Zhao X, *et al.* Surface-enhanced Raman scattering active Ag NPs with enzyme mimicking activities for measuring Glucose and lactate in living tissues, ACS Nano 2017; 11(6): 5558-66.
[http://dx.doi.org/https://doi:10.1021/acsnano7b00905]

[126] Reshmi R, Sugunan S. Superior activities of lipase immobilized on porous and hydrophobic clay support: characterization and catalytic activity study. J Mol Catal, B Enzym 2013; 97: 36-44.
[http://dx.doi.org/10.1016/j.molcatb.2013.04.003]

[127] Murphy JM, Zhang Q, Young SN, *et al.* A robust methodology to subclassify pseudokinases based on their nucleotide-binding properties. Biochem J 2014; 457(2): 323-34.

[http://dx.doi.org/10.1042/BJ20131174] [PMID: 24107129]

[128] Liang JF, Li YT, Yang VC. Biomedical application of immobilized enzymes. J Pharm Sci 2000; 89(8): 979-90.
[http://dx.doi.org/10.1002/1520-6017(200008)89:8<979::AID-JPS2>3.0.CO;2-H] [PMID: 10906721]

[129] Kim J, Chang JY, Kim YY, Kim MJ, Kho HS. Effects of molecular weight of hyaluronic acid on its viscosity and enzymatic activities of lysozyme and peroxidase. Arch Oral Biol 2018; 89: 55-64.
[http://dx.doi.org/10.1016/j.archoralbio.2018.02.007] [PMID: 29475188]

[130] Jaeger KE, Eggert T. Enantioselective biocatalysts optimized by direct evolution, Current Opinion in Biotechnology. 2004; 15: pp. (4) 305-13.
[http://dx.doi.org/https://doi:10,1016/j.copbio.2004.06.007]

CHAPTER 9

Nanomaterials and Immune System: Interactions

Abstract: Drug delivery systems, vaccination, and diagnostic imaging are the main aspects that enhance the effectiveness of human health and safety. This system protects an organism against microbes, viruses, parasites, and allergens. The immune system represents the level of the ontogenic and phylogenetic development of a biosystem. The degree of efficiency to protect against infection varies in different organisms. Overall, the functional mechanism of the immune system is complicated. Any molecule, or a pathogen entering the human body or a biosystem, has to face various components of the immune system. It is imperative to understand the concept of the interactions between nanomaterials and the components of the immune system. This understanding will improve, improvise, and elevate the degree of clinical translation of nanomedicine in the field of human health and safety. There seems to be an enormous scope of studies related to the intricacies of interaction occurring at the bio-nano-interface. These efforts will guide to design the rational nanomaterials that are either fabricated or synthesized with specific targets. The study of the modes or the patterns involved during the interactions between nanomaterials and the immune system can maintain the appropriate defense system of the individual against infections, xenobiotics, and any foreign molecule. This chapter deals with the applications of nanomaterials in the delivery system, competence of nanomaterials concerning the immune system, immunomodulation, immunosuppression, immunostimulation, and interactions between various nanomaterials and the components of the immune system.

Keywords: Immunomodulation, Immunosuppression, Immunostimulation, Immunological Memory, Nanomaterials.

INTRODUCTION: OVERVIEW-IMMUNE SYSTEM IN HUMANS, NANOTECHNOLOGY, NANOSCIENCE, AND NANOMATERIALS

All organisms have some defense system and specialized mechanism dedicated to the protection against the derogative impacts of microbes, viruses, and parasites, and any agent (xenobiotics) that causes sickness or any form of infection. This protective system exhibits the ontogenic and phylogenic development in different organisms. The mechanism and functionality of the immune system are complicated. Despite enormous research in the field of immunology, still, there exist some lacunae concerning the mechanism involved during the interactions between nanomaterials and components of the immune system. Further, this mechanism becomes more ambiguous because of the wide variety of nanomate-

Yogendrakumar H. Lahir & Pramod Avti

rials. Nanomaterials are the products of nanoscience either fabricated or formulated, as per the targeted applications in pharmaceutical, imaging, biomedical implants, and other related applications.

The blood cells like monocytes, platelets, leukocytes, and dendritic cells in tissues, and macrophages in the lungs, tend to engulf the internalized nanoparticles involving phagocytosis among humans and the vertebrates. The fluid components, like plasma proteins, opsonins, and related complements, also get involved in the interactions between nanomaterials, immune cells, and proteins. These interactions can influence the uptake, biodistribution, and the clearance of the internalized nanomaterials. This response can cause a disturbance in the distribution of nanomaterials and divert them from the target tissue and organ [1]. The immune system has two sub-components, namely, innate and adaptive resistant components. The main features of the innate component (network) are non-specificity and cell-mediated responses. Its humeral parts are present in body fluids. An appropriate exposure causes immediate remedial reactions that may not have immunological memory. The features of adaptive immune system concern with the pathogen and are antigen-specific in function. There is a lag time between exposure and duration of maximal response. The exposure leads to immunological memory. There is a complex conglomerating functionality between protein and the cells that regulate the short term and long term efficacies of the immune system. The innate immunity stands for fast and broad response spectrum affectivity. It utilizes proteins present in the blood, body fluids, and tissues.

Cells, such as macrophages, dendritic cells, neutrophils, mast cells, and natural killer cells participate in innate immunity [2, 3]. Innate immunity is the first line of defense among human beings and is non-specific. Its function depends on the Pattern Recognition Receptors (PRPs) that identifies a broad and conserved molecular pattern of the target. This innate immune system recognizes the foreign bodies that enter in a biosystem. It is also responsible for proinflammatory responses [2]. The adaptive immunity has a high degree of specificity, but slow functionality and antibodies identify the antigen. The cells like B-cells and T-cells play their role in antigen-antibody interaction [2, 3].

Nanomaterials activate the complement system-a component of the innate immune system, significantly. It is a complex system comprising of about thirty proteins; some of these are functionally protease enzymes, and the proteolytic dissociation induces these proteins. These all have a functional role that involves binding interaction with the surfaces of the foreign body. During the process of activation of complement, the third complement protein referred to as (C3), undergoes proteolytic cleavage, resulting in the formation of two opsonic complements called (C3b) and (iC3b)]. Macrophages identify nanoparticles or administered materials. The distributed materials in the human body bind with this opsonic complement [C3b] and [iC3b], involving corresponding complement receptors 1 and 3 present in the recipient. Nanoparticles follow three primary pathways, namely classical, lectin, and alternative pathways during their cellular internalization. As the activation of complement system progresses, anaphylatoxins (also referred to as biologically active complement peptides namely, C4a, C3a, C5a), are the action of the product of proteolytic enzyme causing cleavage of the original molecules (C3, C4, and C5). These anaphylatoxins induce the release of histamine from immune cells, like mast cells. These immune cells also participate in allergic reactions. The histamine increases the permeability of vascular cells and even during the contraction of smooth muscle. Anaphylatoxins promote some of the physiological conditions, like distressed breathing, acute changes in blood pressure, chest pain, declined output from heat and cardiac arrest, *etc.*, (Fig. **1**) [4 - 6].

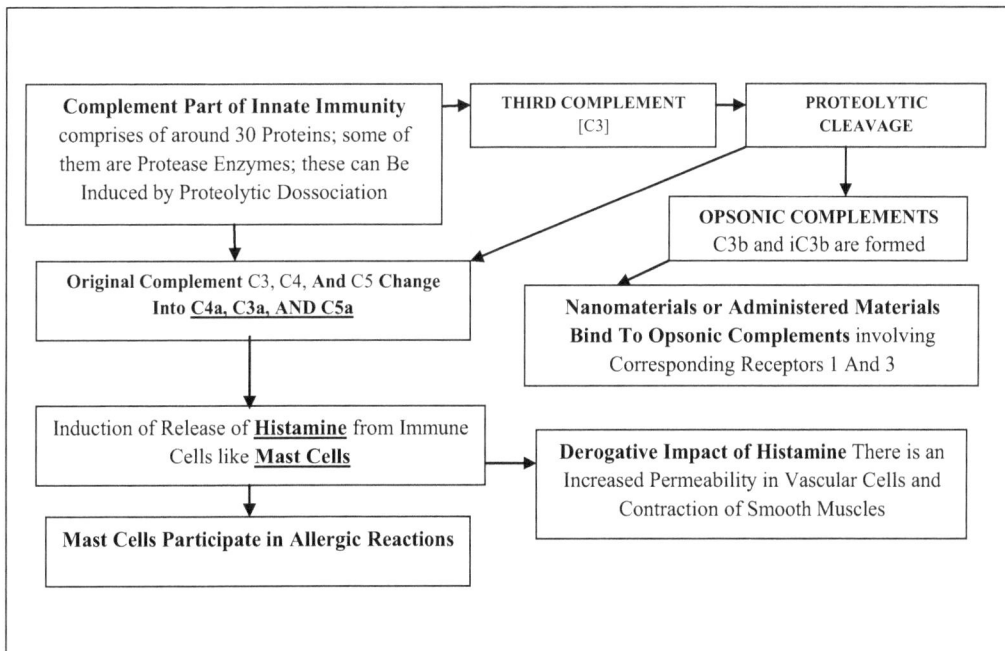

Fig. (1). The flow chart representing the probable process of activation of complement peptides (also called biological active complement peptides).

Nanoscience, nanotechnology, and their products are applicable in every industry, biomedical and pharmacological field. The nanomaterials, either in pristine or modified forms or fabricated forms, are utilized. It becomes imperative to know about the interactions between nanomaterials and the immune system because mammals act as an experimental model. The immune system identifies xenobiotics, any foreign molecule that internalize the body, and this process applies to the agents that carry infection or pathogens, carriers for drugs, genetic materials, or any biomolecules. Thus, nanomaterials involved in such applications must avoid the interactions with the immune system or its components for their successful performance. The nanomaterials have specificity concerning inhibition or to enhance the immune response in the recipient. The successful fabricated or engineered nanomaterials to be used as a drug carrier must be non-immuno-toxic and immuno-compatible. They should not either induce an immune response or get destroyed or get eliminated by the components of the immune system [7]. The manipulation of nanomaterials concerning their physicochemical properties is an essential aspect because these influence the immune system or its components. These modifications also relate to the immunity-related cells or biomolecules. These modulate immunoreactions and avoid undesirable or toxic interactions. These intentions are safeguarding to the host and prevent unpleasant responses like infection, malignancies, inflammation or necrosis, *etc.* These efforts help immunological functions, such as meeting the challenges of the host body while responding to internal and or external agents.

Nanomaterials can cause immunostimulatory or immunosuppressive responses in the host. The vaccine efficacy, anti-tumoral effects are desirable while the hypersensitivity reactions, inflammation, anaphylaxis are not. Anaphylaxis is a severe allergic reaction; it is instantaneous and can lead to death. The symptoms of anaphylaxis include itchiness, rashes, throat swelling, and low blood pressure. These symptoms appear as a result of an insect bite, food, or medication. These conditions are not desirable during the process of the immunostimulation. The treatment of inflammatory disorders, autoimmune diseases, prevention of allergic responses, acceptance of transplants, *etc.*, are helpful while low responses of the body to infected and cancerous cells, myelosuppression, and thymic suppression, are undesirable. The applications of nanomaterials may relate to vaccines, therapeutics for inflammation, or autoimmune disorders. Fig. (**2**) represents the probable interactions of nanomaterials with the immune system.

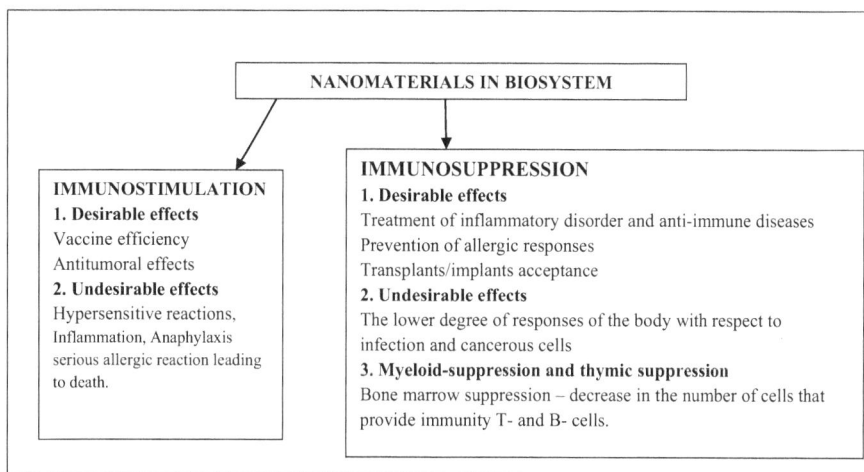

Fig. (2). Probable interactions of nanomaterials with the immune system.

POTENTIAL ADVANTAGES OF NANOMATERIAL BASED DELIVERY SYSTEM

Nanomaterial-based delivery systems are readily modulated as per the need and appear to be better options as compared to the customary modes. The following are the advantages of using nanomaterials as a carrier of drugs.

1. Site-specific delivery of drugs, peptides, and genes,
2. Improved *in vivo* and *in vitro* stability,
3. Reduced side effect profile.

PROBABLE FLAWS OR UNDESIRABLE INTERACTIONS THAT APPEAR DURING THE APPLICATIONS OF NANOMATERIAL BASED DELIVERY SYSTEMS

Not every interaction needs to be advantageous or beneficial regarding nanomaterials. These can pose a problem because the preferred modified nanomaterials are in use as per the requirements. The following are the probable hurdles one may face while using nanomaterials for the delivery system.

1. The phagocytic cells, like macrophages and other immune cells, identify the nanomaterials in use for delivery systems and interplay with these nanomaterials. This interaction results in immunostimulation or immunosuppression and promotes inflammation or autoimmune disorders.

2. This system enhances the susceptibility of the host to either infection or cancer.
3. There is a possibility of wrong recognition of nanoparticle or nanomaterials as foreign by the immune system. It causes multiple immune responses against nanomaterials, eventually leading to toxicity.
4. There are chances that miss-identification declines therapeutic efficiency [8].

The components of the immune system recognize nanomaterials as self or non-self, this state is of great interest, at least, in the drug delivery system, biomedical implants, and biosensors. In this context, the properties like size, shape, hydrophobicity or hydrophilicity, and steric effect, the specific coating on nanoparticles undergo suitable change to enhance the degree of biocompatibility of specific nanomaterials. Polyethylene glycol or other types of polymer are preferred materials for coating nanomaterials, and these provide a hydrophilic environment in a biosystem. These polymers are useful to treat nanomaterials, and this treatment helps them to avoid immunological recognition. The immune system must recognize the polyethylene glycol-coated liposomes and polymer shielded nanomaterials because these modified nanomaterials change the pharmacokinetic profile of subsequently administered dose.

IMMUNOMODULATION

Immunomodulation plays an essential role in immunotherapy. It is imperative to understand the mechanism involved in immuno-modulations. This understanding improves the designing and formulation of nanomaterials for their applications as carriers, clinical and diagnostic agents. The process of immunomodulation is vital as it deals with providing signals that either stimulate or inhibit the function of T-cells. The cells, like dendritic cells, present antigens, and are called antigen-presenting cells. These play a crucial role during immunomodulation. The immune response depends on the process and degree of activation or inhibition of T-cells. The dendritic cells and antigen-presenting cells carry antigen, immunostimulatory adjuvant, immunoregulatory agents, and reach the specific action site to activate T-cells during the stimulation of adaptive immunity. These T-cells are from lymph nodes or spleen. Dendritic cells or antigen-presenting cells get associated with many signals during the activation of T-cells. These signals are concerned with (i)-presentation of antigen, (ii)-co-stimulatory molecules, (iii)-cytokines, and iv-recognition of a specific antigen. These all are related to the antigen-specific T-cell receptor. The receptor of the antigen-specific T-cell recognizes the antigen peptide among the major histocompatibility complex or of the dendritic cells or antigen-presenting cell. The recognition of antigen is an essential aspect of immunotherapy and immunomodulation. If the presented antigen combines with the binding site of the co-stimulatory molecule present on the dendritic cell or antigen-presenting cell, then the paracrine delivery of cytokine from dendritic cell to T-cell takes place successfully. This process induces the process of activation of antigen-specific T-cells [9].

NANOMATERIALS ARE COMPETENT AGENTS TO INFLUENCE THE IMMUNE SYSTEM

Immunotherapy is revolutionizing the current therapy related to cancer and autoimmune diseases. Such nano-bio-molecular studies show impressive outcomes during their clinical trials. The development of innovative nano-bio material is the focus during the interactions between such nanomaterials and the immune system, in adaptive and innate aspects of the immune system. Biomaterials display either immuno-simulative or immunosuppressive behavior [10]. When developed innovative nanomaterial makes contact with the immune system, it induces tolerance response or sequential regulatory T-cell response. It confirms that there is no initiation of defense towards any harmless agent. If confronting agent is a pathogen, macrophages, neutrophils, *etc.*, engulf them and remove from the system. T-cells produce proinflammatory cytokines such as interferon (IFN-α) or interleukins (IL-12). All these help the process of phagocytosis. Other defensive responses include prime parameters, like cytokines such as IL-4, IL-5, and IL-13 along with different immune cells like basophils, eosinophils, mast cell, anti-inflammatory macrophages (M2), T-helper cells (T-H2) cells. This mode of defensive response takes place during parasitic infection, wound healing, chronic inflammatory conditions, asthma, and allergy [11]. The surface of nanomaterials readily gets modified involving covalent bonding with specific ligands during the interactive behavior between the nanomaterials and the components of the immune system. These ligands include antibodies, their fragments, allergens, peptides nucleic acids as immunostimulatory agents. The typical examples of immunostimulatory agents are CpG-DNA, small inhibitory RNAs (si-RNA), aptamer (oligonucleotides), carbohydrates, and biomolecules, like Vit-D, or Toll-like receptors (TLRs) ligands. The conjugated forms of nanomaterials are useful in the delivery process. These conjugated forms enhance the cellular uptake involving target cell, mucosal adhesion, penetration or retention, immunostimulatory response, or effects that result in modulation. These modified nanomaterials also bring about the optimization of safety profile, specificity, and efficiency of vaccine candidates [11].

INFLUENCE OF NANOMATERIALS ON IMMUNOSTIMULATION, IMMUNOSUPPRESSION

Immunostimulation is a phenomenon during which unintended or inappropriate antigen-specific or non-specific activation of the components of the immune system takes place. This activation is the result of (i) unintended immunogenicity of biomolecule involving both antibody and cellular immune response with respect to foreign molecules in the body of a recipient, (ii) adjuvant, *i.e.*, activation of any component of the immune system concerned with the specific antigen or an additional portion which comes in contact either externally or internally to the body of the recipient. During immune stimulation, the immune responses are different. The histopathological response is noncritical, and humoral responses are critical; cellular responses depend on the role of a specific cell, *i.e.*, T-Cells critical, natural killer cells, and granulocytes are not applicable. Host resistance is noncritical, but the signs of illness must be observed critically [12].

Nanomaterials need a concerned investigation to evaluate their immunostimulatory potentials concerned with the stimulation of innate or adaptive immune responses. The related aspects examined, including cytokine secretion, induction of antibody response involving immunogenicity and complement immune system, *etc*. The antigenicity of nanoparticles is still in its infancy. The nanomaterials, like fullerene derivatives, gold colloids, cationic polyamidoamine, and polypropylene imine dendrimers, do not exhibit particle specific immune response, even in the presence of potent adjuvants. Nanomaterials like dendrimers and fullerene are not antigenic. When these bind to proteins, like bovine serum albumin (BSA) become susceptible to B-cell for phagocytosis, these cells digest their protein components within themselves. B-Cells exhibit BSA peptides on human leukocytes antigens molecules and binds to T-helper cells. This interaction activates B-cell and T-helper cells. T-lymphocytes produce cytokines. Plasma B-cells help to direct the antibodies to nanoparticles [8].

Nanomaterials act as an immunostimulant or immunosuppressant agents. This behavior illustrates in the case of nanomaterials that are in use in treatment. These act as an agent to elevate the efficacy of vaccine and cancer treatment. These can result in an undesirable impact on the immunostimulatory process. Such consequences take place during micro and nano delivery modes of trafficking agents. These enhance cellular uptake of the therapeutic drugs as nano-formulation and also in case of processing of antigens. The specific formulation follows particle characteristics; as a result, these devices are addressing the preferred concomitant delivery of antigens and have immuno-potential [13].

RNA-interference is biological processes in which RNA molecules inhibit gene expression or translation process. This inhibition process targeted RNA molecule gets neutralized. The RNA is a specifically endogenous biomolecule and is applicable in targeting leukocytes [14]. Cyclin-D1 (Cy-D1) is an original molecule that regulates the cell cycle. It is beneficial in cancer-related studies. During the interaction between nanomaterials, antibodies with B-7 integrin (B-7I), nanoformulations stabilize with CyD1, and small interfering RNA (siRNA). This practice indicates that this interactive act as a potential anti-inflammatory target. Its application in a similar mode can help to target the desired molecule, cell, or tissue using si-RNA. This interplay acts as an appropriate mode in therapeutic applications [15]. The rabies virus glycoprotein conjugates with paclitaxel and forms spherical nanoformulations having a size of about 140 nm. This spherical formulation (rabies virus glycoprotein) is useful for anti-glioma therapy, and moves across the blood-brain barrier (BBB) and reaches the affected cells. This formulation is helpful in the case of Tumor-Associated Macrophages [16].

Immunosuppression is a state that represents the declined effectiveness of the functionality of the immune system in the body. This condition arises because of the suppressive impacts of the components of the immune system. This process can be the result of defective, adverse reaction of a drug, immunosuppressive agent, or concerned with the development of some specific conditions or some clinico-physiological malfunctioning, during a particular treatment. During immunosuppression, the ability to fight infection or immunodeficiency declines. This clinico-physiological condition may develop deliberately or being careless. Induction of immune–tolerance due to nanomaterials is a type of desirable state of immunosuppression. This state functions in two ways (i) it declines the defense of the host against infection or cancer or any biomedical implant, (ii) it may elevate the benefits of treatment for allergies and autoimmune diseases and prevent or reduce the chances of rejection of any biomedical implant or devices. There are few reports on immunosuppression and nanomaterials, but reports on the induction of inflammation and nanomaterials in the recipients. Induction of immune–tolerance response because of nanomaterials can be regarded as a type of desirable immunosuppression [1, 8, 17].

There are four types of allergic reactions based on the COOMBS and GELL classification.

Type-I: IgE mediates this type of reaction, and it acts immediately.
Type-II: This type of reaction mediates involving by IgG and IgM complement; it is cytotoxic-antibody dependent.
Type-III: IgG Complement neutrophils mediate the third type. It relates to complex immune disease.
Type-IV: It is a delayed typed hypersensitivity cell-mediated immune memory response and antibody independent in nature; T-cells mediate this type.

Generally, these allergies are due to environmental conditions, consumption of food items, medicines, *etc*.

Nanomaterials loaded with water-soluble fullerene derivatives like polyhydroxy C60, Poly (D-L-Lactide –co-glycolide) chitosan, poly (lactic acid), poly-methyl vinyl-ether-co-malic anhydrate, dendrosomes are active suppressive agents for Type-I and Type-II allergy. Generally, these allergies are due to environmental conditions and food consumption [8].

Nanoparticles bind to proteins present in the blood of the recipient, and the resultant product has different physicochemical properties. These nanoproducts, with the modified features, modify the related biological responses and their biodistribution in the recipient. Such nanoproducts do not cause fluctuations in the complex biotic systems. Since this form is non-toxic; it enhances its biodistribution [1]. Carbon nanotubes, water-soluble derivatives of fullerene-like polyhydroxy C60, allergen loaded PLGA, chitosan, polylactic acid, and poly (methyl vinyl ether-co-maleic anhydride, *etc.*, these nanomaterials undergo investigated concerning inflammation. Poly (D-L-Lactide-co-glycolide (PLGA) prevents inflammation. The toxicity of these substances causes immunosuppression of T- Cells. Immunosuppressive agents like tetrachloro-dibenzo-p-dioxin, cadmium, corticosteroids, and radiation, *etc.*, impair the development and functions of T-cells [17]. Carbon nanotubes (CNTs) cause suppression in the service of B-cells. Alveolar macrophages produce transforming Growth-Factor-β (TGF-β) and these acts as a critical factor for the immunosuppression.

Poly (D-L-Lactide-co-glycolide (PLGA) modifies nanomaterials act as a carrier for glucocorticoids for treating the inflamed joints. These prevent the immuno-suppressive properties of drugs in the form of small molecules. These modified nanomaterials release the steroid in a controlled manner, and this controlled release elevates the efficacy of modified nanomaterials and improves the impact [18, 19]. Multiwalled carbon nanotubes when the inhaled result in the suppression of the functioning of B-Cells and TGF-β (a product of alveolar macrophages). Cyclooxygenase enzyme gets activated in the spleen in response to signals from the lungs. These indicate suppression of the immune system in the recipient [20]. In breast cancer patients, Abraxane as nanoparticle colloidal suspension containing paclitaxel causes grade 4 neutropenia (a condition called myelosuppression leading to the decline in several neutrophils). When liposomes loaded with Clodronate-antiosteoporotic drug-like Cladronic acid/Di-Sodium clodronate) protect the lungs of swine, against endotoxins [21].

Nanowires, nanotubes, nanoparticles, cantilevers, micro-nano-arrays, techniques are helpful in the clinical diagnosis. This process involves biomarkers; needs less amount of test sample and give precise and quality detection during the early stages of a clinical condition. The nanomaterials increase the efficacy of drugs and decrease the severity of the adverse effects of therapeutics. The application of such nanomaterials is helpful in reinvestigations concerning pharmaceutically suboptimal molecules. Earlier, their functionality was obscure, but currently, these are important for their clinical aspects. The surfaces of these nanomaterials are coated or biochemically modify conveniently as per the requirements. These changes enhance the degree of biocompatibility, bioavailability, and adherence to biomolecular entities. Nanomaterials provide optimal mechanical support in the form of scaffolding, and this support regulates the spatiotemporal aspects of biomolecular substances that enhance the respective functionality. Immunotherapy has revolutionized in the treatment during cancer, other autoimmune diseases, and tissue engineering. These are more prevalent at clinical stages and have shown impressive results. This tendency of nanomaterials encourages the development of innovative nano-biomaterials that interact with the various components of the immune system. The nanomaterials exhibit immunosuppressive, immunostimulatory, or immunoneutral impacts on the immune system of the biosystem. Nanomaterials are applicable in the fields of clinical treatments in case of autoimmune diseases, allergic conditions, cancer, and tissue engineering, *etc* [10, 22].

The avidity and modulations in the immune responses play an essential role during cancer immunotherapy. Even during the development of vaccine adjuvant, inflammation, and disease response, the molecular inflections are significant. RNA plays a vital role in such conditions. It plays a functional role during immunomodulation because it regulates the shape of the transition of the conformations of RNA. These conformations involve triangle, square, or pentagon shapes. When immunoadjuvants incorporate, there is an impact on the cytokines, TNF-α, and interleukin-6 (IL-6). During this effect, the RNA strand in the polygonal conformation tends to stretch. During this stretching, the interior angle extends from 60° to a maximum of 108°. This interaction causes self-assembled conformational changes. The RNA polygon regulates unnoticeable induced effects [23]. Macrophages have RNA nanoparticles concerning size, shape, the threshold of payload, *etc.*, [23].

INTERACTIONS BETWEEN METAL, METAL OXIDE NANOPARTICLES, AND IMMUNE SYSTEM

The physicochemical features of nanomaterials, like size, shape, surface charge, surface modification, hydrophobicity, hydrophilicity, steric properties dependent on the type of coatings and modes of administration, *etc.*, influence their interactions with the immune system. These parameters affect the degree and patterns of interactions and biocompatibility of these nanoparticles. The interaction between the nanoparticles and cell, cell organelles, biomolecules like proteins, DNA, *etc*., involves biophysical dynamics. The typical interactions result in the formation of the protein corona, particle wrapping, intracellular uptake, and processes like biocatalysis, *etc*. These interactions can be biocompatible or non-biocompatible. The involve activation phase transformation, the release of energy, fluctuations in molecular conformation, or dissolution at the bio-nano-interface [24]. Administration of lower than the threshold amount the recipient gives moderate responses, but these responses become complicated because of the aggregation or agglomeration of nanomaterials. It also results in relevant biomolecular modifications [25]. Interaction between cobalt alloy particles and cells induce interleukins like IL-1β, TNF-α, and IL-8. Toll-like receptors (TLRs) function like sensors and signals for an innate immune response during this interaction. Pathway related to IL-8 becomes nonfunctional when the transduction of signal to TLRs gets interrupted. The prevention of interplay between the respective antibodies and TLR plays a significant role in this process. The cobalt microparticles (0.5 -2.0 μm) also prevent the activation pathway to release IL-8; this interplay causes activation of the neutrophils, and these cells exhibit phagocytosis and remove the cobalt microparticles. Possibly, this mechanism involves similar to that engaged during the activation of IL-8. This mechanism induces aggregation and agglomeration of cobalt microparticles in the epithelial cells [26, 27].

The inflammatory responses are different due to the different sizes of silver nanoparticles and the coating present on them. This response also depends on the stain of the experimental models. Smaller silver nanoparticles (20nm and 110nm; 0.1mg/k; coated with polyvinylpyrrolidone and citrate) induce a higher degree of proinflammatory response in terms of neutrophil and eosinophil influx in the case of Brown Norway rats, but in the case of Sprague-Dawley rates, only the neutrophil influx shows similar induction. The difference in proinflammatory response is because of polyvinylpyrrolidone and citrate treatment, and this response declines in both the strains of rats. These silver nanoparticles increase permeability, causing oxidative stress and activate the release of cytokines KC, IL-13, and IgE from the blood alveolar epithelial barrier. The degree of severity becomes higher in the experimental models treated with 20 nm silver nanoparticles in comparison to those treated with bigger sized silver nanoparticles 110nm [28].

During normal biodegradation of durable metal and other nanomaterials may not break down or dissociate into their sub-molecular entities within the recipient. These nanomaterials remain in the particulate state right from the stage of administration to biodistribution, and either accumulate or eliminate form the system. Mostly, durable nanoparticles exhibit accumulation in the mononuclear phagocyte system (MPS). This system includes cells that exhibit phagocytosis, and are present in different tissues as Kupffer's Cells in the liver, dendritic cells in lymph nodes, macrophages in lungs and B-Cells in blood, *etc.* The durable nanomaterials accumulate in these cells because of their phagocytic nature. This accumulation depends on parameters like size, shape, concentration, and presence or absence of coating on the nanomaterials. The exposure of gold nanoparticles (10 nm; 10mg/kg) and silver nanoparticles (50nm; 5mg/kg) accumulate in mononuclear phagocytic cells during eight weeks of exposure. This mode of exposure does not change cytokine activation, but exposure for 8weeks of silica nanoparticles (50nm; 10mgkg) induces mild inflammation along with clustering (less than moderate) of neutrophils, macrophage, and lymphocytes. These nanoparticles cause an increase in the amount of globulin along with TNF-α and KC-GRO from the 2nd and 3rd week onwards [29]. Coating of nanoparticles using polyethylene glycol or other similar types of compound, help these nanoparticles to bypass immune recognition. The PEG specific antibodies formation eliminates these coated nanomaterials. These aspects are likely to influence the efficacy of safe interaction of therapeutics. Nanoparticles are fabricated to induce direct immunostimulation of antigen-represent-cell or as an agent for the delivery antigen to a specific cell (immune). Such nanoparticles induce an immune response [8].

The degree of affinities, and stoichiometry between proteins and nanomaterials, and the detailed kinetics involved regulating the biodistribution of nanomaterials in a biosystem. These functional parameters influence the association and dissociation of proteins and nanoparticles. These control the biological responses to specific nanoparticles. The albumin, IgG, IgM, fibrinogen, and apolipoproteins form a protein corona with iron oxide nanoparticles. The binding of proteins with the nanoparticles involves pre-incubation with bulk plasma/serum or pre-incubation of an individual protein or attaching specific protein. This binding of protein concerns the rate of macro-phagocytosis. Biodistribution of nanoparticles depends on the state and type of coatings on them. The number of proteins binds with uncoated nanoparticles when these come in contact with human plasma. Coated nanoparticles with polyethylene glycol, bind with fewer proteins while nanoparticles coated with polysorbate materials bind with only selected proteins. This behavior indicates selective binding between nanoparticles and plasma proteins depending on the specific coating [30].

The intravenously and intraperitoneally administered nanoparticles encounter the cellular and biochemical components of body fluid. In vascular bloodstream, the immune cells like monocytes, platelets, leukocytes (white blood cells) help the uptake of administrated nanoparticles. When these reach, organs like liver, lymph nodes, spleen, Kupffer's cells, dendritic cells, macrophage, and B-cell uptake these nanomaterials. Further, nanoparticles also show adsorption of specific plasma protein (opsonins) on to their surface. There are three prime modes of resultant processes that occur in immune cells and tissues. These are hemolysis (damage of red blood cells); thrombogenicity (tendency to induce and promote and activate clot formation), and complement activation (related to the removal of a pathogen from the body and complement-receptor-mediated-phagocytosis) [1]. Gold (Au), silver (Ag), silver oxide (AgO), cobalt oxide (CoO), cerium oxide (CeO_2), iron oxide (Fe_2O_3) coated with citrate, $-OH$ ions, tetramethylammonium and uncoated silver oxide (AgO), cobalt oxide (CoO), cerium oxide (CeO_2), iron oxide (Fe_2O_3) undergo immuno-modulatory interactions. This immuno-modulatory affects the cell line of lung epithelial cells, bronchial epithelial cells, human acute monocytic leukemia, human colon carcinoma cells, and human primary monocytes. These nanoparticles induce cytotoxicity, change the pH and osmolarity of the medium, and reduce the rate of metabolic activity. Different cytokine promoters and TNF-α dependent promoter, motivate immunomodulation. The three inflammatory factors like IL-1β; sICAM-1and CXCL1/GRO activate after the induction of IL-1BETA. Another factor, the activity of the chemo-kinase CXCl4/PF4 also increases. The unexpected modulation of cell activation occurs, but these do not induce direct cytotoxicity. The experimental cells exhibit a varying degree of sensitivity to the respective endotoxicity. Most of the immune cells like monocytes, macrophages, show sensitivity towards endotoxins, specifically monocytes, macrophages [31]. The stannum oxide (SnO_2) cerium oxide (CeO_2) and iron oxide (Fe_3O_4) nanoparticles, these enter coelom of sea urchin, *Paracentrotus lividus.* These nanoparticles internalize and cause agglomeration in coelomocytes. The immune cells present in the body cavity coelom of the sea urchin. During later stages of treatment, the activity of cholinesterase declines; this behavior indicates the blood intoxication. The internalized nanoparticles change the morphology of cell organelles transmembrane, Golgi body, and endoplasmic reticulum. The stress-related proteins like hsp-70 and grp-78 and their expression decline [32]. The peripheral human blood lymphocytes interact with metal oxides like ZnO, CeO_2, TiO_2, and Al_2O_3 the degree of adhesion of the molecules changes and so are the chemokine receptor type 4 (CXCR4), the aluminum oxide (Al_2O_3) and titanium oxide (TiO_2) induce T-cell proliferation (basophils and oxidative explosion), this involves leukocytes. These metal oxides are not inert but have the potential to change the degree of adhesion molecules and chemokine receptor that influences immune response [33].

Evaluation of LD50 is of practical significance [34]. In the case of TiO_2, Al2O$_3$ nanoparticles value for LC50 is 55µg/ml, and it activates the immune system. Human T-cells and metal oxide nanoparticles like cerium oxide (CeO_2), titanium dioxide (TiO_2), aluminum oxide (Al_2O_3) (10 -100µg/ml) and ZnO (5-50µg/ml) after incubation for 24 h along show increased cell proliferation increases due to the release of anti-inflammatory cytokines. The expression of antiapoptotic genes also elevates. CeO$_2$ nanoparticles exhibit a safe profile with gentle activation of signaling proteins. Under these conditions, the pattern of gene expression and cytokines remains within the safer limits. Zinc oxide nanoparticles increase the expression of xenobiotic stress, level of pro-inflammatory cytokines, and apoptosis [35].

The surface properties of the metal and metal oxide nanomaterials and the nature of the proteins control the protein interactions between them and also the process of corona formation. This feature also associates with the coating of different materials. The surface coating changes the behavior and applications such as nanomaterials, and their interplay with immune interactions, and play significant roles. The biocompatibility and biodistribution are the two aspects that assist such interplays in the body of the recipient. Mostly, protein-coated nanomaterials induce an immune response. The macrophages-a component of the immune system, responds to the different types of stimuli. This ability of macrophages shows anti-inflammatory or pro-inflammatory tendencies concerning with their origin, and the intensity of the stimuli. The macrophageal identification of nanomaterials and the foreign molecules induce the manifestation of immune receptors like toll-like receptors (TLRs), nucleotide oligomerization domain-like receptors (NLRs), Fc-Rc receptors of antibody, scavenger receptors, MARCO, SR-Al, integrins, and carbohydrates receptors. Fc Rs receptors are the fragments crystallizable region of antibodies. Investigations related to macrophages and receptors will enhance the understanding of the mechanism involved in the interactions between nanomaterials and the immune system [36].

The types of protein coatings that form corona on the nanomaterials play a functional role during recognition processes. When nanomaterials are present in lower concentrations, the interactive secondary modified protein and cytokine interleukin-6 (IL-6) help in the process of recognition while at higher amounts, the cytokine interleukin like IL-1β, IL-6 and IL-10 play significant roles in recognition process [37]. Macrophages uptake silica-coated superparamagnetic iron oxide nanoparticles (SPIONs) and change the protein corona formation. SPIONs coated with dextran, do not undergo cellular internalization [37, 38].

INTERACTIONS BETWEEN CARBON NANOMATERIALS AND IMMUNE SYSTEM

The interactions between carbon nanomaterials and components of the immune system depend on features like structure, ability to get modified, and other standard physicochemical characteristics, *etc*. These features ensure their possible use in the products of day to day consumption. The recognition processes between components of the immune system and different carbon-based nanomaterials are not well understood. Carbon-based nanomaterials trigger an inflammatory response. Single-walled carbon nanotubes induce macrophages to exhibit strong chemokine response, but, graphene oxide does not. Observed profiles of the studies concerning cytokines and chemokines are helpful in the cytological investigations. During the process of production of chemokine, toll likes receptors also get affected [39].

Carbon nanotubes have relatively more applications in fields like biomedical and biological sciences because of their physicochemical features. These nanomaterials can move across most of the physiological barriers. Carbon nanotubes are applicable as sensors, diagnostic, therapeutic, imaging tools, and tissue engineering, *etc*. Carbon nanotubes are insoluble in water but soluble in organic solvents, and this feature helps in their convenient functionalization and is converting them into biocompatible products. These are preferred options as carriers for biomolecules, drugs, radioactive elements, and specific proteins. Biomolecules and organic molecules are either covalently or non-covalently graft or adsorb on their surfaces [40].

Carbon nanotubes exhibit high potential as a component in the development of vaccines and other related products of immunotherapy. The parameters like higher aspects of the ratio of volume to area, functional flexibility of surface chemistry, controlled structural and morphological abilities; enhance their utility in his field. These are suitably modified to function in a precisely targeted manner. As a result of these modifications, carbon nanotubes can carry antigens across the cell membrane or biological barriers. Despite many beneficial aspects, the functionality of carbon nanotubes causes a pro-inflammatory response [41, 57]. The single-walled carbon nanotubes functionalized with 1.3 dipole acryl, adsorbed with phospholipids, polyethylene glycol, and after amidation, *etc*., do not change their functions [42].

The components of the innate immune system identify carbon nanotubes as a non-self entity. Carbon nanotubes interact with phagocytes. The surface properties of carbon nanotubes and the surface receptors on the phagocytes play an essential role during the recognition process and immune response. The soluble components help in their clearance from the biosystem. There is an interaction between carbon nanotubes and the soluble complements of the innate immune system because of their chemical nature. This feature acts as the bases of activation of the soluble complement of the natural immune system. Their interactions with the immune system are complex and are unavoidable. The masked carbon nanomaterials with some specific group or polymeric compound behave differently in comparison to those to the pristine form. Their overall interaction is of great interest and applies to the rationalized designing of nanomaterials [42]. Modification or modulation of the surface of carbon nanotubes is responsible for the inflammatory response, and it is likely to induce phagocytosis. Phagocytes help in the process of uptake of carbon nanotubes when soluble complement gets deposited on phagocytes. It activates the receptor pathway and results in a decline of proinflammatory response. This process involves complement-receptor bonding with macrophage and B-cell [3, 41]. The surface of carbon nanotubes functionalized with polyethylene glycol interacts with serum complement protein (C1q). This interaction leads to the target recognition and initiates the complex pathway of the complement-related to the immune system, specifically in the case of C1q. The multiwalled carbon nanotubes coated with polyethylene glycol, show relatively less reactivity with C1q, but this reactivity increases with carbon nanotubes bonded covalently (grafted) with polyethylene glycol [43].

The internalization of carbon nanotubes is related to the degree of their binding ability with the complement present in body fluids, like blood. Carbon nanotubes activate complement through the classical pathway, and sometimes, the alternative path is applicable. The degree of activation of complement depends on the surface properties of carbon nanotubes. Multiwalled carbon nanotubes functionalized with alanine or ε-caprolactam exhibit about a 75% increase, while the same multiwalled carbon nanotubes show a decline of about 95% in the activation process during the classical pathway, in comparison to pristine multiwalled carbon nanotubes. Changes in the surface chemistry of carbon nanotubes play a functional role. The degree of activation of the pathways depends on the degree of complement C1q binding with carbon nanotube. This binding factor is an important parameter and is related to the degree of hemolytic and direct protein binding process. The binding ability of carbon nanotubes with complement proteins like C3b, iC3b influences the cellular internalization of carbon nanotubes [44]. Double-walled carbon nanotubes (DWCNT) (10 to 50 µg/ml; exposure time (6h, 24h, 48h, 72h) exhibit non-toxic impacts towards THP-1 monocytes (human). The baseline index increases indicating improved cell adhesion during the early phase of exposure. The double-walled carbon nanotubes aggregate and localize in the cytoplasm of the THP-1 monocytes, and cause an increase in the pro-inflammatory response. When cell surface proteins get modified, it reflects on the ability of double-walled carbon nanotubes. This behavior induces the innate immune system [45].

The graphene is another carbon-based nanomaterial and is applicable in biomolecular, biomedical, and pharmaceutical fields [39, 42, 46]. The cellular functioning such as TGF-β, Smad, and Bcl-2 production relate to the cellular functions and the regular pathway controls it under normal conditions. TGF-α (transforming growth-α path) concern with cellular growth, cellular differentiation, apoptosis, and cellular homeostasis; the Smad pathway relates to the cell cycle and the gene expression, while the Bcl-2 rout refers to lymphoma condition. The JNK kinase is a signaling pathway a first signaling cassette of the mitogenic-activated protein kinase-MAPK [47]. Graphene and graphene oxide induces the production of proinflammatory cytokines like IL-1, IL-6, IL-10, and TNF-α because these activate Toll-like-receptor (TLRs) in macrophages and lymphocytes. Graphene activates TGF-β, Smad, Bcl-2 pathways, and JNK kinase and as a result, induces apoptosis in macrophages. Graphene causes a local inflammatory response and activates the granuloma in parenchyma tissues [48]. The pristine graphene and functionalized graphene (75 µg/ml-1; 72h of exposure) both are hemocompatible specifically to red blood cells, platelets, and pathways of coagulation. These do not cause changes in minerals involved during the expression of cytokines related to peripheral blood mononuclear cells [49].

Smaller sheets of graphene oxide induce a relatively stronger innate and adaptive immune response in comparison to the larger sheets of graphene. During this interaction, cytokines IL1β and TNFα are released involving T-cells and monocytes. These observations indicate that graphene oxide is a good option for applications that include modulation of functionality under the immune system [42]. Graphene oxide nanoparticles (500µg/ml) induce cytotoxicity in the murine macrophage cell line and whole blood cell line. The mitogens, like lipopolysaccharide and phytohaemagglutinin, produce cytotoxic effects. There is no inflammatory response in the absence of these mitogens in the test cell lines. The lipopolysaccharide and phytohaemagglutinin initiate the synthesis of interleukin-6 and interferon-γ in the murine macrophage cell line and the whole blood cell line. In the presence of graphene oxide nanoparticles, synthesis of interleukin-6 and interferon-γ do not take place. This behavior reflects on the ability of graphene oxide nanoparticles to modulate the biomarkers of the immune system [50].

Fullerene is another preferred option as a suitable biomolecular carrier. During their interactions, specific properties such as spatial, dynamics of orientation and surface properties, *etc.*, play a significant role. Its ability to scavenge most of the free radicals in the biosystem makes it a suitable carrier for biomolecules [51]. It is easy to functionalize the Buckyball. The functionalized fullerene oxides with anionic moieties cause hemolysis of blood. The negatively charged moieties under this group bring about hemolysis while positive charged groups/moieties result in hemolysis as per the intensity of these groups on fullerene [1, 52].

The interactions between fullerene and the immune system are useful in biomedical and biomolecular fields. The fullerene C60 derivatives complexed with ovalbumin initiate the interplay between ova-specific IgE and IgG, but this is not so with ovalbumin alone. Fullerene C60 derivatives suppress the synthesis of interleukin-2 (IL-2) by lymphocytes. The lymphocytes produce less IL-2 during their interplay with fullerene C60. The CD-4 type of lymphocytes and T-cell activates in the presence of antiCD-3 and anti-CD-28 antibodies, but the fullerene C60 derivatives minimize the production of IL-2 (interleukin-2). These interactions do not take place when these cells interact with antioxidants. These observations exhibit the suppressive effect of fullerene derivatives on acquired immune. Fullerene activates the cells concerned with the innate immune system and produces various cytokines. These cytokines have antiviral influence (IFN-γ, TNF-α), provide stimulatory signals (IL-12, IL-18), and initiate inflammation (IL-1α/β, IL-6) or produce chemo-attractants. These engage inflammatory cells at the location of the infection, and the antigen initiates the interactions with T-cells. There is a link between the innate and adaptive immune systems. When T-cells get activated The T- cells produce IL-2 on activation, induce clonal proliferation, and develop effectors. IlL-21 activates antigens and cells of innate immunity directly involving B-cells. On activation, the cells of innate immunity become differentiated as plasma cells that form antigen or specific antibodies [53].

Fullerene derivatives, like C60 pyrrolidine (C60-P) and C60-polyhydroxylated fullerene [(C60 (OH) 36)] suppress the activity of acquired immunity. T-cell receptor and antibody production by B-cell activate affecting T-cells. These cells are under the stimulatory influence of anti-CD40/IL-4. There is a decline in the production of ovalbumin-specific antibody, but ovalbumin response by T-cells is not affected when these stimulated cells to interact with (C60-P) and C60 (OH36), (C60-P) [54]. Bis-malonic acid derivatives of fullerene, like derivative-51, reduce the influence of interleukin-33 but activate the effect of IL-6 in the mast cells derived from bone marrow. These nanomaterials are functional intracellularly and extracellularly [55, 56, 58]. When cells become necrotic, they release intracellular interleukin-33. It sends a signal to alert other immune cells while extracellular interleukin-33 brings about the coordination between immune defense and repair mechanism. It also activates the differentiation among helper T-cells facilitating adaptive immune response [59].

INTERACTIONS BETWEEN DENDRIMERS AND IMMUNE SYSTEM

Dendrimers are three-dimensional spherical polymeric molecules. These are conveniently formulated and are suitable carriers of drugs or any other nanomaterials that traps within its branching structure. This structural feature provides ample surface area and loci either at the core or at the peripheral region, for appropriately lodging various biomolecules, pharmaceutical formulations, and other related materials as cargo. These nanomaterials are among the most suitable options to carry the cargo to the specific site in the biosystem. The surface to volume ratio extends up to $1000 m2/g$. This dimension gives the approximate idea of the area that can be available to carry the cargo. Dendrimers are individually perfect symmetrical structures. The dendrimers are polymeric and monodispersible nanomaterials. These can be conveniently functionalized at their core component as well as at their peripheral zone because of their unique architecture and the binding sites available. The incorporation of a specific group within the interior of the dendrimer is related to parameters like the design of monomer, mode of synthesis, properties of the type of backbone of the dendrimer. Further, during the processes of incorporating specific group/s, the main focus is on the adherence to the covalent modifications [60]. Small guest molecules readily get set within the inner zone of dendrimers. The dendrimers based on olefin metathesis exhibit similar behavior. The metathesis is a biochemical reaction that involves the exchange of different kinds of molecules parts and forms other types of molecules. The ruthenium-based Grubbs-Catalys--immunostimulant is an alkylated pyrene, and imidazolinone gets cross-metathesis [61]. The selective proliferation of natural killer cells takes place at the termini of phosphate dendrimers during their interaction with human mononuclear cells from peripheral blood. These natural killer cells play a crucial role during the anticancer immune process [62]. A change of the proinflammatory response occurs when immune cells come in contact with the dendrimer glucosamine, and this interplay influence the chemokines, such as macrophage inhibitory proteins (MIP-1K), Iβ, interleukins IL-8 and cytokines, like tumor necrosis factor (TNF-α), IL-1β, IL-6). The quantitative estimations of m-RNA help to evaluate the proinflammatory inhibitory responses [63]. Dendrimers having terminal carboxyl groups can release cytokines and influence the phenotypes and other functional aspects of cells. Dendrimers having azobisphonate groups, change the phagocytic behavior of human monocytes phenotypically. Polyamidoamine (PAMAM) having hydroxylated groups can prevent or reduce the binding efficacy of cell adhesion molecules. Polyamidoamine dendrimers with $-NH_2$, -OH and –COOH groups at their termini decline the activities due to anti-inflammatory agents like induced nitric oxide and cyclooxygenase [64, 65]. Dendrimers influence the functions of monocytes and macrophages mononuclear phagocytes. These cells exhibit varied phenotypes that respond to various stimuli. Lipopolysaccharide stimulates the translocation of nuclear-factor-kappa-β (NF kβ) and the polarization of the macrophages to the classical activation pathway. The lipopolysaccharides also induce the release of tumor necrosis factor (TNF-α), interleukins like IL-1, IL-6 proinflammatory cytokines. These molecules also affect the expression of cyclin-dependent kinase-86 [66].

INTERACTIONS BETWEEN QUANTUM DOTS AND IMMUNE SYSTEM

Quantum Dots are particular nanosized semiconductor particles. Their optical and electrical features are not the same as that of their respective bulk forms. Different types of quantum dots emit light with specific frequency when exposed to electricity or visible light. These emitted frequencies are controlled by modulating the size of the quantum dots and are of multifaceted applications [67, 68]. Quantum dots are also considered to be as a single object having a particular quantum state of a particle with a specific potential. Such particle shows a tendency to remain localized in one or more regions of the space at this potential. The most remarkable properties of quantum dots include size (9-12nm), their Gaussian emission spectra; these spectra are narrow; these nanomaterials possess tunable dimensions, and ability to excite at a single wavelength. They are exceptionally bright because of the significant absorption extinction coefficients and yield high fluorescent quantum yield. It is possible to discern the emission from a single quantum dot by the human eye under a fluorescence microscope. Quantum dots are inorganic; hence, they are photochemically strong and have relatively more resistance to photobleaching. This feature ensures extended dynamic imaging. Quantum dots blink, *i.e.*, single dot shows intermittent fluoresce. This tendency helps their observation. This feature is of great help to investigate the dynamics of a single protein [69]. Quantum dots incorporate with biological moieties. When a linker arm includes between the quantum dot and the biomolecule moiety, it makes it convenient to study the dynamics of a single protein. This behavior ensures the steric liberty of the resultant molecule, and specifically, is more suitable to the smaller biomolecules like small peptide. Quite often, quantum dots exhibit non-specific binding with varied biomolecular specimens. This aspect is observed more specifically during cell imaging, tissue staining, or *in vivo* investigations. This non-specific binding poses a problem also, and one has to be skeptical about it. This bright and blinking molecule or particle is one of the prime options to facilitate the investigation of protein at the single-molecule level. It is conveniently used to understand the molecular mechanism involved during protein binding, signaling, and regulatory aspects of metabolic activities [70, 71]. Quantum dots are suitable optional tools useful in cell labeling, specifically immune cells, as intracellular sensors, deep tissue imaging, tumor detector, sensitizers, and photodynamic therapy. These nanomaterials influence cell division but without cytokines response concerning CD4+T lymphocytes. These features help the peritoneal macrophages that are activated by bacteria and their cellular functions [72]. Quantum dots support cell proliferation, specifically, the epithelial cells of the gastrointestinal tract, keratinous cells of the skin, germinal cells, and lymphocytes. Peritoneal macrophages produce cytokines and chemokines only when quantum dots interact with quantum dots. The CpG motif, *i.e.*, it is oligodeoxynucleotides-short single-stranded synthetic DNA containing cytosine triphosphate deoxynucleotide (C) then guanine triphosphate deoxynucleotide (G) and p indicate phosphodiesters. This CpG motif in the unmethylated state and can act as immunostimulant [72, 73]. Such peritoneal macrophages and dendritic cells get activated when treated with quantum dots that are conjugated with such a CpG motif involving Toll-like receptors-9 [72].

Quantum dots can influence the inflammatory response. The transcription complex regulates inflammation; it can be inductive or repressive. Quantum dots affect this transcriptional process. These wonder nanomaterials are potential agents to modulate immune response using members of cytokines, like interleukins IL-1 and interferon-γ. Interferon γ-a cytokine is related to innate and adaptive immune responses. This interferon concerns the activation of macrophages and the induction of major compatible complex that constitutes a molecular expression, and different interleukins upregulate. The processes of interleukin up-regulation and down-regulation depend on the concentration of quantum dots. The pro-inflammatory inducing interleukin IL-6 upregulates (11.94 times) during eight hours of exposure of quantum dots (60nM) at the same time anti-inflammatory interleukin IL-10 get downregulated (0.13 times) [69]. The CD-80 and CD-86 are the pro-inflammatory markers that appear in clusters in the murine macrophage cells (RAW264-7). Both the CD-80 and CD-86 pro-inflammatory markers increase in these cells (RAW264.7) during this interplay with quantum dots (1-10nM), and, after their cellular internalization, the pro-inflammatory state develops in these cells. Quantum dots also induce apoptosis and necrosis; the necrotic cells formed exhibit declined fluorescence [74]. Quantum dots pose a threat to health and safety because these nanomaterials behave in a toxic manner causing oxidative stress and inflammation. In addition to these adverse impacts, the subcellular get structures affect adversely.

Quantum dots induce endoplasmic reticulum that gets structurally and functionally disturbed. These also cause dysfunction of mitochondria, rupture of the lysosome, and damage of DNA. Autophagy and apoptosis are of common observation, at least, during exposure of cells and tissues to quantum dots. The quantum dots induce pyroptosis, which involves the complex of inflammasomes and sections of interleukin-1β and interleukin-18. Interleukin-18 is a member of gasdermin-D proteins, and, mostly expressed in epithelial tissues. It is functionally important as a regulator of proliferation and differentiation of epithelial cells. These are expected to be tumor suppressive [75]. Signaling pathways like toll-like receptor-4/myeloid differentiation of primary response-88, nuclear-factor-kappa-light-chain-enhancer of activated B-cells are affected when cells get exposed to quantum dots. The other routes like nucleotide-binding, oligomerization-domain receptors like NOD-like receptor, (NACHI), N-terminal pyrin domain (PYI), and C-terminal caspases-recruitment domain (PYD), Levinerich-repeat-receptor kinase (LRR) concern with the various normal functioning of cell also affects under the influence of quantum dots [76].

INTERACTIVE BEHAVIOR OF MESOPOROUS NANOMATERIALS AND IMMUNE SYSTEM

The zeolites are microporous materials and consist of aluminosilicate minerals. These are useful as commercial absorbent, adsorbent, and catalyst, in industrial and academic fields. Microporous matrix possesses pores size up to 2 nm while pores in the mesoporous form are with 2-50 nm, but, pores in case of the macro-porous model have pore size with range 50 to 75 nm. Some of the standard mesoporous matrices include oxides of niobium, tantalum, titanium, cerium, and tin, *etc*. The activated carbon is the natural and the most common mesoporous material. Generally, mesoporous silica materials are very suitable products for catalytic-immobilization purposes. Their suitability and applications depend on their specific features such as large surface area, mechanical, thermal and chemical stability, enhanced uniform and tunable pore distribution, size, higher capacity for adsorption, a systematic and systemic network of pores that help diffusion for substrate and products during catalytic reaction and immobilization. Solid-supported catalysts are of higher significance in academics as well as in industrial applications because of their properties which promote catalytic reactivity, recyclability, and selectivity remarkably increased contact area and overall yield. The mesoporous natures of solid matrix assist multi-functionality and it ensures the immobilization of multiple catalysts [77]. Porous nanomaterials are the most suitable matrix for loading biological payload like small molecules, peptides, protein, antibodies, and cells. The related degree of loading and releasing of these payloads and kinetics of cargo and releasing are conveniently regulated concerning the properties of pores fabricated in mesoporous nanomaterials [9].

Silica mesoporous nanomaterials induce Th-1 and Th-2 immune responses. The formation of interferon-γ, bacterial activation activity of macrophages, and also induction of B-cell to produce opsonized antibody characterizes Th-1 immune response, and this is cell-mediated immunity. The release of interleukin-5 describes the Th-2 immune response, and it induces eosinophils. It also produces interleukin-4 that facilitates the activation of B-cell. The mesoporous silica nanoparticles enhance the proliferation of lymphocytes, interferon-γ, interleukin-2, and interleukin-4. Once lymphocytes get activated, they secrete interleukin-10. These nanomaterials also increase the antibody titer of IgG, IgG1, IgG2a, IgM, and IgA. These silica mesoporous nanoparticles elevate the level of CD4+ and CD8+. The number of T-cells in bone marrow, spleen, and lymph node also increase. Immunoadjuvant porous silicon-based nano vaccines are biocompatible within a reasonably long range of concentrations. These nanoporous materials help the expression of co-stimulatory signals and the formation of cytokines related to proinflammatory responses [78]. Mesoporous carbon hollow nanospheres suitable carriers and have a higher transporting capacity for vaccine proteins. These are safe adjuvant to boost immune responses. These porous carbon hollow nanospheres have an average size of 200 nm as diameter, average pore size 15 nm, and volume of the pore as $2.85cm3/g^{-1}$ exhibit higher loading capacity, $1040\mu g/ml^{-1}$ of OVA model antigen. This OVA carrier show regulated the release of OVA (ovalbumin), safe profile to healthy cells, and high delivery efficiency concerning macrophages. The OVA-loaded nanomaterials show monodispersibility. The invaginated mesostructured hollow carbon spheres give three times more IgG response in comparison to generally used as adjuvant-Quil A. This fabricated meso-nanomatrix results in a higher concentration of IgG but not IgG2. This nature reflects on the functionality of T-helper -2 (Th-2) responses. The elevated amount of interferon-γ and interleukin-4 also indicate the stimulation of T-helper-1 and T-helper-2 immune response because of the fabricated OVA-loaded mesoporous carbon hollow nanospheres in comparison to Quil (conventional adjuvant) [79].

SOME OF THE ENGINEERED NANOMATERIALS INVOLVED WITH IMMUNE SYSTEM

There are many fabricated or engineered nanomaterials that are useful in biological and biomedical fields. These act as drug carriers, biosensors, clinical investigations, academics, and lots of other industrial applications. The nanomaterials applicable in medical sciences and come under the preview of the immune system directly or indirectly, and these are biodegradable and biocompatible.

POLYMERIC NANOPARTICLES

Polymeric nanoparticles assemble at nanoscale delivery devices. In this class, biodegradable polymer, dendrimers, and micelles are common. These formulations are beneficial as vaccine carriers and the encapsulating antigen in polymer PLGA and polylactides, *etc.* These polymeric nanomaterials are advantageous to sustain and regulate the release of the molecular cargo, protection of the encapsulated antigens from the harsh or adverse ambient environment in the recipient body; and the enzymatic degradation in the shape of a recipient, provide targeted delivery with the attachment of ligaments, and provide probable adjuvant effects. The strategies are useful for their fabrication. The prime focus is on the modulation of their physicochemical properties that enhance their functionality. Smaller polymeric nanomaterials can move across quickly through lymphatic tissue and accumulate in a lymph node. These reside in dendritic cells at least for some time. Other properties, such as charge, surface chemistry, *etc.*, are easily manipulated to gain therapeutic benefits [80].

NANOLIPOSOMES

Nanoliposomes are the liposomal nanomaterials consists of one or more bilayers of amphipathic lipids molecules; lipids have hydrophobic and hydrophilic properties and enclose one or more aqueous compartments. These nanomaterials exhibit two types of immunological perspective (i)- the design of liposomes facilitates the expression of the immune response to an antigen enclosed within; (ii)- these liposomes are coated to avoid immune recognition. In recent times, the developed liposome-based assemblies are beneficial for therapeutic vaccines against HIV, malaria, hepatitis-A, influenza, prostate cancer, colorectal cancer, *etc.* Cationic liposomes have a relatively higher potential in generating an immune response than anionic or neutral liposomes. Antigen delivery to dendritic cells and other targeting moieties like antibodies are improved and improvised. This behavior helps the attachment of liposome to antigens or antibodies as per the need or immunological situation [81].

NANOEMULSIONS

Nanoemulsions are nanosized materials. An emulsion is a thermodynamically unstable system. These unstable systems are stabilized using an emulsifier. These emulsions consist of at least two immiscible liquids; one is dispersed and forms a dispersed phase as globules and other fluid and acts as the continuous phase. Traditionally emulsions are used in case of vaccines. These nanosized emulsions can permeate the nasal mucosa. These carry antigen to the antigen-presenting cells relatively more effectively in comparison to large-sized emulsions. Such nanoemulsions concern with intranasal vaccines and are undertrials in cases of hepatitis-B, HIV, influenza, and anthrax. These show better performance and excellent results. N B-1008-a seasonal influenza vaccine is one such similar product, and it is under therapeutic trials. Nano-HBsAg is a hepatitis vaccine and its assembly nanoemulsion consists of 400 nm oil-in-water emulsion with surface antigen (HBs-Ag) that is a recombinant hepatitis-B virus. Other aspects of this nano-assembly include systemic IgG, mucosal IgA, and the ability to express strong antigen-specific immune responses. Such nanoformulations exhibit advantages such as secure handling, the convenience of administration, longer shelf-life at a higher temperature, and improved storage facilities at room temperature [81].

SOLID-LIPID-NANOMATERIALS

Solid-lipid-nanoparticles are another kind of nanomaterials that are applicable in the field of immunology. These are fabricated or assembled as colloidal particles of the lipid matrix. This lipid is solid at physiological temperature. Such nanoformulations are encapsulated, hydrophobic, or hydrophilic molecules and have a size ranging around 400 nm. These modules correlate with the lower degree of immunostimulatory efficacy; for example, antisense oligo-deoxyribose nucleotides G3139 show a lesser degree of immunostimulatory efficiency. Even, the antitumor activity is low in comparison to encapsulated SLN-encapsulated G3139. The smaller particle size of nanomaterials is responsible for its higher degree of immunostimulatory and antitumoral activity [82].

There is a great scope for the specifically designed nanomaterials in the biomedical and industrial fields. Such nanomaterials exhibit controlled interplay with the components of the immune system and can cause unwanted toxic impacts in the biosystem. The biomedical tools and imaging agents are of significance; even industrial and biological catalytic processes involve the engineered nanomaterials [83, 84, 85].

The immunogenicity represents the potential of a substance that can induce an immune response. When any generated antibody recognizes an antigen, this indicates antigenicity. These aspects need a thorough understanding of the nanomaterials prior to their applications in nanomedicine [86]. It is essential to find the appropriate balance between the overall cellular uptake of nanomaterials and their ability to deliver the specific payload as required [87]. There are many benefits of nanotechnology and also involve high risk like oxidative stress, inflammation, and tumor formation, etc. Some interactions of hybrid engineered nanomaterials and biomolecules, like carbon-bridge like structure and bio-corona, may elaborate immune response. These interactions affect the immune response also. The understanding of the bio-physicochemical interactions at bio-nan--interface is essential as it can help in elaborating biocompatibility, biodispersibility, and effects of engineered nanomaterials on the immune system [88].

CONCLUSION

Most organisms have a defense mechanism for protection against microbial, viral infection, and parasitic invasion, and any other agent (xenobiotics) that causes sickness or any form of health issues. The nanomaterials after the internalization, specifically, the human body, interact with the components of body fluids such as blood cells like monocytes, platelets, leukocytes and dendritic cells in tissues, and macrophages in the lungs, tend to engulf these particles due to phagocytosis. The components of body fluids like plasma proteins, opsonins, and other complement components also get involved in the interaction between nanomaterials, immune cells, and proteins, *etc*. The exposure leads to immunological memory. There is a complex conglomerating functionality between protein and the cells that regulate the short and long terms efficacy of the immune system. The nanomaterials, either in pristine form or engineered forms, are applicable in biomedical, biological, biotechnological, molecular sciences, *etc*. The preferred experimental models are mostly mammals; this makes it easy to extrapolate the observations with a human being. Nanomaterials can activate the innate immune system called the complement system and the adaptive immune system. Immunostimulation and immunomodulation play essential roles in the success of immunotherapy. The nanomaterials act as drug carriers and other applications. Immunomodulation is a critical aspect that can influence the immune system. The understanding of the mechanism involving nanomaterials and immunomodulation helps to improve the degree of utility of nanomaterials about immunological treatment. Nanomaterials are either immunostimulant or immunosuppressant and this aspect concerns their therapeutic significance. The common nanomaterials such as metal and metal oxide, carbon nanotubes, graphene, fullerene, and quantum dots, and mesoporous nanomatrix, *etc*., influence the immune system of an organism. The antigenicity concerning nanomaterials is not well established. The different derivatized forms or primitive forms of nanomaterials do not exhibit specific immune response even in the presence of potent adjuvant. Some of them, when conjugating with some polymer or specific biomolecules, do exhibit strong antigenic properties. The nanomaterials are in use in various fields; it becomes imperative to judicially check the immunological aspects of the nanomaterials in different therapeutic products for human and veterinary use.

REFERENCES

[1] Dobrovolskaia MA, Aggarwal P, Hall JB, McNeil SE. Preclinical studies to understand nanoparticle interaction with the immune system and its potential effects on nanoparticle biodistribution. Mol Pharm 2008; 5(4): 487-95.
[http://dx.doi.org/10.1021/mp800032f] [PMID: 18510338]

[2] Luo YH, Chang LW, Lin P. Metal-based nanoparticles and the immune system: activation, inflammation and applications Biomedical Research International 2015; 12.
[http://dx.doi.org/http://dx.doi.org/10.1155/2015/143720] [PMID: 143720]

[3] Pondman KM, Salvador-Morales C, Pandya B, Sim RB, Kishor U. Interactions of innate immune system with carbon nanotubes. Nano Scale Horizons 2017.
[http://dx.doi.org/10.1039/C6NH00227G]

[4] Moghimi SM, Andersen AJ, Ahmadvand D, Wibroe PP, Andresen TL, Hunter AC. Material properties in complement activation. Adv Drug Deliv Rev 2011; 63(12): 1000-7. https:// doi.org
[http://dx.doi.org/10.1016/j.addr.2011.06.002] [PMID: 21689701]

[5] Moghimi SM, Farhangrazi ZS. Nanomedicine and the complement paradigm. Nanomedicine (Lond) 2013; 9(4): 458-60.
[http://dx.doi.org/10.1016/j.nano.2013.02.011] [PMID: 23499667]

[6]   Moghini SM, Simberg D. Complement activation turnover on the surface of nanoparticles Nano Today 2017; 15: 8- 10.
[http://dx.doi.org/10.1016/j.nanotod.2017.03.001]

[7]   Boraschi D, Costantino L, Italiani P. Interaction of nanoparticles with immunocompetent cells: nanosafety considerations. Nanomedicine (Lond) 2012; 7(1): 121-31.
[http://dx.doi.org/10.2217/nnm.11.169] [PMID: 22191781]

[8]   Zolnik BS, González-Fernández A, Sadrieh N, Dobrovolskaia MA. Nanoparticles and the immune system. Endocrinology 2010; 151(2): 458-65.
[http://dx.doi.org/10.1210/en.2009-1082] [PMID: 20016026]

[9]   Jeong M, Kim H, Park JH. Porous materials for immune modulation. Open Material Science 2018; 4(1): 1-14.
[http://dx.doi.org/10.1515/oms-2018-0001]

[10]   Fontana F, Figueiredo P. Bauleth- Ramos T, Correia A, Santos H A, Immunostimulation and and immunosuppression: Nanotechnology on the brink. Small Methods 2018; 2(5): 1700347.
[http://dx.doi.org/10.1002/smtd.201700347]

[11]   Himly M, Mills-Goodlet R, Geppert M, Duschl A. Nanomaterials in the context of type-2 immune responses-fears and potentials. Front Immunol 2017; 8: 471.
[http://dx.doi.org/10.3389/fimmu.2017.00471] [PMID: 28487697]

[12]   FDA. 1999.https://www.fda.gov/RegulatoryInformation/Guidances/ucm080495.htm

[13]   De Temmerman ML, Rejman J, Demeester J, Irvine DJ, Gander B, De Smedt SC. Particulate vaccines: on the quest for optimal delivery and immune response. Drug Discov Today 2011; 16(13-14): 569-82.
[http://dx.doi.org/10.1016/j.drudis.2011.04.006] [PMID: 21570475]

[14]   Goldsmith M, Mizrahy S, Peer D. Grand challenges in molding the immune response with RNA. Nanomedicine 2011; 6: 10.
[http://dx.doi.org/10.2217/nmm.11.162]

[15]   Peer D, Park EJ, Morishita Y, Carman CV, Shimaoka M. Systemic leukocyte-directed siRNA delivery revealing cyclin D1 as an anti-inflammatory target. Science 2008; 319(5863): 627-30.
[http://dx.doi.org/10.1126/science.1149859] [PMID: 18239128]

[16]   Zou L, Tao Y, Payne G, *et al.* Targeted delivery of nano-PTX to the brain tumor-associated macrophages. Oncotarget 2017; 8(4): 6564-78.
[http://dx.doi.org/10.18632/oncotarget.14169] [PMID: 28036254]

[17]   Martin A. Physical pharmacy: physical-chemical principles in the pharmaceutical science. 4thEd., Williams Wilkins. Baltimore Publication, MD 1993.

[18]   Higaki M, Ishihara T, Izumo N, Takatsu M, Mizushima Y. Treatment of experimental arthritis with poly(D, L-lactic/glycolic acid) nanoparticles encapsulating betamethasone sodium phosphate. Ann Rheum Dis 2005; 64(8): 1132-6.
[http://dx.doi.org/10.1136/ard.2004.030759] [PMID: 15695536]

[19]   Mundargi RC, Babu VR, Rangaswamy V, Patel P, Aminabhavi TM. Nano/micro technologies for delivering macromolecular therapeutics using poly(D,L-lactide-co-glycolide) and its derivatives. J Control Release 2008; 125(3): 193-209.
[http://dx.doi.org/10.1016/j.jconrel.2007.09.013] [PMID: 18083265]

[20]   Mitchell LA, Lauer FT, Burchiel SW, McDonald JD. Mechanisms for how inhaled multiwalled carbon nanotubes suppress systemic immune function in mice. Nat Nanotechnol 2009; 4(7): 451-6.
[http://dx.doi.org/10.1038/nnano.2009.151] [PMID: 19581899]

[21]   FDA. Guidance for industry: liposome drug products. Food and Drug Administration 2002.

[22]   Shi J, Votruba AR, Farokhzad OC, Langer R. Nanotechnology in drug delivery and tissue engineering: from discovery to applications. Nano Lett 2010; 10(9): 3223-30.

[http://dx.doi.org/10.1021/nl102184c] [PMID: 20726522]

[23]   Khisamutdinov EF, Li H, Jasinski DL, Chen J, Fu J, Guo P. Enhancing immunomodulation on innate immunity by shape transition among RNA triangle, square and pentagon nanovehicles. Nucleic Acids Res 2014; 42(15): 9996-10004.
[http://dx.doi.org/10.1093/nar/gku516] [PMID: 25092921]

[24]   Nel AE, Mädler L, Velegol D, *et al.* Understanding biophysicochemical interactions at the nano-bio interface. Nat Mater 2009; 8(7): 543-57.
[http://dx.doi.org/10.1038/nmat2442] [PMID: 19525947]

[25]   Petrarca C, Clemente E, Amato V, *et al.* Engineered metal based nanoparticles and innate immunity. Clin Mol Allergy 2015; 13(1): 13.
[http://dx.doi.org/10.1186/s12948-015-0020-1] [PMID: 26180517]

[26]   Horev-Azaria L, Kirkpatrick CJ, Korenstein R, *et al.* Predictive toxicology of cobalt nanoparticles and ions: comparative *in vitro* study of different cellular models using methods of knowledge discovery from data. Toxicol Sci 2011; 122(2): 489-501.
[http://dx.doi.org/10.1093/toxsci/kfr124] [PMID: 21602188]

[27]   Potnis P, Dutta DK, Wood SC. Toll-like receptor 4 signaling pathway mediates proinflammatory immune response to cobalt-alloy particles Cell Immunology 2013; 282(1): 53-65.
[http://dx.doi.org/10.1016/j.cellimm 2013.04.003]

[28]   Seiffert J, Hussain F, Wiegman C, *et al.* Pulmonary toxicity of instilled silver nanoparticles: influence of size, coating and rat strain. PLoS One 2015; 10(3): e0119726.
[http://dx.doi.org/10.1371/journal.pone.0119726] [PMID: 25747867]

[29]   Weaver JL, Tobin GA, Ingle T, *et al.* Evaluating the potential of gold, silver, and silicon nanoparticles to saturate mononuclear phagocytic system tissues under repeated dosing conditions, Particle and Fibre Toxicology 2017; 14: 25.

[30]   Aggarwal P, Hall JB, McLeland CB, Dobrovolskaia MA, McNeil SE. Nanoparticle interaction with plasma proteins as it relates to particle biodistribution, biocompatibility and therapeutic efficacy. Adv Drug Deliv Rev 2009; 61(6): 428-37.
[http://dx.doi.org/10.1016/j.addr.2009.03.009] [PMID: 19376175]

[31]   Oostingh GH, Casals E, Italiani P, *et al.* Problems and challenges in the development and validation of human cell-based assays to determine nanoparticles induced immunomodulatory effects Particle and Fibre Toxicology 2011; 8.
[http://dx.doi.org/10.101186/1743-8977-8/1/8]

[32]   Falugi C, Aluigi MG, Chiantore MC, *et al.* Toxicity of metal oxide nanoparticles in immune cells of the sea urchin. Mar Environ Res 2012; 76: 114-21.
[http://dx.doi.org/10.1016/j.marenvres.2011.10.003] [PMID: 22104963]

[33]   Lozano-Fernández T, Ballester-Antxordoki L, Pérez-Temprano N, *et al.* Potential impact of metal oxide nanoparticles on the immune system: The role of integrins, L-selectin and the chemokine receptor CXCR4. Nanomedicine (Lond) 2014; 10(6): 1301-10.
[http://dx.doi.org/10.1016/j.nano.2014.03.007] [PMID: 24650882] DOI.

[34]   Lahir YK. Principles and applications toxicology. Bangalore, India: See Kay Publications 2013; pp. 040-560. ISBN: 978-81-924169-5-3.

[35]   Simon-Vazquez R, Loozano-Fernandez T, Devila-Grana A. Gonzelez-Fernandez, Metal oxide nanoparticles interact with immune cells and activate different responses. Int J Nanomedicine 2016; 2016(11): 4657-68.
[http://dx.doi.org/10.2147/IJN.S110465] [PMID: 27695324]

[36]   Borgognoni CF, Kim JH, Zucolotto V, Fuchs H, Riehemann K. Human macrophage response to metal oxide nanoparticles: A review. Artif Cells Nanomed Biotechnol 2018; 46: 694-703.
[http://dx.doi.org/10.1080/21691401.2018.1468767]

[37]   Borgognoni CF, Mormann M, Qu Y, *et al.* Reaction of human macrophages on protein corona covered TiO_2 nanoparticles. Nanomedicine (Lond) 2015; 11(2): 275-82.
[http://dx.doi.org/10.1016/j.nano.2014.10.001] [PMID: 25461290]

[38]   Vogt C, Pernamalm M, Kohonen P, *et al.* Proteomics analysis reveals distinct corona composition on magnetic nanoparticles with different surface coating implication for interaction with primary human macrophages. PLoS One 2015; 10(10): e129008.
[http://dx.doi.org/10.1371/journal.pone.0129008]

[39]   Mukherjee SP, Bondarenko O, Kohonen P, *et al.* Macrophage sensing of single-walled carbon nanotubes via Toll-like receptors. Sci Rep 2018; 8(1): 1115.
[http://dx.doi.org/10.1038/s41598-018-19521-9] [PMID: 29348435]

[40]   Battigelli A, Ménard-Moyon C, Da Ros T, Prato M, Bianco A. Endowing carbon nanotubes with biological and biomedical properties by chemical modifications. Adv Drug Deliv Rev 2013; 65(15): 1899-920.
[http://dx.doi.org/10.1016/j.addr.2013.07.006] [PMID: 23856410]

[41]   Oostingh GH, Casals E, Italiani P, *et al.* Problems and challenges in the development and validation of human cell-based assays to determine nanoparticles induced immunomodulatory effects. Particle and Fibre Toxicology 2011; 8.
[http://dx.doi.org/10.101186/1743-8977-8/1/8]

[42]   Orecchioni M, Bedognetti D, Sgarrella F, Marinol M, Bianco A. Impact of carbon nanotubes and graphene on immune cells. Journal of Translational Medicine 2014; 12: 138.
[http://dx.doi.org/10.1186/1479-5876-12-138]

[43]   Belime A, Gravel E, Brenet S, *et al.* Mode of polyethylene glycol coverage on carbon nanotubes affects the binding of innate immune protein C1q. J Phys Chem B 2018; 122(2): 757-63.
[http://dx.doi.org/10.1021/acs.jpcb.7b06596] [PMID: 28915042]

[44]   Recognition of carbon nanotube by the human innate immune system RayBak-Smith M J. 2011.
https://Labofnano.gmu.edu/wp-content/uploads/2017/02/Raybak-Smith_Pondman.pdf

[45]   Samir D, Bachelet C, Severine M-R, *et al.* Biological effects of double-walled carbon nanotubes on the innate immune system: an in-vivo study on THP-1human monocytes. Toxicology 2016; 365: 1-8.
[http://dx.doi.org/10.1016/j.tox.2016.07.019]

[46]   Lahir YK. Graphene and graphene-based nano-drug delivery: a critical review.Applications of Targeted Nano-Drugs and Delivery Systems. Elsevier PublicationElsevier 2018.

[47]   Vlahopoulos S, Zoumpourlis VC. A key modulator of intracellular signaling Biochemistry Mosc 2004; 69(8): 844-54.
[http://dx.doi.org/10.1023/B BIRY.0000040215.02460.45]

[48]   Dudek I, Skoda M, Jarosz A, Szukiewicz D. The molecular influence of graphene and graphene oxide on the immune system under *in vitro* and *in vivo* conditions. Archivum Immunologiae et therapies experiments 2016; 64(3): 195-215.
[http://dx.doi.org/10.1007/s00005-015-0369-3]

[49]   Sasidharan A, Panchakarla LS, Sadanandan AR, *et al.* Hemocompatibility and macrophage response of pristine and functionalized graphene. Small 2012; 8(8): 1251-63.
[http://dx.doi.org/10.1002/smll.201102393] [PMID: 22334378]

[50]   Lategan K, Alghadi H, Bayati M, de Cortalezzi MF, Pool E. Effects of graphene oxide nanoparticles on immune system biomarkers produced by RAW 264.7 and human whole blood cell cultures. Nanomaterials (Basel) 2018; 8(2): 125.
[http://dx.doi.org/10.3390/nano8020125] [PMID: 29495255]

[51]   Lahir YK. Impacts of fullerene on the biological system, Clinical Immunology Endocrinology and Metabolic Drugs 2017; 4(1): 48- 57 .
[http://dx.doi.org/10.2174/2212707004666171111351624]

[52] Bosi S, Feruglio L, Da Ros T, *et al.* Hemolytic effects of water-soluble fullerene derivatives. J Med Chem 2004; 47(27): 6711-5.
[http://dx.doi.org/10.1021/jm0497489] [PMID: 15615520]

[53] Petrovic D, Seke M, Branislava S, Djordjevic A. Application of anti/pro-oxidant fullerene in nanomedicine along with fullerene influence as the immune system Journal of Nanomaterials 2015; 11.
[http://dx.doi.org/https://dx.doi.org/10.1155/2015/565638] [PMID: 565638]

[54] Hirai T, Yoshioka Y, Udaka E, *et al.* Potential suppressive effects of two C60 fullerene derivative on acquired immunity Nanoscale Research Letters 2016; 11: 449.
[http://dx.doi.org/10.1086/s11671-016-1663-7]

[55] Miller AM. Role of IL-33 in inflammation and disease. J Inflamm (Lond) 2011; 8(1): 22.
[http://dx.doi.org/10.1186/1476-9255-8-22] [PMID: 21871091]

[56] Funakoshi-Tago M, Miyagawa Y, Ueda F, *et al.* A bis-malonic acid fullerene derivative significantly suppressed IL-33-induced IL-6 expression by inhibiting NF-κB activation. Int Immunopharmacol 2016; 40: 254-64.
[http://dx.doi.org/10.1016/j.intimp.2016.08.031] [PMID: 27632703]

[57] Fadel TR, Fahmy TM. Immunotherapy applications of carbon nanotubes: from design to safe applications. Trends Biotechnol 2014; 32(4): 198-209.
[http://dx.doi.org/10.1016/j.tibtech.2014.02.005] [PMID: 24630474]

[58] Castro E, Garcia AH, Zavala G, Echegoyen L. Fullerene in biology and medicine Journal of Materials Chemistry 2017. https://www.sesres.com/wp.contents

[59] Martin NT, Martin MU. Interleukin 33 is a guardian of barriers and a local alarmin. Nat Immunol 2016; 17(2): 122-31.
[http://dx.doi.org/10.1038/ni.3370] [PMID: 26784265]

[60] Hecht S. Functionalizing the interior of dendrimers, synthetic challenges, and applications. J Polym Sci A Polym Chem 2003; 41(8): 1047-58.
[http://dx.doi.org/10.1002/pola.10643]

[61] Catherine O, Fréchet JMJ. Incorporation of functional guest molecules into an internally functionalizable dendrimers through olefin metathesis. Macromolecules 2005; 38(15): 6276-84.
[http://dx.doi.org/10.1021/ma050818a]

[62] Griffe L, Poupot M, Marchand P, *et al.* Multiplication of human natural killer cells by nanosized phosphonate-capped dendrimers. Angew Chem Int Ed Engl 2007; 46(14): 2523-6.
[http://dx.doi.org/10.1002/anie.200604651] [PMID: 17300122]

[63] Hayder M, Fruchon S, Fournié JJ, Poupot M, Poupot R. Anti-inflammatory properties of dendrimers per se. ScientificWorldJournal 2011; 11: 1367-82.
[http://dx.doi.org/10.1100/tsw.2011.129] [PMID: 21789472]

[64] Dernedde J, Rauch A, Weinhart M, *et al.* Dendritic polyglycerol sulfate as multivalent inhibitors of inflammation. PANS 2010; 107(46): 19679-84.
[http://dx.doi.org/10.1073/pnas.1003103107]

[65] Avti PK, Kakkar A. Dendrimers as anti-inflammatory agents. Braz J Pharm Sci 2013; 49(special issue)www.scielo.br/pdf/bjps/v49nspe/a06v49nspe.pdf

[66] Posados I, Romero Castillo L, Brahmi EL, *et al.* Neutral high generation phosphorous dendrimers inhibit the macrophage-mediated inflammatory response *in vitro* and *in vivo*. PANS-USA 2017; 114(37): E7660-9.
[http://dx.doi.org/10.1073//pnas.1704859114]

[67] Sabaeian Mohammad, Khaledi-Nasab Ali. Size-dependent inter-sub-band optical properties of dome-shaped in As/GaAAs quantum dots with wetting layer Applies optics 2012; 51(18): 4176-85.

[http://dx.doi.org/10.1364/AO.51.004176]

[68]   Khaledi-Nasab Ali. Kerr nonlinearity due to inter-sub-band transitions in a three- level InAs/GaAs quantum dot: the impact of a wetting layer on dispersion curves. J Opt 2014; 16(5): 055004.
[http://dx.doi.org/10.1088/2040-8978/16/5/055004]

[69]   Romoser AA, Chen PL, Berg JM, *et al.* Quantum dots trigger immunomodulation of the NFκB pathway in human skin cells. Mol Immunol 2011; 48(12-13): 1349-59.
[http://dx.doi.org/10.1016/j.molimm.2011.02.009] [PMID: 21481475]

[70]   Pons T, Mattoussi H. Investigating biological processes at the single molecule level using luminescent quantum dots. Ann Biomed Eng 2009; 37(10): 1934-59.
[http://dx.doi.org/10.1007/s10439-009-9715-0] [PMID: 19521775]

[71]   Pinaud F, Clarke S, Sittner A, Dahan M. Probing cellular events, one quantum dot at a time. Nat Methods 2010; 7(4): 275-85.
[http://dx.doi.org/10.1038/nmeth.1444] [PMID: 20354518]

[72]   Hoshino A, Hanada S, Manabe N, Nakayama T, Yamamoto K. Immune response induced by fluorescent nanocrystal quantum dots *in vitro* and *in vivo*. IEEE Trans Nanobioscience 2009; 8(1): 51-7.
[http://dx.doi.org/10.1109/TNB.2009.2016550] [PMID: 19304501]

[73]   Weiner GJ, Liu HM, Wooldridge JE, Dahle CE, Krieg AM. Immunostimulatory oligodeoxynucleotides containing the CpG motif are effective as immune adjuvants in tumor antigen immunization. Proc Natl Acad Sci USA 1997; 94(20): 10833-7.
[http://dx.doi.org/10.1073/pnas.94.20.10833] [PMID: 9380720]

[74]   Gladovaskay A, Gerard VA, Nosov M, *et al.* The interaction of quantum dots with RAW 2647 cells nanoparticle quantification uptake kinetics and immune response study RSC Adv. 2015; 5.
[http://dx.doi.org/10.1039/c5ea04233j] [PMID: 43350]

[75]   Liu X, Zhang Z, Ruan J, *et al.* Inflammasome-activated gasdermin D causes pyroptosis by forming membrane pores. Nature 2016; 535(7610): 153-8.
[http://dx.doi.org/10.1038/nature18629] [PMID: 27383986]

[76]   Wang Y, Tang M. Dysfunction of various organelles provokes multiple cell death after quantum dot exposure. Int J Nanomedicine 2018; 13: 2729-42.
[http://dx.doi.org/10.2147/IJN.S157135] [PMID: 29765216]

[77]   Lai Cheng Y. Mesoporous silica nanomaterials applications in catalysis, J Thermodyn Catal 2013; 5(1)1000e124. https://dx.doi.org/10

[78]   Fontana F, Shahbazi MA, Liu D, *et al.* Multistaged nanovaccines based on porous silicon @ acetal Dextran @ cancer cell membrane for cancer immunotherapy. Advanced Materials 2017; 29(7): 1603239.
[http://dx.doi.org/10.1002/adma.20 1603239] [PMID: 1603239]

[79]   Jambhrunkar M, Yu M, Zhang H, *et al.* Pristine mesoporous carbon hollow spheres as safe adjuvants induce excellent Th-2 biased immune response Nano Research 2018; 11(1): 370-82.

[80]   Eshete M, Bailey K, Nguyen TDT, Santosh A, Choi S-O. Interaction of immune system protein with PEGylated and un-PEGylated polymeric nanoparticle Advances in nanoparticles 2017; 6(3): 103-13.
[http://dx.doi.org/10.4236/anp.2017.63009]

[81]   Zamani P, Momtazi-Borojeni AA, Nik ME, Oskuee RK, Sahebkar A. Nanoliposomes as the adjuvant delivery systems in cancer immunotherapy. J Cell Physiol 2018; 233(7): 5189-99.
[http://dx.doi.org/10.1002/jcp.26361] [PMID: 29215747]

[82]   Pan X, Chen L, Liu S, Yang X, Gao JX, Lee RJ. Antitumor activity of G3139 lipid nanoparticles (LNPs). Mol Pharm 2009; 6(1): 211-20.
[http://dx.doi.org/10.1021/mp800146j] [PMID: 19072654]

[83]   Lovelyn C, Attama AA. Current status of nanoemulsions in drug delivery. J Biomater Nanobiotechnol 2011; 2: 626-39.
[http://dx.doi.org/10.4236/jbnb.2011.225075]

[84]   Martínez Sánchez MC, Corma Canós A. Inorganic molecular sieves: Preparation, modification and industrial application in catalytic processes. Coord Chem Rev 2011; 255(13-14): 1558-80.
[http://dx.doi.org/10.1016/j.ccr.2011.03.014]

[85]   Xu L. Modulation of the immune system by fullerene and graphene derivatives Biomedical applications and toxicology of carbon nanomaterials 2016; 213-30.
[http://dx.doi.org/10.1002/9783527692866]

[86]   Ilinskaya AN, Dobrovolskaia MA. Understanding the immunogenicity and antigenicity of nanomaterials: Past, present and future. Toxicol Appl Pharmacol 2016; 299: 70-7.
[http://dx.doi.org/10.1016/j.taap.2016.01.005] [PMID: 26773813]

[87]   Fadeel B. Hide and seek: Nanomaterials interactions with the immune system. Front Immunol 2019; 10: 133.
[http://dx.doi.org/10.3389/fimmu.2019.00133] [PMID: 30774634]

[88]   Bonner JC, Brown JM. Introduction. In: Interactions of nanomaterials with the immune system. Springer International Publishing, 2020.
[http://dx.doi.org/10.101007/978-3-030-33962-3]

CHAPTER 10

Broad Spectra of Applications Based Interactions of Nanomaterials

Abstract: Advancements in the nanoscience, nanotechnology and material science involve the principles of fundamental physical, chemical, and biological sciences. These scientific advancements have opened a vast horizon for understanding the mechanisms of the physiology of life and the environment. This progress has also provided suitable materials and appropriate methodology concerning the developments. Nanomaterials are the bridge between the atomic and molecular and bulk form of matter. These nanoscaled materials are modified, fabricated, and reach most of the biological targets in life forms. As a result, these become useful materials for applications in medical sciences, industrial processes, health care, and home security. The biological components, such as cells and tissues in the biosystem and nanomaterials, interact amicably with restrictions concerning their physicochemical features. Nanomaterials exhibit biocompatibility and bioavailability within the physiological environment. This ability is the primary basis of their applications in almost every sphere of investigation, diagnosis, and treatment of ailments. Enzyme technology, DNA and RNA technology, tissue engineering, military, and communications, energy, and many industrial processes are the fields where different nanomaterials are useful and provide beneficial and desired results. In this chapter, various potential application based interactions related to medical sciences, biomolecular investigations, biotechnology, genetic engineering, tissue engineering, environmental aspects, military, *etc.*, have been envisaged.

Keywords: Antibacterial nanomaterials, Antifungal nanomaterials, Antiviral nanomaterials, Bladder and cartilage implants, Drug delivery, Nanomaterials and defense, Nanomaterials as biocomposites, nanomaterials as neural, nanotechnology, Nanomaterials for tissue culture, Vascular tracheal implants,.

OVERVIEW

The nanoscience, nanotechnology, and materials sciences are the basis for the ever-increasing pace of development. These developments meet the challenges related to industrial, health, agricultural, natural, manufacturing, energy, digital expression and communication, neuronal, and cognitive fields. These make provisions for long-lasting technologies that take care of the problems related to these fields and also digital currencies, hydrogen energy storage, brain to brain in-

Yogendrakumar H. Lahir & Pramod Avti

terface and robotics, *etc*. These fields reflect on many interdependent technologies that influence the society, economy, and the environment. The advancements in the area of nanoscience, nanotechnology, and materials science are the fast-paced changes in materials science. The materials science perceives the matter to be recognizable as functional materials, next-generation materials, and self-assembling materials. The substances that function on the borrowed principles from biological aspects and adopt a new behavior pattern accordingly are considered to be functional materials. Super light and active materials get affected or react in accordance to the environmental changes and generally culminate as smart materials. These original materials come under next-generation substances; can stimulate the components of the environment, the central users. Materials that exhibit self-assembly behavior constitute a large scale, very precise, able to improve their properties concerning strength, tear resistance, conductivity, *etc*., and are included in next-generation substances [1].

Maters get organized in different types and this organizing concerns with the current advancements in materials science. Smart materials are the material with one or more specifically modified properties. The external changes or stimuli like stress, electric or magnetic field, temperature, pH, moisture, *etc*., help in such modifications during the designing of these materials. Thermo bimetals self-regulate and consume energy for a more extended period. Furthermore, these bimetals are activated thermally and can make glass to shade after being exposed to sunlight. Superomniphobic materials can float on aqueous fluid and repel oil. These behave like water bugs on the surface of the water pond. Auxetic contents change their thickness under force. These become thicker in the perpendicular direction to the direction of force applied. These materials either have hinge-like structures, or some of their parts act like a hinge, which gets flexed under stretch. Such auxetic materials are suitable for packaging, pads for knee and elbow, robust shock absorbing material, sponge materials for mops, and materials for body armor, *etc*. Aerogels are ultra-light porous materials, and they are modifications from the gel after replacing the liquid present by air or gas. Aerogels have extremely low density, conductivity, and give polystyrene (Styrofoam) feel when touched. These materials are useful to improve thermal insulation, as chemical absorbents to clean the spills, these are suitable materials for electrochemical supercapacitors, and shock absorbers. Biomaterials are derived from natural sources or synthesized in the laboratory. These materials are appropriate to replace or increase the inherent functionality of the organ or body part. These are the potential materials to improve the drug delivery system and degree of acceptance of graft among transplants. Graphene consists of only carbon elements. The arrangement of the carbon atoms in a typical hexagonal pattern, forming a single sheet with thickness, equals the thickness of one atom. These materials are very light but strong and similar to graphite [2]. These nanomaterials are suitable for the use of low-cost solar cells and display screens in inexpensive mobile devices. These are suitable to store hydrogen for cars powered by the fuel cell, like biomedical and chemical sensors, ultracapacitors and faster-charging batteries, *etc*. Graphene nanomaterials are advantageous for use in integrated transistors. These form the functional components of actuators, nanoelectromechanical systems (the devices that integrate electrical and mechanical systems functionally). These systems also include pumps or motors and good options for physical, chemical, and biological sensors.

Nano-factories are the devices that act as nanomachines. These interact with reactive molecules involving mechanosynthesis and form correct assembled structures of organic products. These products follow a specific arrangement of the macroscopic sized pattern with atomic precision. A material product with large scale assembly exhibit specific behavior, *i.e.*, in an organized system, the components form an organized structure with its local parts in a particular sequence. Thus, the bottom-up approach results in the formation of 3D sequenced structures like DNA, RNA, and protein that come in this category. Substances under metamaterial have specifically and precisely arranged geometrical shaped materials and influence light and sound unconventionally. Such materials are useful in aerospace devices, devices to monitor fractures, and to manage smart solar power devices. Self-healing materials are referred to as intelligent materials work as a biological system, and incorporate a repair mechanism helping the healing of damaged tissues. These materials include polymers, ceramics, and can rectify intrinsic damage caused due to the devices. These materials also reduce the impact of degradation, enhance life span, and are cost-effective [1, 3]

VARIOUS APPLICATIONS OF NANOMATERIALS

Tremendous advancement has taken place in all spheres of life due to the innovative technologies in recent times. Improvements in the fields of nanotechnology and materials science have played significant roles in these advancements. These two disciplines provide methodologies and suitable materials for specific applications for such developments. Nanomaterials are aptly modified or fabricated those help involvements in different investigations and brought successful and useful applications in industries, health care, and home security in addition to the cosmetics and pharmaceutical advancements.

Nanomaterials for Drug Delivery and Biomedical Applications

Drug delivery is the fundamental aspect of health care and safety; it delivers genes, biomolecules, and drugs to a specific site in a biosystem. Parameters like an increased area of the reactants that offers a more significant number of atoms and molecules that interact, bind, adsorb, and facilitate the delivery process.

Nanomaterials act as efficient agents to carry drugs, probes, genes, peptides, RNA, DNA, *etc.*, successfully. Drugs and the corresponding nanomaterials are either engineered or fabricated as per the required to act as carriers. The nanosized substances can reach the extracellular matrix, cells, and cell organelles easily because they promptly move across most of the biological barriers. Nanomaterials are selective application in cancer therapy, tissue engineering, treating inflamed tissue, or any other set target within a biosystem. Nanomaterials increase the efficacy of the passive targeting of drugs concerning intracellular macrophages, those present in the liver or spleen. The altered and specially modified nanomaterials avoid the first-pass stage of hepatic metabolism that involves hydrophobic polymers. The specially modified nanomaterials enhance the life of drugs and cargo in the circulatory system; these ensure the release of the shipment and complete the intended interaction at the site [4 - 6].

Silica, gold, and other heavy metal nanoparticles, quantum dots, nanocrystals, dendrimers, and the concerned contrast agents, *etc.*, make them useful for the drug delivery system, diagnostic and sensing agents. The fabricated and modified forms exhibit the ability to avoid a multi-drug-resistance system present in a biosystem or otherwise. This capability makes them capable of drug carriers. The polymeric nanomaterials, nanoliposomes, dendrimers, and micelle, conjugate with ligands or encapsulate drugs.

Their basic functionalities favor to carry and deliver the drug. The lipospheres are the stable and solid lipid nanoparticles or nanospheres. These nanomaterials are the products of mono-, di-, triglycerides, fatty acids, waxes, and their combinations; these are fabricated using an appropriate surfactant. These nanomaterials are solid at the physiological temperature (37°C), and their diameter is smaller than 1000 nm. Such formulations are useful in the case of hydrophilic materials and suitable drug vehicles as emulsions [6, 7]. The polymer and biological polysaccharides, chitosan, cyclodextrin, dextran, and other polymers, having biodegradable core-shell, are suitable options as drug vehicles and for controlled drug release [6, 8].

Dendrimers are branched polymeric molecules having a diameter (2.5 to 10.0 nm) and many functional groups. These nanomaterials are natural, synthetic, and exhibit monodispersibility. The dendrimers are peptides, glycodendrimers, poly propylamine, and polyethyleneimine types for biochemical aspects. These nanomaterials are the most suitable materials as drug vehicles because these are biocompatible and monodispersible in biological fluids [9]. Nanoshells are nanomaterials that have silica core and thin metallic coating. These are convenient options for targeting tissues immunologically, as imaging agents, and as drug carriers. These are useful in the investigations concerning micrometastasis of tumors, diabetic therapy, and diagnostic agents for immunoassay of the whole blood [10].

Carbon-based nanomaterials such as carbon nanotubes, fullerene, and graphene are useful options for biomedical applications. These are hollow tubular, spherical or ellipsoidal, or sheet-like nanomaterials, and are readily modified and fabricated as biocompatible nanomaterials that show a high degree of biodistribution. Generally, these incorporate with carboxyl or ammonium group, the resultant products act are capable vehicles for antiviral drugs, antibiotics, and agents for anticancer drugs [2, 11 - 15]. Nanopores are wafer-like nanomaterials having about 20 nm diameter and enormous distribution of pores. Through these pores, molecules like oxygen, glucose, and insulin, can move across. Such nanomaterials are used to protect the transplanted tissues from the adverse effects of the immune system of the recipient. These nanopores are helpful substances in transplanting β-cells of the pancreas, as enclosures in these nanopores meeting the pertinent requirements. These are applicable in DNA sequencing and purine and pyrimidine differentiation [6, 16].

Quantum dots are semiconductors in nature, measuring around less than 10 nm and exhibit fluorescence when stimulated by light. These nanomaterials readily conjugate with biomolecules and serve as practical biomarkers and as sensitive probes. These are useful in imaging the stages of tumor formation and also help in planning concern treatment. These conjugate with molecules that have an amphiphilic triblock copolymer layer, this form elevates their degree of biocompatibility, and remain in circulation emitting bright signals even at different wavelengths. This ability ensures their role as useful tools for imaging and tracking as agents. Thus, these help in detecting multiple tumors and cancer with higher sensitivity. These also conjugate with antibodies and related biomolecules; therefore, these are potential tools in the nanomedicine field [6, 17, 18].

Parkinson's disorder (PD) and Alzheimer's disorder (AD) are the major neurodegenerative disorders. These disorders are the representations of both sporadic and familial loss due to the progressive degenerative process of loss of neuronal tissue. Parkinson's disease affects around 1% of individuals that are 60 years old or above. The rate of derogative progress of this pathological condition declines under treatment but cannot be prevented (at least in the current scenario). Parkinson's disorder exhibits two major primary neuropathological conditions, (i) loss of the pigmented dopaminergic neurons from Substantia nigra and (ii) appearance of the Lewy bodies and Lewy neuritis. Lewy bodies are eosinophilic inclusion present in the cytoplasm of the pigmented neuron, and each Lewy body has a specific halo around it. Since these cytoplasmic inclusions contain polymerized alpha-synuclein, hence, the PD is regarded as synucleinopathy or α-synucleinopathy [19]. Alzheimer's disease is a neurodegenerative disease that initiates slowly but becomes serious in due course of time, resulting in 60% to 70% of cases of dementia. The cause for this ailment is obscure, and, since many genes are involved, it is a genetically related ailment. Other factors may be the history of head injury, depression or hypertension, *etc*. Formation of plaque and tangles in the brain may add to the severity of this disease [20, 21].

The genetic cause plays a significant role in the occurrence of Alzheimer's disorder (AD). The gene concerned with AD is one of the forms of apolipoprotein E (APOE) gene, and it is the most probable cause of the risks of AD. Apolipoprotein E (APOE) genes are present as different alleles, namely APOEε2, APOEε3, and APOEε4; APOEε2 and these are not common. It may prevent this ailment, but during later age, it can become derogative. APOEε3 is a common risk factor for AD but is neutral functionally. In the case of Alzheimer's disease, the alleles of ApoE ε 4 belong to the Apo E gene family. ApoEε4 allele of the Apo- E gene is present on chromosome number 19. This allele is associated with the risk of the onset of AD during a younger age. The role of ApoEε4 in AD is considered to be concerned with the formation of amyloidal plaque because it is a carrier of β-amyloid. It can lead to either deficiency or failure of phosphorylation of Tan when ApoEe4 fails to bind to Tan protein. This condition causes the generation of neurofibrillary tangles. The TAN protein-N-β-Nany Dopamine hydrolase enzyme catalyzes the hydrolysis of N-β-antidopamine, which is a primary precursor of brown cuticle and plays vital roles during the metabolism of the brain neurotransmitter, cycling of dopamine and histamine] [21, 22]. One of the conventional treatments of AD includes the replacement or modification of either neurotransmitter or the enzyme involved. The steps involved are primarily of symptomatic significance. The neurotransmitters and enzymes are acetylcholinesterase inhibitor; the nerve growth factors, cholinesterase inhibitors, antioxidants, amyloid β targeted drugs, electron secretase inhibitors, a vaccine against β- amyloid, *etc*., play significant roles in such treatment [21 - 23].

The nanotherapeutics approaches help address neurological disorders. Generally, polymeric nanotherapeutics undergoes surface functionalization to enable them to move across the blood-brain barrier. The brain tissue consists of the neuron, astrocytes, oligodendrocytes, *etc*. Neural stem cells interact with nanoparticles like nanotubes, nanofibers, and these help in the formation of nano-scaffolds for the targeted tissue. The modification of nanoparticles is done using polymer chemicals to improve the delivery system and to gain entry into the target tissues [24, 25]. The therapeutic drugs like antioxidants, anti-inflammatory agents, immunomodulatory agents, growth factors, genes, SiRNA antimicrobial drugs, *etc*., are the therapeutic options for treating nervous disorders. The compatibility of drug-polymer and its mode of crossing the blood-brain barrier are the two major parameters that guide the rationale designing of the nanomaterials. The variables like surface properties, size, the chemistry of nanomaterials, and, the drug carrier, feasibility to move across the blood-brain barrier, and, the site of action in the central nervous system, need careful consideration during the fabrication of nanotherapeutics for the said purpose [24 - 27].

Some of the degradable polyanhydrides and polyethers exhibit tunable erosion profiles and chemical aspects that are readily modified to attain sustainable payload release profiles. Polybutylcyanoacrylate as nanoform coated with surfactant polysorbate-80 is suitable to carry drugs that are administered intravenously like dalargin, doxorubicin, lope amide, tubocurarine, *etc.* The use of surfactant and coating of polysorbate helps the drug to form conjugate to ensure its movement across the blood-brain barrier [25, 28 - 31]. Polylactic acid (PLA) and polylactic-CO-glycolic acid (PGLA) come under polyethers, and these are FDA approved products for human use [32]. These products are relatively less toxic, undergo cellular metabolism, and exhibit better biodegradability [32, 33]. Generally, most of the polyesters show erosion because ester bonds rapidly burst during the release of the payload, and the molecular weight of polymers changes from this release of drug payload. The behavior of the drug released can change if the hydrophobic materials like gelatin or chitosan replace the coating of PLGA. The rate of degradation increases in the microenvironment, specifically, cellular microenvironment having a pH range of 1.5 to 3.6 pH (acidic range). This condition interferes with the dynamics of therapeutic payload release [34 - 37].

Polyanhydrides exhibit a higher degree of biocompatibility and are formulated especially for drug delivery to the central nervous system. Further, the formulation, related to polyanhydrides and the respective implantable wafer system is useful to treat Alzheimer's disorder and brain cancer. Sebacic acid and 1.3.bis (p-carboxyphenoxy) propane are the most commonly used as drug delivery agents. These polyanhydrides metabolizes into biocompatible forms, and after that, these are readily eliminated from the biosystem [38, 39]. Polyethers are synthetic and naturally inspired products that are useful in polymeric drug delivery. Polyethylene-glycol (PGA) and polypropylene glycol (PPG) are applicable as triblock pluronic [PEG] n-[PPG] m–[PEG] together with other polymers. Naturally derived polymers like chitosan, a cationic polysaccharide are promising drug delivery vehicles [40, 41].

Nanomaterials as Antifungal Agents

Fungal infections frequently occur during medical procedures or because of unhygienic conditions of post-operative infection. These infections contribute significantly towards morbidity and mortality in the patient, specifically immunologically compromised individuals. These individuals need intensive care, during antifungal treatment, especially when treatment involves broad-spectrum antifungal agents. A solution of silver nitrate nanoparticles (1%) exhibits antifungal activity and inhibits the growth of most of the yeast, fungal, and mold strains. The mode of interaction of Ag nanoparticles as an antifungal agent includes the disruption of cell membrane potential. The TEM technique reveals the formation of a pit like structure on the surface of the cell membrane of Candida albicans. The creations of pits or pore-like structures lead to the death of the fungus. The possible mechanism involves (i) disruption of cell membrane, (ii) protein oxidation, (iii) DNA damage, (iv) interruption of the electron transport system, and (v) rise in the concentration of ROS (reactive oxygen species) (Fig. **1**) [42, 43].

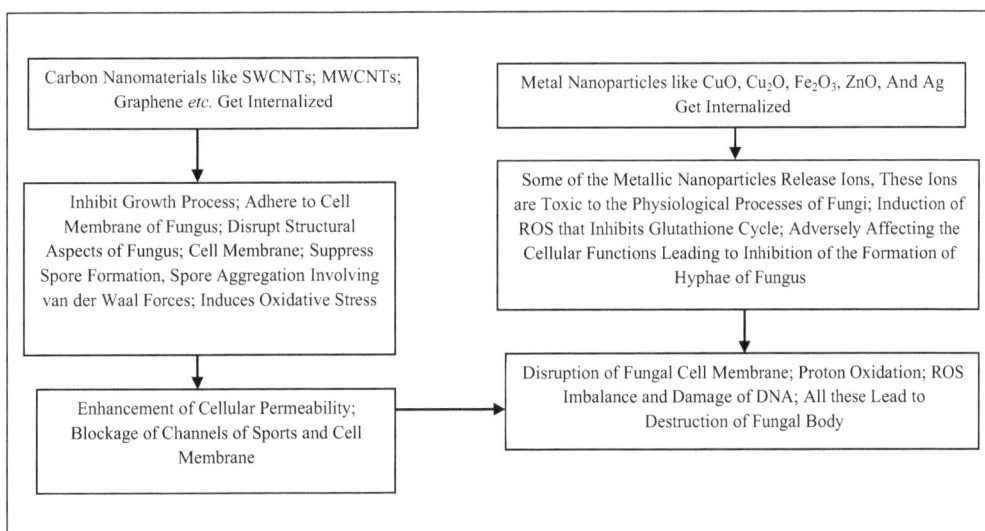

Fig. (1). Schematic representation of antifungal impacts of nanomaterials.

Synthesized ZnO nanoparticles, using a green approach involving egg white, are a suitable antifungal agent. These nanoparticles enter cells involving either diffusion or endocytosis. Once internalized, these interfere with energy production at mitochondria and promote the production of ROS. ZnO nanoparticles release Zn++ and the ROS, and move across the nuclear membrane and cause DNA damage. As a result, irreversible chromosomal damage, apoptosis, and cell death or repair of DNA, *etc.*, is impaired [44]. The cetyl trimethyl ammonium bromide and isopropyl alcohol coated copper nanoparticles exhibit intense antifungal activity against *Phoma destructive, Curvularia lunata, Alternaria alternata, and Fusarium oxysporum.* These nanoparticles have good efficacy as disinfectants, more so, in poultry and animal husbandry [45].

Nanomaterials as Antiviral Agents

A viral infection is worldwide and causes health, social, and economic implications. Viral infection treatment is a considerable challenge. Such challenges pose two main impediments, namely, maintaining the viability of host cells and targeting the disease-causing virus.

Metal nanoparticles with core-materials and ligand shells act as antiviral agents. Non-toxic doses of silver nanoparticles are antiviral against HIV-1. The human cytomegalovirus (CMV) is a herpes virus and causes significant health problems in neonatal and immunologically compromised individuals. A monoclonal antibody formed against cytomegalovirus, surface glycoprotein (GB), and chemically conjugated with gold nanoparticles, acts as antiviral agents. It becomes distinct as a potential antiviral agent against cytomegalovirus. Gold nanoparticles treated with amino-thiol-phenol and glutaraldehyde spacer are active agents for covalent immobilization of monoclonal anti-GB-antibody on its surface. These nanoparticles develop the ability to block the replication of the virus. The virus induces cytopathogenic effects and spreads in the cell culture without causing cytotoxicity. Higher concentrations of glycoprotein form a coat on the surface of virus particles, and this coat prevents the entry of the virus into the cells. The lower amounts of glycoproteins block the development of later stages in the life cycle of the virus. As a result, cell treated with glycoproteins gains resistance to cytomegalovirus infection. The infected cells selectively leys on exposure to a specific wavelength of the laser in the presence of gold nanoparticles bound to glycoprotein, *i.e.*, the infected cells undergo nanophotothermolysis [46, 47]. CD-4 is a cluster of differentiated-4-glycoprotein and is present on the surface of immune cells like T- cells, monocytes, macrophages, and dendritic cells. The CD-4 protein encodes the CD4 gene, and this protein acts as a receptor for HIV.

Gold nanoparticles prevent CD-4 dependent iron-binding, fusion, and infectivity. These particles also act as active viricidal agents against both cell-free virus and cell-associated virus. Gold nanoparticles prevent or inhibit post-entry stages of life cycles of HIV-1. These nanoparticles are useful as anti-HIV-agents. The multiple copies of amphiphilic sulfate ended ligand coated on the gold nanoparticles bind to the HIV, it envelops glycoprotein group 120 (gp120), and inhibit the infection of HIV to T-cells at a nanomolar concentration *in vitro*. The binding of nanoparticles with gp120 depends on the ligand density attached. The higher density of ligand results in a higher degree of binding and a lower degree of density causes low binding. Nanoparticles based presentations of multiple ligands create a higher local concentration of binding molecules. The concentrated binding facilitates the targeted biological interaction [48].

Silver nanoparticles (Ag 10 nm diameter and AgNO3 50 nm) cause a decline in extracellular HBV DNA formation in the model cell line HepAD 38 cells up to more than 50%. These nanoparticles cast relatively less impact on those cells that have closed circular DNA at the same time, restricting the formation of intracellular HBV RNA. During this interaction, the molar ratio of silver is around 1:50. Nanoparticles exhibit an increased affinity for binding with HBV DNA. Silver nanoparticles show the ability to interrupt the formation of HBV RNA and virions outside the cell. It appears that there is a direct interaction between the nanoparticles and the double-stranded DNA of HBV, and it exhibits antiviral activity [49, 50].

Most of the nano viricides are polymeric micelles consisting of the polymeric single chemical chain having ligands attached covalently, which specify the virus target. The antiviral spectrum of the nano viricide drug is determined. The specificity of the set of ligands attached to the chain, in addition to other functionally essential aspects, helps to determine the antiviral spectrum of the nano viricide assigned. The formulation of nano viricides looks for specific virus type, engulf or encapsulate the virus particle, to neutralize the effectiveness, destabilize it, and possibly dismantle it. These nano viricides are suitable options to destroy the viral genome and results in their destruction. There is another option of keeping an active pharmaceutical component in the core of the nano viricides [46].

The physicochemical properties of nano viricides exhibit the structural flexibility of the polymers that have well defined nano-particulate materials. Such nanomaterials show an ability to modulate their fundamental aspects following the physicochemical properties. The characterization of nanomaterials is in terms of molecular weight average number of ligand per chain. Since it is not possible to manufacture single molecular weight species, the manufactured nanoviricides have a specific molecular weight. The molecular weight of the resultant product depends on the technique used for measurement. The single molecular chain with different molecular sizes and only polymer chemistries enable the secure attachment of ligand for blocking the open sites. The amphiphilic materials exhibit self-assembly, and, because of this feature, there is a restriction on many procedures related to molecular weight distribution. Since, most of these are soluble in organic, aqueous, and intermediate solvents, these pose problems for fractionalization. It is not accessible to characterize the non-particulate materials, using scanning electron microscopy, transmission electron microscopy, and atomic force microscopy. These viricides fabricate with relation to target specific virus and do not cause adverse metabolic impacts on host. The fabricated viricides are safe and biodegradable within the body. These do not result in a significant generation of mutagens because of their unique design. It is possible to fabricate viricides with high potential to either inactivate or destroy substantial numbers of viruses and minimize the viral threats, and it is possible by using targeting ligand against the desired set of viral pathogens. This procedure is relatively at low cost for their development, manufacture, and distribution [46, 47]. The flow chart representing nanomaterials as antiviral agents is shown as (Fig. **2**).

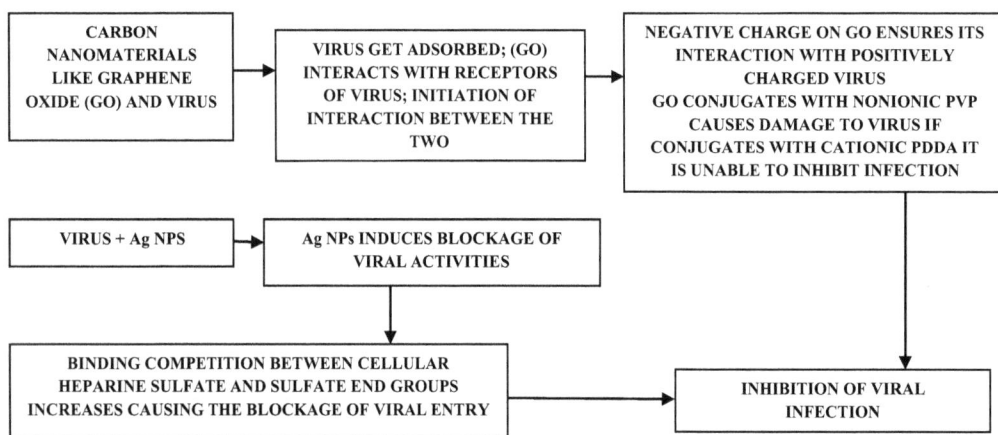

Fig. (2). The flow chart representing nanomaterials as antiviral agents.

Nanomaterials as Antibacterial Agents

Microbes/bacteria have specific features that ensure their survival or sustainability. The prime feature is its ability to resist drug interaction. Some microbes can form a biofilm; this biofilm ensures their durability and successful microbial lifestyle. Bacteria are among some of the smaller organisms and have all the cellular organelles that facilitate their growth, reproduction, respiration, survival, and other processes that support their life. Even though the size of microbes is tiny (average diameter is around 10^{-3} mm) but these organisms have an active osmotic barrier-cell membrane and cell wall. It also resists or tolerates mechanical stress. A thick layer of peptidoglycan is present bacteria that act as a basis of their differentiation. This layer interacts with basic dye crystal violet. Hans Christian Gram, around 130 years ago, demonstrated this feature around 130 years ago. Those bacteria which interact with Crystal Violet dye and stain are Gram-positive Bacteria (Gram +ve). Bacteria under this category possess 20 to 80 nm thick layer of peptidoglycan. It is present as the outermost layer. Those bacteria that do not stain with Crystal Violet dye are Gram-negative bacteria (Gram−). Bacteria in this group have a layer of peptidoglycan (around 8 nm in thickness). Some bacteria exhibit another structural feature in addition to the peptidoglycan layer, which is called a capsule. This layer is protective against toxic hydrophobic molecules because it acts as a barrier to detergent molecules. This layer facilitates and promotes adherence to other bacteria and the host tissue surface. Cariogenic bacteria like mutant Streptococci and Lactobacilli (oral bacteria) have a well-developed polysaccharide capsule that exhibits a very high degree of adherence to other bacteria and the enamel of teeth. This extracellular polysaccharide capsule is mainly composed of dextran and levan. (Levans come under fructans –polymers of fructose that form non-structural carbohydrate, *i.e.*, levans, these links with each other and form supermolecules under natural conditions). The levans can also be synthesized by enzymes dextransucrase and levansucrase, respectively. These enzymes are present either near or on the surface of the cell membrane of these organisms.

Once these bacteria get settled in an aggregated manner, they form a layer of polysaccharides around them. This layer favors bacterial interaction and results in the formation of biofilm. This feature is ubiquitous in the case of bacteria present in our oral cavity. This film lodges many bacteria, and the assembled form that is a dental plaque. Such bacteria also adhere to dental devices or other medical devices or damaged tissues by forming a biofilm. This feature is unique and beneficial for these microorganisms. This biofilm provides vital mechanical protection and is a favorable site of rich nutrients for oral bacteria. This biological adaption helps the microbes to develop overall resistance towards chemicals or antibiotics. The degree of favorable conditions is related to the degree of oxidative stress in the structural aspects of biofilm. The biofilm is like a community that provides space and nutrients for the sustenance of microbial life and the most appropriate means for developing a persistent infection. Biofilm can include single or multiple species of bacteria. There can be up to 500 species or more inhabiting biofilm [51 - 55]. The schematic representation of derogative inter-actions between bacteria and nanoparticles is shown in Fig. (**3**).

Biofilm is the most appropriate site for sustaining microbial colonization. Colonization is a complex process. It depends on (i) modulation of nutrients, (ii) microbial interactions, (iii) survival conditions, and (iv) nature of early colonizers. The colony changes once it has attained the climax of relatively stable conditions. Frequent changes in the ambient environmental conditions of biofilm favor the development of a specific organism and promote a resistant species of bacteria in it [51, 54]. Interactive correlation between biofilm, medicine (antibiotic), and the bacterial cell wall is dependent on (i) the ability of cell wall to act as a site of interactions between the cell wall and antibiotic, (ii) the degree of interplay and duration of bacteria, (iii) the immunological efficiency of the host, (iv) the minimum inhibitory concentration of the drug involved, increases 10 to 103 times after bacteria in the biofilm get well organized. Under these conditions, the bacteria develop either resistance or offer many modes to resist the interaction of drugs. (iii) The biofilm is the representation of a highly successful and well-established association involving cell-cell and cell-surface interactions, and are well organized, cumulative, and coordinated in nature. Any chemotherapeutic drug or agent has to overcome such intricacies of biofilm before it reaches and interacts with the organism. Inappropriate and widespread use of antimicrobial agents that have no controlled resources concerning the infection stimulates growth. These features favor the development of a resistant strain of bacteria. There exist a massive lacuna between the interacting drug-resistant strain of the bacteria and related infectious disease. This aspect is one of the prime foci for researchers concerning health professionals. Oral bacteria are more resistant to chlorohexidine (it is one of the lethal drugs to oral bacteria) present in the matured biofilm and nutrient-limited biofilms. A developed oral biofilm contains more drug-resistant species of oral bacteria. Similarly, pulmonary biofilm exhibits co-ordinational features of the respective bacteria in the lungs, and this acts as a source of cystic fibrosis [56].

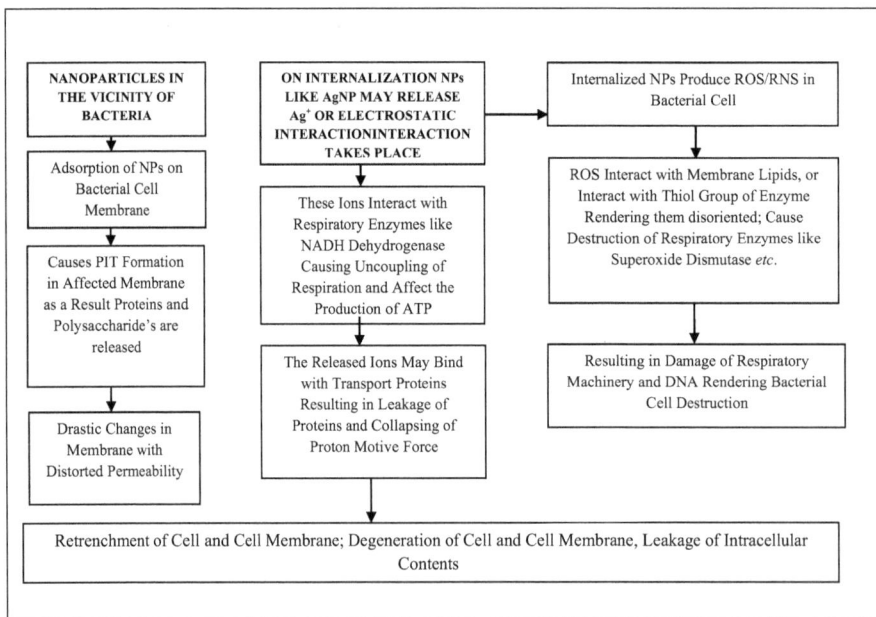

Fig. (3). Schematic representation of derogative interactions between bacteria and nanoparticles.

Overview of the Mechanism Involved During Antimicrobial Activity of Metal and Metal Oxide Nanomaterials

The prospective nanomaterials should avoid rejection because of the efflux from the bacteria. The modified or fabricated antimicrobial nanomaterials tolerate the adverse effects of the biochemical and physiological environment within the microbes and retain its antibacterial activity effectively. The engineered nanomaterials should produce reactive species of oxygen or nitrogen that damage the cell or its organelles. The nanomaterials can either interrupt or render microbial activities like the energy transduction process, enzyme cascade, and genetic transformations, *etc.* The colistin (polymixins B and E) t damages the microbial cell membrane. These antibiotics are in use against infections caused by Gram-negative bacteria. This antibiotic effectively prevents the disease due to the multidrug-resistant microbe, Pseudomonas aeruginosa, or carbapcnemases. This antibiotic is less effective in the case of Gram-positive bacteria, but when combined with trimethoprim; its efficiency enhances the broader microbial spectrum. These antibiotics interact with the cell membrane, damaging vital barriers of the cell membrane. Another mode of interaction forms holes or pores in the microbial cell membrane [57].

Nanomaterials, release ions, and these ions can act derogatively. Such ions damage the interacting microbes and also with proteins. Cadmium, zinc, and silver nanoparticles release metallic ions like Cd++, Zn++, and Ag+. Silver ions form a sparingly soluble salt solution to attack bacterial cells. For example, Cl^- inhibits cellular respiration when it precipitates as silver chloride within the cytoplasm of the affected cell. The silver nanoparticles exhibit intense antimicrobial action against Gram-negative bacteria like *E. coli.* These nanoparticles also release Ag+ ions, and these penetrate the bacterial cell wall and interfere with their metabolic activities. Cadmium and zinc ions bind with sulfur-containing proteins of the cell membrane and severely influence cellular permeability. A deficient concentration of Ag^+ (10^{-9} mol/l) affects the DNA derogatively, specifically inhibiting DNA replication.

Nanomaterials interrupt the electron transport system; induce protein oxidation and the collapse of the microbial cell membrane. Most of the nanoparticles have a positive charge, and the cell membrane has a negative charge, and this feature acts as antibacterial nature, but the related mechanism is debatable. Silver ions disrupt the respiratory enzymes bound to the membrane and also the efflux bombs of ions as a result, the microbial cell dies. Efflux Bomb-is an active efflux of ions-represents, the movements of neurotransmitters, toxic metabolites, antibiotics, and sudden shifts of ions in large amounts. This feature is concerned with antibiotic resistance and involves an energy-dependent mechanism. Some efflux systems are also drug-specific, while some may accommodate drugs. It contributes to the process of multidrug resistance (MDR).

Nanomaterials induce the generation of reactive species of oxygen. Oxygen is one of the effective oxidative agents, best electron acceptors during respiration, but it is lethal in bacterial respiration. A triplet of oxygen is the ground state of oxygen molecules and is a common cytotoxic allotropic form of oxygen. The singlet oxygen is dioxygen or dioxide-gaseous inorganic chemical ($1O_2$) and represents a quantum state in which all electrons are spin paired but kinetically is unstable at ambient temperature exhibiting a lower rate of decay. The formation of singlet oxygen leads to the peroxidation of cellular molecules like proteins and lipids. It causes undesirable cellular oxidation and also promotes spontaneous cytotoxic molecules. Technically along with peroxidation, H_2O_2 is also produced during decomposition, redox reaction, and other organic reactions or as a precursor of other peroxide compounds. During respiration, burst H_2O_2 forms, and it consumes oxygen and forms free radicals. Thus, the production of free OH^- is the basis of H_2O_2 action. The oxidation of DNA, protein, and membrane lipids is the result of interaction with H_2O_2 within the cell. The ROS affects the integrity of the bacterial cell, and as a result, bacterial cells lose its ability of surface adherence, interbacterial communication, and physiological functionalities decline [58].

ROS produced due to nanomaterials cause a decline in cellular adhesion among bacteria. Microbes try to fight or restrict the production of ROS and produce superoxide dismutase enzyme that neutralizes oxidative stress. The microbes cope with oxidative stress involving (i) SOXRS system that deals with the superoxide production and (ii) OxyR system that handles the hydrogen peroxide production. Both of these systems repair the damaged bacterial cell and regulate the reducing conditions. A healthy working cascade of events makes contact between nanoparticles and the organisms and cause oxidation of respiratory enzymes. As a result, there is a ROS production, degradation of DNA, and disruption of bacterial physiology. Commonly, aluminum, silver, gold, magnesium oxide, titanium oxide, and zinc oxide nanoparticles exhibit antibacterial activities [58 - 64].

The Carbon-Based Nanomaterials and Antibacterial Activity and The Mechanism Involved

The carbon nanotubes, single-walled carbon nanotubes, fullerene, and graphene oxide have antimicrobial activity. Some features of carbon-based nanomaterials like size, shape, conformation, and the available bonds, influence their interaction with microbes. The integrity of cell membrane, morphology, metabolism, become dysfunctional during the interaction between carbon-based nanomaterials and bacteria. The antimicrobial activity carbon-based nanomaterials are the function of their composition, surface modification, intrinsic features, and the type of interacting microbes. The physical contact between the carbon nanomaterials and the bacteria adversely affect the structural integrity of bacterial cell membrane while oxidative stress has relatively less impact [65 - 70].

Microbial drug resistance is a common observation all across the globe and encountered during the treatment of microbial infection. All these conditions offer opportunities in this field to search for some feasible alternatives. Nanotechnology has shown the potential to solve such problems. The carbon nanomaterials are good biocompatible options to address microbial drug resistance. These nanomaterials cause mechanical damage to microbes, influence oxidative stress, photothermal and photocatalytic impact, extraction of lipids, inhibition of microbial metabolism, isolation of microbes undergoing wrapping these organisms, and inducing synergistic has implications on bacteria. The carbon nanomaterials function individually or in combination with other antimicrobial materials. The carbon nanomaterials are safe and effective antimicrobial agents [71].

The single-walled carbon nanotubes (1-3μm) and multi-walled carbon nanotubes (0.5 -2μm – short and less than 50μm long) have the antimicrobial ability towards a wide range of microbes. Their functionalized forms with hydroxy and carboxyl radicals also have antibacterial activity. This behavior indicates that size, groups used during their surface modifications, and the type of interacting microbe plays a significant role during their antimicrobial activity. These nanomaterials wrap around the microbial wall or membrane by piercing or enclosing the organism, and after that release, the intracellular contents like DNA, RNA, causing the loss of membrane potential resulting in total disorientation of the entrapped body. This interaction is dependent on the diameter, size, length, and surface modification of the carbon nanotubes. Single-walled carbon nanotubes are more effective in comparison to multiwalled carbon nanotubes. These nanomaterials are a potential hazard to probiotic microbes [61].

Graphene is single-layered carbon-based nanomaterials having compactly arranged carbon atoms. Graphene oxide and reduced graphene are the two water-dispersible forms of graphene; these inhibit the growth of Escherichia coli and induce minimum cellular toxicity [72]. The dispersion of graphene nanoplatelets in a polymeric solution undergoes pressurized gyration the fibers become polymethyl methacrylate.

The amount of graphene nanoplatelets is one of the prime factors that regulate its beaded porous nature and the diameter of the fibers formed. The percentage of graphene nanoplatelets present varies from 2 to 8%. The fabricated fibers using graphene platelet concentration (2-4%) promote microbial growth, but the higher amount graphene nanoplatelets above 8% show vigorous antibacterial activity [73].

The Fullerenes are the cage-like nanoparticles and are known to exhibit stronger antimicrobial activity for a broader microbial spectrum. Its antimicrobial activity depends on its unique cage morphology that provides conjugation *via* π-electrons. This feature helps in the absorption of light, and that results in the production of reactive oxygen species. During this interplay, the fullerene becomes excited and quickly releases a lower triplet of oxygen with a longer exciting life; thus, it induces a photochemical transfer pathway and superoxide anion formation. These radicals have a short reactive duration and exhibit one or more unpaired excited electrons in their atomic or molecular state. Thus, there is a higher degree of formation of ROS that is lethal to microbes. These ROS trigger maximum damage to biomolecules like lipids, proteins, and nucleic acids, *etc* [61, 72, 74 - 76].

Quantum dots also exhibit concentration-dependent antimicrobial activity. The quantum dots bind with the bacteria and disrupt the functionality of the microbial antioxidative system, down-regulate the genes related to antioxidation, and severely declines the enzymatic activities that participate in antioxidation. As a result of these activities, oxidation of proteins, lipids, do not permit the growth of the colony and render the microbe as inactive or dead [50]. This ability of quantum dot plays a significant role during their fabrication concerning specific microbial issues.

Graphene conjugated quantum dots maintain their basal plan, which resembles the idea of graphene oxide (GO-QGD). These conjugates have restricted antibiotic impact, *i.e.*, it can readily kill Staphylococcus aureus and also concerned with persisters that support antibiotic effect. These conjugates are not capable or effective against microbes like *Bacillus subtilis, E. coli, and Pseudomonas aeruginosa.* The essential factor that seems to be responsible for its restricted antimicrobial activity is the matching of the Surface-Gaussian-Curvature and its contact with the cell membrane of the interacting microbe. The first step involves physical contact with the cell membrane, followed by the disruption of the cell envelope of the bacterium [77].

The polyamidoamine dendrimers increase the degree of solubility of antibiotics (erythromycin and tobramycin) and as a result, the distribution, the bioavailability of antibiotic drugs, and the effectivity of antimicrobial medications increase. Dendrimers exhibit limited bactericidal impacts. This impact is not strong enough to ensure antibiotic activity. This aspect offers enormous scope for further investigations [78]. The N-methyl pyrrolidine dendrimers (heterocyclic 1, 2, 3, 4) exhibit different degrees of growth restrictions concerning *B. subtilis, K. pneumonia, E. coil,* and some particular species of Streptococcus at varying concentrations [79].

NANOCOMPOSITES-A PRODUCTS OF NANOTECHNOLOGY AND THEIR BIOCOMPATIBILITY

Nanocomposites are the modification of nanomaterials and have multiple applications. The nanocomposite materials are one, two, or three-dimensional substances, and are less than 100 nm (at least one dimension). Mostly, these structures have nanoscale repeat distance and in between different phases also. The nanocomposites are porous media, colloids, gels, copolymers that show a solid combination of bulk material (as a matrix) with nano-dimensional aspects or phases that show different properties. There are differences in the structure and their chemistry and have unusual properties like mechanical, electrical, thermal, optical, electrochemical, catalytic properties of nanocomposites; these properties are different from the component materials that constitute it [80, 81]. The functions of nanocomposites restrict to the size; less than 5 nm is suitable for catalytic activity, less than 20 nm is useful for making hard magnetite material soft; less than 50 nm is capable of causing changes in the refractive index while less than 100 nm can attain superparamagnetism, mechanical strengthening and to restrict the dislocation movements of the matrix. There are many natural examples of nanocomposites, like shells of mollusks — vertebral bone, pumice stone, Abalone shell, *etc.* Nanocomposite materials exhibit an exceptional reinforced phase and another dimensional ratio. The reinforcing material is made up of particles like minerals, sheets, fibers like carbon nanotubes, electrospun fibers, *etc.* There exists a large area of interface between the matrix and reinforcement components. Further, the matrix and reinforcement phase is in order of the magnitude of composites. The properties of the material of the configuration are significantly affected in the vicinity of the reinforcement [80].

The polymer nanocomposites have the properties concerning the local chemistry, degree of thermoset curve, polymer chain mobility, polymer chain conformation, degree of polymer chain ordering or crystallinity, *etc.*, are adjustable. Generally, when carbon nanotubes interact with nanocomposites and their properties like electrical, thermal conductivity improves. Even other properties such as optical, dielectric, heat resistance, mechanical features like stiffness, strong aversion to wear and tear and damage, *etc.*, also get better, although the mass fraction, (the percentage by weight of nanocomposites) remains low [82].

Applications of Synthetic Biopolymer Composites for Tissue Engineering Scaffolds

Tissue engineering concerns with the formulation of biological substitutes that are useful the repair or replaces defective, dysfunctioning, or structurally deformed tissues. These products are biocompatible and have adjustable kinetics of biodegradation. Synthetic polymer nanocomposites are three-dimensional structures and are good options for scaffolds in tissue engineering. The conventional nanocomposites are polylactic acid (PLA), polyglycolic acid (PGA), polylactic acid-co-glycolic acid (PLGA), poly ε-caprolactone (PCL), *etc*. These nanocomposites incorporate with apatite components, carbon nanostructures, and nanoparticles. Apatite materials are minerals having phosphate refers to hydroxylapatite, fluorapatite, and chlorapatite and their crystals have higher concentrations of OH^-, F^-, and Cl^-, respectively. These additives enhance (i) mechanical properties, (ii) degree of cellular adhesion, and (iii) rate of proliferation of cells. These composite materials are in use and worth study their interactions with stromal cells and biopolymer interfaces. These materials are acceptable (at least) during experimental stages of cell culture analysis involving cell-scaffold interaction with mesenchymal stem cells *in vitro* [83].

Chitosan-Halloysite nanotubes nanocomposites act as a scaffold for tissue engineering and are developed using the combined solutions mixing and freeze-drying techniques. These composite materials are suitable for application as a scaffold in tissue engineering. Chitosan and Halloysite nanotubes (HNTs) have hydrogen bonding and electrostatic attraction between them. The spectroscopic and morphological analyses confirm these types of bondings. These materials are cytocompatible and have high compressive strength, thermal-stability, suitable porosity suitable for tissue engineering in comparison to the chitosan. These are also appropriate nanocomposite options as drug or gene carriers [84, 85].

Applications of Bacterial Cellulose-Collagen Nanocomposite for Bone Tissue Engineering

This nanocomposite is a product of the interaction between bacterial cellulose and collagen modification that involves glycine esterification and cross-linking. This cross-linking is brought about by employing 1-ethyl-3-(3-dimethyl aminopropyl) carbo-di-imide. Spectroscopic and X-ray diffraction techniques can characterize these. These techniques also describe those nanomaterials which incorporate as components of the nanocomposite. This nanocomposite is biocompatible and the most suitable materials for osteoblast and appropriate for bone tissue engineering [86].

Applications of Nanomaterials as Biosensors

A biosensor is an analytical device that can convert a biological response into a detectable, usable, and informative output as an electrical signal. The signal obtained converts into either digital or graphic representation. Quite often, biosensor involves in the detection of concentrations of the analytes in addition to other parameters of biological interest. These devices investigate whole-cell metabolism, ligand binding, and antigen-antibody interactions, *etc*. The field of biosensors is expanding very fast, with around 60% of the annual growth rate. Major thrust area includes the health care industry, diabetes, blood pressure, obesity, some types of cancer, food industry, cosmetics, pharmaceuticals, material science, and other industries. Research and development in the field of biosensors have become multidisciplinary science that includes basic principles of biochemistry, bioreactors, physical chemistry, electrochemistry, electronics, software, biophysics, nanoscience, and bionanotechnology. A useful and suitable biosensor should be biocompatible, capable, quick, reliable, accurate, precise, and capable of providing reproducible results, linear over the useful analytical range without dilution or concentration. It must be free from electrical noise. Other functional qualities include energetic catalytic activities, stability, tendency to cause minimum surface fouling [87]. The device should be easy to handle even by non-technical individuals. The nanoparticles, nanotubes, quantum dots, dendrimers, and other nanomaterials, along with the fabricated, are useful for detection and other functional components of bio and chemical sensors. Nanomaterials exhibit specific physicochemical qualities and interactive abilities that make them suitable options for the structural and functional components of biosensors [12]. These devices are more advantageous over otherwise used as sensors because these provide faster sampling, reliable information about the metabolic fate of analytes, drugs and their interactions, dosages (if needed). These also provide information related to the detection of the degree of recovery during various remedial treatments. These nanodevices have some limitations concerning the electrochemical or electrical activity of analytes, complexities due to the interfering species or molecules that may be present in varying concentrations in comparison to the analytes, electrode fouling, charges that are likely to cause background drift and lack of knowledge of the specific detection, *etc* [88].

One of the prime functional parameters of biosensors is miniaturization. This feature is utilized in all most all sensors to increase their sensitivity, efficiency, and degree of detecting analytes up to pg/ml concentration and in a concise time rather instantly [88 - 90]. The biocatalyst is highly specific for the analysis and stable under normal storage conditions, specifically, in the case of colorimetric enzyme techniques and dipsticks techniques. It must exhibit a high degree of stability, ability to carry out large numbers of assays, and to provide reproducible results. Biosensing relates to molecular recognition and this aspect is a prerequisite for the transducer to attain useful functionality referred to as receptors. These may be in the form of biological molecules concerning biosensors. These biomolecules are able either to receive and give a specialized molecular, functional, and detectable response. The receptors (responding biomolecules) are the products or byproducts or intermediates molecules of specific physiological or biochemical or metabolic processes/pathways and biochemical targets [91]. Some of the most common receptors are enzymes, antibodies, nucleic acids, lectin hormones, cell structures, ligands, *etc*. Enzymes are among the most appropriate biomolecules considered as analytic reagents. The categories of enzymes that are useful during biosensing are oxidoreductases that regulate oxidative and reduction reactions, transferases that are related to transfer molecular groups, hydrolases that result in hydrolytic cleavages of bigger molecules and ligases ensuring binding between two related molecules, *etc*. The factors that influence the functioning of enzymes are like loading of catalysts, pH, temperature, concentration, any cofactors if involved [13, 92].

Antibodies are specialized and specific biomolecules having a distinct and unique arrangement of amino acids. Their structural and functional efficacies act as the basis of analytical technology, immune- technology, and immuno-sensors. The variable selectivity of antibodies acts as an efficient tool that helps a wide range of biomolecular reactions and immune-based assays. Recombinant techniques are suitable to solve the issues where the formation of antibodies, two or more antibodies, or fragments of antibodies make the detection problematic [13, 93 - 95]. Nucleic acids are the biomacromolecules concern to the genetics and protein synthesis. These play a prime role in clinical analysis, genotoxic investigations, identification of antibodies, and the related detections. Biosensors react with nucleic acids and are genosensors. The basic principle involved is concerned with the ability of precise recognition of complementary base pairs in a given nucleic acid. The prime components of DNA sensors are sensing elements, microlaser, and generator of signals. The sensor component fits in the strand of DNA, *i.e.*, it exactly matches with the position of nucleotide or matches with fluorescent that releases the respective signal [13, 96]. The aptamers are handy tools to detect non-nucleic acid targets like proteins, cells, viruses, microbes, and other small molecules, including dyes, metal ions, *etc* [97]. The aptamers have an exceptional affinity for non-nucleic acid target moieties and bind with those molecules that act as suitable biomolecules to recognize elements for biosensor or genosensors [98]. Their affinity for the target guides the movement of the aptamers-nanoparticles conjugate to the site and in the detection of cancerous tissues [99].

Epigenetic molecules are helpful as functional receptors because these are the products of genetic changes due to either turning off or turning on some base pairs. These genetic changes are the result of chemical interactions like methylation of DNA or modifications in histones proteins. Such changes may not occur when silencer proteins are active, *i.e.*; no change takes place in the sequence of nucleotides or expression of the gene. During epigenetic gene regulation, cellular differentiation, migration, morphogenesis, the transformation of stem cellular totipotency to pluripotency of cell line in embryo, *etc.*, are accomplished. The activation of the gene takes place while epigenetic regulation is underway [100 - 102].

Cell organelles play a prime role in different pathways that retain cell physiology in totality. These organelles either acquire or synthesize most of the needed metabolites involving specific signals as per the demand of the cell, thereby ensuring the entirety of structural and functional stability of the cell, and a healthy state of homeostasis establishes between the cell and its ambient environment, including the extracellular matrix [12]. Cell organelles such as mitochondria, vacuoles, lysosomes, *etc.*, are at the research targets to investigate the potentials of the organelles in maintaining specific cellular functionality even during drug delivery, developmental stages, senescence and cellular death [103, 104]. Cells and tissues are also suitable receptors because these perform coordinated and cooperative functional approach while surviving in a specific physiological environment. These receptors are useful in monitoring metabolic fluctuations, stress, xenobiotics, temperature, pH, organic and inorganic metabolic derivatives, *etc* [91, 105]. Microbes, fungi, eukaryotic cells and tissues of plants are different, but these also respond to the various stimuli, and, strive to survive, and, to rejuvenate. Viable and non-viable cells are useful to detect different molecular bioreceptors [13].

The prime aim for developing nano biosensors should have functional features like capable, fast, reliable, sensitive, accurate, easy to handle, and the ability to analyze large numbers of samples in a short time. Metal and metal oxide nanoparticles like gold, silver, palladium, platinum, rhodium, copper, cobalt, zinc, either in pristine form or as combined form (mostly oxide form), are suitable materials for biosensors [90, 106]. Magnesium ferrite and other metal oxide nanoparticles (15 nm) detect dopamine electrochemically (7.7×10^{-8}) [107]. ZnO nanoparticles exhibit electro-catalytic nature and are biocompatible. These are effective in the process of immobilization of enzymes [108]. ZnO and ZnO/polyglycine are the potential detectors for dopamine, ascorbic acid, uric acid in the serum samples. These nanomaterials show high electrolytic behavior that is very effective in analytic chemistry and biosensors concerning the biosystem [109]. Metal-organic framework (MOF)-crystalline molecular materials exhibit a higher degree of porosity, structural tunability, higher thermal, and chemical stability. Any substance with these features is suitable for storage, separation, catalysis, sensing, and bioimaging [13, 110]. Nanowires show specific surface properties either in pristine form or in conjugated forms have the potential to interact with chemical and biomolecules in the biosystem. The formation of such conjugates depends on the chemical binding influencing the conductance of nanowires. This feature of conductance is susceptible, real-time, quantitative, and also suitably applicable to the development of biosensors or bio-medical-sensor. Silicon nanowires field-effect transistor is a useful tool as biosensors because these are label-free and highly sensitive. These sensors exhibit specific selectivity, better functionality, and real-time [13, 102].

Porous nanomaterials are also suitable materials to be used in biosensors. These specific materials exhibit very high surface area to volume ratio, ability to get fabricated conveniently as per the need, a close resemblance to the scaffolding, a controllable molecular template, and facilitate the growth of solid materials. Features like thickness, refractive index, and pore size modify according to the set target. These materials are most suitable for electrochemical, optical sensing devices that are very sensitive for the detection of biomolecules like glucose, DNA, antibodies, microbes, *etc* [13, 92, 111]. Magnetic nanoparticles play a very significant role in diagnostics related to biological and medical fields. These can be readily fabricated suitably and are useful as biosensors in different forms like single domain or superparamagnetic (Fe_3O_4), gerigite (Fe_3S_4), mega mite (γ-Fe_2O_3) and types of ferrites (MeO-Fe_2O_3. These forms of ferrite are adjustable, and if needed, the metal (Me) of these ferrites can be replaced by Ni, Co, Mg, Zn, or Mn as per the need. Nanobiosensors with such magnetic nanomaterials identify biomolecules and interact also. This behavior helps in the identification, separation, and detection of biomolecules in a given sample [13, 112].

Another suitable nanomaterial for biosensors is optic fibers. These are flexible, transparent composed of glass and plastic. Their core is transparent and enclosed in cladding materials (cladding material acts as either covering material or a coating). The core component retains light within while cladding materials have a lower refractive index that helps the retention of light within core. Biosensors based on optical fibers are cost-effective, efficient, accurate, and are conveniently handled for the detection of biological samples like cells, proteins, DNA, microbes, *etc* [13].

Carbon nanomaterials like single-walled and multiwalled nanotubes, graphene, fullerene, carbon nanowires, *etc.*, are quite suitable materials to be used in bio and chemical sensors. These nanomaterials act as a bio-electronic interface at the nanoscale that helps to identify cells and tissues. These are sensitive enough to pick up signals emitted during the detection. These nanomaterials conjugate with other nanomaterials like porous nanomaterials and help in the specific fabrication of nanomaterials for particular biomedical and biological applications [13, 15, 113 - 115].

Quantum dots are also suitable materials for biosensors because of their properties like broad continuous absorption spectra wavelength (ranging between UV and visible light), a lower degree of photobleaching, a better degree of photochemical stability, colloidal, and, noncrystalline semiconductor nature. These nanomaterials have a better ability to interact with biomolecules like proteins, peptides, antibodies, DNA, RNA, and ligands involving covalent binding; hence, they are useful as labeling for biomolecules [13, 116, 117].

The dendrimers are biocompatible, stable macromolecules, and exhibit appropriate biodistribution and accessibility to the targeted analytes or tissue and do not influence by the biochemical and physiological fluctuation occurring in a biosystem. Their fabrication regulates the target analytes or the specific tissues; these are also suitable materials to detect cytokines and are useful as biomarkers. The dendrimers are used for multiple micro-based assays, conductometric immune sensors, and field-effect biosensors [118 - 120].

Applications of Nanomaterials for Implants, Prosthesis and Tissue Engineering

Implants are the devices that promote life by enhancing structural and functional aspects, and expectancy of the body and its parts. These implants may be for short term or life-long use. There are some primary requirements for the success of the implants and nanomedical devices. The nano-implants and the materials used for them should be biocompatible and mechanically, chemically, and biologically stable. It should exhibit thermal and electrical conductivity and diffusibility. These should have a minimum of zero degrees of rejection in the body of the recipient. If the degree of rejection is higher, then there are higher chances of failure of the deviceswhich causes harmful impacts in the recipient and need immediate removal from the biosystem. The ready to use tools must be either neutral to the physiological processes of the tissues like growth, regeneration, and repairs and the related biochemical and physiological activities. If these devices are reactive, then, these must not interfere biochemically at molecular and atomic levels in the biosystem or the body of the recipient. The materials to be used can be either coated or treated with suitable polymer or any relevant biomolecules; thereby, increasing their sustainability of the device. Nanotechnology guides to fabricate biocompatible, antifungal and antimicrobial, anticorrosion, and free of contamination nanodevices or nano-implants [121].

When the device in use is a synthetic vascular graft, then there is a need to check the deposition of biological materials and to monitor the development of occlusion or blockage in the system under consideration. These should be able to withstand the physical stress caused due to the body fluids of the biosystem. The device should be very light but stronger and sturdier concerning its shape, size, and usability. The device should be conveniently monitored and controlled or participate (without interfering) or be neutral to the growth processes of the tissue. The nanodevices should reduce excessive pain or irritation, degree of inflammation, and allow the regular developmental, growth, and healing processes. The nanodevice should not interfere with the procedural biomimetic cure of the particular chronic and degenerative diseases under treatment. Such devices should induce tissue engineering and its applications, along with degenerative conditions, with the help of specific medicines [121].

Applications of Nanomaterials for Bone and Dental Implants

The nanomaterials help regulate the process of osseointegration. The nanophase materials having particular biocompatibility polymers concerning the extracellular matrix are a beneficial option in the successful regulation of osseointegration. The nanomaterial should provide a porous surface with a large surface area and functional mechanical integrity that improves the degree of cellular adhesion, cellular spreading, and migration. Zirconia-toughened alumina complex implants have potential with a reasonably good life span (Booker European Union Countries), which are better options for bone and dental implants. Angstrom media is a nanoparticulate material based synthetic bone and is successful in replacing the severely damaged bone. Human bone consists of calcium and phosphate composites called hydroxyapatite. These materials resemble each other and are helpful in orthopedic surgeries.

When nanofibers are involved, there is an increase and promotion of the matrices that induce the growth of living cells incorporating the growth factors. As a result, the appropriate degree of stimulation is available as per the need of the treatment; these nanomaterials enhance osteointegration and ensure good binding with orthopedic materials of the implant. The juxtaposed bone should be able to match the physical features of the surrounding bone without disturbing the ambient physiological aspects. Otherwise, there are chances of ejection, rejection, or induction of adverse conditions in the recipient. The dental implant having TiO_2 nanoparticles with nanohydroxyapatite coating induces the regeneration of nerves, and cause no cytomorphological changes during culture experiments. This ability of the coated TiO_2 feature is beneficial to increase the degree of biocompatibility. The growth factors and biomolecules immobilized on to the implant increase the rate of growth and osteointegration. TiO_2 nanoparticles alone cannot meet the requirements for bone tissue growth but along with other nanomaterials like Ti_6Al_4V, Co Cr Mo alloy, nano tote with a coating of $CaPO_4$, bone void filler and nanomaterials having zirconia nanoparticles, nanoceramics, and nanomaterials with hydroxyapatite materials, *etc.*, are useful in the bone tissue engineering. The tantalum coated hydroxyapatite at nanophase causes better growth of bone tissue [122 - 124].

Applications of Nanomaterials as Cartilage Implants

Cartilage cannot repair itself because of its cells, and the matrix does not grow. Polylactic-co-glycolic-acid (PLGA)/hydroxyapatite composite activates and increases the growth of chondrocytes attachment, proliferation, and other responses that increase the tensile strength of the cartilage tissue to its double than the standard value. A self-assembled peptide-hydrogel as a scaffold to chondrocytes of the cartilage under consideration helps the repairs and regeneration. The anodized titanium having nanosized pores helps to increase the degree of cell adhesion and migration of chondrocytes. [The anodized metal has a protective oxide layer by an electrolytic process in which the metal forms anode]. The carbon nanotubes are an excellent option to grow chondrocytes; on electric impulse applied to the composite; it accelerates the rate of formation and growth of chondrocytes when an electrical stimulation [125 - 127].

Applications of Nanomaterials as Oesophageal, Tracheal and Bladder Implants

Tissue engineering technique also caters to the need to fabricate suitable implants for tissues of the organs like gastro-intestine, trachea, and urinary bladder, *etc.* There are postoperative side effects due to the implants in the tissues of the trachea and esophagus. These side effects are scarring, inflammation, constriction at the site of implant, loss of regular functional aspect of the epithelial barrier, and a higher rate of mortality. Under normal conditions, there is a need to replace the tissue by an autologous tracheal part of the gastrointestinal segment or allograft. The cell sheet engineering technique is vital from the structural and functional aspects, and it promotes the growth of epithelial cells, and, replacement of damaged cells, involving non-invasive procedure. Tracheal replacements in the case of rabbits, there have been successful migration and growth of implanted cells on to the host trachea and the epithelial cell sheet formed to the luminal side. Nanostructured polymers like PGAL, PLA, and poly-ether-urethane (PU) facilitates the cancerous bladder cells to enhance surface roughness and growth. The electrospun polystyrene scaffolds promote the growth of smooth bladder cells. Nanoscaffolds made from fibrinogen and cellulose promotes cellular adhesion and proliferation of bladder cells. The electrospun technique helps to produce the PGA coated with self-assembled peptide- amphiphilic molecules and nanofibers, these nanofibers facilitate the growth of bladder cells [128 - 131].

Applications of Nanomaterials as Vascular Implants and Stents

The complex, multilayered structure of blood vessel offers a complex challenge to design a suitable fabricated device. The cell sheet engineering technique is beneficial to produce three-dimensional tissue. This technique involves the seeding of cells, cell sheet formation, use of hydrogels for cellular growth, and the electrospun polymers to produce nanofibers to form a suitable scaffold to create 3D tissue. A mixture of collagen and elastin is used to the production of the fibers involves a combination of collagen and elastin in fabricating a three-layered tubular structure. Magnetic nanoparticles labeled smooth muscle cells undergo seeding process on to the luminal surface of the tubular-shaped collagen membrane leading to the completion of a suitable vascular graft. Magnetic force induces the active participation of labeled smooth muscle cells, and to improve the seeding efficiency, elevates up to 90% [132 - 134]. Polydioxanone (PDO/PDS) is colorless, and crystalline biodegradable polymers are useful. The polymers are suitable as suture materials, and even for vascular graft engineering. Mechanical features of PDO nanofibers closely resemble the natural soft tissues like collagen and elastin. The polycaprolactone (PCA) and polylactic acid (PLA) both are suitable substances for the fabrication of the nanofibers. Their mixtures enhance the flexibility, elasticity that maintains the tensile strength of the PLA fibers. As a result, these fibers act as an ideal nanomaterial for the vascular scaffold, and, become able to withstand the blood pressure due to blood flow. The stent is the most widely used devices in vascular intervention. It is a thin tubular device implanted on to arteries to support the blood vessel (angioplasty). The material used to fabricate stent is stainless steel, cobalt-chromium alloy. Recently, NiTinol (an alloy of Ni and Titanium) is also used to make stent. The probable side effects include venous neointimal hyperplasia, aggressive growth of smooth muscle cells in the vascular wall at the site or nearby region, restenosis (a condition called restenosis) this can lead to thromboses. Nanostructured Ti and Co/Cr/Mo show an increase in endothelial cells loaded with polymeric magnetic nanoparticles; these cells are the target at the surface of the stent. Other options for fabricating stent are hydroxylapatite and alumina. Nanoscaled sol-gel coatings reduce the sludge accumulation in the case of the biliary Teflon stent [92, 132].

Applications of Nanomaterials as Neural Implants

Neurogenesis is a complicated and challenging process, and so are the neural implants. Carbon nanotubes and nanofibers are the most suitable materials for neural implants. These promote mammalian neuron cellular growth, migration, proliferation, and reduction of scar tissue formation. Nanoscaffolds are a biodegradable polymer, and these help axonal regeneration and scaffolding peptides formation that exhibits self-assembly. These fibers mimic the natural extracellular matrix and get converted into natural amino acids, and this ensures the reduction in the rejection of the neural implant. Biocompatible self-assembled monolayer having 16-mercapto-hexadecanoic acid (MHA) and Endo-thiolated-poly (3-(2-Ethylhexyl) thiophene mixed in different proportion on gold nanoparticles. This product forms a coat on the neuron electrode; these elevate the degree of biocompatibility but reduce the impedance. Fabricated nanofibrils and neuronal prosthetic devices having [poly (3, 4-dioxyethelenethiophene) PEDOT] coatings improve conductivity and also reduction in impedance. Thus, this feature makes these nanomaterials preferred options for neural implants and neuro biosensors. The nanowire array is useful as a contact site for neural stimulation. These are an ideal conductor for electrical stimulation concerning the brain. The single-crystal silicon wafers that act as micromachine do not damage the neighboring tissue at the site of implanting silicon as an insulator wafer in the brain for deep brain stimulation of the subthalamic nucleus. This type of stimulation using small devices offers excellent possibilities in the treatment of degenerative diseases like Parkinson's and Alzheimer's, blindness, and spinal injuries [135 - 137].

Role of Nanomaterials in the Varied Aspects of Defense

Nanomaterials have broad applications in military applications like defense arena, aerodynamics, mobility, stealth technology that make the detection by radar and sensors difficult. Other applications involve sensing, power generation, management, smart materials, and structures. Nanomaterials also play a significant role in the process of resilience and recovering from difficulties. Areas like battlespace, related information, and signal processing autonomy, the robustness of materials, threat detection, novel electronic detection, and display and interface system in various syntheses of materials. Nanomaterials play a significant role in the field of evolving vision and related capabilities. Thus, nanotechnology and nanoscience are of the functional significance of modern military warfare and defense systems. These materials have impacts on military personnel, hardware, and nature of the war on the whole.

The structural and functional aspects of micro-radar for personal use and crewless miniaturized vehicles involve nanomaterials and the basics of nanotechnology. Nanomaterials form the specific technical components of the thermal IR sensors with enhanced sensitivity, portable and wearable inertial and positional, motional and accelerative sensors, miniaturized and healthy sensitive vision camera system, *etc.* Biochemical sensors (remote operation and handheld); health monitoring sensors (embedded and continuous original); self-healing and self-repair materials; conditions monitoring of equipment ammunition, military weapons, *etc.*; drug and nutraceuticals form the necessary items in military functioning.

Delivery sensors and systems; wireless-secured Radio-Frequency-links (RF-Linked) between sensors and the equipment; enhanced surface-treated against corrosion, hardware, and frictionless coatings; protective clothing for army personnel, stealth coats; smart skin materials, adaptive camouflage; impact-resistant the Rheo-fluid system, reactive nano-armor composites, bioactive textiles, and electronic textiles, *etc.*, all utilized to great extent in the working of the military during peace and war times. The application of functionalized nanomaterials in various modalities like stealth and camouflage light-weight, and with Radiofrequency linked materials, are well evident. The nanomaterials are made to be highly adaptive structures with specific hardness and impact resistance. The range of such structures includes nylon-6 Montmorillonite-nanocomposites that have a high tensile strength of 40%, tensile modules with elasticity 68% flexural strength up to 60%, flexural modulus (bending up to 120°) distortion temperature up to 65°C to 152°C and improved flame retardant properties. The boron, arsenic, fluorine, and cobalt nanomaterials are branched polyisocyanates, and these emit colors within the visible range. These also have the switch off and on the facility to reflect specific color only. These nanomaterials are superior abrasion-resistant, antireflectant, tailored refractive indices, anti-corrosive, and protection from chemical corrosion, self-cleaning abilities. Pilingtoris active surface-the surface having a coating of chemically bonded on to the outer surface; this coating is designed to absorb U V light from solar radiation. This process of absorption causes a reaction on the surface; as a result, the link between the surface and dirt gets loosened and breaks. This behavior does not allow the dust to settle on the affected surface. When it rains, the coating causes rainwater drops to sheet off the surface of the glass. It not only washes away the loosened dirt but also prevents the formation of droplets that form streaks on glass or window panes or surface. Most of these nanocomposites materials are 60% lighter and twice resistant to denting and scratching [138 - 140].

Nanovaccines and Nanomaterials

Vaccines are useful in the treatment of diseases like malaria, acquired immunodeficiency syndrome, and tuberculosis. Currently available vaccines are deficient, specifically; in stimulating the immune system completely and mostly do not remain stable in vivo. These vaccines cause toxic impacts, need a cold chain, and repeated administration to be effective [141-Ghebi and Darroudi, 2019). Nanotechnology provides a feasible option in this field. Nanomaterials are better and convenient agents as carriers and adjuvants for vaccines. The materials at nano dimensions are quite similar to the pathogens and this feature plays a significant role during the stimulation of the immune system effectively. As a result, the humoral immune response improves. The nanovaccines exhibit better stability and biodistribution. The shelf life is also more than the traditional vaccines. Other advantages of nanovaccines are either reduction or avoidance of booster dose, maintenance of cold chain, and improved ability to create active targeting. Nanovaccines show better results during the treatment of rheumatoid arthritis, AIDS, and chronic autoimmune syndrome [141]. Nanomaterials are preferred options for incorporating with less virulent microbes used as vaccines. The resultant products, the nanovaccines, are better in performance in comparison to the traditional vaccines because nanovaccines have better biocompatibility and biodegradability [142].

The CD8 cells are a type of lymphocytes and are cytotoxic. These cells are also known as cytotoxic-lymphocytes, T-Killer cells, cytotoxic T-cells, CD8+T-cells, or killer T-cells. Most of the cytotoxic T-cells can interact with T-cell receptors and identify a specific antigen. Antigens bind to the I-MHC class of molecules and get exposed after coming to the surface with the help of MHC-1 molecules. The T-cell recognizes them. If the T-cell receptor is specific for this antigen, it forms a complex with it and kills the cell. The glycoproteins *i.e.*, CD-8s facilitate the binding between T-cell receptor and MHC-1 molecule; hence, the T-cells are called CD8+T-cells [143]. It is tedious to get a response of robust CD8+T-cell during innate stimulation because this response requires the Spatio-temporal reorganization of antigen. This process represents the antigen-presenting cells during stimulation of the innate immune component. Nanovaccines are effective agents to modulate and enhance the degree of adaptive immunity and even reduce the toxicity [144].

Nanomaterials, such as virus-like particles, polymeric and protein nanomaterials, and liposomes, are suitable delivery agents for vaccine antigens and adjuvants. Such combinations provide improved stability of antigen, targeted delivery, and extended-release duration. Such a delivery agent offers extended flexibility in designing and programming the required nanomaterials and immune response. An antigen and adjuvant need encapsulation within the nanomaterials or to be coated on to nanomaterials to attain their optimum effect [145]. Some of the commonly available nanovaccines are liposomes, viral-like proteins (VLPs), self-assembling proteins, and cell membranes coated with nanoparticles. Liposomes having liposome-polycationic-DNA nanoparticles are applicable to deliver DNA vaccines [146]. Liposome having PLGA nanoparticles, associated with lipid-antigenic nanovaccines, are efficient malaria vaccines [147]. Cancer cell membranes with lipid-coated onto polymeric nanoparticles involving liposomes are good agents to deliver anticancer vaccines [148]. Genetically modified virus-like proteins (VLPs) are useful carriers against antiviral protection and avian retrovirus having gag fusion proteins are effective delivery agents for intracellular proteins [145, 149, 150]. Hallow vault protein is one of the self-assembling proteins and helps in suppressing the proliferation of lung cancer cells [151]. Cell membranes associated with nanoparticles, such as the cell membrane of gastric epithelium coated with PLGAs is suitable to load antibiotics and are convenient to use for antibacterial treatment. Gold nanoparticles coated bacterial cell membranes effectively elevate the antibacterial immunity [145, 152, 153].

CONCLUSION

Nanomaterials and nanotechnology can meet the challenges concerning energy, medical sciences, biomolecular investigations, biotechnology, genetic engineering, tissue engineering, environmental aspects, *etc*. The nanomaterials cause useful as well as harmful impacts on biotic and abiotic components of the environment, industrial development, and all those fields, wherever these materials are in use. Their use must be selective and judicial, with utmost cautiousness. These include finances, investments, human profit-making tendencies, and state, local and federal government regulatory authorities. There is a need to establish the most suitable and distinct clear risk control policy and this policy must execute all the activities related to the production, storage, distribution, procurement, and management, the health of personnel involved, regulatory direction, and specific codes on the small scale and large scale production of nanomaterials. Any preferred technology must be applied cautiously, and continuous research and investigations should run parallel to the applications to minimize their derogative impacts. This attitude will help in the sustenance of the environment, biota, and the society.

REFERENCES

[1]   Policy Horizons–Canada. Nanotechnology and materials science 2013. www.horizons.gc.ca/sites/ default/files/uploaded_media/pdf-version218_bw_354kb-1p.pdf

[2]   Lahir YK. Graphene and graphene-based nanomaterials are suitable for drug delivery, In: Applications of Targeted Nano-Drugs and Delivery Systems. 50 Hampshire Street: Elsevier Publication 2018; pp. 157-90.

[3]   Doris K. 2012.http://www.theverge.com/2012/10/27/3562340/doris-kim

[4]   Sahoo SK, Parveen S, Panda JJ. The present and future of nanotechnology in human health care. Nanomedicine (Lond) 2007; 3(1): 20-31.
[http://dx.doi.org/10.1016/j.nano.2006.11.008] [PMID: 17379166]

[5]   Chakraborty G, Seth N, Sharma V. Nanotechnology: Clinical, toxicological, social, regulatory and other aspects of nanotechnology. J Drug Deliv Ther 2013; 3(4): 138-41.

[7]   Ogawara K, Un K, Tanaka K, Higaki K, Kimura T. *In vivo* anti-tumor effect of PEG liposomal doxorubicin (DOX) in DOX-resistant tumor-bearing mice: Involvement of cytotoxic effect on vascular endothelial cells. J Control Release 2009; 133(1): 4-10.
[http://dx.doi.org/10.1016/j.jconrel.2008.09.008] [PMID: 18840484]

[8]   Chan JM, Zhang L, Yuet KP, *et al*. PLGA-lecithin-PEG core-shell nanoparticles for controlled drug delivery. Biomaterials 2009; 30(8): 1627-34.
[http://dx.doi.org/10.1016/j.biomaterials.2008.12.013] [PMID: 19111339]

[9]   Albertazzi L, Gherardini L, Bondi M, *et al.* In vivo distribution and toxicology of PAMAM dendrimers in the central nervous system depend on their surface chemistry . Molecular Pharmaceutics 2013; 10(): 249-60.
[http://dx.doi.org/dx.doi.org.10.1021/mp.300391w]

[10]   Kherlopian AR, Song T, Duan Q, *et al.* A review of imaging techniques for systems biology. BMC Syst Biol 2008; 2: 74.
[http://dx.doi.org/10.1186/1752-0509-2-74] [PMID: 18700030]

[11]   Reilly RM. Risk of nanotechnology in nuclear medicine. J Nucl Med 2007; 48(7): 1039-42.
[http://dx.doi.org/10.2967/jnumed.107.041723] [PMID: 17607037]

[12]   Lahir YK. A dynamic component of tissue- Extracellular matrix: a structural and adaptive Approach. Biochem Cell Arch 2015; 15(2): 331-47.

[13]   Lahir YK, Samant M, Dongre PM. The role of nanomaterials in the development of biosensors. Global Journal of Bioscience and Biotechnology 2016; 5(2): 146-63.

[14]   Lahir YK. Impacts of fullerene on a biological system, Clinical Immunology Endocrinology and Metabolic Drugs 2017; 4(1): 48-57.
[http://dx.doi.org/10.2174/2212707004666171111351624]

[15]   Lahir YK. Impacts of metal and metal oxide nanoparticles on reproductive tissues and spermatogenesis. J Exp Zool India 2018; 18(2): 594-608.

[16]   Yadav A, Ghine M, Jain DK. ISSN 2240-3379. Nano-medicine based delivery system Journal of Advanced Pharmacy Education and Research 2011; 1(4): 201-13.

[17]   Amiot CL, Xu S, Liang S, Pan L, Zhao JX. Near-infrared fluorescent materials for sensing of biological targets. Sensors (Basel) 2008; 8(5): 3082-105.
[http://dx.doi.org/10.3390/s8053082] [PMID: 27879867]

[18]   Gangrade SM. Nanocrystal-a way for carrier-free drug delivery. Pharm Buzz 2011; 6: 26-31.

[19]   Braak H, Ghebremedhin E, Rüb U, Bratzke H, Del Tredici K. Stages in the development of Parkinson's disease-related pathology. Cell Tissue Res 2004; 318(1): 121-34.
[http://dx.doi.org/10.1007/s00441-004-0956-9] [PMID: 15338272]

[20]   Burns A, Steve I. Alzheimer's disease, British Medical Journal, 2009.
[http://dx.doi.org/10.1136/bmj.b158]

[21]   Ballard C, Gauthier S, Corbett A, Brayne C, Aarsland D, Jones E. Alzheimer's disease. Lancet 2011; 377(9770): 1019-31.
[http://dx.doi.org/10.1016/S0140-6736(10)61349-9] [PMID: 21371747]

[22]   Rostagno A, Holton JL, Lashley T, Revesz T, Ghiso J. Cerebral amyloidosis: amyloid subunits, mutants and phenotypes. Cell Mol Life Sci 2010; 67(4): 581-600.
[http://dx.doi.org/10.1007/s00018-009-0182-4] [PMID: 19898742]

[23]   Todd S, Barr S, Roberts M, Passmore AP. Survival in dementia and predictors of mortality: a review. Int J Geriatr Psychiatry 2013; 28(11): 1109-24.
[http://dx.doi.org/10.1002/gps.3946] [PMID: 23526458]

[24]   Kreuter J, Petrov VE, Kharkevich DA, Alyautdin R. Influence of the type of surfactant on analgesic effects induced by the peptide dalargin after delivery across the blood-brain barrier using surfactant coated nanoparticles. J Control Release 1995; 49: 81-7.
[http://dx.doi.org/10.1016/S0168-3659(97)00061-8]

[25]   Mallapragada SK, Brenza TM, McMillan JM, *et al.* Enabling nanomaterial, nanofabrication and cellular technologies for nanoneuromedicines. Nanomedicine (Lond) 2015; 11(3): 715-29.
[http://dx.doi.org/10.1016/j.nano.2014.12.013] [PMID: 25652894]

[26]   Alyautdin R, Khalin I, Nafeeza MI, Haron MH, Kuznetsov D. Haron MH, Kuznetsov D, Nanoscale

drug delivery system and the blood-brain barrier International Journal Nanomedicine 2011; 9: 795-811.
[http://dx.doi.org/10.2147/IJN.S5236]

[27]   Zhang TT, Li W, Meng G, Wang P, Liao W. Strategies for transporting nanoparticles across the blood-brain barrier. Biomater Sci 2016; 4(2): 219-29.
[http://dx.doi.org/10.1039/C5BM00383K] [PMID: 26646694]

[28]   Kreuter J, Alyautdin R, Kharkevich DA, Ivanov AA. Passage of peptides through the blood-brain barrier with colloidal polymer particles nanoparticle Brain Research 1997; 171-4.

[29]   Alyautdin RN, Petrov VE, Langer K, Berthold A, Kharkevich DA, Kreuter J. Delivery of loperamide across the blood-brain barrier with polysorbate 80-coated polybutylcyanoacrylate nanoparticles. Pharm Res 1997; 14(3): 325-8.
[http://dx.doi.org/10.1023/A:1012098005098] [PMID: 9098875]

[30]   Alyautdin RN, Tezikov EB, Ramge P, Kharkevich DA, Begley DJ, Kreuter J. Significant entry of tubocurarine into the brain of rats by adsorption to polysorbate 80-coated polybutylcyanoacrylate nanoparticles: an in situ brain perfusion study. J Microencapsul 1998; 15(1): 67-74.
[http://dx.doi.org/10.3109/02652049809006836] [PMID: 9463808]

[31]   Steiniger SC, Kreuter J, Khalansky AS, *et al.* Chemotherapy of glioblastoma in rats using doxorubicin-loaded nanoparticles. Int J Cancer 2004; 109(5): 759-67.
[http://dx.doi.org/10.1002/ijc.20048] [PMID: 14999786]

[32]   Gelperina S, Maksimenko O, Khalansky A, *et al.* Drug delivery to the brain using surfactant-coated poly(lactide-co-glycolide) nanoparticles: influence of the formulation parameters. Eur J Pharm Biopharm 2010; 74(2): 157-63.
[http://dx.doi.org/10.1016/j.ejpb.2009.09.003] [PMID: 19755158]

[33]   Gunatillake PA, Adhikari R. Biodegradable synthetic polymers for tissue engineering European Journal cells and Materials 2003; 5: 1-16.
[http://dx.doi.org/10.22203/Ecm.v005a01]

[34]   Determan AS, Wilson JH, Kipper MJ, Wannemuehler MJ, Narasimhan B. Protein Stability in presence of polymer degradation products: consequences of controlled release formulations Biomaterials 2006; 27(17): 3312-20.
[http://dx.doi.org/10.1016/j.biomaterials.2006.01.054]

[35]   Tosi G, Costantino L, Rivasi F, *et al.* Targeting the central nervous system: in vivo experiments with peptide-derivatized nanoparticles loaded with Loperamide and Rhodamine-123. J Control Release 2007; 122(1): 1-9.
[http://dx.doi.org/10.1016/j.jconrel.2007.05.022] [PMID: 17651855]

[36]   Budhian A, Siegel SJ, Winey KI. Controlling the *in vitro* release profiles for a system of haloperidol-loaded PLGA nanoparticles. Int J Pharm 2008; 346(1-2): 151-9.
[http://dx.doi.org/10.1016/j.ijpharm.2007.06.011] [PMID: 17681683]

[37]   Wohlfart S, Khalansky AS, Gelperina S, *et al.* Efficient chemotherapy of rat glioblastoma using doxorubicin-loaded PLGA nanoparticles with different stabilizers. PLoS One 2011; 6(5): e19121.
[http://dx.doi.org/10.1371/journal.pone.0019121] [PMID: 21573151]

[38]   Lesniak MS, Upadhyay U, Goodwin R, Tyler B, Brem H. Local delivery of doxorubicin for the treatment of malignant brain tumors in rats. Anticancer Res 2005; 25(6B): 3825-31.
[PMID: 16312042]

[39]   Attenello FJ, Mukherjee D, Datoo G, *et al.* Use of Gliadel (BCNU) wafer in the surgical treatment of malignant glioma: a 10-year institutional experience. Ann Surg Oncol 2008; 15(10): 2887-93.
[http://dx.doi.org/10.1245/s10434-008-0048-2] [PMID: 18636295]

[40]   Batrakova EV, Kabanov AV. Pluronic block copolymers: evolution of drug delivery concept from inert nanocarriers to biological response modifiers. J Control Release 2008; 130(2): 98-106.

[http://dx.doi.org/10.1016/j.jconrel.2008.04.013] [PMID: 18534704]

[41]   Kim JY, Choi WI, Kim YH, Tae G. Brain-targeted delivery of protein using chitosan- and RVG peptide-conjugated, pluronic-based nano-carrier. Biomaterials 2013; 34(4): 1170-8.
[http://dx.doi.org/10.1016/j.biomaterials.2012.09.047] [PMID: 23122677]

[42]   Falkiewicz-Dulik M, Macura AB. Nanosilver as a substance for biostabilizing footwear materials in the foot mycosis prophylaxis. Mikologia Lekarska 2008; 15: 145-50.

[43]   Gajbhiye M, Kesharwani J, Ingle A, Gade A, Rai M. Fungus-mediated synthesis of silver nanoparticles and their activity against pathogenic fungi in combination with fluconazole. Nanomedicine (Lond) 2009; 5(4): 382-6.
[http://dx.doi.org/10.1016/j.nano.2009.06.005] [PMID: 19616127]

[44]   Shoeb M, Singh BR, Khan J, *et al.* ROS dependent Anti-candida activity of ZnO nanoparticles synthesized by using albumen as biotemplate. Advanced Nature Science: Nanoscience and Nanotechnology 2013; 4: 035015.
[http://dx.doi.org/10.1088/2043-6262/4/3/035015]

[45]   Kanhed P, Birla S, Gaikwad S, *et al. In vitro* antifungal efficacy of copper nanoparticles against selected crop pathogenic fungi. Mater Lett 2014; 115: 13-4.
[http://dx.doi.org/10.1016/j.matlet.2013.10.011]

[46]   Di Gianvincenzo P, Marradi M, Martínez-Avila OM, Bedoya LM, Alcamí J, Penadés S. Gold nanoparticles capped with sulfate-ended ligands as anti-HIV agents. Bioorg Med Chem Lett 2010; 20(9): 2718-21.
[http://dx.doi.org/10.1016/j.bmcl.2010.03.079] [PMID: 20382017]

[47]   Lara HH, Ayala-Nuñez NV, Ixtepan-Turrent L, Rodriguez-Padilla C. Mode of antiviral action of silver nanoparticles against HIV-1. J Nanobiotechnology 2010; 8: 1-9.
[http://dx.doi.org/10.1186/1477-3155-8-1] [PMID: 20145735]

[48]   Bernadette M. Inhibiting of cytomegalovirus infection and photo thermolysis of infected cells using bioconjugate gold nanoparticles, Scientific Reports 2014; 4: 5550. www.nanoviricde .com/Antiviral-Therapeutics-Technologies

[49]   Lu Z, Li CM, Bao H, Qiao Y, Toh Y, Yang X. Mechanism of antimicrobial activity of CdTe quantum dots. Langmuir 2008; 24(10): 5445-52.
[http://dx.doi.org/10.1021/la704075r] [PMID: 18419147]

[50]   Lu L, Sun RW, Chen R, *et al.* Silver nanoparticles inhibit hepatitis B virus replication. Antivir Ther (Lond) 2008; 13(2): 253-62.
[PMID: 18505176]

[51]   Donlan RM. Biofilms: microbial life on surfaces. Emerg Infect Dis 2002; 8(9): 881-90.
[http://dx.doi.org/10.3201/eid0809.020063] [PMID: 12194761]

[52]   Hall-Stoodley L, Costerton JW, Stoodley P. Bacterial biofilms: from the natural environment to infectious diseases. Nat Rev Microbiol 2004; 2(2): 95-108.
[http://dx.doi.org/10.1038/nrmicro821] [PMID: 15040259]

[53]   Karatan E, Watnick P. Signals, regulatory networks, and materials that build and break bacterial biofilms. Microbiol Mol Biol Rev 2009; 73(2): 310-47.
[http://dx.doi.org/10.1128/MMBR.00041-08] [PMID: 19487730]

[54]   Lopez D, Vlamakis H, Kolter R. Cold Spring Harbor- Perspective in Biology 2010; 2: 7.
[http://dx.doi.org/ 10.1101/cshperspect.a000398] [PMID: a000398]

[55]   Aggarwal S, Stewart PS, Hozalski RM. Biofilm cohesive strength as the basis for biofilm recalcitrance: Are bacterial biofilms over-designed? Microbiology Insights 2016; 8s2: 29-32.
[http://dx.doi.org/10.4137/MBI.S31444]

[56]   Shen Y, Stogici S, Haapasalo M. Antibacterial efficacy of 'Chlorohexidine against bacteria in different

biofilms at different stages of developments. J Endod 2011; 37: 657-61.
[http://dx.doi.org/10.1016/j.joen.2011.02.007] [PMID: 21496666]

[57]   Aruguete DM, Kim B, Hochella MF Jr, *et al.* Antimicrobial nanotechnology: its potential for the effective management of microbial drug resistance and implications for research needs in microbial nanotoxicology. Environ Sci Process Impacts 2013; 15(1): 93-102.
[http://dx.doi.org/10.1039/C2EM30692A] [PMID: 24592430]

[58]   Emami-Karvani Z, Cheharzi P. Antibacterial activity of zinc oxide nanoparticles on Gram-positive and Gram-negative bacteria. Afr J Microbiol Res 2011; 5(12): 1368-73.

[59]   Azam A, Ahmed AS, Oves M, Khan MS, Habib SS, Memic A. Antimicrobial activity of metal oxide nanoparticles against Gram-positive and Gram-negative bacteria: a comparative study. Int J Nanomedicine 2012; 7: 6003-9.
[http://dx.doi.org/10.2147/IJN.S35347] [PMID: 23233805]

[60]   Usman MS, El Zowalaty ME, Shameli K, Zainuddin N, Salama M, Ibrahim NA. Synthesis, characterization, and antimicrobial properties of copper nanoparticles. Int J Nanomedicine 2013; 8: 4467-79.
[http://dx.doi.org/10.2147/IJN.S50837] [PMID: 24293998]

[61]   Chen H, Wang B, Gao D, *et al.* Broad-spectrum antibacterial activity of carbon nanotubes to human gut bacteria. Small 2013; 9(16): 2735-46.
[http://dx.doi.org/10.1002/smll.201202792] [PMID: 23463684]

[62]   Chen Q, Xue Y, Sun J. Kupffer cell-mediated hepatic injury induced by silica nanoparticles *in vitro* and in vivo. Int J Nanomedicine 2013; 8: 1129-40.
[http://dx.doi.org/10.2147/IJN.S42242] [PMID: 23515466]

[63]   Besinis A, Peralta TD, Handy RD. The antibacterial effects of silver, titanium oxide and Silicon dioxide nanoparticles on Streptococcus mutants using a suite of bioassays. Nanotoxicology 2014; 8(1): 1-16.
[http://dx.doi.org/10.3109/17435390.2012.742935] [PMID: 23092443]

[64]   Zarei M, Jamnejad A, Khajehali E. Antibacterial effect of silver nanoparticles against four foodborne pathogens. Jundishapur J Microbiol 2014; 7(1): e8720.
[http://dx.doi.org/10.5812/jjm.8720] [PMID: 25147658]

[65]   Kang S, Pinault M, Pfefferle LD, Elimelech M. Single-walled carbon nanotubes exhibit strong antimicrobial activity. Langmuir 2007; 23(17): 8670-3.
[http://dx.doi.org/10.1021/la701067r] [PMID: 17658863]

[66]   Vecitis CD, Zodrow KR, Kang S, Elimelech M. Electronic-structure-dependent bacterial cytotoxicity of single-walled carbon nanotubes. ACS Nano 2010; 4(9): 5471-9.
[http://dx.doi.org/10.1021/nn101558x] [PMID: 20812689]

[67]   Yong C, Mamount J, Tang Y, Yang L. Antimicrobial activity of single-walled carbon nanotubes: length, effect, Langmuir 2010; 26(20): 16013-9.
[http://dx.doi.org/10.1021/la103110g]

[68]   Hajipour MJ, Fromm KM, Ashkarran AA, *et al.* Antibacterial properties of nanoparticles. Trends Biotechnol 2012; 30(10): 499-511.
[http://dx.doi.org/10.1016/j.tibtech.2012.06.004] [PMID: 22884769]

[69]   Gurunathan S, Han JW, Dayem AA, Eppakayala V, Kim JH. Oxidative stress-mediated antibacterial activity of graphene oxide and reduced graphene oxide in Pseudomonas aeruginosa. Int J Nanomedicine 2012; 7: 5901-14.
[http://dx.doi.org/10.2147/IJN.S37397] [PMID: 23226696]

[70]   Murray AR, Kisin ER, Tkach AV, *et al.* Factoring-in agglomeration of carbon nanotubes and nanofibers for better prediction of their toxicity versus asbestos. Part Fibre Toxicol 2012; 9: 10.
[http://dx.doi.org/10.1186/1743-8977-9-10] [PMID: 22490147]

[71] Anand A, Unnikrishnan B, Wei SC, Chou CP, Zhang LZ, Huang CC. Graphene oxide and carbon dots as broad-spectrum antimicrobial agents - a minireview. Nanoscale Horiz 2019; 4(1): 117-37.
[http://dx.doi.org/10.1039/C8NH00174J] [PMID: 32254148]

[72] Hu W, Peng C, Luo W, *et al.* Graphene base antibacterial paper, American Society of Chemistry. Nano 2016; 4(7): 4317-23.
[http://dx.doi.org/10.1021/nn101097]

[73] Matharu RK, Porwal H, Ciric L, Edirisinghe M. The effect of graphene-poly(methyl methacrylate) fibres on microbial growth. Interface Focus 2018; 8(3): 20170058.
[http://dx.doi.org/10.1098/rsfs.2017.0058] [PMID: 29696090]

[74] Naddeo J, Ratti M, O'Malley SM, *et al.* Antibacterial properties of nanoparticles: A comparative review of chemically synthesized and laser-generated particles. Adv Sci Eng Med 2015; 7(12): 1044-57.
[http://dx.doi.org/10.1166/asem.2015.1811]

[75] Prasad K, Lekshmi GS, Ostrikov K, *et al.* Synergic bactericidal effects of reduced graphene oxide and silver nanoparticles against Gram-positive and Gram-negative bacteria. Sci Rep 2017; 7(1): 1591.
[http://dx.doi.org/10.1038/s41598-017-01669-5] [PMID: 28484209]

[76] Grinholc M, Nakonieczna J, Fila G, *et al.* Antimicrobial photodynamic therapy with fulleropyrrolidine: photoinactivation mechanism of Staphylococcus aureus, *in vitro* and in vivo studies. Appl Microbiol Biotechnol 2015; 99(9): 4031-43.
[http://dx.doi.org/10.1007/s00253-015-6539-8] [PMID: 25820601]

[77] Hui L, Huang J, Chen G, Zhu Y, Yang L. Antibacterial property of graphene quantum dots (both sources material and bacterial shape matter). ACS Appl Mater Interfaces 2016; 8(1): 20-5.
[http://dx.doi.org/10.1021/acsami.5b10132] [PMID: 26696468]

[78] Winnicka K, Wroblewski M, Wieczorek P, Sacha PT. The effects of polyamidoamine (PAMAM) dendrimers on antibacterial activity with different solubility. Molecules 2013; 18: 8607-17.
[http://dx.doi.org/10.3390/molecules18078607] [PMID: 23881050]

[79] Perumal R, Thirunayanan A, Sebastian R. Synthesis and antibacterial activity of novel N-methyl pyrrolidine dendrimers via [3+2] cycloaddition, Proceedings of National Academy of Science, Sect A Physical Science 2014; 84(3): 371-9.
[http://dx.doi.org/ 10.1007/s40010-013-01210-6]

[80] Kamigaito O. What can be improved by nanometer composites? Journal of Japan Society Powder Metals 1991; 38(3): 315-21.
[http://dx.doi.org/10.2497/jjspm.38.315]

[81] Zhiting T, Hu H, Sun Y. A molecular dynamics study of effective thermal conductivity in nanocomposites. International Heat Mass Transfer 2013; 61: 577-82.
[http://dx.doi.org/10.1016/j.ijheatmasstransfer.2013.02.023]

[82] Ajayan PM, Schadler LS, Braun PV. Nanocomposites science and technology. Wiley 2003. ISBN -3-527-30359-6.

[83] Okarmoto M, John B. Synthesis biopolymer nanocomposites for tissue engineering scaffolds. Prog Polym Sci 2013; 38(10-11): 1487-503.
[http://dx.doi.org/10.1016/j.progpolymsci.2013.06.001]

[84] Liu W, Zhao Q-F, Liu JY, *et al.* Environmental and biological influence on the stability of silver nanoparticles. Chin Sci Bull 2011; 56(10): 2009-15.
[http://dx.doi.org/10.1007/s11434-010-4332-8]

[85] Liu M, Wu C, Jiao Y, Xiong S, Zhou C. Chitosan-halloysite nanotubes nanocomposite scaffolds for tissue engineering. J Mater Chem B Mater Biol Med 2013; 1(15): 2078-89.
[http://dx.doi.org/10.1039/c3tb20084a] [PMID: 32260898]

[86]   Sask S, Teixeira L N, de Oliveira PT, *et al.* Nanocomposites for bone tissue engineering Journal Material Chemistry 2012; 22: 22102-12.
[http://dx.doi.org/10.1039/C2J 33762B]

[87]   Xu GK, Qian J, Hu J. The glycocalyx promotes cooperative binding and clustering of adhesion receptors. Soft Matter 2016; 12(20): 4572-83.
[http://dx.doi.org/10.1039/C5SM03139G] [PMID: 27102288]

[88]   Lahir YK. ISSN 2056-3779 . Role and adverse effects of nanomaterials in food technology Journal of Toxicology and Health 2015; 2
[http://dx.doi.org/10.7243/2056-3779-2-2]

[89]   Tothill IE. Biosensors and nanomaterials and their application for mycotoxin determination. World Mycotoxin J 2011; 4(4): 361-74.
[http://dx.doi.org/10.3920/WMJ2011.1318]

[90]   Sagadeven S, Periswamy M. Recent trends in nano-bio-sensors. Cell Calcium 2014; 26(5): 193-200.

[91]   Lahir YK. Apoptosis: A biological phenomenon. Biochem Cell Arch 2010; 12(2): 237-48.

[92]   Sweetman MJ, Shearer CJ, Shapter JG, Voelcker NH. Dual silane surface functionalization for the selective attachment of human neuronal cells to porous silicon. Langmuir 2011; 27(15): 9497-503.
[http://dx.doi.org/10.1021/la201760w] [PMID: 21678982]

[93]   Krishna G, Schulte J, Cornell BA, Pace RJ, Osman PD. Tethered bilayer membrane containing ion reservoir selectivity and conductance. Langmuir 2003; 19: 2294-305.
[http://dx.doi.org/10.1021/la026238d]

[94]   Kramer K, Mahlkhecht G, Bertold H. 6-Molecular antibody technology for biosensors and Bioanalytics, Handbook of biosensors and biochips Two6. Wiley Sons Ltd 2008.
[http://dx.doi.org/10.1002/97804006156.hbb007]

[95]   Zhang GJ, Ning Y. Silicon nanowire biosensor and its applications in disease diagnostics: a review. Anal Chim Acta 2012; 749: 1-15.
[http://dx.doi.org/10.1016/j.aca.2012.08.035] [PMID: 23036462]

[96]   Borgmann S, Schulte A, Neugebauer S, Schuhmann W. Advances in electrochemical Science and Engineering. Weinheim: Wiley –VCH, Verlag Gmbh and & CO 2011. ISBN: 978-3-527-32885-7.

[97]   Strehlitz B, Nikolaus N, Stoltenburg R. Protein detection with aptamer biosensors. Sensors (Basel) 2008; 8(7): 4296-307.
[http://dx.doi.org/10.3390/s8074296] [PMID: 27879936]

[98]   Nikolaus N, Strehlitz B. DNA-aptamers binding aminoglycoside antibiotics. Sensors (Basel) 2014; 14(2): 3737-55.
[http://dx.doi.org/10.3390/s140203737] [PMID: 24566637]

[99]   Reinemann C, Strechlitze B. Aptamer-modified nanoparticles and their use in cancer, Diagnostics and Treatment Swiss Medical Weakly. 2014; p. 144.
[PMID: W13908]

[100]   Reik W. Stability and flexibility of epigenetic gene regulation in mammalian development. Nature 2007; 447(7143): 425-32.
[http://dx.doi.org/10.1038/nature05918] [PMID: 17522676]

[101]   Mohammad Z. Recognition receptors in biosensors. Spring link 2010.ISBN: 978-1-4419-0919.

[102]   Mohd Azmi MA, Tehrani Z, Lewis RP, *et al.* Highly sensitive covalently functionalised integrated silicon nanowire biosensor devices for detection of cancer risk biomarker. Biosens Bioelectron 2014; 52: 216-24.
[http://dx.doi.org/10.1016/j.bios.2013.08.030] [PMID: 24060972]

[103]   Rangel R, Guzman-Rojas L, Lucia G. Combinatorial targeting and discovery of ligand-receptor in

organelles of a mammalian cell, Nature Communications 2012; 3.
[http://dx.doi.org/10.38/ncomms.1773] [PMID: 788]

[104] Deatherage BL, Cookson BT. Membrane vesicle release in bacteria, eukaryotes, and archaea: a conserved yet underappreciated aspect of microbial life. Infect Immun 2012; 80(6): 1948-57.
[http://dx.doi.org/10.1128/IAI.06014-11] [PMID: 22409932]

[105] Samant MP, Lahir YK, Chitre AV, Kale A. Review- Influence of nanomaterials on cellular stress and cellular behavior, Advances in Clinical Toxicology. 2018; 3: pp. (3)13-140.

[106] Sau TK, Rogach AL. Nonspherical noble metal nanoparticles: colloid-chemical synthesis and morphology control. Adv Mater 2010; 22(16): 1781-804.
[http://dx.doi.org/10.1002/adma.200901271] [PMID: 20512953]

[107] Reddy S, Kumaraswamy BE, Chandra U, *et al.* Synthesis of MgFe2O4 nanoparticles and MgFe2O4/CPE for electrochemical investigation of dopamine. Anal Methods 2011; 3: 2792-6.
[http://dx.doi.org/10.1039/c1ay05483j]

[108] Dubey KK. Kamaldeep, Kumar D, Optimization of ZnO nanoparticles synthesis to Fabricate glucose sensor. Advances in Applied Science Research 2012; 3(5): 3081-8.

[109] Reddy S, Kumaraswamy BE, Aruna S, *et al.* Preparation of NiO/ZnO hybrid nanoparticles, for electrochemical sensing of Dopamine. Chemical Sensors 2012; 2: 7.

[110] Lei J, Qian R, Ling P, Cui I, Ju H. Design and sensing applications of metal-organic framework composites Transactions A C trends in Analytic Chemistry. 2014; 2014; pp. 71-8.
[http://dx.doi.org/10.1016/j.trac.2014.02.012]

[111] Ju H, Zhang X, Wang J. Biosensor based on porous materials, nanobiosensors: Principles, development and application, Biological and Medical Physics, Biomedical Engineering. Springer Business Media 2011; LLC.
[http://dx.doi.org/10.1007/978-1-4419-9622-0_6:175-205]

[112] Zhang X, Guo Q, Cui D. Recent advances in nanotechnology applied to biosensors. Sensors (Basel) 2009; 9(2): 1033-53.
[http://dx.doi.org/10.3390/s90201033] [PMID: 22399954]

[113] Chun TK, Timothy LK, Michael JS, Li YY. Highly stable porous silicon-carbon composites as label-free optical biosensors, ACS. Nano Lett 2012; 6(12): 10546-54.

[114] Kruss S, Hilmer AJ, Zhang J, *et al.* Carbon nanotubes optical biomedical sensors,Advances in Drug Delivery Research Adv Drug Deliv Res 2013; 65(15): 1933-50.
[http://dx.doi.org/10.1016/jaddr.2013.07.015]

[115] Gayathri SB, Kamraj P, Arthanareeswari M. Multi-Walled Carbon Nanotubes based Purine Electrodes for Electrochemical Detection of Benzene and its Derivatives using Differential Pulse Voltammetry Int J Multidiscip Curr Res 2014; 2: 211-7. http://ijmer.com

[116] Frasco MF, Chaniotakis N. Semiconductor quantum dots in chemical sensors and biosensors. Sensors (Basel) 2009; 9(9): 7266-86.
[http://dx.doi.org/10.3390/s90907266] [PMID: 22423206]

[117] Ung TD, Pham ST, Tran KC, Diuh DK, Nguyen L. CdTe QDs for an application in Life science, Advances in Natural Science Nanoscience Nanotechnology Adv Nat Sci: Nanosci Nanotechnol 2010; 1(4): 045009.
[http://dx.doi.org/10.1008/2043045009]

[118] Thi TP, Thi KCT, Quang LN. Temperature-dependent photoluminescence study of InP/ZnS QDs. Adv Nat Sci: Nanosci Nanotechnol 2011; 2: 025001.
[http://dx.doi.org/10.1088/2043-6262/2/2/025001]

[119] Thi D T, Thi KCT, Thu NP, *et al.* CdTE and CdSe quantum dots Synthesis, characterization, and application in agriculture. Adv Nat Sci: Nanosci Nanotechnol 2012; 3(4): 043001.

[http://dx.doi.org/10.10882043-6262/3/4/043001]

[120]   Araque E, Villalonga R, Gamella M, Martínez-Ruiz P, Reviejo J, Pingarrón JM. Crumpled reduced graphene oxide-polyamidoamine dendrimer hybrid nanoparticles for the preparation of an electrochemical biosensor. J Mater Chem B Mater Biol Med 2013; 1(17): 2289-96.
[http://dx.doi.org/10.1039/c3tb20078g] [PMID: 32260882]

[121]   Stevens B, Yang Y, Mohandas A, Stucker B, Nguyen KT. A review of materials, fabrication methods, and strategies used to enhance bone regeneration in engineered bone tissues. J Biomed Mater Res B Appl Biomater 2008; 85(2): 573-82.
[http://dx.doi.org/10.1002/jbm.b.30962] [PMID: 17937408]

[122]   Sivolella S, Stellini E, Brunello G, *et al.* Silver nanoparticles in alveolar bone surgery devices. J Nanomater 2012. 975842: 12 pages.
[http://dx.doi.org/10.1155/2012/975842]

[123]   Yi H, Ur Rehman F, Zhao C, Liu B, He N. Recent advances in nano scaffolds for bone repair. Bone Res 2016; 4(4): 16050.
[http://dx.doi.org/10.1038/boneres.2016.50] [PMID: 28018707]

[124]   Priyadarsini S, Mukherjee S, Mishra M. Nanoparticles used in dentistry: A review. J Oral Biol Craniofac Res 2018; 8(1): 58-67.
[http://dx.doi.org/10.1016/j.jobcr.2017.12.004] [PMID: 29556466]

[125]   Oseni AO, Crowley C, Boland MZ, Butler PE, Seialian AM. Cartilage tissue engineering the application of nanomaterial of nanomaterials and stem cell technology, IntechOpen 2011. Dimensions ID.
[http://dx.doi.org/10.5772/22453] [PMID: 1028285806]

[126]   Cao Z, Dou C, Dong S. Scaffolding biomaterial for cartilage regeneration. J Nano 2014; 8.
[http://dx.doi.org/http:dx.doi.org/10.1155/2014/489128] [PMID: 489128]

[127]   Limongi T, Tirinato L, Pagliari F, *et al.* Fabrication and applications of micro-nano structured devices for tissue engineering. Nano-Micro Letters 2017; 9-1.

[128]   Parekh A, Cigan AD, Wognum S, Heise RL, Chancellor MB, Sacks MS. Ex vivo deformations of the urinary bladder wall during whole bladder filling: contributions of extracellular matrix and smooth muscle. J Biomech 2010; 43(9): 1708-16.
[http://dx.doi.org/10.1016/j.jbiomech.2010.02.034] [PMID: 20398903]

[129]   Barbara R, Jane M, Zhang B, *et al.* The Fibrotic response to implanted biomaterials: Implications for tissue engineering, regenerative medicine, and tissue engineering - cells and biomaterials. 2011. ISBN: 978-953-307-663-8.

[130]   Baiguera S. Urbani, Gaudio CD, Tissue-engineered scaffold for effective healing and regeneration: Reviewing orthopedic studies. BioMed Res Int 2014; 398069.
[http://dx.doi.org/10.1155/398069]

[131]   Saksena R, Gao C, Wicox M, de Mel A. Tubular organ epithelialisation. J Tissue Eng 2016; 7: 1-16.
[http://dx.doi.org/10.1177/2041731416683950] [PMID: 28228931]

[132]   Bennett MR. In-stent stenosis: pathology and implications for the development of drug eluting stents. Heart 2003; 89(2): 218-24.
[http://dx.doi.org/10.1136/heart.89.2.218] [PMID: 12527687]

[133]   Rhee JW, Wu JC. Advances in nanotechnology for the management of coronary artery disease. Trends Cardiovasc Med 2013; 23(2): 39-45.
[http://dx.doi.org/10.1016/j.tcm.2012.08.009] [PMID: 23245913]

[134]   Mahsa B, Marzieh M, Jerry JWS, Mohammad R. Nanomaterial coatings on the stent surfaces. Nanomedicine 2016; 11: 10.
[http://dx.doi.org/10.2217/nnm-2015-007]

[135] Gibney E. Injectable brain implant spies on individual neurons. Nature 2015; 522(7555): 137-8.
[http://dx.doi.org/10.1038/522137a] [PMID: 26062488]

[136] Krucoff MO, Rahimpour S, Slutzky MW, Edgerton VR, Turner DA. Enhancing nervous system recovery through neurobiology, neural interface training, and neuro-rehabilitation. Front Neurosci 2016; 10: 584.
[http://dx.doi.org/10.3389/fnins.2016.00584] [PMID: 28082858]

[137] Marchesan S, Ballerini L, Prato M. Nanomaterials for stimulating nerve growth. Science 2017; 356(6342): 1010-1.
[http://dx.doi.org/10.1126/science.aan1227] [PMID: 28596325]

[138] Reynold JG, Hart BR. Nanomaterials and their applications to defense and home security Journal of the Minerals, Metals and Materials 2004; 56(1): 36-9.
[http://dx.doi.org/10.1007/s11837-004-0270-8]

[139] Lele A. Role of nanotechnology in defense. Strategic Analysis 2009; 33(2): 229-41.
[http://dx.doi.org/10.1080/09700160802518700]

[140] Navanni NK, Sinha S. Nanotechnology, Defense applications. Stadium Press LLC 2013.ISBN: 10.1-62699-005-0.

[141] Gheibi HSM, Darroudi M. A novel approach in immunization Journal of Cell Physiology 2019; 234(8): 12530-8.
[http://dx.doi.org/10.02/jcp.28120]

[142] Yadav HSK, Dibi M, Mohammad A, Srouji AE. Nanovaccines formulation and application: a Review. Journal of Drug Delivery Science and Technology 2018; 44: 380-7.
[http://dx.doi.org/doi,org/10.1016/j.jddst.2018.01.015]

[143] Gao GF, Jakobsen BK. Molecular interactions of coreceptor CD8 and MHC class I: the molecular basis for functional coordination with the T-cell receptor. Immunol Today 2000; 21(12): 630-6.
[http://dx.doi.org/10.1016/S0167-5699(00)01750-3] [PMID: 11114424]

[144] Luo M, Samandi LZ, Wang Z, Chen ZJ, Gao J. Synthetic nanovaccines for immunotherapy. J Control Release 2017; 263: 200-10.
[http://dx.doi.org/10.1016/j.jconrel.2017.03.033] [PMID: 28336379]

[145] Vijayan V, Mohapatra A, Uthaman S, Park IK. Recent advances in nanovaccines using biomimetic immunomodulatory materials. Pharmaceutics 2019; 11(10): 534.
[http://dx.doi.org/10.3390/pharmaceutics11100534] [PMID: 31615112]

[146] Li S, Rizzo MA, Bhattacharya S, Huang L. Characterization of cationic lipid-protamine-DNA (LPD) complexes for intravenous gene delivery. Gene Ther 1998; 5(7): 930-7. https://www.stockton-press.cp.uk/gt
[http://dx.doi.org/10.1038/sj.gt.3300683] [PMID: 9813664]

[147] Moon JJ, Suh H, Polhemus ME, Ockenhouse CF, Yadava A, Irvine DJ. Antigen-displaying lipid-enveloped PLGA nanoparticles as delivery agents for a Plasmodium vivax malaria vaccine. PLoS One 2012; 7(2): e31472.
[http://dx.doi.org/10.1371/journal.pone.0031472] [PMID: 22328935]

[148] Yang R, Xu J, Xu L, *et al.* Cancer cell membrane-coated adjuvant nanoparticles with mannose modification of effective anticancer vaccine. ACS Nano 2018; 12(6): 5121-9.
[http://dx.doi.org/10.1021/acsnano.7609041]

[149] Kaczmarczyk SJ, Sitaraman K, Young HA, Hughes SH, Chatterjee DK. Protein delivery using engineered virus-like particles. Proc Natl Acad Sci USA 2011; 108(41): 16998-7003.
[http://dx.doi.org/10.1073/pnas.1101874108] [PMID: 21949376]

[150] Wu CY, Yeh YC, Chan JT, *et al.* A VLP vaccine induces broad-spectrum cross-protective antibody immunity against H5N1 and H1N1 subtypes of influenza A virus. PLoS One 2012; 7(8): e42363.

[http://dx.doi.org/10.1371/journal.pone.0042363] [PMID: 22879951]

[151] Bai H, Wang C, Qi Y, *et al.* Major vault protein suppresses lung cancer cell proliferation by inhibiting STAT3 signaling pathway. BMC Cancer 2019; 19(1): 454.
[http://dx.doi.org/10.1186/s12885-019-5665-6] [PMID: 31092229]

[152] Angsantikul P, Thamphiwatana S, Zhang Q, *et al.* Coating nanoparticles with gastric epithelial cell membrane for targeted antibiotic delivery against *Helicobacter pylori* infection. Adv Ther (Weinh) 2018; 1(2): 1800016.
[http://dx.doi.org/10.1002/adtp.201800016] [PMID: 30320205]

[153] Gao W, Fang RH, Thamphiwatana S, *et al.* Modulating antibacterial immunity via bacterial membrane-coated nanoparticles. Nano Lett 2015; 15(2): 1403-9.
[http://dx.doi.org/10.1021/nl504798g] [PMID: 25615236]

SUBJECT INDEX

A

Absorbance 29, 223, 238
  hypochromic 238
Acids 23, 25, 77, 78, 79, 82, 117, 122, 141, 153, 154, 156, 157, 162, 195, 196, 201, 202, 208, 225, 265, 277, 279, 283, 286, 307, 349, 352, 357, 358, 337
  6-amino penicillanic 277
  16-mercapto-hexadecanoic 358
  arginyl glycocylate aspartic 156
  ascorbic 77, 352
  cellular retinoic 279
  complex molecule N-acetyl neuraminic 162
  folic 154, 265
  humic 195, 196, 202
  hyaluronic 157
  kojic 23
  lactic 307
  malolactic 265
  mercaptoundecanoic 79
  polyglycolic 349
  polyinosinic 208
  polylactic 82, 307, 337, 349, 357
  polylactic acid-co-glycolic 349
  silicone phosphoric 122
  sulphonic 78
  tannic 141
  uric 352
Acquired immunodeficiency syndrome 359
Actin 152, 156, 166, 167, 168, 169
  assembly 152
  network acts 169
  polymerization of 169
Action 73, 82, 115, 140, 154, 249, 287, 289, 344
  concentration-dependent toxic 115
  enzymatic 154, 249
  homeostatic 82
  intense antimicrobial 344
  neutral response 73
  proton pumping 140
  synergistic 287, 289

Acute phase proteins (APPs) 186
Adaptive resistant components 299
Adenocarcinoma 112
Adhesion 10, 11, 106, 116, 157, 159, 160, 163, 166, 167, 304, 311, 345, 349, 355, 356
  cellular 157, 160, 345, 349, 355, 356
  mucosal 304
  physical 106
ADP-ribose polymerase 168
Affinity capillary electrophoresis techniques 204
Albumin145, 153, 192
  -bound nanoparticles 145
  receptors 153, 192
Alkaline phosphatase activities 280, 283
Alzheimer's disorder 335, 336, 337
Ampere's law 33
Analytical 30, 31, 209
  microchip 209
  techniques 30, 31
Antigens 137, 149, 185, 186, 187, 299, 303, 305, 313, 316, 321, 322, 323, 360
  classical pathway 149
  encapsulated 321
  encapsulating 321
  function-associated 137
  vaccine 360
Anti-HIV-agents 339
Anti-immune diseases prevention 302
Antimicrobial activity 345, 346, 347
  concentration-dependent 347
  restricted 347
Apolipoproteins 83, 204, 336
Apoptosis 74, 109, 110, 145, 168, 187, 221, 312, 315, 319, 338
  mitochondrial 168
Application of immobilization 278
  in food technology 278
  of enzyme in biofuel technology 278
Applications and significance of protein corona 197
Apurinic DNA hydrolyzes 225
Arterial natriuretic peptides secretion 158

www.ingramcontent.com/pod-product-compliance
Lightning Source LLC
Chambersburg PA
CBHW050800220326

41598CB00006B/75